T0331110

NEAR EARTH OBJECTS, OUR CELESTIAL NEIGHBORS:
OPPORTUNITY AND RISK

IAU SYMPOSIUM No. 236

COVER ILLUSTRATION: (66391) 1999 KW$_4$ and its satellite.

Radar-derived model of near-Earth binary asteroid (66391) 1999 KW4 (Ostro et al. 2006, Science 314, 1276-1280, and Scheeres et al. 2006 Science 314, 1280-1283). This object was discovered in 1999 by MIT's LINEAR Project and was imaged extensively by radar with Arecibo and Goldstone in May 2001 during the object's 13-lunar-distance pass by the Earth, its closest until 2036. The 1.5-kilometer-diameter primary (Alpha) is an unconsolidated gravitational aggregate with a spin period 2.8 h, bulk density 2 g/cm^3, porosity 50%, and an oblate shape dominated by an equatorial ridge at the object's potential-energy minimum. The 0.5-km secondary (Beta) is elongated and probably is denser than Alpha. Its average orbit about Alpha is circular with a radius 2.5 km and period 17.4 h, and its average rotation is synchronous with the long axis pointed toward Alpha, but librational departures from that orientation are evident. The binary system is in a dynamically excited state due to the effects of close passes by the Sun and probably the Earth. Excitation of the mutual orbit stimulates complex fluctuations in that orbit and in each component's rotation, inducing the attitude of Beta to have large variation within some orbits and to hardly vary within others. Alpha's proximity to its rotational stability limit suggests an origin from spin-up and disruption of a loosely bound precursor within the past million years. KW4 is the best characterized Potentially Hazardous Asteroid as large as a kilometer. NASA/JPL image.

Steve Ostro, JPL

This image has been selected as an illustration of the very significant progress in the state of the art with Near Earth Objects (NEO) science. Among the recent steps forward we can cite: the availabilty of an enormously increased inventory of objects (this object was discovered only in 1999), the possibility of imaging (not only with space missions but also with radar), the observational evidence as well as the theroretical understanding of asteroid shapes, rotation dynamics and binary dynamics, the extremely accurate orbit determinations allowing to assess the specific impact risk from each object. They have upgraded the status of NEO science to a major contributor to the astrophysical understanding of our solar system (and potentially also of other ones).

The Editors (A. Milani, G.B. Valsecchi & D. Vokrouhlický)

IAU SYMPOSIUM PROCEEDINGS SERIES

2006 EDITORIAL BOARD

Chairman

K.A. VAN DER HUCHT, IAU Assistant General Secretary
SRON Netherlands Institute for Space Research
Sorbonnelaan 2, NL-3584 CA Utrecht, The Netherlands
K.A.van.der.Hucht@sron.nl

Advisors

O. ENGVOLD, IAU General Secretary, *Institute of Theoretical Astrophysics, University of Oslo, Norway*
E.J. DE GEUS, *Netherlands Foundation for Research in Astronomy, Dwingeloo, The Netherlands*
M.C. STOREY, *Australia Telescope National Facility, Australia*
P.A. WHITELOCK, *South African Astronomical Observatory, South Africa*

Members

IAUS233
V. BOTHMER, *Universitäts Sternwarte, Georg-August-Universität, Göttingen, B.R. Deutschland*
IAUS234
M.J. BARLOW, *Department of Physics and Astronomy, University College London, London, UK*
IAUS235
F. COMBES, *LERMA, Observatoire de Paris, Paris, France*
IAUS236
A. MILANI, *Dipartimento di Matematica, Università di Pisa, Pisa, Italia*
IAUS237
B.G. ELMEGREEN, *IBM Research Division, T.J. Watson Research Center, Yorktown Heights, NY, USA*
IAUS238
V. KARAS, *Astronomical Institute, Academy of Sciences of the Czech Republic, Praha, Czech Republic*
IAUS239
F. KUPKA, *Max-Planck-Institut für Astrophysik, Garching-bei München, B.R. Deutschland*
IAUS240
W.I. HARTKOPF, *U.S. Naval Observatory, Washington D.C., USA*
IAUS241
A. VAZDEKIS, *Instituto de Astrofísica de Canarias, La Laguna, Tenerife, Canary Islands, Spain*

INTERNATIONAL ASTRONOMICAL UNION

UNION ASTRONOMIQUE INTERNATIONALE

NEAR EARTH OBJECTS, OUR CELESTIAL NEIGHBORS: OPPORTUNITY AND RISK

PROCEEDINGS OF THE 236th SYMPOSIUM OF THE INTERNATIONAL ASTRONOMICAL UNION HELD IN PRAGUE, CZECH REPUBLIC AUGUST 14–18, 2006

Edited by

ANDREA MILANI
University of Pisa, Pisa, Italy

GIOVANNI B. VALSECCHI
IASF-INAF, Roma, Italy

and

DAVID VOKROUHLICKÝ
Charles University, Prague, Czech Republic

CAMBRIDGE
UNIVERSITY PRESS

Shaftesbury Road, Cambridge CB2 8EA, United Kingdom

One Liberty Plaza, 20th Floor, New York, NY 10006, USA

477 Williamstown Road, Port Melbourne, VIC 3207, Australia

314–321, 3rd Floor, Plot 3, Splendor Forum, Jasola District Centre, New Delhi – 110025, India

103 Penang Road, #05–06/07, Visioncrest Commercial, Singapore 238467

Cambridge University Press is part of Cambridge University Press & Assessment, a department of the University of Cambridge.

We share the University's mission to contribute to society through the pursuit of education, learning and research at the highest international levels of excellence.

www.cambridge.org
Information on this title: www.cambridge.org/9780521863452

© International Astronomical Union 2007

This publication is in copyright. Subject to statutory exception and to the provisions of relevant collective licensing agreements, no reproduction of any part may take place without the written permission of Cambridge University Press & Assessment.

First published 2007

A catalogue record for this publication is available from the British Library

ISBN   978-0-521-86345-2   Hardback

Cambridge University Press & Assessment has no responsibility for the persistence or accuracy of URLs for external or third-party internet websites referred to in this publication and does not guarantee that any content on such websites is, or will remain, accurate or appropriate.

Table of Contents

Part 2. THE METEOR/ASTEROID IMPACT TRANSITION

Part 3. ROTATION, SHAPES AND BINARIES

Part 4. SURFACES AND COMPOSITION

Part 5. SURVEYS: ORBIT DETERMINATION AND DATA PROCESSING

Part 6. SURVEYS: OBSERVATORIES AND THEIR PERFORMANCES

Part 7. CURRENT AND FUTURE MISSIONS TO NEOs

Part 8. IMPACT MONITORING AND RISK ESTIMATES

Part 9. IAU AND GOVERNMENTS ROLE IN THE NEO PROBLEM

EPILOGUE

Preface

Near Earth Objects (NEOs), and more seldom Czech lands, played occasionally important role in discovering the Universe and our place in it. As an example in 1577 Tycho Brahe, using his own observations and those of a Czech nobleman Thaddaeus Hagecius (Tadeáš Hájek), determined that a comet discovered earlier that year moves in translunar zone. This was in contrast with a scholastic view, hold since Aristotle through the whole Middle Ages, that all ephemeral phenomena occur in the Earth atmosphere up to lunar sphere only, beyond which extends an eternal and unchangeable Universe. The same Hagecius, personal physician of emperor Rudolph II, made later Brahe move to Prague in 1599, attracting also young Johannes Kepler. For short 13 years Prague became one of the two most important astronomy centers in the world.

Today, both Prague and NEOs hold a more modest role in shaping our fundamental views on Universe. Yet the amount of research devoted to NEOs seems to significantly grow over the few last decades. NEOs visit the neighborhood of our planet making thus available detailed observations of objects as small as metres across by various techniques including optical, infrared and radar observations. Thus the smallest observed NEOs today have about the same size as meteorite precursors; meteorites themselves are the cheapest samples of the Universe in our laboratories and lot of effort is being done to put them into a broader context in the Solar system. NEOs proximity to the Earth allows low-energy transfer orbits for spacecraft, on the other hand this results in the threat of a possible collision. With their sources ranging from the immediate region beyond Mars up to Oort cloud, studies of NEOs contribute importantly into our overall understanding of the Solar system and its origins. Recognizing all these facts the Prague Symposium was the first meeting about minor objects in the Solar system to be held within the frame of an IAU General Assembly. The present book attempts to collect the most important contributions in the Symposium programme.

As much as Prague was honored to become a venue of the IAU General Assembly, Czech astronomers active in planetary science research were pleased to contribute in organization of the Symposium. I would like to thank, also on behalf of the SOC co-chairs (Andrea Milani and Giovanni B. Valsecchi), all members of the LOC and SOC that made the preparations and later the meeting itself very successful. Obviously the same goes to all participants who contributed to a friendly and scientifically profitable atmosphere during these 5 days in August 2006. Finally, we also thank all referees that had to work fast to make this volume published in a short time after the Symposium.

David Vokrouhlický

THE SCIENTIFIC ORGANIZING COMMITTEE

- G.B. Valsecchi (Co-chair, INAF, Italy)
- A. Milani (Co-chair, Univ. Pisa, Italy)
- D. Vokrouhlický (Co-chair, Charles Univ., Czech Rep.)
- G. Consolmagno (Vatican Obs., Vatican)
- S. Isobe (NAO, Japan)
- Z. Knežević (Belgrade Obs., Serbia)
- I. Mann (Univ. Munster, Germany)
- D. Morrison (NASA Ames, USA)
- P. Pravec (Ondřejov Obs., Czech Rep.)
- H. Rickman (Uppsala Obs., Sweden)
- H. Scholl (OCA, France)
- T. Spahr (MPC, USA)
- E. Tedesco (Univ. New Hampshire, USA)
- I.P. Williams (Univ. London, UK)
- D.K. Yeomans (NASA JPL, USA)

Acknowledgements

The symposium was sponsored and supported by the IAU Divisions III (Solar System) and I (Fundamental Astronomy); and by the IAU Commissions No. 20 (Positions and motions of minor planets, comets and satellites), No. 7 (Celestial Mechanics and Dynamical Astronomy), No. 15 (Physical Studies of Comets, Minor Planets and Meteorites), No. 16 (Physical Study of Planets and Satellites) and No. 22 (Meteors, Meteorites and Interplanetary Dust).

The Local Organizing Committee of the IAU XXVI General Assembly acted as local organizing committee also for this Symposium.

Prologue

Near Earth Objects, our Celestial Neighbors: Opportunity and Risk
Proceedings IAU Symposium No. 236, 2006 © 2007 International Astronomical Union
A. Milani, G.B. Valsecchi & D. Vokrouhlický, eds. doi:10.1017/S1743921307002980

236 years ago...

Giovanni B. Valsecchi

IASF-Roma, INAF, via Fosso del Cavaliere 100, I-00133 Roma (Italy)
email: giovanni@iasf-roma.inaf.it

Abstract. Some of the problems related to Near Earth Objects (NEOs), like orbit determination and ephemeris computation, are not new, and had to be dealt with since the beginning of NEO astronomy. The latter practically started with the discovery of Comet D/1770 L1 Lexell, that passed very close to the Earth in 1770; studies of the chaotic dynamics of this exceptional object continued well into the XIXth century. At the end of the XXth century there has been a renewal of interest in NEOs, as attested by IAU Symposium 236.

Keywords. orbit determination; multiple solutions; close approaches

1. A close passage

Astronomers started to deal with objects coming very close to the Earth in 1770, when Charles Messier discovered, on the night between 14 and 15 June, the comet that is now known under the name of comet D/1770 L1 Lexell. The comet was heading towards the Earth; within a few days from the discovery it became visible to the naked eye, reaching the second magnitude. The minimum distance from the Earth was reached on 1 July, at about six times the lunar distance, and in a few more days the comet disappeared due to its proximity to the Sun; Pingré computed an ephemeris for its recovery based on a parabolic orbit, as it was then customary, and Messier was able to see the comet again starting from the beginning of August. The comet then remained under observation until the beginning of October.

It soon became clear that, although the ephemeris by Pingré had allowed the recovery of the comet in August, a parabolic orbit could not account for the entire set of observations. Prosperin tried to use three parabolas to fit separately the observations of June, August, and September, but this was evidently unsatisfactory, and unjustifiable from the point of view of Celestial Mechanics.

2. Lexell's work

The solution to this problem was found by Lexell, who showed that the comet was on an elliptical orbit with a period of five and a half years, far shorter than the 76 years of the other case then known, that of comet Halley; his solution accounted well for the entire set of observations. Messier was puzzled by this finding, and asked why the comet had not been observed before, given its short orbital period and its small perihelion distance; to this, Lexell answered that in May 1767 the comet and Jupiter had been very close to each other, and that the action of the gravity of the giant planet had greatly transformed the orbit of the comet. In fact, before 1767 the comet had a much larger perihelion distance, implying that it could not become very bright, thus explaining the fact that it had not been observed before.

Lexell also found that the orbit of the comet in 1770 was nearly resonant with that of Jupiter; in fact, in the time it took the comet to make two revolutions about the Sun, the giant planet would make one revolution. As a consequence, in 1779 the comet would encounter Jupiter again, at a distance even closer than in 1767, and would be expelled from the inner solar system into an orbit of large perihelion distance and period, that would make it invisible again for the telescopes of the time. The comet, in fact, was not observed again in 1782, as it should have been if it had remained in its 1770 orbit. Lexell's reconstruction of what had happened in 1767, as well as his prediction of what would happen in 1779, became generally accepted and the comet, although discovered by Messier, now brings Lexell's name. It can be said that the work of Lexell started the modern understanding of the dynamics of small solar system bodies.

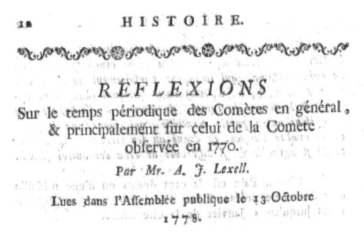

Figure 1. The Latin title of Lexell (1777b); in English it may be translated as "Conjecture about where in the sky the comet of 1770, in its next return to perihelion, has to be looked for from our Earth".

It is interesting to note that one of the current problems of NEO astronomy, that of predicting the future position of an Earth-approaching body from a short observed arc, was a major concern also for Lexell, as testified by the title of Lexell (1777b), reproduced in Figure 1.

HISTOIRE.

RÉFLEXIONS

Sur le temps périodique des Comètes en général, & principalement sur celui de la Comète observée en 1770.

Par Mr. A. J. Lexell.

Lues dans l'Assemblée publique le 13 Octobre 1778.

Figure 2. The French title of Lexell (1778b); in English it may be translated as "Considerations about the periods of comets in general, and principally on that of the comet observed in 1770".

The title of another of Lexell's papers, that of Lexell (1778b), reproduced in Figure 2, reflects one of the major worries of current NEO astronomy, i.e. that of communicating to fellow scientists and to the interested public about NEO matters, given the potential implications in terms of hazard (the idea that Earth approaching comets could be on orbits of period much shorter than that of P/Halley could have been rather hard to accept at the end of the XVIIIth century).

3. Enter LeVerrier: Line of Variations, chaos and all that...

In mid XIXth century, at about the same time of the celebrated work that led to the discovery of Neptune, LeVerrier reconsidered the orbital problems posed by Lexell's comet. LeVerrier critically examined the available observations, and among them identified a subset that he trusted; he then tried to compute an accurate orbit for the comet, taking also into account the gravitational action of the Earth. After many computations, described in detail in LeVerrier (1844), LeVerrier (1848) and LeVerrier (1857), he concluded that the best-fit orbit of the comet was poorly constrained and that all what could be done was to identify a line in the space of orbital elements —what we would nowadays call the Line of Variations (LoV), see Milani *et al.* (2005a)— in which the point corresponding to the true orbit most probably lies. As described by Milani *et al.* (2005b), current impact monitoring software robots exploit the same concept.

Thus, LeVerrier expressed the six orbital elements of comet Lexell as functions of a single unknown parameter, that he called μ; he also showed that the observations could be used to find the possible range of variation of μ, since outside a certain range, the path of the comet on the sky would have been measurably different from the observed one. Figure 3 shows an excerpt from LeVerrier (1857), in which the LoV is given as a straight line, parametrized by μ, in the space of orbital elements.

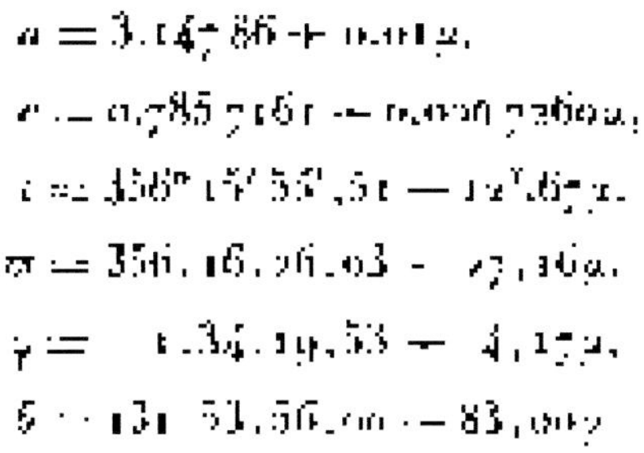

Figure 3. The Line of Variations introduced by LeVerrier for comet Lexell: from top to bottom, semimajor axis, eccentricity, mean longitude at epoch, longitude of perihelion, inclination and longitude of node.

LeVerrier computed the post-1770 time evolution of orbits lying on the LoV, so as to obtain a global view of all the possible outcomes, as done much later by Carusi *et al.* (1982); a comparison of the orbits examined in the latter paper with those given in Figure 3 reveals that, in fact, the uncertainty affecting LeVerrier's orbit solution was mostly in the magnitude of the heliocentric velocity vector, since the orbits computed by Carusi *et al.* only differed from each other for that quantity, and they appear to lie, in orbital elements space, along a line almost coincident with LeVerrier's LoV.

The computations showed, among other things, that the comet could approach Jupiter extremely closely in 1779, as close as less than three and a half radii of the planet from its centre; nevertheless, the comet could not become a satellite of Jupiter, not even temporarily, for any allowed value of μ. The range of post-1779 orbits included the possibility, for the comet, to leave the solar system on a hyperbolic orbit. The reason for this wide range of possible outcomes was the extreme sensitivity of the subsequent evolution to the precise value adopted for μ; this sensitivity is a crucial part of the modern concept of chaos, and in fact Le Verrier's computations probably represent the first instance of this concept in scientific literature.

Figure 4 shows the largest perturbations computed by LeVerrier, that are very close to his assumed nominal orbit .

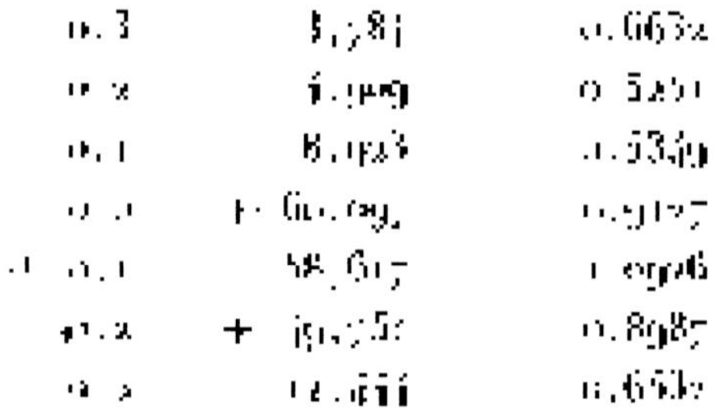

Figure 4. A small excerpt from LeVerrier's computations: for the values of μ in the left column, the corresponding post-1779 values of semimajor axis and eccentricity of the orbit of comet Lexell.

4. NEO studies have a long tradition

After a long hiatus in the second half of the XIXth century and in the first three quarters of the XXth, the subject is again attracting the attention of astronomers. Some of the basic ideas have not changed; however, new observational and computational techniques have allowed us to make significant progress in NEO studies. We have learnt many things in the process, and these Proceedings aim at presenting a comprehensive view of the current status and achievements of NEO studies.

References

Carusi, A., Kresáková, M. and Valsecchi, G. B. 1982, *Astron. Astrophys.* 116, 201

LeVerrier, U. J. 1844, *Comptes rendues de l'Academie des sciences* 19, 982

LeVerrier, U. J. 1848, *Comptes rendues de l'Academie des sciences* 26, 465

LeVerrier, U. J. 1857, *Annales de l'observatoire de Paris* 3, 203

Lexell, A. J. 1777a, *Acta Academiae Scientiarum Imperialis Petropolitanae* I, 332

Lexell, A. J. 1777b, *Acta Academiae Scientiarum Imperialis Petropolitanae* II, 328

Lexell, A. J. 1778a, *Acta Academiae Scientiarum Imperialis Petropolitanae* I, 317

Lexell, A. J. 1778b, *Acta Academiae Scientiarum Imperialis Petropolitanae* II, 12

Milani, A., Sansaturio, M. E., Tommei, G., Arratia, O. and Chesley, S. R. 2005a, *Astron. Astrophys.* 431, 729

Milani, A., Chesley, S. R., Sansaturio, M. E., Tommei, G. and Valsecchi, G. B. 2005b, *Icarus* 173, 362

Near Earth Objects, our Celestial Neighbors: Opportunity and Risk
Proceedings IAU Symposium No. 236, 2006
A. Milani, G.B. Valsecchi & D. Vokrouhlický, eds.
© 2007 International Astronomical Union
doi:10.1017/S1743921307002992

A tribute to the life and work of George W. Wetherill: Some reflections of his career at DTM

Douglas O. ReVelle

Los Alamos National Laboratory, Los Alamos, New Mexico, USA
email: revelle@lanl.gov

George Wetherill and I worked together as scientific collaborators when I was a post-doctoral fellow in 1977-1978 at the Department of Terrestrial Magnetism (DTM) of the Carnegie Institution of Washington (CIW) in Washington, D.C. We worked on problems of meteoroids interacting in Earth's atmosphere along with Richard McCrosky at Harvard College Observatory and Zdeněk Ceplecha at the Ondřejov Observatory in Czechoslovakia and also with Sundar Rajan who had already arrived at DTM from the University of California at Berkeley before me.

George had just returned from being the Chairman of the Planetary and Space Science Department at UCLA and while there was on the governing board that proposed the creation of the IGPP (Institute of Geophysics and Planetary Physics) at the National Laboratories including Los Alamos where I have been ever since January 1994. His direct intent and interest in my being an Astronomy postdoctoral fellow at DTM was evident from even before I arrived at DTM.

One day I phoned him to confirm that everything was prepared for my arrival at DTM and he acted very odd to me on the phone. It turns out that Dr. Vera Rubin was in his office lobbying for a extragalactic astronomer postdoctoral position, but George had already made up his mind to have me at DTM so he let Vera have her say completely and then told her I was coming and that was that. Later on that day he just called me back as if the other conversation hadn't happened and told me that I should plan to arrive at DTM very soon. I found out about my officially being called an Astronomy postdoctoral fellow about a year later when Professor Bart Bok came to give a seminar on the origin of Bok Globules. I was listed on an official postdoctoral list that had recently been assembled as being an Astronomy postdoc much to my surprise since all of my degrees are in Atmospheric and Oceanic Sciences or Aeronomy and Planetary Atmospheres.

At least in those days DTM was very lively and always being visited by dignitaries such as Bok and many others. Pictures on the wall of Niehls Bohr were evident in the staff library and office (of the director's secretary) which I loved to use which had many wooden book shelves and wonderful sets of books and journals. Many times we would have 3-4 seminars in one week for the scientific staff's fields of Classical and Solar System Astronomy, Geophysics and Seismology, Geochemistry and of Nuclear Physics.

George was always mostly very friendly, but also very intense. When we met for the last time years later at the Cornell ACM in 1999 (see below), he had changed quite a bit and was much more social and willing to have several beers together, etc. His first wife, Phyllis, had passed away and I think he really missed the "good" old days together with the solar system Astronomy postdoctoral fellows, etc. at DTM. He almost never took breaks from his almost 12 hour days at DTM and rarely drank coffee or even water or participated in the "lunch" club that had been formed there since the original DTM chef and his staff had left for more fertile pastures (the chef just cost too much to keep on in the 1960's and beyond). One day while we were playing volleyball in preparation for the big game against the geophysical lab (which has since moved onto the same site at 5241 Broad Branch Road near Chevy Chase Circle in Washington, D.C.), George came out of his office and made a rare appearance on his veranda to the entire postdoctoral group. He politely reminded us of the need to go back to work and not just to play games, but it was all in good fun and we certainly worked harder after the great exercise as I remember it or at least I know that I worked even harder.

One day in the Spring of 1978 George and other senior DTM staffers asked me to show our new secretary Maura around DTM to help get her oriented to the buildings and offices. In order to illustrate George's full intensity, I recall a drinking fountain episode that occurred at DTM as Maura began working for the Carnegie Institution of Washington (this story was also told by me in front of his many DTM colleagues at George's DTM retirement party many years later). George's office was on the southern end of the third floor of DTM. There were standard drinking fountains present only on the first and third floors at the top of the main hallway through the three story structure. Maura and I had just examined the room where "glass slides" were still being made and were going up to the third floor from the first floor, but while we briefly stood at the landing between the second and third floors, we observed the following: George got up from his desk and walked directly toward the stairwell and proceeded down to the second floor. He walked right by the drinking fountain on his own floor on the way down the stairs. When he reached and had passed us on the second floor he proceeded to bend over at the spot where the drinking fountain would have been found on the first and third floors and "pushed" the nonexistent button down on the nonexistent drinking fountain and decidedly drank until he was completely satisfied from the nonexistent fountain. Maura looked directly at me and immediately said "That's my new boss?" George's response to my comments at his retirement dinner in D.C. several years ago was "Well, I am here and I am still ok anyway."

As yet another illustration of his scientific intensity, I will relate the story of his dosages of "Wetherillitis" to me (as my wife Ann liked to call them since I always came home completely drained mentally on those inevitable days), our monthly 12 hour meetings with no breaks in 1977-1978. Again the intensity of his purpose was thoroughly evident in these very intense meetings. George had the habit of dragging out all that his collaborators had been working on for the past month or so at such meetings. We would first make a list of topics that we were both working on and where our various lines of thought intersected. The various meteoroid related topics that we usually discussed are listed later on. After many many tiring hours of converging towards a consensus of where we should be heading scientifically over the next month or so we would finally begin to relax and be positive about what had already been done. On one occasion late in the day George's first wife, Phyllis, arrived at our meeting place in his office. I was finally summarizing everything for the day's efforts and was almost finished as she entered the room. George almost instinctively jumped up and put on his coat and was out the door in a flash and I was still talking. They left almost as quickly as she had entered and I slowly rose and put on my jacket and gathered my things for my long ride home to Culpeper, Virgina (some 85 miles away from D.C.). After another day of "Wetherillitis", I almost looked forward to these long drives! We were nonetheless able to get an enormous number of things done during these sessions and finally wrote two major scientific papers together as well as submitted and subsequently presented numerous meeting abstracts (at the DPS of the AAS, the Meteoritical Society, AGU, ASA, etc.).

George helped me immensely in my interactions with Zdeněk Ceplecha and the staff of the Ondrejov Observatory through a U.S. National Academy of Sciences Scientific

Exchange grant for 8 weeks in the summer of 1980. Dr. Zdeněk Ceplecha, the Chief of the Interplanetary Matter Department for many years at the Ondrejov Observatory, was even then a very well known world leader in the science of meteors. Since this all happened before email and the multitude of personal computers that we have today at our disposal, numerous and very lengthy letters were being written back and forth across the Atlantic often taking 4-8 weeks each way due to the Communist government in Czechoslovakia.

George recognized the vital interaction that needed to take place between Ceplecha and myself and saw to it that this connection was fostered. This work culminated in the Spring of 1983 when Zdeněk was awarded the U.S. National Academy of Sciences Prize, the George P. Merrill Award, and Zdeněk was invited to receive the prize at their offices in Washington D.C. Again with George's help I was made financially able to come to the award ceremony and be with Ceplecha during his time in the U.S. when great honors were being bestowed upon him.

During his early and teenage years George was infatuated with the daughter of the great classical meteor physicist Dr. Charles P. Olivier while growing up in his native Philadelphia. In fact, his fundamental reason for turning to our science of Planetary and Solar System problems were fundamentally instilled by this need to impress the teenage daughter back in the 1920's and 1930's. She of course was not impressed by her father's pursuits much to George's dismay. He retold this story to me at the ACM at Cornell in 1999 while still asking if anyone knew of her whereabouts even then.

Another illustration of George's scientific intensity, in case you still may doubt it, was during our excursion boat trip together with the numerous ACM scientists at Belgirate, Italy in 1993, when George and I spent one and a half hours totally alone onboard the large cruise boat that had been hired for the day on Lake Maggiore in Northern Italy near the Swiss border (we were together for this long time with the exception of the barebones crew of the boat and their mascot dog). Everyone else had departed the boat to shop and sightsee on a well-known tourist excursion island. George and I discussed meteor science intensely during this lengthy period. We never even knew everyone got off the boat until they all returned. Several people later got back on and laughed heartily when they saw that we were both still there and at the very same seats for all that time (and still drinking a fine European beer) discussing what else, but our first loves, science (this tremendous laughter included outbursts from the late Mayo Greenberg and several others as well)!

George is perhaps most well known in our scientific community for his absolutely pioneering effort on discovering the theoretical origins of the terrestrial planets and their formation processes and the huge role played by Jupiter in its protection of the terrestrial planets. Nonetheless all the problems that I collaborated on with George involved our knowledge of meteors and their composition and structure and their orbital associations and their properties as deduced by directly studying their entry into Earth's atmosphere. As an honor for all of his very hard and pioneering efforts, I believe that we as a community should consider bestowing his name on a region of our Solar System, such as that from the outer edge of the asteroid belt to Jupiter's orbit, as the Wetherill zone.

Topics that we collaborated on together while I was at DTM included (from 1977 to 1981, which included return summer visits after my final departure to Arizona in late 1978):

(*a*) Single-body entry modeling to determine initial masses (through ablation effects) and the initial velocities and orbits of meteorites (with Sundar Rajan);

(*b*) Luminous efficiency of meteorites (with Sundar Rajan);

(*c*) Theoretical predicted mass loss comparisons to cosmic ray tracks measurements in meteorites (with Sundar Rajan);

(*d*) Identification of meteorites with bright meteors;

(*e*) Large bolide influx rate determined using microbarographs;

(*f*) Studies and identification of meteorites from beyond Jupiter;

(*g*) Impacts of extraterrestrial bodies on the Earth and the Moon.

Some of this work appeared as articles in the Annual Report of the Director of DTM, subsequently became formal papers in Icarus and in the Arizona Comets book of 1981 (L. Wilkening editor), and many of the ideas and concepts were presented at multiple scientific conferences including the Meteoritical Society in both Canada and in Europe (with the latest in 1983).

During some years, George as the DTM Director would pick a poem to add to the mix of papers contained in the Annual Reports that were done back then. It seems appropriate that we should add one of his favorite poems to this tribute to his life and times. In the section entitled Meteorites and Meteoroids in the DTM Annual Report for 1980-1981 you can still find the provocative poem:

.... the might
Of earth-convulsing behemoth, which once
Were monarch beasts, and on the slimy shores
And weed-overgrown continents of earth,
Increased and multiplied like summer worms
On an abandoned corpse, till the blue globe
Wrapped deluge round it like a cloak, and they
Yelled, gasped and were abolished; or some God
Whose throne was in a Comet, passed and cried
"Be not!"and like my words they were no more.

Shelley (1820)

When I finally received an offer of university employment at a Physics and Astronomy Department in Arizona the summer of 1978 George's first response to me was: "We can sue them!" and George was referring to the fact that they had offered me employment long after the traditional deadline of May 1st for hiring faculty for the next academic year.

George even offered me a staff position in meteoroids and meteoritics at DTM in 1999 while we were at the ACM at Cornell which would have been quite nice. Very reluctantly I told him that I couldn't even think of that offer as an option because it wasn't possible to make a living in the U.S.A. doing meteoroid research at DTM since I needed the NSF (US National Science Foundation) or some similar organization for my complete funding support. When I first arrived at DTM individual support funding was provided by interest monies on the Carnegie endowment. Later after equipment grants to the NSF were routinely written and funded for technical staff support equipment, the CIW Board of Trustees decided to do the same for CIW staff salaries, a fact which greatly limited my chances to return to DTM.

George ultimately was extremely successful and triumphant in his scientific career. Like most of us, he also had tragedies including the loss of his son. Nevertheless he should be remembered as a great scientist who foresaw the possibilities of the future and made things happen. He will be missed by all of us who knew him well.

Part 1

Population Models and Transport Mechanisms

Near Earth Objects, our Celestial Neighbors: Opportunity and Risk
Proceedings IAU Symposium No. 236, 2006
A. Milani, G.B. Valsecchi and D. Vokrouhlický, eds.
© 2007 International Astronomical Union
doi:10.1017/S1743921307003018

Mutual geometry of confocal Keplerian orbits: uncertainty of the MOID and search for virtual PHAs

Giovanni F. Gronchi, Giacomo Tommei and Andrea Milani

Department of Mathematics, University of Pisa, Largo B. Pontecorvo 5, 56127 Pisa, Italy
email: gronchi@dm.unipi.it

Abstract. The Minimum Orbit Intersection Distance (MOID) between two confocal Keplerian orbits is a useful tool to know if two celestial bodies can collide or undergo a very close approach. We describe some results and open problems on the number of local minimum points of the distance between two points on the two orbits and the position of such points with respect to the mutual nodes. The errors affecting the observations of an asteroid result in uncertainty in its orbit determination and, consequently, uncertainty in the MOID. The latter is always positive and is not regular where it vanishes; this prevents us from considering it as a Gaussian random variable, and from computing its covariance by standard tools. In a recent work we have introduced a regularization of the maps giving the local minimum values of the distance between two orbits. It uses a signed value of the distance, with the sign given to the MOID according to a simple orientation property. The uncertainty of the regularized MOID has been computed for a large database of orbits. In this way we have searched for Virtual PHAs, i.e. asteroids which can belong to the category of PHAs (Potentially Hazardous Asteroids) if the errors in the orbit determination are taken into account. Among the Virtual PHAs we have found objects that are not even NEA, according to their nominal orbit.

Keywords. MOID, Potentially Hazardous Asteroids, orbit determination

1. Introduction

The *orbit distance* between two Keplerian orbits with a common focus is useful to know if two celestial bodies moving along these orbits can collide or undergo a very close approach. If the orbit distance is large enough there is no possibility of such an event, at least during the time span in which the Keplerian solutions are a good approximation of the real orbits. This distance is called MOID (*Minimum Orbit Intersection Distance*) in the literature; this acronym was introduced in (Bowell & Muinonen 1994) with the definition of *Potentially Hazardous Asteroids* (PHA), an asteroid with MOID $\leqslant 0.05$ AU and absolute magnitude $H \leqslant 22$.

Two confocal Keplerian orbits can get close at more than a pair of points, for example near both the mutual nodes, thus it is useful to compute all the local minima of the *Keplerian distance function d*, distance between two points on the two orbits as a function of the two anomalies along the orbits (see Figure 1), not only the absolute minimum. We compute these values as the stationary points of the function d^2, squared to be smooth also in case of *orbit crossing*, when the distance can be zero (see Section 2).

When a new celestial body is detected and its observations are enough, the orbit of the body can be determined by means of a least squares fit. Moreover the uncertainty in the determination of the nominal orbit produced by the errors in the observations can be represented by a covariance matrix (see, e.g., Milani 1999). The errors in the orbit determination also affect the computation of the MOID, and it is important to estimate

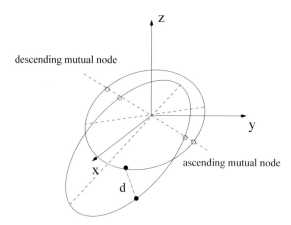

Figure 1. Two confocal orbits and the distance d for two selected values of the anomalies. The apsidal lines and the mutual nodal lines (intersection of the two orbital planes) are also drawn.

the resulting uncertainty. In other cases, the covariance of some function of the orbital elements can be computed by a covariance propagation formula (see Section 3), but in the case of the MOID the possibility of orbit crossings produces a singularity in this computation. An additional difficulty is that the uncertainty of a non-zero but small orbit distance may allow negative values of the distance, that are meaningless. Both these problems prevent us from computing a meaningful confidence interval just when the MOID can be small or vanishing.

We use the results of (Gronchi & Tommei 2007), in which we regularize the local minima of d as maps of the orbit configurations according to an intuitive geometric rule. Using these regularized maps we can compute a meaningful confidence interval for the local minimal distance maps also when they vanish (see Section 4). With this new algorithm we compute the covariance of the regularized maps for a large database of orbits with covariance matrix, and we search for *Virtual PHAs*, i.e. asteroids which can belong to the PHA category if the errors in the orbit determination are taken into account (see Section 5). Among the Virtual PHAs we have found objects whose nominal orbits are not even *Near Earth Asteroids* (NEA), that is have a nominal perihelion $q > 1.3$ AU.

A mathematical theory to compute the uncertainty of the orbit distance is relevant to applications such as to produce an observation priority list for follow up of NEAs based on the possible orbit distance with the Earth. We briefly comment on this in Section 6.

2. Stationary points of the squared Keplerian distance function

2.1. *Computation of the stationary points of d^2*

There are several papers in the literature on the computation of the minimum points of d (e.g. Sitarski 1968; Hoots 1984; Dybczynski *et al.* 1986). Recently some algebraic methods to compute all the stationary points of d^2 have been introduced, using the *Gröbner bases* (Kholshevnikov & Vassiliev 1999) and the *resultant theory* (Gronchi 2002; Gronchi 2005). They are both based on a polynomial formulation of the problem. The resultant method is used to compute the MOID and all the stationary points of d^2 (also for cometary orbits) by the orbit determination software Orbfit† and in the NEODyS website‡. In

---

† A free software, released under a GNU Public License software, by the Orbfit consortium lead by the University of Pisa. To download see `http://newton.dm.unipi.it/orbfit`

‡ The *Near Earth Objects Dynamic Site*, maintained at the Universities of Pisa and Valladolid, `http://newton.dm.unipi.it/neodys` and `http://unicorn.eis.uva.es/neodys`

Table 2.1 and Figure 2.1 we show the result of the computation of the stationary points for the Near Earth Asteroid 1991 TB$_2$. The data for all the examples in this paper are from the NEODyS site for the NEOs and from JPL ephemerides for the planets.

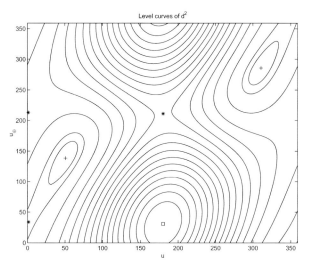

Figure 2. Level curves of d^2 for the asteroid 1991 TB$_2$ in the plane of the eccentric anomalies u (for the asteroid), $u_\oplus$ (for the Earth). Asterisks (*) are saddle points, squares (□) are maxima, crosses (+) are minima.

u	$u_\oplus$	d	type
50.73	138.52	0.11	MINIMUM
310.72	285.75	0.14	MINIMUM
1.50	33.86	0.56	SADDLE
1.15	212.86	1.45	SADDLE
180.21	210.96	2.61	SADDLE
180.05	30.90	4.61	MAXIMUM

Table 1. Stationary points for asteroid 1991 TB$_2$, giving the values of the eccentric anomalies (in degrees), the values of the distance d (in AU) and the type of stationary point.

The algebraic formulation of the problem has two main advantages: (a) allows to search for all the solutions using the efficient methods of modern Computational Algebra (Cox 1992; Bini 1997); (b) allows to give a bound for the maximum number of stationary points, as discussed below.

2.2. *Mutual geometry of confocal Keplerian orbits*

How many stationary points of d^2 may exist? And among them, how many local minima? In (Gronchi 2002) we found that the stationary points are at most 16 for the case of two ellipses and at most 12 if one orbit is circular, except for very particular cases with infinitely many stationary points. We do not have a proof that these are the optimal bounds. By a large number of numerical experiments we have found cases with at most 12 stationary points of d^2 and at most 4 local minima (see Gronchi 2002; Gronchi 2005).

In Table 2 we show a statistics of the stationary and minimum points of the distance function d between points on the orbit of the Earth and on the orbit of a NEA. On August 4, 2006 the NEODyS database contained the orbits of 4102 Near Earth Asteroids.

NEAs with 4 stationary points:	1829	NEAs with 1 minimum point:	1830
NEAs with 6 stationary points:	2244	NEAs with 2 minimum points:	2267
NEAs with 8 stationary points:	28	NEAs with 3 minimum points:	5
NEAs with 10 stationary points:	1	NEAs with 4 minimum points:	0

Table 2. Statistics with NEODyS database. Total number of NEAs: 4102.

Note that most mutual orbit configurations from this database have 2 local minima of d, among 6 stationary points. This is the most intuitive case, with a simple geometry. Moreover at least one maximum point must always exist, for topological reasons (d is a continuous function defined on a compact set, more precisely a 2–torus). There are also several cases with only 1 local minimum of d: surprisingly enough, the asteroid 2000 DK_{79} shows 2 maximum and 1 minimum point, that is not intuitive at all. This explains the difference of one unit between the NEAs with 4 stationary points and the NEAs with only 1 minimum. From a classical mathematical theory (see Milnor 1963) there is a simple relation among the stationary points, apart from special cases, in which the Hessian matrix of d^2 evaluated at the stationary points is degenerate. Generically

NUMBER of MIN. + MAX. POINTS = NUMBER of SADDLE POINTS .

The only NEA so far discovered with 10 stationary points is 2004 LG; this asteroid and 1997 US_9, 2004 BU_{58}, 2004 XM_{14}, 2005 NK_1 have 3 minimum points. No real asteroid has been found so far with 4 minimum points, although we know this case is possible from numerical experiments conducted with large sets of fictitious orbits.

Is there a geometric method to locate the minimum points along the orbits? When there is a crossing between the orbits (MOID=0) the minimum point of d corresponds to a mutual node. Is it always true that at least a local minimum point of d is close to a mutual node? The answer is negative: we can find examples, like asteroid 1991 TB_2 (see Figure 2.1 and Table 2.1), with two minima, both far enough from the mutual nodes. We can understand such cases arising from orbit configurations with low mutual inclination.

3. Uncertainty of the MOID

In this section we describe a method to define a meaningful uncertainty for all the local minima of d (see Gronchi & Tommei 2007), and in particular for the MOID, taking into account the uncertainty of the orbit of the asteroids. The second orbit is always the one of the Earth, so that its uncertainty is negligible. Nevertheless the method would allow to take into account also the uncertainty of the second orbit.

3.1. *Uncertainty of the orbit*

The observations of a celestial body are affected by errors, producing an error in the determination of the orbit. Let (E, v) be a set of orbital elements: E describes the geometric configuration of the orbit and v is a parameter along the trajectory. For example we can use the 5 Keplerian elements $E = (a, e, I, \Omega, \omega)$, and the true anomaly as v. For cometary orbits we can use the perihelion distance q in place of a.

Gauss' method gives us a nominal orbit (E^*, v^*), solution of a least squares fit, together with its uncertainty. The uncertainty is represented by the 6×6 *covariance matrix*

$$\Gamma_{(E,v)} = \begin{pmatrix} \Gamma_E & \vdots \\ \cdots & \Gamma_v \end{pmatrix},$$

which is the inverse of the *normal matrix*

$$C_{(E,v)} = \left[\frac{\partial \Xi}{\partial(E,v)}\right]^t \left[\frac{\partial \Xi}{\partial(E,v)}\right],$$

where Ξ is the vector of the observational residuals. Note that the 5×5 sub-matrix Γ_E gives the *marginal covariance* of the five elements E, independently from the value of the sixth one v, and $C^E = \Gamma_E^{-1}$ is the *marginal normal matrix*.

One way of representing the uncertainty of the orbital elements E is by means of *confidence ellipsoids*, an appropriate approximation whenever the least squares problem is quasi-linear and the uncertainty is moderate (see Milani 1999). By truncating the expansion of the target function (sum of squares, suitably weighed, of the residuals) to order 2 it is possible to represent the region in the orbital elements space where the target function does not increase above the minimum by more than a given value σ^2 as

$$Z_E(\sigma) = \{E \mid (E - E^*)^t \, C^E \, (E - E^*) \leqslant \sigma^2\},$$

that is, as an ellipsoid in the space of the geometric configurations E. If the eigenvalues of the covariance matrix Γ_E are too large this representation is inaccurate, because the truncation to order 2 is a poor approximation.

3.2. *The minimal distance maps and their singularities*

Let E be the vector of 5 orbital elements representing the orbit configuration of the asteroid and v the anomaly, and let $(E_\oplus, v_\oplus)$ be the orbit of the Earth, supposedly known with negligible errors.

Given the two orbit configuration $(E, E_\oplus)$, the Keplerian distance function can be regarded as a map $V = (v, v_\oplus) \mapsto d(E, V)$, with $E_\oplus$ as fixed parameters. Moreover the number of stationary points of $V \mapsto d(E, V)$ is constant for E in a neighborhood $\mathfrak{U}$ of E^*; actually this number can change only if the 2×2 Hessian matrix of the function $V \mapsto d(E^*, V)$ is degenerate. For each configuration E in $\mathfrak{U}$ we consider the minimum points $V_h(E)$ and we define the maps

$$d_h(E) = d(E, V_h(E)) \qquad \text{local minimal distance};$$

$$d_{min}(E) = \min_h d_h(E) \qquad \text{orbit distance (MOID)}.$$

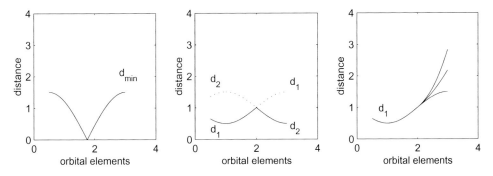

Figure 3. The singularities of the maps d_h and d_{min} are shown in the figure: from left to write we draw a sketch of the problems described in (i), (ii), (iii).

In Figure 3 we show the singularities of d_h and of d_{min}:
  (i) d_h and d_{min} are not differentiable where they vanish;

(ii) in a neighborhood of an orbit configuration E^*, two local minima can exchange their role as absolute minimum: then d_{min} can lose its regularity even without vanishing;

(iii) when a bifurcation occurs the definition of the maps d_h may become ambiguous. Note that this ambiguity does not occur for the d_{min} map. The bifurcation phenomena can occur only where the Hessian matrix of $d^2(E, V)$ is degenerate.

3.3. *Computation of the uncertainty of d_h and d_{min}*

The errors in the orbit affect the computation of the local minima of d: we want to estimate the size of this effect. Let us consider the orbit distance map d_{min}; the same method can be applied to compute the uncertainty of the minimal distance maps d_h.

For a given $(E^*, E_\oplus)$, the nominal orbit configuration E^* being endowed with its covariance matrix Γ_E, we can compute the covariance of $d_{min}(E^*)$ by a linear propagation of the matrix Γ_E (Jazwinski 1970)

$$\Gamma_{d_{min}(E)} = \left[\frac{\partial d_{min}}{\partial E}(E^*) \right] \Gamma_E \left[\frac{\partial d_{min}}{\partial E}(E^*) \right]^t. \tag{3.1}$$

The possibility of crossings between the orbits produces a singularity in this computation because the partial derivatives $\partial d_{min}/\partial E(E^*)$ do not exist when $d_{min}(E^*) = 0$, e.g., when the two orbits in the configuration $(E^*, E_\oplus)$ intersect each other.

An additional problem is that the uncertainty of a non–zero but small orbit distance may allow meaningless negative values of the distance.

Note that we are interested in knowing the uncertainty just when the orbit distance can be small or vanishing, that is when a collision or a close approach is possible. Thus an algorithm exploiting the classical covariance propagation to compute the uncertainty of the MOID is available only when it is not very useful. As a result of this deeply rooted mathematical difficulty, so far all the authors discussing the distribution of the MOID of asteroid orbits (e.g., the PHA population) have given up the use of the available orbit uncertainty information, with the only exception of the paper discussed in the next subsection.

3.4. *An approximation of the MOID*

An approximation of the MOID has been proposed in Wetherill (1967) by using the straight lines tangent to the orbits at the mutual nodes and taking the distance between these lines as two approximations of the local minima. (See the comments by ReVelle on Wetherill's biography in this volume.)

In Bonanno (2000) the uncertainty for the approximated MOID is computed by using Wetherill's approximation, as a distance between two straight lines. This distance can be given a sign, in accordance with the sign of the nodal distance (positive if the asteroid and the Earth mutual nodal points and the Sun are in this order, negative otherwise). This regularizes the approximated MOID function and allows to use the covariance propagation formulae of Section 3.3.

This approach gives useful approximate results in many cases, but has the following problems:

1) if the mutual orbital inclination I_M is zero the mutual nodes are not defined;

2) the minimum points can be located far from the mutual nodes; they also can be close to only one node;

3) the approximations of the local minima at the mutual nodes cannot be more than two while there are known cases with up to four local minimum points.

4. Regularization of the minimal distance maps

In Gronchi & Tommei (2007) a regularization of the maps d_h, d_{min} is introduced. Let us take into account the map d_{min}, the same method can also be applied to d_h. It is possible to make d_{min} locally analytic even where its value is zero, simply by changing its sign according to some properties of the orbit configuration.

The idea of the regularization can be illustrated by a simple example. Let us consider the positive function, defined on the whole plane, $f(x, y) = \sqrt{x^2 + y^2}$ and the function $\tilde{f}$, defined on a smaller domain by $\tilde{f}(x, y) = sign(x) \, f(x, y)$. The directional derivative of f in $(x, y) = (0, 0)$ does not exist for every choice of the direction. The regularized function $\tilde{f}$, extended by continuity to the origin $(0, 0)$, has all the directional derivatives in $(x, y) = (0, 0)$. How to extend such method to the problem at hand is discussed below.

4.1. *Geometric definition of the regularization*

Let τ_1, τ_2 be the tangent vectors to the orbits at the minimum point and let Δ_{min} be the vector joining the two tangency points ($|\Delta_{min}| = d_{min}$). If $\Delta_{min} \neq 0$ and τ_1 is not parallel to τ_2 we can define the nonzero vector $\tau_3 = \tau_1 \times \tau_2$. Due to the stationary points properties Δ_{min} is parallel to τ_3.

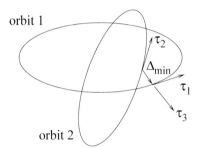

Figure 4. The orientation of the two parallel vectors Δ_{min}, τ_3 is the key to define a regular map $\tilde{d}_{min}$ by simply changing the sign of d_{min} on selected configurations E.

We define the regularized map $\tilde{d}_{min}$ by setting $|\tilde{d}_{min}| = d_{min}$ and choosing the sign $+$ for $\tilde{d}_{min}$ if Δ_{min} and τ_3 have the same orientation, the sign $-$ otherwise. This sign is well defined, with the only exception of the cases in which τ_1 and τ_2 are parallel.

Then we extend the definition domain to most crossing orbits setting $\tilde{d}_{min} = 0$ if $d_{min} = 0$. The orbit configurations with parallel tangent vectors to minimum points are also excluded from the definition domain even if they are not crossing points.

The resulting map $E \mapsto \tilde{d}_{min}(E)$ is locally analytic almost everywhere (see Gronchi & Tommei 2007), including in a neighborhood of most orbit configurations E such that $d_{min}(E) = 0$. In particular the partial derivatives can be computed as

$$\frac{\partial \tilde{d}_{min}}{\partial E_k}(E^*) = \left\langle \hat{\tau}_3(E^*), \frac{\partial \Delta}{\partial E_k}(E^*, V_{min}(E^*)) \right\rangle \qquad k = 1 \dots 5 \qquad (4.1)$$

where $V_{min}(E^*)$ is the absolute minimum point and $\Delta(E, V)$ is the vector joining the points corresponding to v and $v_\oplus$ on the orbit of the asteroid and of the Earth respectively.

Thus it becomes possible to use for the smooth function $\tilde{d}_{min}(E)$ the standard covariance propagation formula, applicable only to differentiable functions, including the really interesting low MOID cases.

4.2. *Covariance of $\tilde{d}_{min}$*

For each configuration $(E^*, E_\oplus)$, with covariance matrix Γ_E, we can compute the covariance of $\tilde{d}_{min}(E^*)$ as

$$\Gamma_{\tilde{d}_{min}(E)} = \left[\frac{\partial \tilde{d}_{min}}{\partial E}(E^*) \right] \Gamma_E \left[\frac{\partial \tilde{d}_{min}}{\partial E}(E^*) \right]^t \qquad (4.2)$$

by using the smooth partial derivatives of eq. (4.1). The *standard deviation*, defined as

$$\sigma_{min}(E^*) = \sqrt{\Gamma_{\tilde{d}_{min}(E^*)}},$$

allows us to define an *uncertainty interval* for $\tilde{d}_{min}(E^*)$. If we assume that the minimal distance $\tilde{d}_h(\tilde{\mathcal{E}})$ is a Gaussian random variable, there is a high probability ($\sim 99.7\%$) that its value is within the interval $[\tilde{d}_{min}(E^*) - 3\sigma_{min}(E^*), \tilde{d}_{min}(E^*) + 3\sigma_{min}(E^*)]$. This statement needs to be taken with some caution. It is necessary to check that the singular case (τ_1 parallel to τ_2) does not occur at E^* and is not even within the confidence ellipsoid. We need to check the variance of the determinant of the Hessian matrix to look for possible bifurcations of the stationary points. Last but not least, the propagation of the covariance by the linear formula of eq. (4.2) may be mathematically consistent, but to consider $\tilde{d}_{min}(E)$ as a Gaussian random variable is a good approximation only if the function $\tilde{d}_{min}$ is quasi-linear, when the uncertainty on E is small.

In this paper we are using only this quasi-linear approximation, thus the results are meaningful only when the distance between the nominal orbit configuration E^* and the ones corresponding to low MOID is small. This implies that the estimate of the probability is only approximate, and this approximation is useful only for comparatively high probabilities of being a PHA. Thus in the next Section we will present the results of the numerical tests on real asteroids by giving the probabilities approximated to 1% and by neglecting entirely the cases with probabilities < 1%. We will also apply this method only to asteroids for which there are enough data to compute an orbit with low to moderate uncertainty.

5. Search for Virtual NEAs and Virtual PHAs

We have computed the uncertainty of all the local minima of d using the orbit of the Earth and a large database of asteroid orbits, each with a covariance matrix representing its uncertainty. These orbits have been computed using the astrometric and photometric data made public by the Minor Planet Center (MPC) on March 2006, and they are divided into quality classes, according to their Arc–Type (see Milani *et al.* 2006).

The regularized distance maps $\tilde{d}_h$, $\tilde{d}_{min}$, and the perihelion distance q can be considered as Gaussian random variables, thus for each asteroid we can decide if it is a *Virtual NEA* or a *Virtual PHA* according to the following definitions:

1) a *Virtual NEA (VNEA)* is an asteroid that has a nonzero probability of being a NEA, i.e. having the perihelion distance $q \leqslant 1.3$ AU;

2) *Virtual PHA (VPHA)* is an asteroid that has a nonzero probability of being a PHA, i.e. having MOID $(= |\tilde{d}_{min}|]) \leqslant 0.05$ AU and absolute magnitude $H \leqslant 22$.†

Because of the limitations discussed above, we report only the cases in which the probabilities are $\geqslant 1\%$. The computation of such probabilities are explained below.

---

† We are not taking into account the uncertainty of the absolute magnitude; it depends on the uncertainty in the radial geocentric distance and in a deeper analysis it should be considered.

5.1. *Probability of being a NEA*

According to the standard formulae for the probability that a Gaussian random variable belongs to a given interval, we can compute the probability that for a given nominal two–orbit configuration E^* the perihelion distance $q = q(E^*)$ is less than 1.3 AU as

$$\mathcal{P}\left(q \leqslant 1.3\,\text{AU}\right) = \frac{1}{\sqrt{2\pi}} \int_{z_1}^{z_2} \exp(-z^2/2)\,dz \qquad (5.1)$$

with $z_i = (x_i - q(E^*))/\sigma_q(E^*)$ for $i = 1, 2$, where x_i are the extrema of the interval

$$[x_1, x_2] = [0, 1.3] \cap [q(E^*) - 3\sigma_q(E^*), q(E^*) + 3\sigma_q(E^*)],$$

and $\sigma_q(E^*)$ is the standard deviation of q, defined by

$$\sigma_q(E^*) = \sqrt{\Gamma_{q(E^*)}}, \qquad \Gamma_q(E^*) = \left[\frac{\partial q}{\partial E}(E^*)\right] \Gamma_E \left[\frac{\partial q}{\partial E}(E^*)\right]^t.$$

5.2. *Probability of being a PHA*

We describe the computation of the probability that a local minimum value d_h of d is less than 0.05 AU. We stress that the use of the regularized minimal distances $\tilde{d}_h$ is essential for this purpose: for small nominal values of d_h also negative values of the distance have to be admissible to perform the computation and the function measuring the minimal distance has to be regular also when it vanishes. The regularized map $\tilde{d}_h$ fulfills these properties and can be regarded as a random variable.

The probability that $\tilde{d}_h$ belongs to $[-0.05\,\text{AU}, 0.05\,\text{AU}]$ is

$$\mathcal{P}\left(|\tilde{d}_h| \leqslant 0.05\,\text{AU}\right) = \frac{1}{\sqrt{2\pi}} \int_{z_1}^{z_2} \exp(-z^2/2)\,dz \qquad (5.2)$$

with $z_i = (x_i - \tilde{d}_h(E^*))/\sigma_{\tilde{d}_h}(E^*)$ for $i = 1, 2$, where x_i are the extrema of the interval

$$[x_1, x_2] = [-0.05, 0.05] \cap [\tilde{d}_h(E^*) - 3\sigma_{\tilde{d}_h}(E^*), \tilde{d}_h(E^*) + 3\sigma_{\tilde{d}_h}(E^*)],$$

and $\sigma_{\tilde{d}_h}(E^*)$ is the standard deviation of $\tilde{d}_h$, defined by

$$\sigma_{\tilde{d}_h}(E^*) = \sqrt{\Gamma_{\tilde{d}_h(E^*)}}, \qquad \Gamma_{\tilde{d}_h}(E^*) = \left[\frac{\partial \tilde{d}_h}{\partial E}(E^*)\right] \Gamma_E \left[\frac{\partial \tilde{d}_h}{\partial E}(E^*)\right]^t.$$

5.3. *Results of the computation*

In this section we show first the asteroids that are not present in the "official" list of NEAs given by the Minor Planet Center (MPC)†, updated to August 4, and indeed have a non–negligible probability of being NEA or even PHA, we display the probability that they belong to such classes. Then we show the asteroids in the "official" list of NEA, that are not PHAs according to their nominal orbit, but are VPHAs.

The "official" lists of NEA and PHA are compiled by the MPC by using the elements of the nominal least squares orbits computed by the MPC. We recompute the least squares orbit, obtaining a slightly different result because of some different orbit determination algorithms and because we use of a different observations error model (see Carpino *et al.* 2003). Then we compute the uncertainty of both the perihelion q and the (regularized) MOID. In Tables 3 to 8 we show the VNEAs and VPHAs that are not present in the list given by the MPC. These asteroids are divided into classes according to their Arc–Type (see Milani *et al.* 2006).

† It is maintained at `http://cfa-www.harvard.edu/iau/NEO/TheNEOPage.html`

VNEA	q	RMS	prob
1994 RD	0.4955	0.00122	100%
2002 ON$_4$	1.2989	4.18×10^{-6}	100%
2002 PT$_{140}$	1.2936	9.10×10^{-6}	100%
2003 UW$_{26}$	1.2959	1.29×10^{-5}	100%
2004 SB	1.3028	0.00359	22%
1999 JS$_6$	1.3176	0.00831	2%

Table 3. VNEAs not present in the list of NEAs given by the MPC: Arc–Type $\geqslant 5$

VPHA	dist	RMS	H	prob
1994 RD	-0.020	1.72×10^{-4}	17.21	100%
1994 RD	-0.034	1.61×10^{-4}	17.21	100%

Table 4. VPHAs not present in the list of NEAs given by the MPC: Arc–Type $\geqslant 5$

First we take into account the best determined orbits, i.e. Arc–Type $\geqslant 5$. In Table 3 there are asteroids with a nominal perihelion distance very close to the boundary of the NEA class (i.e. 1.3 AU). This is a consequence of the use of an arbitrary boundary, not corresponding to a gap in the population: comparatively small differences in the nominal orbits computed by us and by the MPC result in a different classification.

We also find a very peculiar case: asteroid 1994 RD is present in the table as a NEA with probability 100% and a nominal value of $q < 0.5$ AU; moreover it is present in Table 4 as a PHA at both local minima of d with probability 100%. Further investigations led us to the following conclusion: indeed 1994 RD has been identified with 1991 AQ, and later numbered as (85182). It is present in the NEODyS website as (85182) 1991 AQ. Then why the name 1994 RD is present in the database provided by the MPC? The answer is that a sort of *administrative error* occured: the arc of 1994 RD, made by observations done on September 2,4,5,6,7,11 1994, is not present in the observation file for (85182) which contains the observations of September 8,17,18,22,17,19 and October 3,4, 1994.

VNEA	q	RMS	prob
2000 GM$_{146}$	1.044	0.020	100%
2005 QN$_{87}$	1.281	0.048	65%
2006 DV$_{62}$	1.230	0.001	59%
2004 AT$_1$	1.503	0.564	36%
2004 BD$_{11}$	1.305	0.012	35%
2003 FF$_{42}$	1.701	1.306	28%
2005 JK$_{173}$	1.647	2.079	22%
2004 VT$_{16}$	1.315	0.017	18%
2002 VF$_{118}$	1.478	0.178	16%

Table 5. VNEAs not present in the list of NEAs given by the MPC: Arc–Type $= 4$. We list only the asteroids with probability of being a NEA $> 10\%$.

The results for orbits with Arc–Type 4 are described in Table 5. The asteroid 2000 GM$_{146}$ has probability 100% of being a NEA with nominal $q < 1.05$ AU. This asteroid has been observed for 2 nights only, and it has not even an orbit computed by the MPC. We have found some VPHAs with Arc–Type 4 (see Table 6), but the probabilities are low.

In Tables 7, 8 we show our results for less well determined orbits: these have Arc–Type=3, but are still considered reliable enough for claiming discovery of the corresponding asteroids, according to (Milani *et al.* 2006).

VPHA	dist	RMS	H	prob
2004 AT_1	-0.654	1.146	16.34	3%
2003 FF_{42}	-0.714	1.308	15.51	3%
2005 JK_{173}	-0.812	1.820	15.04	2%
2002 VF_{118}	-0.634	0.278	17.06	1%
2004 BS_{159}	-1.560	0.732	17.06	1%

Table 6. VPHAs not present in the list of NEAs given by the MPC: Arc–Type = 4.

VNEA	q	RMS	prob
2005 UO_{497}	1.044	0.084	100%
2005 TQ_{79}	1.112	0.034	100%
2002 TO_{301}	0.947	0.040	100%
2001 FN_{91}	0.974	0.007	100%
2000 TF_2	1.289	0.007	94%
2002 EP_{150}	0.928	0.254	93%
2004 XZ_{44}	1.044	0.216	88%
2005 PX_{21}	1.072	0.215	85%
2005 JC_{78}	0.959	0.394	80%
2005 YX_{231}	1.187	0.183	73%
2001 SG_{340}	1.233	0.202	63%
1999 CG_{130}	0.844	0.704	63%
2003 LD_3	1.053	0.618	61%

Table 7. VNEAs not present in the list of NEAs given by the MPC: Arc–Type = 3. We list only the asteroids with probability of being a NEA > 60%.

Note that asteroid 2005 UO_{497}, present as a sure VNEA and as a VPHA with 37% probability, has a constrained orbit (see Milani *et al.* 2005). Another interesting case is 2001 FN_{91}, that is both a VNEA and a VPHA with probability 100%. The cases with the highest probability of being a PHA, namely 2001 FN_{91}, 1999 CG_{130} and 2005 UO_{497}, do not have an orbit computed by the MPC. This indicates that their orbit determination with the available data is not a trivial task. For 1999 CG_{130} there are no photometric data, leaving doubts on the nature of the object; note that in this case the minimum distance could be less than 0.05 AU near both nodes.

VPHA	dist	RMS	H	prob
2001 FN_{91}	-0.025	0.001	21.39	100%
1999 CG_{130}	0.050	0.037	–	50%
1999 CG_{130}	-0.016	0.077	–	48%
2005 UO_{497}	-0.062	0.076	20.28	37%
2005 JC_{78}	-0.084	0.134	17.52	24%
2005 PX_{21}	0.101	0.176	17.02	19%
2002 EP_{150}	0.133	0.098	17.85	17%
2004 XZ_{44}	0.089	0.284	19.01	13%
2006 DA_{72}	-0.067	0.311	15.67	13%
2005 YX_{231}	-0.197	0.184	19.52	12%
2006 DZ_{65}	0.088	0.353	18.70	11%
2005 RE_{26}	-0.134	0.067	13.29	10%

Table 8. VPHAs not present in the list of NEAs given by the MPC: Arc–Type = 3. We list only the asteroids with probability of being a NEA > 10%.

In Table 9 we display the VPHAs, found in the MPC list of NEAs updated to August 4, 2006, which are not PHA according to their nominal orbit.

VPHA	dist	RMS	H	prob
1994 XG	0.063	0.030	18.58	33%
2006 FW$_{33}$	0.066	0.111	20.12	30%
2000 VZ$_{44}$	-0.052	0.003	21.03	25%
2006 FW$_{33}$	0.108	0.115	20.12	22%
2006 KT$_{67}$	0.111	0.145	19.59	20%
2006 CD	-0.142	0.155	20.46	17%
1999 UZ$_5$	0.055	0.004	21.87	12%
1984 QY$_1$	0.179	0.084	14.16	6%
2006 OV$_5$	0.192	0.090	19.02	6%
2000 RK$_{12}$	0.056	0.004	21.27	5%

Table 9. VPHAs (down to probability > 5%) in the "official" list of NEAs, that are not PHAs according to their nominal orbit.

6. Conclusions and future work

We have done some progress in the understanding of the mutual geometry of two confocal Keplerian orbits: this seems to be a fairly difficult problem and there are still some interesting geometric features to investigate. The regularization of d_h and d_{min} explained above allows to define a meaningful uncertainty of the local minima of d even if this uncertainty leads to negative values of the distance. Moreover the orbit crossing singularity is removed, except for the tangent crossing case. We have computed the uncertainty of the local minima of d for several thousands of asteroids, whose orbit uncertainty is not too large, and for all the known NEAs. By this computation we have found some VPHAs among asteroids that are not even NEAs, according to their nominal orbits. The effect of the nonlinear terms, neglected in the determination of the orbit uncertainty and in the covariance propagation formula, still needs to be investigated. We expect that if the orbit of an asteroid is poorly determined, like for objects whose observations form an arc with Arc–Type 2, then the uncertainty of $\tilde{d}_h$ and $\tilde{d}_{min}$ might need to be computed with a method taking properly into account the nonlinearity of the orbit determination. This is the target for our future work.

References

Bini, D. A. 1997, *Numer. Algorithms* 13, 179

Bonanno, C. 2000, *Astron. Astrophys.* 360, 416

Bowell, E. & Muinonen, K., 1994, in: T. Gehrels (ed.), *Hazards due to comets & asteroids*, (Tucson: The University of Arizona Press), p. 149

Carpino, M., Milani, A. & Chesley, S. R. 2003, *Icarus* 166, 248

Cox, D., Little, J. & O'Shea, D. 1992, *Ideals, Varieties and Algorithms*, Springer-Verlag

Dybczynski, P. A., Jopek, T. J. & Serafin, R. A. 1986, *Celest. Mech. & Dynam. Astron.* 38, 345

Gronchi, G. F. 2002, *SIAM Journ. Sci. Comp.* 24, 61

Gronchi, G. F. 2005, *Celest. Mech. & Dynam. Ast.* 93, 297

Gronchi, G. F. & Tommei, G. 2007, *DCDS-B*, 7, 755

Hoots, F. R. 1984, *Celest. Mech. & Dynam. Astron.* 33, 143

Jazwinski, A. H. 1970, *Stochastic processes and filtering theory* , Academic Press

Kholshevnikov, K. V. & Vassiliev, N. 1999, *Celest. Mech. & Dynam. Astron.* 75, 75

Milani, A. 1999, *Icarus* 137, 269

Milani, A., Sansaturio, M. E., Tommei, G., Arratia, O. & Chesley, S. R. 2005, *Astron. Astrophys.* 431, 729

Milani, A., Gronchi, G. F. & Knežević, Z. 2006, *Earth, Moon & Planets*, in press

Milnor, J. 1963, *Morse theory*, Princeton University Press

Sitarski, G. 1968, *Acta Astron.* 18, 171

Wetherill, G. W. 1967, *J. Geophys. Res.* 72, 2429

Near Earth Objects, our Celestial Neighbors: Opportunity and Risk
Proceedings IAU Symposium No. 236, 2006
A. Milani, G.B. Valsecchi & D. Vokrouhlický, eds.
© 2007 International Astronomical Union
doi:10.1017/S174392130700302X

On the Lyapunov exponents of the asteroidal motion subject to resonances and encounters

Ivan I. Shevchenko

Pulkovo Observatory of the Russian Academy of Sciences, Pulkovskoje ave. 65-1,
St. Petersburg 196140, Russia
email: iis@gao.spb.ru

Abstract. In theoretical as well as practical issues of the asteroidal hazard problem, it is important to be able to assess the degree of predictability of the orbital motion of asteroids. Some asteroids move in a virtually predictable way, others do not. The characteristic time of predictability of any motion is nothing but the Lyapunov time (the reciprocal of the maximum Lyapunov exponent) of the motion. In this report, a method of analytical estimation of the maximum Lyapunov exponents of the orbital motion of asteroids is described in application for two settings of the problem. Namely, the following two types of the motion are considered: (1) the motion close to the ordinary or three-body mean motion resonances with planets, and (2) the motion in highly eccentric orbits subject to moderately close encounters with planets. Whatever different these settings may look, the analytical treatment is universal: it is performed within a single framework of the general separatrix map theory. (Recall that the separatrix maps describe the motion near the separatrices of a nonlinear resonance.) The analytical estimates of the Lyapunov times are compared to known numerical ones, i.e., to known estimates obtained by means of numerical integration of the orbits.

Keywords. Chaos, Lyapounov exponents, resonances, close approaches

---

1. Introduction

Assessment of the degree of predictability of the orbital motion of asteroids and other potentially hazardous objects is one of the most complicated aspects of the asteroidal hazard problem. Some asteroids move in a virtually predictable way, others do not. Whipple (1995) wrote: "The existence of a significant population of extremely chaotic Earth-crossing asteroids must be factored into the thinking about the potential hazard posed by these objects. An asteroid with a Lyapunov time of 20 years may be considered as an example. If the initial error in its position is 100 km (a very optimistic assumption) then that error will grow to one Earth radius in 83 years and to an Earth–Moon distance in 165 years. Assessments of the threat from specific objects like this can be made for only short spans of time."

Generally, the estimation of the Lyapunov exponents is one of the most important tools in the study of chaotic motion (Lichtenberg & Lieberman (1992)), in particular in celestial mechanics. The Lyapunov exponents characterize the mean rate of exponential divergence of trajectories close to each other in phase space; in the Hamiltonian systems, nonzero Lyapunov exponents indicate chaotic character of motion, while the maximum Lyapunov exponent equal to zero signifies regular (periodic or quasi-periodic) motion. The Lyapunov time (quantity reciprocal to the maximum Lyapunov exponent) gives the characteristic time of predictable dynamics.

The development of methods of numerical computation of the Lyapunov exponents has more than a thirty year history (see reviews in Froeschlé (1984), Lichtenberg & Lieberman (1992)). On the contrary, methods of analytical estimation of the Lyapunov exponents started to be developed only recently (Holman & Murray (1996), Murray & Holman (1997), Shevchenko (2000a), Shevchenko (2002), Shevchenko (2004a)).

In this report, a method of analytical estimation of the maximum Lyapunov exponents of the orbital motion of asteroids is described in application for two settings of the problem. Namely, the following two types of the motion are considered: (1) the motion close to the ordinary or three-body mean motion resonances with planets, and (2) the motion in highly eccentric orbits subject to moderately close encounters with planets. Whatever different these settings may look, the analytical treatment is universal: it is performed within a single framework of the general separatrix map theory.

The analytical estimates of the Lyapunov times are compared to known numerical ones, i.e., to known estimates obtained by means of numerical integration of the orbits.

2. The model of perturbed resonance

Under general conditions (Chirikov (1977), Chirikov (1979), Lichtenberg & Lieberman (1992)), a model of nonlinear resonance is provided by the Hamiltonian of the nonlinear pendulum with periodic perturbations:

$$H = \frac{\mathcal{G}p^2}{2} - \mathcal{F}\cos\varphi + a\cos(\varphi - \tau) + b\cos(\varphi + \tau). \tag{2.1}$$

The first two terms in Eq. (2.1) represent the Hamiltonian H_0 of the unperturbed pendulum; φ is the pendulum angle (the resonance phase angle), p is the momentum. The periodic perturbations are given by the last two terms; τ is the phase angle of perturbation: $\tau = \Omega t + \tau_0$, where Ω is the perturbation frequency, and τ_0 is the initial phase of the perturbation. The quantities $\mathcal{F}$, $\mathcal{G}$, a, b are constants. The frequency of the pendulum small-amplitude oscillations $\omega_0 \equiv (\mathcal{F}\mathcal{G})^{1/2}$.

An example of section of phase space of the Hamiltonian (2.1) at $\tau = 0 \bmod 2\pi$ is shown in Fig. 1 ($\Omega = 5$, $\omega_0 = 1$, $a = b$, $\varepsilon \equiv \frac{a}{\mathcal{F}} = 0.5$). This is a chaotic resonance triplet.

The motion near the separatrices of Hamiltonian (2.1) is described by the so-called separatrix algorithmic map (Shevchenko (1999)):

$$\text{if } w_n < 0 \text{ and } W = W^- \text{ then } W = W^+,$$
$$\text{if } w_n < 0 \text{ and } W = W^+ \text{ then } W = W^-;$$
$$w_{n+1} = w_n - W \sin\tau_n,$$
$$\tau_{n+1} = \tau_n + \lambda \ln\frac{32}{|w_{n+1}|} \quad (\bmod\ 2\pi); \tag{2.2}$$

with the parameters

$$\lambda = \frac{\Omega}{\omega_0}, \tag{2.3}$$

$$W^+(\lambda, \eta) = \varepsilon\lambda\left(A_2(\lambda) + \eta A_2(-\lambda)\right),$$
$$W^-(\lambda, \eta) = \varepsilon\lambda\left(\eta A_2(\lambda) + A_2(-\lambda)\right), \tag{2.4}$$

$\varepsilon = \frac{a}{\mathcal{F}}$, $\eta = \frac{b}{a}$. The Melnikov–Arnold integral $A_2(\lambda)$ is given by the relation

$$A_2(\lambda) = 4\pi\lambda\frac{\exp(\pi\lambda/2)}{\sinh(\pi\lambda)}, \tag{2.5}$$

see (Chirikov (1979), Shevchenko (1998b), Shevchenko (2000b)).

The quantity w denotes the relative (with respect to the separatrix value) pendulum energy: $w \equiv \frac{H_0}{\mathcal{F}} - 1$. The variable τ retains its meaning of the phase angle of perturbation.

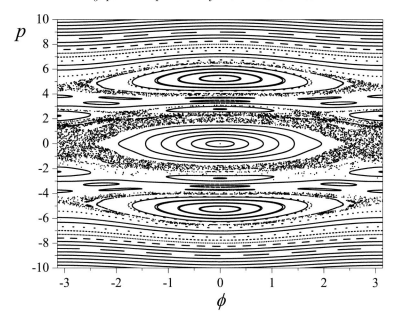

Figure 1. An example of a chaotic resonance triplet.

One iteration of map (2.2) corresponds to one half-period of pendulum's libration or one period of its rotation.

If $a = b$ (symmetric case), the separatrix algorithmic map reduces to the well-known ordinary separatrix map

$$w_{i+1} = w_i - W \sin \tau_i,$$
$$\tau_{i+1} = \tau_i + \lambda \ln \frac{32}{|w_{i+1}|} \quad (\text{mod } 2\pi), \tag{2.6}$$

written in the present form in (Chirikov (1977), Chirikov (1979)); the expression for W (Shevchenko (1998b), Shevchenko (2000b)) is

$$W = \varepsilon \lambda \left(A_2(\lambda) + A_2(-\lambda) \right) = 4\pi \varepsilon \frac{\lambda^2}{\sinh \frac{\pi \lambda}{2}}. \tag{2.7}$$

Formula (2.7) differs from that given in (Chirikov (1979), Lichtenberg & Lieberman (1992)) by the term $A_2(-\lambda)$, which is small for $\lambda \gg 1$. However, its contribution is significant for λ small (Shevchenko (1998b)), i.e., in the case of adiabatic chaos.

An equivalent form of Eqs. (2.6), used, e.g., in (Chirikov & Shepelyansky (1984), Shevchenko (1998a)), is

$$y_{i+1} = y_i + \sin x_i,$$
$$x_{i+1} = x_i - \lambda \ln |y_{i+1}| + c \quad (\text{mod } 2\pi), \tag{2.8}$$

where $y = w/W$, $x = \tau + \pi$; and

$$c = \lambda \ln \frac{32}{|W|}. \tag{2.9}$$

The applicability of the theory of separatrix maps for description of the motion near the separatrices of the perturbed nonlinear resonance in the full range of the relative frequency of perturbation, including its low values, was discussed and shown to be legitimate in (Shevchenko (2000b)).

The half-width y_b of the main chaotic layer of the separatrix map (2.8) in the case of the least perturbed border of the layer is presented as a function of λ in Fig. 1 in (Shevchenko (2004a)). The observed dependence follows the piecewise linear law with a transition point at $\lambda \approx 1/2$. This transition takes place not only in what concerns the width of the layer, but also in other characteristics of the motion, in particular, in the maximum Lyapunov exponent. The clear sharp transition at this point manifests a qualitative distinction between two types of dynamics, "slow" and "fast" chaos.

3. The method of analytical estimation of Lyapunov times

In (Shevchenko (2000a), Shevchenko (2002)), a method for estimation of the maximum Lyapunov exponent of the chaotic motion in the vicinity of separatrices of perturbed non-linear resonance was derived in the framework of the separatrix map theory. Following the general approach (Shevchenko (2000a), Shevchenko (2002)), we represent the maximum Lyapunov exponent L of the motion in the main chaotic layer of system (2.1) as the ratio of the maximum Lyapunov exponent L_{sx} of its separatrix map and the average period T of rotation (or, equivalently, the average half-period of libration) of the resonance phase φ inside the layer. For convenience, we introduce a non-dimensional quantity $T_{sx} = \Omega T$. Then the general expression for L is

$$L = \Omega \frac{L_{sx}}{T_{sx}}. \tag{3.1}$$

The quantity $T_L \equiv L^{-1}$, by definition, is the Lyapunov time.

We consider four generic resonance types: the fastly chaotic resonance triplet, fastly chaotic resonance doublet, slowly chaotic resonance triplet, slowly chaotic resonance doublet (we call them, respectively, the "ft", "fd", "st", "sd" resonance types).

3.1. Fast chaos. Resonance triplet

Consider the case of $a = b$, $\lambda > 1/2$. This means that there is a symmetric triad of interacting resonances (first condition), and chaos is fast (second condition).

The case of the fastly chaotic triad is completely within the range of applicability of the method presented in (Shevchenko (2000a), Shevchenko (2002)). The perturbed nonlinear resonance is modelled by Hamiltonian (2.1) with $a = b$. Following (Shevchenko (2000a), Shevchenko (2002)), we take the dependence of the maximum Lyapunov exponent of the separatrix map (2.8) upon λ in the form

$$L_{sx}(\lambda) \approx C_h \frac{2\lambda}{1 + 2\lambda}, \tag{3.2}$$

where $C_h \approx 0.80$ is a constant (Shevchenko (2004b)).

The average increment of τ (proportional to the average rotation period, or libration half-period) inside the chaotic layer is (Chirikov (1979), Shevchenko (2000a), Shevchenko (2002)):

$$T_{sx}(\lambda, W) \approx \lambda \ln \frac{32e}{\lambda |W|}, \tag{3.3}$$

where e is the base of natural logarithms. From Eq. (3.1), one has for the Lyapunov time for the "ft" resonance type:

$$T_L = \frac{T_{pert}}{2\pi} \frac{T_{sx}}{L_{sx}} \approx T_{pert} \frac{(1 + 2\lambda)}{4\pi C_h} \ln \frac{32e}{\lambda |W|}, \tag{3.4}$$

where $T_{pert} = 2\pi/\Omega$ is the period of perturbation.

3.2. *Fast chaos. Resonance doublet*

The previous analysis of the symmetric case $a = b$ sets a foundation for an analysis of the more general asymmetric case $a \neq b$, since the Lyapunov exponents in the asymmetric case can be found by averaging the contributions of the separate components of the chaotic layer (Shevchenko (2004a)).

Calculation of the average constitutes a complicated problem. In particular, one should know the relative average times of residence of the system in three different components of the layer corresponding to direct rotation, reverse rotation, and libration of the pendulum. The relative times of residence depend on the asymmetry of perturbation. A simple heuristic method of averaging was proposed in (Shevchenko (2000a), Shevchenko (2002)), but rigorous solution is still far from being found.

In view of these difficulties, we consider the limit case of a or b equal to zero. It means that one of the two perturbing resonances simply does not exist, and instead of the resonance triad we have a duad.

If $\lambda > 1/2$, the equality $b = 0$ implies $|W^-| \ll |W^+|$, and, vice versa, $a = 0$ implies $|W^-| \gg |W^+|$. We designate the dominating quantity by W.

Consider first the libration side of the chaotic layer. Then W^- and W^+ alternate (replace each other) at each iteration of the separatrix algorithmic map (2.2). It is straightforward to show that, if W^- or W^+ is equal to zero, the separatrix algorithmic map (2.2) on the doubled iteration step reduces to the ordinary separatrix map (2.6) with the doubled value of λ and the same non-zero value of W. One iteration of the new map corresponds to two iterations of the old one. Since the half-width of the chaotic layer of map (2.6) is $\approx \lambda W$ (Chirikov (1979), Shevchenko (2004a)), the layer's extent in w on the side of librations doubles, it becomes $\approx 2\lambda W$. Note that the parameters λ and W are considered here as independent from each other.

Consider then the circulation sides of the chaotic layer. The side corresponding to reverse (or direct) rotations does not exist, if W^- (or, respectively, W^+) is equal to zero; its measure is zero. The other side, corresponding to direct (or reverse) rotations is described by the ordinary separatrix map (2.6) with the parameters λ, W; its extent in w is $\approx \lambda W$.

The averaged (over the whole layer) value of the maximum Lyapunov exponent is the sum of weighted contributions of the layer components corresponding to the librations, direct rotations and reverse rotations of the pendulum. The weights are directly proportional to the times that the trajectory spends in the components, and, via supposed approximate ergodicity, to the relative measures of the components in phase space. Taking into account the just made estimates of the widths of the chaotic layer's components in the duad case, one can expect that the relative weights of librations and circulations in the "fd" case are respectively 4 and 1.

Hence the formula for the Lyapunov exponent for the "fd" resonance type is

$$L = \frac{\Omega}{\mu_{libr} + 1} \left(\mu_{libr} \frac{L_{sx}(2\lambda)}{T_{sx}(2\lambda, W)} + \frac{L_{sx}(\lambda)}{T_{sx}(\lambda, W)} \right), \qquad (3.5)$$

and

$$T_L = \frac{T_{pert}}{2\pi} \cdot \frac{\mu_{libr} + 1}{\mu_{libr} \frac{L_{sx}(2\lambda)}{T_{sx}(2\lambda, W)} + \frac{L_{sx}(\lambda)}{T_{sx}(\lambda, W)}}, \qquad (3.6)$$

where $\mu_{libr} \approx 4$, and W, L_{sx}, T_{sx} are given by formulas (2.7, 3.2, 3.3).

3.3. *Slow chaos. Resonance triplet*

In the case of $\lambda < 1/2$, the diffusion across the layer is slow, and on a short time interval the phase point of the ordinary separatrix map (2.8) follows close to some current curve. We call this curve guiding. Let us derive an analytical expression for the guiding curve with an irrational winding number far enough from the main rationals. We approximate the winding number by the rationals m/n. Thus $c \approx 2\pi m/n$. Noticing that at an iteration n of the map the phase point hits in a small neighborhood of the starting point, one obtains for the derivative:

$$\frac{dy}{dx} = \frac{1}{nc - 2\pi m} \sum_{k=0}^{n-1} \sin(x + kc) =$$

$$= \frac{1}{nc - 2\pi m} \sin\frac{nc}{2} \operatorname{cosec}\frac{c}{2} \sin\left(x + \frac{n-1}{2}c\right). \tag{3.7}$$

Integrating and passing to the limit $n \to \infty$, one obtains:

$$y = -\frac{1}{2} \operatorname{cosec}\frac{c}{2} \cos\left(x - \frac{c}{2}\right) + \mathcal{C}, \tag{3.8}$$

where $\mathcal{C}$ is an arbitrary constant of integration.

The motion is chaotic only when the curve (3.8) crosses the singular line $y = 0$. Hence the half-width of the chaotic layer is $y_b = \left|\operatorname{cosec}\frac{c}{2}\right|$. Averaging (by taking an integral analytically) the quantity $-\ln|y_{i+1}|$ (equal to $(\langle\Delta x\rangle - c)/\lambda$, Eqs. (2.8)), where y_{i+1} is substituted by y of Eq. (3.8), over the chaotic layer in the derived boundaries, we find the approximate analytical expression for $\Theta \equiv (T_{sx} - c)/\lambda$:

$$\Theta \approx \ln\left|4\sin\frac{c}{2}\right|. \tag{3.9}$$

Then, we need an expression for $L_{sx}(\lambda)$. We explore the λ dependence of L_{sx} in a numerical experiment. At each step in λ (namely, $\Delta\lambda = 0.005$, $\lambda \geqslant 0.005$) we find the value of c corresponding to the case of the least perturbed layer and plot the value of L_{sx}. At $\lambda < 0.3$, the dependence turns out to be practically linear. The linear fit $L_{sx}(\lambda) = a\lambda$ gives $a = 1.01132 \pm 0.00135$, and the correlation coefficient $R = 0.9998$.

We set $L_{sx} \approx \lambda$ for the generic (non-resonant) values of c and for $\lambda < 1/2$. Then, from Eqs. (3.1, 3.9) one has the following approximate formula for the maximum Lyapunov exponent:

$$L \approx \frac{\Omega}{\ln\left|4\sin\frac{c}{2}\right| + \frac{c}{\lambda}}, \tag{3.10}$$

where $c = \lambda \ln\frac{32}{|W|}$ (Eq. (2.9)).

For $\lambda \ll 1$ one has $W \approx 8\varepsilon\lambda$, hence the formula for the Lyapunov time for the "st" resonance type:

$$T_L \approx \frac{T_{pert}}{2\pi} \ln\left|\frac{16}{\varepsilon\lambda} \sin\left(\frac{\lambda}{2}\ln\frac{4}{|\varepsilon|\lambda}\right)\right|. \tag{3.11}$$

3.4. *Slow chaos. Resonance doublet*

Utilizing the approximation of the Melnikov–Arnold integral $A_2(\lambda) \approx 2\pi\lambda + 4$ at $\lambda \ll 1$, $\eta = 0$, one has: $W^{\pm} \approx \varepsilon\lambda(4 \pm 2\pi\lambda) \approx 4\varepsilon\lambda$. So, in the "sd" case, the separatrix

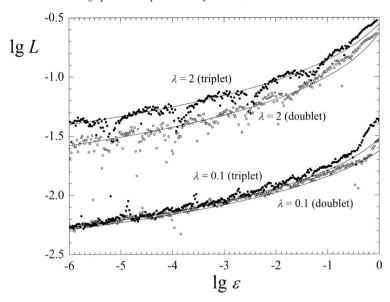

Figure 2. The maximum Lyapunov exponent of the chaotic motion of system (2.1) in dependence on the magnitude of perturbation: the results of direct computation (circles) and the theoretical curves.

algorithmic map (2.2) degenerates to the ordinary separatrix map (2.6) with $W \approx 4\varepsilon\lambda$, i.e., mathematically the case is equivalent to the "st" case, but with a different (halved) value of W.

Following the lines of the previous Subsection, it is then straightforward to write down the formula for the Lyapunov time for the "sd" resonance type:

$$T_L \approx \frac{T_{pert}}{2\pi} \ln \left| \frac{32}{\varepsilon\lambda} \sin \left(\frac{\lambda}{2} \ln \frac{8}{|\varepsilon|\lambda} \right) \right|. \tag{3.12}$$

3.5. *Theory versus numerical experiment*

To check the theory, the Lyapunov exponents of the chaotic motion near the separatrices of Hamiltonian (2.1) have been directly computed by means of the program package (Shevchenko & Kouprianov (2002), Kouprianov & Shevchenko (2003)) utilizing the HQRB method by von Bremen, Udwadia & Proskurowski (1997). The power of this method is far greater than that necessary in the present computation. It allows one to evaluate the full Lyapunov spectrum of a multidimensional system. The spectrum of our system consists of a sole pair of Lyapunov exponents — the maximum one and its negative counterpart. The integration of the equations of motion has been performed by the integrator by Hairer, Nørsett & Wanner (1987). It is an explicit 8th order Runge–Kutta method due to Dormand and Prince, with the step size control. We choose the time unit in such a way that $\omega_0 = 1$.

The results of the computations for $\lambda = 0.1$ and 2 are shown in Fig. 2 (circles). The integration time interval has been chosen to be equal to 10^6. This is sufficient for the computed values of the Lyapunov exponents to saturate in each case. Local wave-like patterns represent prominent features of the constructed dependences; they are conditioned by the process of encountering resonances, while ε changes.

The theoretical dependences are shown in Fig. 2 as solid curves. T_L are given by formulas (3.4, 3.6, 3.11, 3.12); $L = 1/T_L$. Close correspondence is observed between the theory and experimental data for each value of λ.

4. Lyapunov times of the asteroidal motion subject to resonances

The mean motion resonances and secular resonances represent the main classes of the orbital resonances in the motion of asteroids. The mean motion ones represent commensurabilities of the periods of the motion of asteroids and planets. There are two main subclasses of the mean motion resonances: ordinary (two-body) resonances and three-body resonances.

4.1. Ordinary mean motion resonances

The Hamiltonian of the motion of a zero-mass test particle in the gravitational field of the Sun and Jupiter, in the plane of Jupiter's orbit, in the vicinity of a mean motion resonance with Jupiter can be represented in some approximation in the form (Holman & Murray (1996), Murray & Holman (1997)):

$$H = \frac{1}{2}\beta\Lambda^2 - \sum_{p=0}^{q} \phi_{k+q,k+p,k} \cos(\psi - p\omega), \qquad (4.1)$$

where $\beta = 3k^2/a^2$, $\Lambda = \Psi - \Psi_{res}$, $\Psi = (\mu_1 a)^{1/2}/k$, $\Psi_{res} = (\mu_1^2/(k^2(k+q)n_J))^{1/3}$, $\mu_1 = 1 - \mu$, $\omega \equiv -\varpi$ (i.e., ω is minus the longitude of asteroid's perihelion; its time derivative is assumed to be constant); a and e are asteroid's semimajor axis and eccentricity. The integer non-negative numbers k and q define the resonance: the ratio $(k+q)/k$ equals the ratio of mean motions of an asteroid and Jupiter in the exact resonance. The phase $\psi \equiv kl - (k+q)l_J$, where l and l_J are the mean longitudes of an asteroid and Jupiter.

Here the units are chosen in such a way that the total mass (Sun plus Jupiter), the gravitational constant, Jupiter's semimajor axis a_J are all equal to one; $\mu = 1/1047.355$, $\mu_1 = 1 - \mu$. Jupiter's mean longitude $l_J = n_J t$, eccentricity $e_J = 0.048$. Jupiter's mean motion $n_J = 1$, i.e., the time unit equals $\frac{1}{2\pi}$th part of Jupiter's orbital period.

According to Eq. (4.1), the resonance $(k+q)/k$ splits in a cluster of $q+1$ subresonances $p = 0, 1, \ldots, q$. The coefficients of the resonant terms are

$$|\phi_{k+q,k+p,k}| \approx \frac{\mu}{q\pi a_J}\binom{q}{p}\left(\frac{\epsilon}{2}\right)^p\left(\frac{\epsilon_J}{2}\right)^{q-p}, \qquad (4.2)$$

where $\epsilon = ea_J/|a - a_J|$, $\epsilon_J = e_J a_J/|a - a_J|$ (Holman & Murray (1996), Murray & Holman (1997)). The approximation (4.2) is good, if $\epsilon q < 1$ (Holman & Murray (1996)). Besides, the model is restricted to the resonances of relatively high order, $q \geqslant 2$.

The signs of the coefficients $\phi_{k+q,k+p,k}$ alternate with changing p, so, the coefficients with numbers p and $p+2$ are always of the same sign. This means that whatever is the choice of the guiding resonance in the multiplet, its closest neighbors have coefficients of equal signs, and η is always non-negative.

The frequency of small-amplitude oscillations at the subresonance p

$$\omega_0 = (\beta|\phi_{k+q,k+p,k}|)^{1/2} \approx \frac{a_J}{|a - a_J|}n_J\left(\mu_1\mu\frac{4q}{3\pi}\binom{q}{p}\left(\frac{a}{a_J}\right)\left(\frac{\epsilon}{2}\right)^p\left(\frac{\epsilon_J}{2}\right)^{q-p}\right)^{1/2} \qquad (4.3)$$

Table 1. Numerical and analytical estimates of Lyapunov times for ordinary
mean motion resonances, T_L in years, $e = 0.1$

$\frac{k+q}{k}$	λ	$\lg T_L^{num}$ †	$\lg T_L^{theor}$	Res. type
3/1	0.093	3.8–4.3	4.3	sd
5/2	0.192	3.5–3.8	4.1	st
7/3	0.415	3.8–4.2	4.0	st
9/4	0.932	3.9–4.3	4.2	fd*
11/5	1.970	3.9–4.3	4.3	ft*
9/5	0.323	3.6–3.8	3.7	st*
7/4	0.166	3.2–3.3	3.7	st
12/7	0.594	3.6–4.0	3.9	fd*
5/3	0.101	2.5–3.3	3.7	sd
8/5	0.156	2.5–3.3	3.6	st*
11/7	0.264	3.3–3.6	3.5	st*

† Morbidelli & Nesvorný (1999), Holman & Murray (1996)

and the perturbation frequency

$$\Omega = \dot{\omega} \approx \frac{\mu_1 \mu}{2\pi} n_J \left(\frac{a}{a_J} \right)^{1/2} \left(\frac{a_J}{a - a_J} \right)^2 , \tag{4.4}$$

cf. (Holman & Murray (1996), Murray & Holman (1997)). The ratio of Ω and ω_0 gives
the value of λ.

Now we are able to apply the theory developed in Section 3. For comparison, we take
the data on the numerical (based on integrations) values of T_L for the motion near mean
motion resonances from Fig. 1 in (Morbidelli & Nesvorný (1999)) and Fig. 6 in (Holman
& Murray (1996)). The theoretical estimates are made by means of formulas (3.4, 3.6,
3.11, 3.12). Before they are used, the guiding resonance in the multiplet is identified (it
has the maximum value of $|\phi_{k+q,k+p,k}|$), and its two closest neighbors are considered
as the perturbing resonances. Then, the formula is chosen in accord with the resonance
type (fastly chaotic triad "ft", fastly chaotic duad "fd", slowly chaotic triad "st", or
slowly chaotic duad "sd"). If the amplitudes of the neighbors differ from each other less
than twice, the model resonance is considered to be a triad, otherwise a duad. Those
resonances which have $\epsilon q > 1$ are marked in Table 1 by an asterisk.

The analytical maximum Lyapunov exponent estimates are generally in agreement
with the numerical ones. However, some differences can be clearly seen, especially in the
domain of slow chaos. This should be attributed to the imperfectness of model (4.1), and
mainly to the fact that the coefficients $\phi_{k+q,k+p,k}$ are treated as constants. They fix the
frequencies ω_0 of small amplitude oscillations at subresonances, and when the period of
perturbation is large in relation to the period of these oscillations, the variations of ω_0
can have greater dynamical influence.

Also another effect can be of importance. The differential distribution, built by
Shevchenko, Kouprianov & Melnikov (2003) for a representative plane of starting values

(the trajectories were computed by the Wisdom map (Wisdom (1983)) in the planar elliptic restricted three-body problem), demonstrates that the maximum Lyapunov exponent of the trajectories near the 3/1 mean-motion resonance with Jupiter has two, and not one, preferable numerical values: the distribution has a bimodal peak structure. This signifies that there are two distinct domains of chaos in phase space; thus the perturbed pendulum model as applied to this low-order mean motion resonance turns out to be too approximate. This example shows also that generally a closer look at the numerical data may be necessary when comparing it with theory.

Now consider an example of an estimate for a real asteroid, namely (522) Helga. This object is famous to be the first example of "stable chaos" among asteroids (Milani & Nobili (1993)): while its Lyapunov time is relatively small (6900 years), its orbit does not exhibit any gross changes on cosmogonic time scales, according to numerical experiments. It is known to be in the 12/7 mean motion resonance.

Let us apply our method. The necessary data on a, e, the perihelion frequency $g = \dot{\varpi}$ are taken from the "numb.syn" catalogue (Knežević & Milani (2000)) of the AstDyS web service†. T_{pert} is defined by the value of g. We find that the guiding subresonance in the resonance sextet is the third one ($p = 2$), consequently the perturbing neighbors in our model have the numbers $p = 1$ and 3. The quantity $\epsilon q = 0.624 < 1$, so there are no problems with the potential model. The derived separatrix map parameters are: $\lambda = 2.325$, $\eta = 0.812$, consequently the model resonance type is the fastly chaotic triplet "ft". Applying formula (3.4), one has $T_L = 9700$ years. The agreement with the values, obtained in integrations in the full problem, (6900 years (Milani & Nobili (1993)), 6860 years (AstDyS)) should be considered as satisfactory.

4.2. Three-body mean motion resonances

An important role in the orbital dynamics of bodies of the Solar system, in particular asteroids, is played by the so-called three-body resonances (Murray, Holman & Potter (1998), Nesvorný & Morbidelli (1998), Nesvorný & Morbidelli (1999)). In the case of a three-body resonance, the resonant phase is a combination of angular elements of the orbits of three bodies (a test one and two perturbing ones; e.g., an asteroid, Jupiter, and Saturn).

The three-body resonances can be described by the perturbed pendulum model (Murray, Holman & Potter (1998), Nesvorný & Morbidelli (1998), Nesvorný & Morbidelli (1999)). The Hamiltonian of the motion of a zero-mass test particle near a three-body resonance $\{m_J m_S m\}$ with Jupiter and Saturn in the planar-elliptic problem can be expressed, in some approximation, in the following form (Nesvorný & Morbidelli (1999)):

$$H = \alpha S^2 + \sum_{p_J, p_S, p} \beta_{p_J p_S p} \cos \sigma_{p_J p_S p}, \qquad (4.5)$$

where the conjugated to S resonance argument $\sigma_{p_J p_S p} = m_J l_J + m_S l_S + m l + p_J \varpi_J + p_S \varpi_S + p \varpi$ (it is assumed that the time derivatives of l_J, l_S, ϖ_J, ϖ_S are constants), $\alpha = -(3/2) n^2 a_{res}^{-2}$. Analytical expressions $\beta_{p_J p_S p}(e)$ for some important three-body resonances are given in Tables 3–6 in (Nesvorný & Morbidelli (1999)). It is clear from Eq. (4.5) that the three-body resonance $\{m_J m_S m\}$ splits in a cluster of subresonances with various $\{p_J p_S p\}$ combinations.

† http://hamilton.dm.unipi.it/cgi-bin/astdys/

Table 2. Numerical and analytical estimates of Lyapunov times for asteroids in three-body mean motion resonances

Asteroid	Resonance $\{m_J m_S m\}$	λ	T_L^{num} yr †	T_L^{num} yr ‡	T_L^{theor} yr	Res. type
258 Tyche	$2 + 2 - 1$	0.536	35900	–	43100	ft
485 Genua	$3 - 1 - 1$	0.376	6550	6500	35700	sd
1642 Hill	$3 - 1 - 1$	0.643	36100	–	43300	fd
936 Kunigunde	$6 + 1 - 3$	0.624	22200	–	54600	fd
490 Veritas	$5 - 2 - 2$	0.546	10200	8500	9100	fd
2039 Paine-Gaposchkin	$5 - 2 - 2$	0.449	22000	–	6020	sd
3460 Ashkova	$5 - 2 - 2$	0.433	65100	8300	5940	sd

† AstDyS
‡ Nesvorný & Morbidelli (1998), Nesvorný & Morbidelli (1999), Milani, Nobili & Kneževič (1997)

The frequency of small-amplitude oscillations at the subresonance $\{p_J p_S p\}$ is (Nesvorný & Morbidelli (1999)):

$$\omega_0 = 2\pi n (3\beta_{p_J p_S p})^{1/2} a_{res}^{-1}. \tag{4.6}$$

The perturbation frequency Ω is generally an algebraic combination of perihelion frequencies of Jupiter, Saturn and the asteroid. The formula for this combination is defined by the choice of the guiding subresonance (see below). The ratio of Ω and ω_0 gives the value of λ.

We consider the asteroids residing close to the three-body resonances studied in (Nesvorný & Morbidelli (1999)), and utilize the analytical data in Tables 3–6 in (Nesvorný & Morbidelli (1999)) on the coefficients of resonant terms. The theoretical estimates of Lyapunov times are made by means of formulas (3.4, 3.6, 3.11, 3.12). Before they are used, the guiding resonance in the multiplet is identified (it has the maximum value of $|\beta_{p_J p_S p}|$), and its two closest neighbors are considered as the perturbing resonances. Then, the formula is chosen in accord with the resonance type ("ft", "fd", "st", "sd").

On identification of the guiding resonances in the multiplets, it turns out that the three-body resonances under study subdivide in two distinct classes: those for which the perturbation frequency Ω in model (2.1) is equal to $\dot{\varpi} - \dot{\varpi}_J$, and those for which it is equal to $\dot{\varpi}_S - \dot{\varpi}_J$. The resonances $5 - 2 - 2$ and $3 - 1 - 1$ belong to the first class, while $2 + 2 - 1$ and $6 + 1 - 3$ to the second. We use $\dot{\varpi}_J = 4.257''/\text{yr}$ and $\dot{\varpi}_S = 28.243''/\text{yr}$ (Bretagnon (1990)). The data on $\dot{\varpi}$, a, and e are taken from the "numb.syn" catalogue (Kneževič & Milani (2000)) of the AstDyS web service.

The theoretical estimates, obtained in this way, are presented in Table 2. Some of them are in accord with the numerical ones (in particular, in the case of (490) Veritas), others are not. From the fact of disagreement in the cases of (485) Genua and (2039) Paine-Gaposchkin, one can judge that these objects do not, most probably, reside in the chaotic layers of the prescribed resonance multiplets. Thus the analytical estimation of Lyapunov times represents a promising tool for discerning between possible models of chaos in the motion of real asteroids, and, generally, celestial bodies.

5. Lyapunov times of the asteroidal motion subject to encounters

5.1. *Lyapunov exponents of the motion described by the Kepler map*

Consider a map similar to Eqs. (2.8), but with a power-law phase increment instead of the logarithmic one:

$$y_{i+1} = y_i + \sin x_i,$$
$$x_{i+1} = x_i - \lambda|y_{i+1}|^{-\gamma} + c \quad (\text{mod } 2\pi). \qquad (5.1)$$

A number of mechanical and physical models are described by such maps. The case of $\gamma = 3/2$, $c = 0$ corresponds to the Kepler map. It was derived and analyzed in (Chirikov & Vecheslavov (1986), Vecheslavov & Chirikov (1988), Chirikov & Vecheslavov (1989), Petrosky (1986)) in order to describe the chaotic motion of the Halley comet and, generally, the motion of comets in nearly parabolic orbits. The motion model consists in the assumption that the main perturbing effect of Jupiter is concentrated when the comet is close to the perihelion of its orbit. This effect is defined by the phase of encounter with Jupiter. The variable y has the meaning of the normalized full energy of the comet, while x is the normalized time. One iteration of the map corresponds to one orbital revolution of the comet.

Chirikov's constant C_h^{gen} for general separatrix map (5.1) with an arbitrary value of γ is introduced in the same way as one for basic map (2.8): it is the least upper bound for the maximum Lyapunov exponent of the motion in the main chaotic layer of the map. The proper limit can be shown to exist in the same way as it was done in (Shevchenko (2004b)) for the ordinary separatrix map.

By means of linearization of map (5.1) in y it is straightforward to see that the value of y corresponding to the critical value of the stochasticity parameter $K = K_G$ of the approximating standard map is $y_b = (\gamma\lambda/K_G)^{\frac{1}{\gamma+1}}$, while the value of y corresponding to $K = 4$ is $y_p = (\gamma\lambda/4)^{\frac{1}{\gamma+1}}$. The first of these values marks the border of the chaotic layer, while the second one roughly separates mostly "non-porous" and mostly "porous" parts of the layer. The ratio $y_b/y_p \approx 4^{\frac{1}{\gamma+1}}$; hence the contribution of the porous part to the value of the maximum Lyapunov exponent in the layer becomes negligible with γ increasing. This makes the estimation of Chirikov's constant in the case of general separatrix map (5.1) with $\gamma > 0$ more precise than in the case of map (2.8), because the contribution of the porous part, which is small here, is most uncertain.

An expression for Chirikov's constant C_h^{gen} for the general separatrix map is derived from Eq. (9) in (Shevchenko (2004b)) in the same way as in the case of Chirikov's constant C_h for the basic separatrix map (cf. Shevchenko (2004b)), except the change of the variable and the expression for layer's half-width are different. The resulting expression is

$$C_h^{gen}(\gamma) = \frac{K_G^{\frac{1}{\gamma+1}}}{(\gamma+1)\sigma(\gamma)} \int\limits_{K_G}^{\infty} L(K)\mu(K) \frac{dK}{K^{\frac{\gamma+2}{\gamma+1}}}, \qquad (5.2)$$

where

$$\sigma(\gamma) = \frac{K_G^{\frac{1}{\gamma+1}}}{\gamma+1} \int\limits_{K_G}^{\infty} \mu(K) \frac{dK}{K^{\frac{\gamma+2}{\gamma+1}}}. \qquad (5.3)$$

The functions $L(K)$ and $\mu(K)$ (the maximum Lyapunov exponent of the standard map and the measure of the chaotic component in phase space of the standard map, both in dependence on the stochasticity parameter) were computed in (Shevchenko (2004b)).

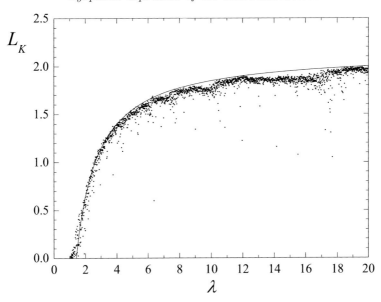

Figure 3. The λ dependence of the maximum Lyapunov exponent of the Kepler map and the curve (5.4).

Performing numerical calculation of integral (5.2), we obtain the value of $C_K \equiv C_h^{gen}(\gamma = 3/2)$, that turns out to be equal to $2.21\ldots$.

Let us check the obtained value $C_K \approx 2.21$ versus a direct computation of the maximum Lyapunov exponent of the Kepler map. In Fig. 3, the computed maximum Lyapunov exponent of map (5.1) is plotted versus λ. The limit $\lambda \to \infty$ of the dependence gives Chirikov's constant; one can see that the computed dependence is in accord with this prediction. All the observed data are below the line $C_K = 2.2$, as expected. A good fit to the computed dependence is given by the function

$$L_K(\lambda) = C_K - \frac{3}{\lambda},\tag{5.4}$$

where, however, $C_K = 2.15$. Therefore, for usage in applications, it is prudent to set $C_K \approx 2.2$.

The derivation of the Lyapunov time estimate for a highly-eccentric object is based on a consideration of the Kepler map as a separatrix map, the unperturbed parabolic trajectory playing the role of the separatrix. Then, the Lyapunov time estimate is given by the relation

$$T_L \approx \frac{T_{orb}}{L_K(\lambda)},\tag{5.5}$$

where T_{orb} is the average orbital period of the object. The lower bound for the Lyapunov time is just

$$T_L \approx \frac{T_{orb}}{C_K}.\tag{5.6}$$

This value of T_L corresponds to the motion with $\lambda \gg 1$.

The cradle cause of dynamical chaos in this setting of the problem lies not in "close encounters", as is often physically interpreted, but in the overlapping of resonances in phase space of the Kepler map. So, the intrinsic source of chaos — the resonant interaction — for the orbital motion of planet-encountering bodies and for the motion of ordinary asteroids is one and the same.

5.2. *Lower bounds for the Lyapunov times of NEAs and comets*

The lower bound for the Lyapunov time of the orbital motion of a planet-encountering body is given by relation (5.6). For the comet 1P/Halley, with the orbital period of 76 years, the lower bound, given by this formula, is about 34 years. Using the data (Chirikov & Vecheslavov (1986), Vecheslavov & Chirikov (1988), Chirikov & Vecheslavov (1989)) on the amplitude of perturbation of the full energy of the comet, one finds $\lambda \approx 1.2 \cdot 10^4 \gg 1$; consequently, the found lower bound for T_L is close to the expected T_L value itself.

It is very probable that relation (5.6) applies for the motion of any long-periodic comet, or any Halley-type comet, or any asteroid in a highly-eccentric orbit with a similar kind of perturbation; i.e., the value of the Lyapunov time of such an object is determined solely by its orbital period.

Whipple (1995) explored the chaotic orbital motion of 175 real asteroids with $q <$ 1.6 AU (in the inner part of the Solar system). He found that the Lyapunov times can be as small as 10 years, due to encounters with the terrestrial planets. The range of values of T_L is 10–20000 years. 34 of these 175 objects are so chaotic, that the errors in determination of their orbits double in less than 70 years.

To our present knowledge, no known asteroid or comet violates the bound (5.6). In this respect, the diagram "Lyapunov time – semimajor axis", constructed by Whipple (1995) (see Fig. 2 in his paper), is of particular interest: one can see that all the considered objects have $T_L \geqslant 10$ years, and, since they all have the semimajor axis $a <$ 3.5 AU, bound (5.6) is in no way violated.

Tancredi (1995), Tancredi (1999) considered the orbital evolution of 145 Jupiter family comets and 307 NEAs (inactive objects with aphelia $Q > 1$ AU and perihelia $q <$ 1.5 AU). He found T_L values in the range 30–200 years for the first group of objects, and 10–300 years, mostly 50–150 years, for the second one. The minimum observed value was ≈ 10 years.

Thus the planet-encountering asteroids and comets are among the most chaotic objects of the Solar system; their Lyapunov times can be as low as several years. This is in accord with the simple estimate (5.6).

6. Conclusions

In this report, we have addressed the problem of predictability of the chaotic asteroidal motion. Up to the present moment, large numerical material on the Lyapunov times of the chaotic asteroidal motion has been accumulated in literature. In overwhelming majority, these estimates were obtained by means of numerical integration. In view of necessity of theoretical explanation of these data, we have presented a method of analytical estimation of the maximum Lyapunov exponents of the orbital motion of asteroids. It is based on the separatrix map theory.

We have considered the chaotic asteroidal motion close to the ordinary and three-body mean motion resonances with planets, and the motion in highly eccentric orbits subject to moderately close encounters with planets. For the case of the mean motion resonances, we have derived simple analytical formulas for the Lyapunov time in four basic resonance type models: the fastly chaotic resonance triad, fastly chaotic resonance duad, slowly chaotic resonance triad, slowly chaotic resonance duad. For the case of highly-eccentric objects subject to moderate encounters with planets, we have derived simple analytical formulas for the Lyapunov time and its lower bound.

The analytical estimates of the Lyapunov times for model and real asteroids have been made and compared to many known numerical ones, i.e., to known estimates obtained by means of numerical integration of orbits, including estimates obtained for real objects in the full problem of motion in the Solar system. In many cases a satisfactory agreement have been observed, that testifies the quality of the corresponding theoretical models of asteroidal motion. On the other hand, the cases of disagreement are even more interesting: they may imply either an imperfectness of the adopted perturbed pendulum model as applied to the considered resonance (as noted in Section 4.1 in relation to the 3/1 mean motion resonance), or an incorrect identification of the guiding resonance. So, one can conclude that the analytical estimation of the Lyapunov times may represent a promising tool for discerning between possible models of chaos in the motion of real asteroids, and, generally, celestial bodies.

Acknowledgements

It is a pleasure to acknowledge the useful comments of a referee. This work was partially supported by the Russian Foundation for Basic Research (project # 05-02-17555) and by the Programme of Fundamental Research of the Russian Academy of Sciences "Fundamental Problems in Nonlinear Dynamics". The computations were partially carried out at the St. Petersburg Branch of the Joint Supercomputer Center of the Russian Academy of Sciences.

References

Bretagnon, P. 1990, *Astron. Astrophys.* 231, 561
Chirikov, B. V. 1977, *Nonlinear Resonance* (Novosibirsk: Izdatel'stvo NGU) (In Russian)
Chirikov, B. V. 1979, *Phys. Rep.* 52, 263
Chirikov, B. V. & Shepelyansky, D. L. 1984, *Physica* D 13, 395
Chirikov, B. V. & Vecheslavov, V. V. 1986, *INP Preprint* 86–184
Chirikov, B. V. & Vecheslavov, V. V. 1989, *Astron. Astrophys.* 221, 146
Froeschlé, Cl. 1984, *Celest. Mech.* 34, 95
Hairer, E., Nørsett, S. P. & Wanner, G. 1987, *Solving Ordinary Differential Equations I. Nonstiff Problems* (Berlin: Springer-Verlag)
Holman, M. J. & Murray, N. W. 1996, *Astron. J.* 112, 1278
Knežević, Z. & Milani, A. 2000, *Preprint*
Kouprianov, V. V. & Shevchenko, I. I. 2003, *Astron. Astrophys.* 410, 749
Lichtenberg, A. J. & Lieberman, M. A. 1992, *Regular and Chaotic Dynamics* (New York: Springer-Verlag)
Milani, A. & Nobili, A. M. 1993, *Celest. Mech. Dyn. Astron.* 56, 323
Milani, A., Nobili, A. M. & Knežević, Z. 1997, *Icarus* 125, 13
Morbidelli, A. & Nesvorný, D. 1999, *Icarus* 139, 295
Murray, N. W. & Holman, M. J. 1997, *Astron. J.* 114, 1246
Murray, N., Holman, M. & Potter, M. 1998, *Astron. J.* 116, 2583
Nesvorný, D. & Morbidelli, A. 1998, *Astron. J.* 116, 3029
Nesvorný, D. & Morbidelli, A. 1999, *Celest. Mech. Dyn. Astron.* 71, 243
Petrosky, T. Y. 1986, *Phys. Letters* A 117, 328
Shevchenko, I. I. 1998a, *Phys. Letters* A 241, 53
Shevchenko, I. I. 1998b, *Physica Scripta* 57, 185
Shevchenko, I. I. 1999, *Celest. Mech. Dyn. Astron.* 73, 259
Shevchenko, I. I. 2000a, *Izvestia GAO* 214, 153 (In Russian)
Shevchenko, I. I. 2000b, *J. Exp. Theor. Phys.* 91, 615 [ZhETP 118, 707]
Shevchenko, I. I. 2002, *Cosmic Res.* 40, 296 [Kosmich. Issled. 40, 317]
Shevchenko, I. I. & Kouprianov, V. V. 2002, *Astron. Astrophys.* 394, 663
Shevchenko, I. I., Kouprianov, V. V. & Melnikov, A. V. 2003, *Solar System Res.* 37, 74 [*Astronomicheskii Vestnik* 37, 80]

Shevchenko, I. I. 2004a, in: G. Byrd *et al.* (eds.), *Order and Chaos in Stellar and Planetary Systems*, ASP Conf. Series, vol. 316, p. 20

Shevchenko, I. I. 2004b, *JETP Letters* 79, 523 [*Pis'ma Zh. Eksp. Teor. Fiz.* 79, 651]

Tancredi, G. 1995, *Astron. Astrophys.* 299, 288

Tancredi, G. 1999, *Celest. Mech. Dyn. Astron.* 70, 181

Vecheslavov, V. V. & Chirikov, B. V. 1988, *Sov. Astron. Letters* 14, 151

von Bremen, H. F., Udwadia, F. E. & Proskurowski, W. 1997, *Physica* D 101, 1

Whipple, A. L. 1995, *Icarus* 115, 347

Wisdom, J. 1983, *Icarus* 56, 51

Near Earth Objects, our Celestial Neighbors: Opportunity and Risk
Proceedings IAU Symposium No. 236, 2006
A. Milani, G.B. Valsecchi & D. Vokrouhlický, eds.
© 2007 International Astronomical Union
doi:10.1017/S1743921307003031

Resonant trans-Neptunian objects as a source of Jupiter-family comets

E. L. Kiseleva† and V. V. Emel'yanenko

Department of Computational and Celestial Mechanics, South Ural University, Russia
email: kleo@susu.ac.ru

Abstract. The dynamical interrelation between resonant trans-Neptunian objects and short-period comets is studied. Initial orbits of resonant objects are based on computations in the model of the outward transport of objects during Neptune's migration in the early history of the outer Solar system. The dynamical evolution of this population is investigated for 4.5 Gyr, using a symplectic integrator. Our calculations show that resonant trans-Neptunian objects give a substantial contribution to the planetary region. We have estimated that the relative fraction of objects captured per year from the 2/3 resonance to Jupiter-family orbits with perihelion distances $q < 2.5$ AU is 0.4×10^{-10} near the present epoch.

Keywords. Trans-Neptunian objects, comets, resonances, migration.

1. Introduction

The dynamical and physical characteristics of short-period comets imply that these bodies originate from the outer Solar system. It was shown by Duncan & Levison (1997) and by Emel'yanenko *et al.* (2004) that the scattered disc trans-Neptunian objects are a source of Jupiter-family comets. On the other hand, Morbidelli (1997), and Nesvorný & Roig (2000) demonstrated that objects from the classical Edgeworth-Kuiper belt moving in the resonance 2/3 with Neptune can also contribute to the population of Jupiter-family comets. The estimate of the transfer rate from the 2/3 resonance to Jupiter-family comets is the main aim of the present paper.

We try to investigate the features of that resonant population which arises during the evolution under planetary perturbations for the age of the Solar system. The long-term motion in the 2/3 resonance was studied by Morbidelli (1997). It was shown that the mechanism of the chaotic diffusion can drive Plutinos out of the 2/3 resonance, after which they are subjected to close encounters with the giant planets and eventually evolve into Jupiter-family comets. However, Morbidelli (1997) studied a uniform distribution of initial orbits. Now we know that the mechanism of Neptune's migration is very important in the formation of the resonant population structure (Malhotra 1993, 1995; Hahn & Malhotra 1999, 2005).

Therefore, we analyze the distribution of resonant objects which arises after the outward transport of objects during Neptune's migration in the early history of the Solar system and the evolution under planetary perturbations for the age of the Solar system. Our integrations are performed using the symplectic integrator by Emel'yanenko (2002). This explicit second-order integrator uses a time transformation which allows to adjust the time-step in correspondence with the distance from the Sun and the magnitude of perturbations. Thus it can handle both high-eccentricity orbits and close encounters with planets.

† Present address: South Ural University, 76 Lenina, Chelyabinsk, Russia, 454080

2. Model of migration

We have tested the hypothesis that Neptune migrated outwards by several AU during the solar system's past and, in so doing, sculpted the pattern of resonance occupation in the Kuiper belt (Malhotra 1995).

The simulation parameters are identical to those in Malhotra (1995). They correspond to a smooth migration of Neptune. We consider four models with different values of initial inclinations and eccentricities. The eccentricities and inclinations of objects are distributed in the ranges (0, 0.001), (0, 0.01), (0, 0.05), and (0, 0.1), respectively, for the four models. Each set contains 1000 initial orbits in the range (27, 42) AU of semimajor axes.

We have found that dynamical features are similar in all models, and they are analogous to results of Malhotra (1995). Therefore, in the following we discuss only computations for the fourth model where the inclination distribution is the broadest, and is more consistent with the observed distribution of the 2/3 resonant population than for the other models.

3. Evolution for the age of the Solar system

We have taken 332 objects located in the 2/3 resonance region from the fourth scenario. The inclinations and eccentricities of these objects lie in the ranges (0.9, 22.1) degrees and (0.01, 0.33), respectively. These orbits have been integrated for 4.5 Gyr, unless objects reach orbits with perihelion distances $q < 2.5$ au or evolve to hyperbolic orbits. We include perturbations from the four outer planets.

The number of particles surviving in the 2/3 resonance after 4.5 Gyr is 44. Thus the 2/3 resonance population after 4.5 Gyr is approximately ten times as small as the original one. From 44 resonant objects, five objects were located in this region at the beginning of the planetary migration, and the others reached this region in the process of migration.

Figure 1 shows the initial and final distributions of semimajor axes and eccentricities for objects in the 2/3 resonance.

4. Evolution to Jupiter-family orbits

Figure 2 shows the number N of particles surviving in the resonance 2/3 as a function of time for the last 2 Gyr of integrations. We have found that 7 objects reach the region $q < 20$ AU for the last billion years of integrations. These objects are cloned 10 times. We study the orbital evolution of these objects until they are captured to $q < 2.5$ au or evolve to a hyperbolic orbit. We then apply the method described in Emel'yanenko *et al.* (2004) to calculate the rate of injection of Jupiter-family comets from the 2/3 resonant population.

We have estimated that the relative fraction of objects captured per year from the 2/3 resonance to Jupiter-family orbits with perihelion distances $q < 2.5$ AU is 0.4×10^{-10} near the present epoch. This is approximately two times as large as the corresponding estimate in the model of Morbidelli (1997).

Figure 3 shows the distribution of Tisserand parameters and inclinations for Jupiter-family comets with $q < 2.5$ AU, captured from the 2/3 resonant population. This distribution is consistent with the observed distribution of Jupiter-family comets. However, this distribution is more concentrated towards high values of Tisserand parameter than that in models studying high-eccentricity trans-Neptunian objects (Emel'yanenko *et al.* 2004) and objects from the Oort cloud (Emel'yanenko *et al.* 2005) as sources of Jupiter-family comets.

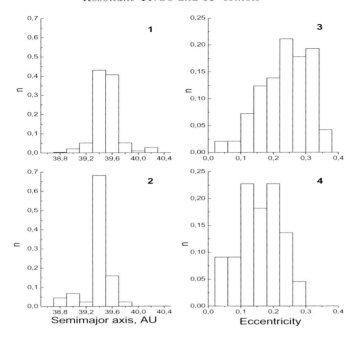

Figure 1. (1) The distribution of semimajor axes for objects in the 2/3 resonance at the end of Neptune's migration. (2) The distribution of semimajor axes for objects in the 2/3 resonance after 4.5 Gyr. (3) The distribution of eccentricities for objects in the 2/3 resonance at the end of Neptune's migration. (4) The distribution of eccentricities for objects in the 2/3 resonant objects after 4.5 Gyr.

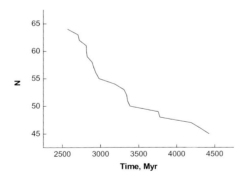

Figure 2. Evolution of the number N of particles survivng in the resonance 2/3 as a function of time.

5. Conclusions

We studied the dynamical connection between resonant trans-Neptunian objects and short-period comets in the model based on the origin of objects by the outward transport during Neptune's migration and the subsequent evolution under planetary perturbations for 4.5 Gyr. Our calculations show that the relative fraction of objects captured per year from the 2/3 resonance to Jupiter-family orbits with perihelion distances $q < 2.5$ AU is 0.4×10^{-10} near the present epoch. This estimate is close to that obtained earlier for scattered disc objects as a source of Jupiter-family comets (Emel'yanenko *et al.* 2004).

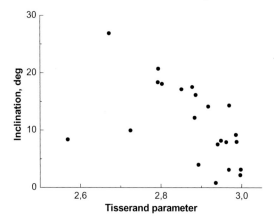

Figure 3. The distribution of Tisserand parameters and inclinations for simulated objects from the 2/3 resonant population when their perihelia first drop below 2.5 AU.

Therefore, the relative number of Jupiter-family comets originating from the 2/3 resonance and the scattered disc should be approximately the same as the ratio of these populations. Thus the contribution of resonant trans-Neptunian objects to the population of Jupiter-family comets is not negligible.

6. Acknowledgments

This work was supported by RFBR Grants 04-02-96042 and 06-02-16512. The authors are grateful to Giovanni Valsecchi for helpful suggestions concerning the style of this paper.

References

Duncan, M.J. & Levison, H.F. 1997, *Science* 276, 1670
Emel'yanenko, V.V. 2002, *Celest. Mech. & Dyn. Astron.* 84, 331
Emel'yanenko, V.V., Asher, D.J. & Bailey, M.E. 2004, *Mon. Not. R. Astron. Soc.* 350, 161
Emel'yanenko, V.V., Asher, D.J. & Bailey, M.E. 2005, *Mon. Not. R. Astron. Soc.* 361, 1345
Hahn, J.M. & Malhotra, R. 1999, *AJ* 117, 3041
Hahn, J.M. & Malhotra, R. 2005, *AJ* 130, 2392
Malhotra, R. 1993, *Nature* 365, 819
Malhotra, R. 1995, *AJ* 110, 420
Morbidelli, A. 1997, *Icarus* 127, 1
Nesvorný, D. & Roig, F. 2000, *Icarus* 148, 282

Near Earth Objects, our Celestial Neighbors: Opportunity and Risk
Proceedings IAU Symposium No. 236, 2006
A. Milani, G.B. Valsecchi & D. Vokrouhlický, eds.
© 2007 International Astronomical Union
doi:10.1017/S1743921307003043

Orbital evolution of short-period comets with high values of the Tisserand constant

N.Yu. Emel'yanenko

Department of Computational and Celestial Mechanics, South Ural University, 76 Pr. Lenina, Chelyabinsk, Russia
email: emel@math.susu.ac.ru

Abstract. The orbital evolution of comets with high values of the Tisserand constant is studied for a time interval of 800 years. Scenarios of dynamical evolution are obtained for 85 comets. Particular features of the orbital evolution of the comets of this class are singled out. The orbits of all comets are tangent to the orbit of Jupiter and have a steadily low inclination. For 80% of comets, the evolution scenario includes a timespan in which the comets move in low-eccentricity orbits. The possibility is analyzed of a change in the Tisserand constant and of a transition of the comet to be controlled by other giant planets.

Keywords. Evolution, short-period comet, cyclic transformation.

1. Introduction

In Jupiter family, about 100 comets are known (Marsden & Williams 2003) whose Tisserand constants satisfy the condition

$$T_s \geqslant 2.9 \qquad (1.1)$$

Nowadays, the orbits of these comets are tangent to the Jovian orbit at the aphelion or at the perihelion. In what follows, such orbits will normally be referred to as A-type (P-type) orbits or denoted as A and P, respectively. Kazimirchak-Polonskaya (1967), Everhart (1973), Carusi & Valsecchi (1980, 1982) and Froeschlé & Rickman (1981) have shown that such orbits are extremely unstable because of frequent encounters with Jupiter.

We study here for a time interval of 800 years the evolution of the comets whose orbits satisfy condition (1.1). Such comets are called the short-period comets of Jupiter family (JFC).

The dynamics of cometary orbits is investigated by numerically integrating the equations of motion. The Everhart method and his RADAU code (Everhart 1974) are used. The calculations are based on the systems of orbital elements borrowed from Marsden & Williams' catalog (2003). Perturbations from nine planets are taken into account (the mass of Mercury being included in the solar mass). If a comet penetrates into the jovicentric sphere of radius $k = 0.08$ AU, perturbations due to the oblateness of Jupiter are taken into account (Emel'yanenko 2003a, 2003b).

The scenarios of the orbital evolution of comets are studied here for the time interval from $t_1 = 2\,300\,000.5$ to $t_2 = 2\,600\,000.5$.

2. Orbital evolution

2.1. *Basic Objectives of the Study*

These are:

1) The search for peculiarities in the orbital evolution of the comets of this class.
2) Possibilities for changes in the Tisserand constant.
3) Possibilities for the transition of comets to being controlled by other giant planets.
4) The presence of persistent tendencies in the variation of cometary orbits in the past and in the future.

2.2. *Analysis of the Orbital Evolution of Comets*

We consider separately three scenarios of preferred variations in the orbital elements of the comets under study.

2.2.1. *Variations in the elements governing the spatial orientation of the orbit*

The performed investigation showed that, to a certain degree, all the comets experience variations in the angular orbital elements over the time interval $t_1 - t_2$. Among all the orientation elements, let us pay particular attention to the inclination of the orbit of the comet. It is well known that different classes of comets have different distributions of the element i. I examined in this study the stability of the distribution of the cometary orbit inclination for the comets of Jupiter's family which obey condition (1.1) and have, by and large, extremely unstable orbits. To this end, I investigated possible transformations of the element i in the approach region, total changes in the inclination over the approach time, and the evolutionary changes in the element i for a time interval that encompasses $100-150$ revolutions of the comet around the Sun.

Near the encounter with Jupiter, the inclination of a cometary orbit can vary over an extremely wide range. For the investigated comets, in the vicinity of the jovicentric distance minima, all the known cometary orbits with the corresponding value of the element i can be found as osculating orbits.

However, the variation of the element i in the encounter region is nearly symmetric with respect to the moment of the achievement of the minimum. Therefore, comets enter the encounter region and leave it at almost the same orbital inclination.

In studying evolutionary changes of the element i, I compared the values of the cometary orbit inclination at the moments t_1 (in the past) and t_2 (in the future).

The comets under study have a steadily low orbit inclination for the investigated time interval. In the past and future, the maximum inclination has that of the orbit of 83P/Russell 1 ($i_1 = 27°$, $i_2 = 23°$).

On the whole, the performed investigation of changes in the cometary orbit inclination shows that the distribution of the element i is stable over the time interval $t_1 - t_2$ for the group of comets of the Jupiter family which satisfy condition (1.1).

Because of the small inclination of the cometary orbit to the ecliptic plane, changes in the two other elements specifying the spatial orientation of the orbital plane virtually do not affect the conditions of subsequent encounters of every comet with Jupiter. The proximity of the aphelion and/or the perihelion of the cometary orbit to the mean radius of the Jovian orbit is the determining factor for the occurrence of these encounters throughout the time interval at hand.

2.2.2. *Variations in the shape of a cometary orbit*

We shall refer to a cometary orbit as a low-eccentricity orbit, or l orbit, if its eccentricity e obeys the condition

$$e \leqslant 0.33 \tag{2.1}$$

as a medium-eccentricity orbit, or m orbit, if

$$0.33 < e \leqslant 0.66 \tag{2.2}$$

and as a high-eccentricity orbit, or h orbit, if

$$e > 0.66 \tag{2.3}$$

Thus, a cometary orbit will hereinafter be denoted by a letter, A or P, with a subscript indicating the orbit eccentricity. For example, the notation A_m corresponds to a comet whose orbit is tangent to the Jovian orbit at the aphelion and has an eccentricity satisfying condition (2.2).

Any change in the shape of a cometary's orbit and in the type of its tangency to the Jovian orbit, which occurs in the course of evolution, will be denoted by an arrow indicating the direction of the change (for example, $A_l \rightarrow P_m$). The evolutionary changes of the cometary orbits studied here are indicated in the first column of Table 1, with the second column giving the number of comets undergoing each evolution.

The investigation of changes in the cometary orbit shapes was aimed at resolving the following questions:
- what were the predominant eccentricities of the comets of Jupiter's family under condition (1.1) for the time interval $t_1 - t_2$,
- how did the eccentricity of a cometary's orbit change near the encounter with Jupiter,
- and what particular features took place in the behavior of the eccentricity in the course of evolutionary transformations of the orbit?

The analysis of the evolution of cometary orbits showed the following: the orbits of all comets satisfy condition (2.1) or (2.2) over the studied time interval. 16 comets are constantly in medium-eccentricity m orbits and three comets, in low-eccentricity P_l or orbits A_l. 7 comets are at all times in low-eccentricity orbits but undergo transitions $A_l \leftrightarrow P_l$. The other 59 comets change the shape of their orbits. All four possible scenarios of shape transformation take place: either the form of a cometary's orbit changes without the reversal of tangency (transitions $A_m \leftrightarrow A_l$, $P_m \leftrightarrow P_l$), or both the form and the type of tangency to the Jovian orbit are changed (transitions $P_m \leftrightarrow P_l$, $P_m \leftrightarrow A_l$).

It is the encounters with Jupiter that directly affect the changes of the element e. Deeply in the sphere of action of Jupiter, the osculating eccentricity values vary over an extremely wide range, up to the hyperbolic value. However, as consequence of the encounter, only two scenarios take place in the behavior of the eccentricity function: the element e either changes (according to the scheme $m \rightarrow l$ or $l \rightarrow m$) or remains unchanged (the orbit of the comet satisfying condition (2.1) or (2.2) both before and after the encounter).

For the comets studied, the main feature of the orbit-eccentricity behavior is the presence of a timespan Δt during which these comets move in P_l or A_l orbits. For 68 comets (80%), the evolution scenario includes at least one such timespan Δt in the interval $t_1 - t_2$.

2.2.3. *Changes in the sizes of cometary orbits*

In considering changes in the cometary's orbit sizes, the principal attention was given to the following questions: what are the orbit sizes of the comets under study in the time interval $t_1 - t_2$ what particular features are related to the orbit sizes of the comets

Table 1. Short-period comets of Jupiter's family with high values of the Tisserand constant. Evolutionary transitions

Evolutionary transitions	N
$P_l \to P_m \to P_l \to P_m$	1
$P_l \to P_m \to P_l \to A_l \to P_l \to A_l$	1
$P_l \to P_m \to P_l \to A_l \to P_l$	1
$P_l \to P_m \to A_l \to A_m \to A_l \to A_m \to P_l \to A_l$	1
$P_l \to P_m \to A_m$	1
$P_l \to A_l \to P_l \to A_l \to A_m \to A_l \to A_m \to A_l \to A_m \to A_l$	1
$P_l \to A_l \to P_l \to A_l \to A_m \to A_l$	2
$P_l \to A_l \to P_l \to P_m$	2
$P_l \to A_l \to P_l$	1
$P_l \to A_l \to A_l \to P_m$	1
$P_l \to A_l \to A_m \to A_l \to P_l \to A_l$	1
$P_l \to A_l \to A_m \to A_l$	1
$P_l \to A_l$	1
P_l	1
$P_m \to P_l \to A_l \to P_l \to A_l \to A_m$	1
$P_m \to P_l \to A_l \to P_l \to A_l$	1
$P_m \to P_l \to A_l$	1
$P_m \to A_l \to P_l \to A_l$	2
$P_m \to A_l \to A_m \to A_l$	1
$P_m \to A_m$	2
P_m	1
$A_l \to P_l \to A_l \to P_l \to A_l \to P_l \to A_l$	1
$A_l \to P_l \to A_l \to P_l \to A_l$	1
$A_l \to P_l \to A_l \to P_l$	2
$A_l \to P_l \to A_l \to A_m \to A_l \to A_m \to A_l$	1
$A_l \to P_l \to A_l \to A_m \to A_l$	1
$A_l \to P_l$	2
A_l	2
$A_l \to A_m \to P_l$	1
$A_l \to A_m \to A_l \to P_l \to A_l \to P_l$	1
$A_l \to A_m \to A_l$	3
$A_l \to A_m \to A_l \to A_m \to A_l$	3
$A_l \to A_m \to A_l \to A_m \to A_l \to A_m$	1
$A_l \to A_m \to A_l \to A_m$	2
$A_l \to A_m$	9
$A_m \to P_l$	1
$A_m \to P_l \to A_l \to A_m \to A_l \to A_m \to A_l \to P_l \to A_l$	1
$A_m \to P_l \to A_m \to A_l$	1
$A_m \to A_l \to P_l \to P_m \to A_m$	1
$A_m \to A_l \to P_l \to A_l \to A_m \to A_l \to P_l \to A_l$	1
$A_m \to A_l \to A_m \to A_l \to P_l \to A_l \to P_l \to A_l$	1
$A_m \to A_l \to A_m \to A_l \to A_m \to A_l$	1
$A_m \to A_l \to A_m \to A_l \to A_m$	1
$A_m \to A_l \to A_m \to A_l$	3
$A_m \to A_l \to A_m$	5
$A_m \to A_l$	1
A_m	14

of a given class, and what are the transformations in the orbit size which occur during encounters with Jupiter?

116P/Wild 4 had the minimum orbit size ($a_{min} = 2.90$ AU) and P/1997 T3 Lagerkvist-Carsenty the maximum size ($a_{max} = 12.03$ AU).

The basic feature related to the sizes of the cometary orbits is their permanent tangency to the orbit of Jupiter over the 800-yr time interval. For the orbits tangent at the perihelion, $q_{min} = 3.71$ AU (P/1997 V1 Larsen) and $q_{max} = 6.74$ AU (P/1999 XN120 Catalina). For the orbits tangent at the aphelion, $Q_{min} = 4.63$ AU P/2001 YX127 LINEAR and $Q_{max} = 6.06$ AU (82P/Gehrels 3).

An estimate of the lifetime of comets in A and P orbits indicates the following. There are 44 comets that move in an A orbit over the entire time interval considered and 2 comets that are constantly in a P orbit (see Table 1). The minimum lifetime is equal to three revolution periods for both A-type (39P/Oterma) and P-type (74P/Smirnova-Chernykh) orbits.

We shall use the above-proposed criteria and notation in analyzing the transformations of orbits which take place at encounters with Jupiter. As it was noted in the discussion of changes in the shape of a cometary orbit, the comets satisfying condition (1.1) have m or l orbits. Therefore, the comets under study can have only A_m, A_l, P_m and P_l orbits.

Changes in the orbit sizes of these comets occur in two ways: either the type of tangency between the orbits of the comet and Jupiter remains invariant, and a close encounter transforms the aphelion distance of P-type comets or the perihelion distance of A-type comets (transitions $P_m \leftrightarrow P_l$, $A_m \leftrightarrow A_l$), or reversal of tangency takes place, and a close encounter with Jupiter transforms both the aphelion and the perihelion distance of the comet (transitions $P_m \leftrightarrow A_m$, $P_l \leftrightarrow A_l$, $P_m \leftrightarrow A_l$, $P_l \leftrightarrow A_m$).

The same comet can undergo both first-type and second-type transitions (see the first column of Table 1).

Even a very close encounter with Jupiter can leave the sizes of the cometary orbit virtually unchanged (for the approach of 147P/Kushida-Muramatsu, $T_p = 2\,391\,334$ JD, $\rho = 0.00107$ AU, $\Delta 1/a = -0.00022$ AU^{-1}).

Let us examine the past and the future of the comets, using Table 1.

Among the comets studied, the number of objects with A orbits shows a decrease in the future and, even more significant, in the past. According to the Table, the nearest past and the future of a certain amount of comets with A orbits demonstrate a redistribution of orbits: these are the objects whose orbits are tangent to the Jovian orbit at the perihelion. They exhibit a great variance in aphelion distances: $Q_{max} = 19.06$ AU (P/1997 T3 Lagerkvist-Carsenty), whereas $Q_{min} = 4.63$ AU (P/2001 YX127 LINEAR).

No concentration of aphelia in the neighborhood of the orbits of Saturn or Uranus is observed.

The exhaustion of A orbits revealed for arbitrarily chosen moments t_1 and t_2 evidences that comets obeying condition (1.1) spend a finite time in short-period orbits.

The current presence of A-type orbits supports the fact that this group is constantly replenished at the expense of comets with P orbits.

The redistribution of orbits in the past and in the future is the result of transitions between A and P type orbits common for the investigated comets: 36 comets (47%) have a total of 85 transitions $A \leftrightarrow P$; 21 comets undergo cascade transitions $P \rightarrow A \rightarrow P$; 21 demonstrate cascade transitions $A \rightarrow P \rightarrow A$; 10 comets show only $P \rightarrow A$ transitions; and 4 comets, only $A \leftrightarrow P$ transitions. It can be suggested that any comet presented in Table 1 has cascade transitions $P \rightarrow A \rightarrow P \rightarrow A \rightarrow ...$ but the encounters with Jupiter responsible for these transitions are beyond the investigated time interval.

Now we summarize some results of the analysis of the orbital evolution of comets.

The comets under consideration have orbits with a steadily low inclination, which are tangent to the Jovian orbit at the aphelion or at the perihelion. The eccentricity of their orbits does not exceed $e = 0.6$. Of these comets, 80% have a timespan Δt in their evolution during which they move in low-eccentricity orbits. Residence in an l orbit increases the instability of the orbit itself and, evidently, of the comet moving in it as well. The instability is due to very efficient encounters with Jupiter which took place during the timespan. Such encounters can be accompanied not only by transitions $A \leftrightarrow P$ but also by multiple minima of the jovicentric-distance function of the comet and by its temporary satellite capture (TSC).

At the same time, only cascade transitions $A \to P \to A \to \dots$ take place for the comets studied, i.e., their orbits evolve during at least 100-150 revolutions about the Sun on a closed cycle limited by two types of tangency to the Jovian orbit. No hyperbolic ejection of a comet is revealed. As shown by Kresák (1977), hyperbolic ejections are impossible for comets with high values of the Tisserand constant.

2.3. *Cyclic Transformations of Cometary Orbits*

Now we shall examine cyclic transformations of the orbits of comets in greater detail.

For the investigated comets, the most substantial of the known orbit-size transformations follow scheme I:

$$P_m \leftrightarrow A_m. \tag{2.4}$$

Such transitions also take place for comets that do not satisfy condition (1.1). They occur in very close approaches of D/1770 L1 Lexell, 76P/West-Kohoutek-Ikemura and 81P/Wild 2. Such encounters are rare, while, as shown by Everhart (1973), the efficiency of more frequent but less close encounters is on the whole small.

The known transitions $A \leftrightarrow P$ of comets which do not satisfy condition (1.1) do not place the comet in a low-eccentricity orbit. Encounters with Jupiter which result in transitions $A \leftrightarrow P$ either leave the comet in a medium-eccentricity m orbit (in a high-eccentricity h orbit) or cause changes in the shape of the cometary orbit according to the scheme $m \leftrightarrow h$. In such encounters, the Öpik model for the change of the heliocentric-velocity vector of the comet is realized. Such encounters result in transitions $A \leftrightarrow P$, if the change corresponds to a transformation of this type. Since very close encounters with Jupiter are rare in occurrence, none of the comets not satisfying condition (1.1) has a cascade $P \to A \to P$ within the time interval $t_1 - t_2$. This interval is too short for investigating the orbital evolution of such comets.

For comets satisfying condition (1.1), four close approaches with transitions $P_m \to A_m$ that follow scheme I (2.4) are known. However, as the analysis of the orbital evolutions shows, the great majority of comets pass in all transitions $A \leftrightarrow P$ either from a low-eccentricity l orbit or to a l orbit; in most cases of the encounters with Jupiter which lead to transitions $A \leftrightarrow P$, the shape of the cometary's orbit changes insignificantly (so that the comet has an l orbit both before and after the encounter).

Hence, for the comets studied, evolutionary transformations $P_m \leftrightarrow A_m$ can be realized by stages, according to the scheme II:

Scheme II (Fig. 1) includes an intervening stage, the transition of the comet to low-eccentricity P_l or A_l orbits.

While a transition by scheme I (2.4) requires a very close approach to Jupiter, each of the transitions of scheme II (Fig. 1) is realized as early as at entering the Jovian sphere of action; moreover, the penetration of the comet into the sphere of unit radius (1 AU) centered at this planet is in many cases sufficient for transitions $P_l \leftrightarrow A_l$.

Thus, the type of tangency of the orbits of Jupiter and a comet that satisfies condition (1.1) can change not only according to scheme I (2.4), but also according to scheme II (Fig. 1). The motion of real comets according to the scheme II (Fig. 1) leads to a fairly fast orbital evolution of the comets of this class even over a 800-yr time interval. This makes it possible to reveal and investigate the basic laws of evolutionary transformations

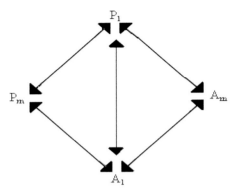

Figure 1. Scheme II.

of cometary orbits considering a time interval shorter than in the case of comets with lower values of the Tisserand constant.

For the comets studied, all the particular transitions of scheme II (Fig. 1) take place. For all the comets, in general, most transitions are realized many times. There are comets that repeatedly undergo the transitions of the central and the right portion of the scheme. Four comets have passed from a P_m to an A_m orbit by scheme II (Fig. 1), choosing one or another sequence of transitions. Note that there are two types of transitions in the scheme which involve only changes in the shape of the cometary orbit (transitions $P_m \leftrightarrow P_l$ and $A_m \leftrightarrow A_l$), as well as the above-discussed transitions $P_l \leftrightarrow A_l$ involving only changes in the orbit sizes. We see that these transitions form an ingredient of scheme II (Fig. 1) and actually take place for real comets (see Table 1).

Scheme II (Fig. 1) includes three basic states of a cometary orbit:

• State A_m. This is a state of relative stability, which is enhanced if the orbit lies within the Jovian orbit. There are 66 comets that passed through state A_m. 14 comets remain in A_m orbits for the whole time interval studied.

• State P_l or A_l. These states were passed by 67 comets. Three comets have P_l or A_l orbits over the entire time interval $t_1 - t_2$. Seven comets undergo only transitions $P_l \leftrightarrow A_l$. The transitions of the central and the right portion of the scheme leave the Tisserand constant virtually unchanged. There are comets that evolve within these two parts of the scheme, undergoing up to seven transitions over the entire time interval at hand.

• State P_m. Only 14 comets passed through state P_m. In the transitions of the left portion of the scheme, the largest change in the Tisserand constant is observed. For the comets listed in Table 1, this is an unstable state. None of them have spent the whole studied time interval in this orbit. The instability is determined not only by Jupiter but also by the influence of other great planets: eight comets have encounters with Saturn and one, with Uranus. At the same time, none of the comets, being in an P_m orbit, broke free of Jupiter's control. Thus, the comets that satisfy condition (1.1) evolve according to scheme II (Fig. 1) for a long time interval that encompasses their 100-150 revolutions about the Sun. Their orbital ellipses pulsate in their sizes, shape, and the type of tangency with the Jovian orbit. The giant planet holds these comets steadily and tightly in its possessions in spite of their "walks" over scheme II (Fig. 1) and "dizzy" jumps according to scheme I (2.4), which is also admissible for them.

3. Discussion

We proposed evolutionary scheme II (Fig. 1) for comets of the Jupiter family with high values of the Tisserand constant. The transformations of the orbits of all the studied comets follows this scheme.

It is beyond doubt that these objects are the most rapidly evolving comets in the Solar System: 85 transitions $A \leftrightarrow P$ and more than 130 transitions $m \leftrightarrow l$ were revealed for them.

In the evolution scenarios of 68 comets (80%), a timespan Δt is present over which the comet moves in a low-eccentricity l orbit, and 12 comets are in l orbits for the 800-yr interval. The presence of this timespan plays an important role in the evolutionary transformations of cometary orbits subject to condition (1.1) and is the main cause of the instability of these orbits.

Low-velocity approaches of these comets to Jupiter are in many cases accompanied by the temporary satellite capture (TSC) of the comet and multiple minima of its jovicentric-distance function. Such encounters result from a similarity between the l orbits and the Jovian orbit for the timespan Δt.

The following evolutionary scenarios are known exclusively for comets with l orbits: comet $\leftrightarrow$ temporary satellite of Jupiter; comet $\rightarrow$ temporary satellite of Jupiter $\rightarrow$ comet disruption $\rightarrow$ many comets. The last factor possibly evidences that l orbits are the primordial orbits of the comets of Jupiter's family which obey condition (1.1). The following arguments in favor of this suggestion can be noted. First, the Tisserand constant of the investigated comets satisfies condition (1.1) for the 800-yr time interval (and for l objects, it satisfies this condition). Second, in the evolution scenarios of 80% of comets, the timespan Δt is present. Third, the above-proposed scheme can be realized (i.e., the entire multitude of the orbits of these comets can be obtained from l orbits). We do not necessarily imply that such comets form immediately in the capture zone. One cannot rule out the possibility of another factor that transforms some comets or other objects into comets of this class. However, what follows is their long-term evolution according to the proposed scheme.

Acknowledgements

This work was supported by RFBR Grant 06-02-16512. The author would like to thank Giovanni Valsecchi for helpful comments.

References

Carusi, A. & Valsecchi, G.B. 1980, *Moon & Planets* 22, 113

Carusi, A. & Valsecchi, G.B. 1982, in: G. Teleki & W. Fricke (eds.), *Sun and Planetary System*, (Dordrecht: D. Reidel) p. 379

Emel'yanenko, N.Yu. 2003a, *Solar System Research* 37/1, 66

Emel'yanenko, N.Yu. 2003b, *Solar System Research* 37/2, 156

Everhart, E. 1973, *Astron. J.* 78, 329

Everhart, E. 1974, *Celest. Mech.* 10, 35

Froeschlé, Cl. & Rickman, H. 1981, *Icarus* 46, 400

Kazimirchak-Polonskaya, E.I. 1967, *Astron. Zh.* 44, 439

Kresák, L. 1977, in: A.H. Delsemme (ed.) *Comets, Asteroids, Meteorites Interrelations, Evolution and Origins* (Univ. of Toledo), p. 313

Marsden, B.G. & Williams G.V. 2003, *Catalogue of Cometary Orbits*

Near Earth Objects, our Celestial Neighbors: Opportunity and Risk
Proceedings IAU Symposium No. 236, 2006
A. Milani, G.B. Valsecchi & D. Vokrouhlický, eds.
© 2007 International Astronomical Union
doi:10.1017/S1743921307003055

Dynamical evolution of Oort cloud comets to near-Earth space

Olga A. Mazeeva†

Department of Computational and Celestial Mechanics, South Ural State University,
Chelyabinsk, 454080, Russia
email: omega@susu.ac.ru

Abstract. The dynamical evolution of $2 \cdot 10^5$ hypothetical Oort cloud comets by the action of planetary, galactic and stellar perturbations during $2 \cdot 10^9$ years is studied numerically. The evolution of comet orbits from the outer (10^4 AU $< a < 5 \cdot 10^4$ AU, a is semimajor axes) and the inner Oort cloud ($5 \cdot 10^3$ AU $< a < 10^4$ AU) to near-Earth space is investigated separately. The distribution of the perihelion (q) passage frequency in the planetary region is obtained calculating the numbers of comets in every interval of Δq per year. The flux of long-period (LP) comets (orbital periods $P > 200$ yr) with perihelion distances $q < 1.5$ AU brighter than visual absolute magnitude $H_{10} = 7$ is ~ 1.5 comets per year, and ~ 18 comets with $H_{10} < 10.9$. The ratio of all LP comets with $q < 1.5$ AU to 'new' comets is ~ 5. The frequency of passages of LP comets from the inner Oort cloud through region $q < 1.5$ AU is $\sim 3.5 \cdot 10^{-13}$ yr^{-1}, that is roughly one order of magnitude less than frequency of passages of LP comets from the outer cloud ($\sim 5.28 \cdot 10^{-12}$ yr^{-1}). We show that the flux of 'new' comets with $15 < q < 31$ AU is higher than with $q < 15$ AU, by a factor ~ 1.7 for comets from the outer Oort cloud and, by a factor ~ 7 for comets from the inner cloud. The perihelia of comets from the outer cloud previously passed through the planetary region are predominated in the Saturn-Uranus region. The majority of inner cloud comets come in the outer solar system ($q > 15$ AU), and a small fraction (~ 0.01) of them can reach orbits with $q < 1.5$ AU. The frequency of transfer of comets from the inner cloud ($a < 10^4$ AU) to the outer Oort cloud ($a > 10^4$ AU), from where they are injected to the region $q < 1.5$ AU, is $\sim 6 \cdot 10^{-14}$ yr^{-1}.

Keywords. Oort Cloud, comets: general, solar system: general.

---

1. Introduction

A constant near-parabolic flux of comets (Marsden & Williams 2005) originating from the Oort cloud (Oort 1950) is observed near the Earth orbit. The distribution of original reciprocal semimajor axes ($\varepsilon = 1/a$) of observed comets shows a narrow peak in the range less than 10^{-4} AU^{-1}. Comets with $\varepsilon < 10^{-4}$ AU^{-1} are dynamically 'new' comets, i.e. presumably on their first passage in the planetary region. The flux of Earth-crossing ($q < 1$ AU) comets with $a > 10^4$ AU and absolute magnitudes $H_{10} < 7$ is 0.2 per year (Bailey & Stagg 1988). Oort (1950) argued that stellar perturbations are responsible for bringing of comets from the solar system cloud into the inner planetary region. Byl (1986) and Heisler & Tremaine (1986) pointed out that the galactic tidal force is a more efficient mechanism of changing the perihelion distances of Oort cloud comets.

Hills (1981) proposed that a massive, unobserved inner cloud of comets with semimajor axes $a < 10^4$ AU exist. Comets of the inner cloud are weakly affected by external perturbations, except in the cases of passing of stars through the inner cloud. Duncan *et al.* (1987) simulated the formation of the Oort cloud and studied its subsequent evolution over an interval $4.5 \cdot 10^9$ yr, assuming that comets formed in the outer planetary region. The resulting spatial density of Oort cloud comets between $3 \cdot 10^3$ AU and

† Present address: South Ural State University, 76 Lenina, Chelyabinsk, Russia.

$5 \cdot 10^4$ AU is $\sim r^{-3.5}$ and roughly corresponds to a distribution of semimajor axes $\sim a^{-1.5}$ (Duncan *et al.* 1987). This implies that 70% of comets have $a < 10^4$ AU. Dones *et al.* (2000) found that the populations of the inner and outer clouds are roughly equal.

Comets which enter into the inner solar system are perturbed by the giant planets. The orbital energy ($\varepsilon = 1/a$) of comets change at every perihelion passage. This process may be viewed as a slow diffusion of $1/a$. The typical energy change $\Delta\varepsilon$ depends from the perihelion distance: $\Delta\varepsilon \approx 10^{-3}$ AU^{-1} for $q < 6$ AU, $\Delta\varepsilon \approx 10^{-4}$ AU^{-1} for $q \approx 10$ AU and $\Delta\varepsilon \approx 10^{-5}$ AU^{-1} for $q \approx 20$ AU (Fernández 1981), and depends less significantly from the inclination. In the region $q < 10$ AU the typical change in energy per perihelion passage due to planetary perturbations is greater than the initial energy of a comet. The region $q < 10$ AU is called the 'loss cone' (Hills 1981).

The aim of this work is to study the dynamical evolution of comets from both the inner and outer Oort clouds to near-Earth space, to estimate the flux of 'new' and long-period comets for all perihelion distances in the planetary region.

2. Model

The dynamical evolution of 10^5 comets of the outer Oort cloud (OOC) and 10^5 comets of the inner Oort cloud (IOC), affected by planetary, galactic and stellar perturbations during $2 \cdot 10^9$ yr, is considered. The evolution of cometary orbits of the outer and inner Oort cloud is investigated separately. The initial semimajor axes (a) of cometary orbits of the outer Oort cloud are distributed according to $a^{-1.5}$ in the range $(1 - 5) \cdot 10^4$ AU (Duncan *et al.* 1987). Our model of the inner Oort cloud includes cometary orbits with semimajor axes distributed according to $a^{-1.5}$ in the interval $(0.5 - 1) \cdot 10^4$ AU.

The number of inner and outer cloud comets with perihelia in the range $(q, q + dq)$ is $N(q)dq$, where $N(q) \sim 1 - q/a$, $q < a$ (Hills 1981). The initial inclinations i are uniformly distributed in $\cos i$. For each initial orbit the argument of perihelion and the longitude of the ascending node are chosen randomly in the interval $(0°, 360°)$.

The perturbations of the four outer planets and the action of Galaxy are taken into account (Emel'yanenko 1999). Stellar perturbations are described by the impulse approximation (Ogorodnikov 1965). The distribution of parameters of stars and the frequency of their passages near the solar system are taken from Heisler *et al.* (1987).

The orbital elements are printed for each comet passage through the planetary region ($q < 35$ AU) with $a > 34.2$ AU. The flux of comets with $\tau < 10^9$ yr and $10^9 < \tau$ (yr) $< 2 \cdot 10^9$ is investigated separately (τ is the time of perihelion passage).

3. Comparison with observations

The comparison was made for observed and hypothetical comets with $a > 8 \cdot 10^3$ AU and $q < 3$ AU. From the Catalogue of Cometary Orbits (Marsden & Williams 2005) were taken the elements of orbits with quality classes 1 and 2, according to the classification proposed by Marsden *et al.* (1978). This excluded comets for which non-gravitational parameters have been determined. The observed and modeled $\varepsilon = 10^4/a$ (a is the original semimajor axis) distributions are shown in Table 1.

The flux of comets with $T < 10^9$ yr in the intervals $\varepsilon(0.5; 0.75)$ AU^{-1} and $\varepsilon(0.75; 1.0)$ AU^{-1} is approximately 1.5 times as large as the observed flux. The computed flux of comets in the range $\varepsilon(1.0; 1.25)$ AU^{-1} (mainly comets coming from the inner cloud) is 8 times larger than the observed flux. The flux of comets with $10^9 < T$ (yr) $< 2 \cdot 10^9$ agrees better with the observations than that with $T < 10^9$ yr (Table 1).

Table 1. The $\varepsilon = 10^4/a$ (a is the original semimajor axis) distributions of observed (Marsden & Williams 2005) and hypothetical comets with $q < 3$ AU. The distributions were normalized so that the number of comes with $\varepsilon(0.25; 0.5)$ AU^{-1} ($2 \cdot 10^4 < a$ (AU) $< 4 \cdot 10^4$) is unity. The flux of comets from the OOC and the combined flux of comets from the OOC and IOC for $T < 10^9$ yr and $10^9 < T$ yr $< 2 \cdot 10^9$ is presented separately, where T is the time of aphelion passage.

$\varepsilon = 10^4/a$ (AU^{-1})	Observed comets	$T < 10^9$ yr		$10^9 < T$ (yr) $< 2 \cdot 10^9$	
		OOC comets	(OOC+IOC) comets	OOC comets	(OOC+IOC) comets
0.00-0.25	0.714	0.994	0.741	0.969	0.878
0.25-0.50	1.000	1.000	1.000	1.000	1.000
0.50-0.75	0.250	0.404	0.348	0.217	0.236
0.75-1.00	0.179	0.258	0.243	0.074	0.135
1.00-1.25	0.036	0.050	0.282	0.039	0.071

We then considered the distributions of original semimajor axes between 10^4 AU and $6 \cdot 10^4$ AU, in intervals of $5 \cdot 10^3$ AU, of the observed and hypothetical comets with $q < 3$ AU. The distribution of OOC comets with $T < 10^9$ yr has a first peak in the interval $a = (4.5-5) \cdot 10^4$ AU, and a second peak in the range $a = (1-1.5) \cdot 10^4$ AU. The combined flux of IOC and OOC comets with $T < 10^9$ yr has a peak in the range $a = (1-1.5) \cdot 10^4$ AU. The distribution of semimajor axes of observed comets with $q < 3$ AU has a peak in the range $(2 - 2.5) \cdot 10^4$ AU, and smaller peaks in the intervals $(4 - 4.5) \cdot 10^4$ AU and $(1 - 1.5) \cdot 10^4$ AU.

The flux of OC comets with $10^9 < T$ (yr) $< 2 \cdot 10^9$ has a first peak in the distribution of semimajor axes in the range $(2 - 2.5) \cdot 10^4$ AU, and second peak in the range $a = (4.5-5) \cdot 10^4$ AU. Thus, our computed distribution of comets with $10^9 < T$ (yr) $< 2 \cdot 10^9$ agrees better with the observations than that with $T < 10^9$ yr.

A possible explanation of the disagreement with the observations for $T < 10^9$ yr is in the adopted initial distribution of perihelion distances of comets (Hills 1981), that corresponds to the distribution caused by a close stellar passage. Weissman (1993) concludes that we are not in a strong shower at present. Thus, hereafter we consider mainly results obtained after 10^9 yr of the evolution of our comet orbits.

4. The flux of dynamically 'new' Oort cloud comets

Comets with original semimajor axes $a > 10^4$ AU entering from the Oort cloud into the planetary system for the first time are conventionally defined as 'new' comets (Oort 1950). Here are also included within the category of 'new' comets some comets with $a < 10^4$ AU coming in the region $q < 35$ AU for the first time, since comets with $a < 10^4$ AU can be injected into the inner planetary system by the action stars passing within 10^4 AU from the Sun.

Figure 1 shows the relative flux of 'new' comets from the outer (Fig. 1a) and the inner (Fig. 1b) parts of the Oort cloud versus perihelion distance. The flux is normalized to its value at $q < 1.5$ AU with $10^9 < \tau$ (yr) $< 2 \cdot 10^9$. The flux of 'new' comets through the outer planetary region is higher than the flux trough the inner planetary region (here defined by $q < 15$ AU).

In the steady-state situation, 'new' comets from the outer Oort cloud (Fig. 1a) come roughly uniformly to the perihelion distance $q < 15$ AU. However, the flux of these comets in the region $15 < q$ (AU) < 31 is ~ 1.7 times as large as the flux with $q < 15$ AU (Fig. 1a). The ratio of the flux of 'new' comets from the outer Oort cloud near the orbits

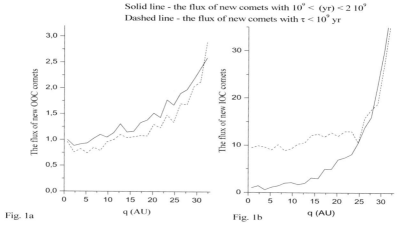

Fig. 1a

Fig. 1b

Figure 1. The relative flux of 'new' comets are coming from the outer (Fig. 1a) and inner (Fig. 1b) Oort cloud in the interval $(q, q + dq)$, where $q \leqslant 31.5$ AU, $dq = 1.5$ AU. The flux is normalized to its value at $q < 1.5$ AU with $10^9 < \tau < 2 \cdot 10^9$, τ is the time of perihelion passage.

of Jupiter $(5 - 6)$ AU, Saturn $(9 - 10)$ AU, Uranus $(19 - 20)$ AU and Neptune $(30 - 31)$ AU to the flux of these comets with $q < 1$ AU is, respectively, 1, 1, 1.4 and 2.4 (Fig. 1a).

The distribution shown in Fig. 1a for $10^9 < \tau$ (yr) $< 2 \cdot 10^9$ is slightly different from that for $\tau < 10^9$ yr. However, the distribution of perihelion distances of 'new' comets from the inner cloud (Fig. 1b) for $10^9 < \tau$ (yr) $< 2 \cdot 10^9$ is very different from that for $\tau < 10^9$ yr. Thus, the steady-state flux of comets cannot come from the inner part Oort cloud uniformly in perihelion distance for the following reasons:

- the galactic potential and stellar encounters remove comets from the inner core of the Oort cloud ($a < 10^4$ AU) mainly to the outside of the planetary system;
- comets of the inner core can be injected in the 'loss cone' only by the action of stars passing within 10^4 AU of the Sun. While the passages of stars through the hypothetical inner cloud are very rare, the passages within the outer Oort cloud are not (Heisler et al. 1987);
- within the region $q < 10$ AU the planetary perturbations move comets quickly, either to hyperbolic orbits, or to orbits that are more tightly bound to the Solar system.

The ratio of the flux of 'new' comets from the inner cloud in the outside planetary region ($15 < q$ (AU) < 31) to that in the region $q < 15$ AU is ~ 7 (Fig. 1b). The flux of comets coming from the inner Oort cloud for the first time near the orbits of Jupiter, Saturn, Uranus and Neptune to that with $q < 1$ AU is 1.4, 1.6, 5.8 and 27.4 respectively (Fig. 1b).

5. The frequency of passages of comets within the planetary region

'New' comets may pass repeatedly through the planetary region. Comets with $a > 10^3$ AU passing more than once within the region $q < 35$ AU are defined as dynamically 'young' comets. Comets with $a < 10^3$ AU, which passed many times through the planetary system, are defined as 'old' comets.

The orbital elements were printed for each perihelion passage within $q < 35$ AU with $a > 34.2$ AU. The frequency $f(q)$ of passages of comets in the interval of perihelion distance $(q, q + dq)$ per year is calculated by

$$f(q) = \frac{n(q)}{N \cdot t} (yr^{-1}),\qquad(5.1)$$

Table 2. The frequency of passages of dynamically new comets (f_{new}), comets with $a > 10^4$ AU ($f_{a>10^4}$), young comets with semimajor axes in the range $(10^3; 10^4)$ AU (f_{young}), long-period comets with $a > 34.2$ AU (f_{LP}), originating from the outer (f_{OOC}) and inner Oort clouds (f_{IOC}) per year in the interval $(q, q + dq)$, where $q \leqslant 31.5$ AU, $dq = 1.5$ AU (10^9 yr $< \tau < 2 \cdot 10^9$ yr).

q (AU)	f_{OOC} (yr^{-1})				f_{IOC} (yr^{-1})			
	f_{new}	$f_{a>10^4}$	f_{young}	f_{LP}	f_{new}	$f_{a>10^4}$	f_{young}	f_{LP}
0.0-1.5	$1.20 \cdot 10^{-12}$	$1.41 \cdot 10^{-12}$	$1.43 \cdot 10^{-12}$	$5.28 \cdot 10^{-12}$	$7.00 \cdot 10^{-14}$	$6.00 \cdot 10^{-14}$	$1.70 \cdot 10^{-13}$	$3.50 \cdot 10^{-13}$
1.5-3.0	$1.06 \cdot 10^{-12}$	$1.50 \cdot 10^{-12}$	$2.47 \cdot 10^{-12}$	$8.98 \cdot 10^{-11}$	$1.00 \cdot 10^{-13}$	$1.00 \cdot 10^{-13}$	$3.80 \cdot 10^{-13}$	$1.46 \cdot 10^{-12}$
3.0-4.5	$1.10 \cdot 10^{-12}$	$1.69 \cdot 10^{-12}$	$5.14 \cdot 10^{-12}$	$3.56 \cdot 10^{-11}$	$4.00 \cdot 10^{-14}$	$1.10 \cdot 10^{-13}$	$7.70 \cdot 10^{-13}$	$4.17 \cdot 10^{-12}$
4.5-6.0	$1.12 \cdot 10^{-12}$	$1.84 \cdot 10^{-12}$	$9.60 \cdot 10^{-12}$	$8.80 \cdot 10^{-11}$	$8.00 \cdot 10^{-14}$	$1.60 \cdot 10^{-13}$	$3.18 \cdot 10^{-12}$	$2.11 \cdot 10^{-11}$
6.0-7.5	$1.23 \cdot 10^{-12}$	$2.66 \cdot 10^{-12}$	$2.54 \cdot 10^{-11}$	$1.58 \cdot 10^{-10}$	$1.00 \cdot 10^{-13}$	$2.60 \cdot 10^{-13}$	$6.90 \cdot 10^{-12}$	$7.38 \cdot 10^{-11}$
7.5-9.0	$1.32 \cdot 10^{-12}$	$4.08 \cdot 10^{-12}$	$3.25 \cdot 10^{-11}$	$4.01 \cdot 10^{-10}$	$1.40 \cdot 10^{-13}$	$4.90 \cdot 10^{-13}$	$3.08 \cdot 10^{-11}$	$4.17 \cdot 10^{-10}$
9.0-10.5	$1.26 \cdot 10^{-12}$	$4.63 \cdot 10^{-12}$	$1.98 \cdot 10^{-11}$	$1.04 \cdot 10^{-9}$	$1.50 \cdot 10^{-13}$	$8.40 \cdot 10^{-13}$	$7.71 \cdot 10^{-11}$	$7.60 \cdot 10^{-10}$
10.5-12.0	$1.36 \cdot 10^{-12}$	$5.78 \cdot 10^{-12}$	$6.84 \cdot 10^{-11}$	$9.04 \cdot 10^{-10}$	$1.20 \cdot 10^{-13}$	$1.37 \cdot 10^{-12}$	$7.76 \cdot 10^{-11}$	$3.96 \cdot 10^{-9}$
12.0-13.5	$1.57 \cdot 10^{-12}$	$8.37 \cdot 10^{-12}$	$5.95 \cdot 10^{-11}$	$2.44 \cdot 10^{-10}$	$1.40 \cdot 10^{-13}$	$1.83 \cdot 10^{-12}$	$1.73 \cdot 10^{-10}$	$3.24 \cdot 10^{-9}$
13.5-15.0	$1.38 \cdot 10^{-12}$	$9.43 \cdot 10^{-12}$	$1.31 \cdot 10^{-10}$	$3.33 \cdot 10^{-10}$	$2.20 \cdot 10^{-13}$	$2.20 \cdot 10^{-12}$	$1.59 \cdot 10^{-10}$	$2.41 \cdot 10^{-9}$
15.0-16.5	$1.40 \cdot 10^{-12}$	$1.04 \cdot 10^{-11}$	$1.02 \cdot 10^{-10}$	$2.81 \cdot 10^{-10}$	$2.10 \cdot 10^{-13}$	$2.80 \cdot 10^{-12}$	$1.54 \cdot 10^{-10}$	$2.52 \cdot 10^{-9}$
16.5-18.0	$1.60 \cdot 10^{-12}$	$1.17 \cdot 10^{-11}$	$6.65 \cdot 10^{-11}$	$5.34 \cdot 10^{-10}$	$3.50 \cdot 10^{-13}$	$3.04 \cdot 10^{-12}$	$1.49 \cdot 10^{-10}$	$2.18 \cdot 10^{-9}$
18.0-19.5	$1.66 \cdot 10^{-12}$	$1.25 \cdot 10^{-11}$	$7.98 \cdot 10^{-11}$	$2.60 \cdot 10^{-10}$	$3.50 \cdot 10^{-13}$	$3.08 \cdot 10^{-12}$	$1.84 \cdot 10^{-10}$	$2.05 \cdot 10^{-9}$
19.5-21.0	$1.82 \cdot 10^{-12}$	$1.36 \cdot 10^{-11}$	$9.36 \cdot 10^{-11}$	$2.16 \cdot 10^{-10}$	$4.90 \cdot 10^{-13}$	$3.08 \cdot 10^{-12}$	$2.72 \cdot 10^{-10}$	$2.69 \cdot 10^{-9}$
21.0-22.5	$1.72 \cdot 10^{-12}$	$1.39 \cdot 10^{-11}$	$5.92 \cdot 10^{-11}$	$1.80 \cdot 10^{-10}$	$5.20 \cdot 10^{-13}$	$3.37 \cdot 10^{-12}$	$4.44 \cdot 10^{-10}$	$2.82 \cdot 10^{-9}$
22.5-24.0	$2.13 \cdot 10^{-12}$	$1.61 \cdot 10^{-11}$	$7.01 \cdot 10^{-11}$	$6.45 \cdot 10^{-10}$	$5.70 \cdot 10^{-13}$	$3.54 \cdot 10^{-12}$	$4.88 \cdot 10^{-10}$	$3.04 \cdot 10^{-9}$
24.0-25.5	$2.01 \cdot 10^{-12}$	$1.68 \cdot 10^{-11}$	$6.88 \cdot 10^{-11}$	$4.46 \cdot 10^{-10}$	$7.40 \cdot 10^{-13}$	$3.71 \cdot 10^{-12}$	$4.83 \cdot 10^{-10}$	$3.63 \cdot 10^{-9}$
25.5-27.0	$2.28 \cdot 10^{-12}$	$1.85 \cdot 10^{-11}$	$5.32 \cdot 10^{-11}$	$1.99 \cdot 10^{-10}$	$9.70 \cdot 10^{-13}$	$3.73 \cdot 10^{-12}$	$3.85 \cdot 10^{-10}$	$5.40 \cdot 10^{-9}$
27.0-28.5	$2.37 \cdot 10^{-12}$	$1.87 \cdot 10^{-11}$	$9.15 \cdot 10^{-11}$	$1.62 \cdot 10^{-10}$	$1.12 \cdot 10^{-12}$	$4.17 \cdot 10^{-12}$	$5.62 \cdot 10^{-10}$	$4.63 \cdot 10^{-9}$
28.5-30.0	$2.61 \cdot 10^{-12}$	$2.06 \cdot 10^{-11}$	$7.22 \cdot 10^{-11}$	$9.73 \cdot 10^{-11}$	$1.59 \cdot 10^{-12}$	$4.48 \cdot 10^{-12}$	$3.78 \cdot 10^{-10}$	$3.51 \cdot 10^{-9}$
30.0-31.5	$2.86 \cdot 10^{-12}$	$1.99 \cdot 10^{-11}$	$8.98 \cdot 10^{-11}$	$1.10 \cdot 10^{-10}$	$2.10 \cdot 10^{-12}$	$4.26 \cdot 10^{-12}$	$3.15 \cdot 10^{-10}$	$2.34 \cdot 10^{-9}$
31.5-33.0	$3.11 \cdot 10^{-12}$	$2.12 \cdot 10^{-11}$	$3.34 \cdot 10^{-11}$	$5.47 \cdot 10^{-11}$	$2.76 \cdot 10^{-12}$	$4.63 \cdot 10^{-12}$	$4.39 \cdot 10^{-10}$	$9.63 \cdot 10^{-10}$

where $n(q)$ is number of perihelion passages of comets in a given interval $(q, q + dq)$, N is the population of hypothetical comets in the inner or outer Oort cloud ($N = 10^5$), t is the investigated time interval ($t = 10^9$ yr). Table 2 shows the computed frequency $f(q)$ of passages of dynamically new comets (f_{new}), new and young comets with a $> 10^4$ AU ($f_{a>10^4}$), young comets with semimajor axes in the range $(10^3; 10^4)$ AU (f_{young}), long-period comets with $a > 34.2$ AU (f_{LP}), originating from the outer (f_{OOC}) and inner cloud (f_{IOC}), in the interval $(q, q + dq)$, where $q < 31.5$ AU, $dq = 1.5$ AU.

Bailey & Stagg (1988) estimated that one comet with $a > 10^4$ AU and $H_{10} < 7$ (diameter in the range $5 - 15$ km) passes perihelion with $q < 1$ AU every 5 years, while Fernández (2002) estimated this frequency at every 3 years. Let $n(1)$ be the number of comets coming in the region $q < 1$ AU per year with $a > 10^4$ AU; then, the population of the Oort cloud may be computed by

$$N_{OC} = \frac{n(1)}{f_{a>10^4}(1) \cdot t}, \qquad (5.2)$$

where $f_{a>10^4}(1)$ is the frequency of passages of Oort cloud comets with $a > 10^4$ AU within $q < 1$ AU, and t is one year. The frequency $f_{a>10^4}(1) = 4.8 \cdot 10^{-13}$ yr^{-1} has been obtained from (5.1), $n(1) = 0.2 - 0.3$ comets per year (Bailey & Stagg 1988; Fernández 2002). The corresponding number of comets in the Oort cloud, including the inner and outer parts, with $H_{10} < 7$ is equal to $N_{OC} = N_{OOC} + N_{IOC} \approx (4.2 - 6.3) \cdot 10^{11}$. The average number is $N_{OC} \approx 5.25 \cdot 10^{11}$ comets with $H_{10} < 7$.

The number $n(1)$ should be 12 times larger for comets with $H_{10} < 10.9$ according to the $10^{\alpha H_{10}}$ dependence of the cumulative distribution of the number of objects brighter than H_{10} (Levison et al. 2002). From this follows that $n(1) = 2.4 - 3.6$ comets with $H_{10} < 10.9$. Then, the population of the Oort cloud is $N_{OC} = N_{OOC} + N_{IOC} \approx (5 - 7.5) \cdot 10^{12}$ comets

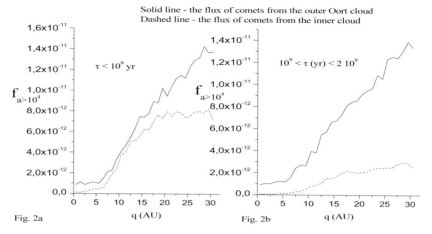

Fig. 2a

Figure 2. The frequency $f_{a>10^4}(q)$ of passages of comets with $a > 10^4$ AU per year in the interval of perihelion distance $(q, q+dq)$, where $q \leqslant 30$ AU, $dq = 1$ AU, for $\tau < 10^9$ yr (Fig. 2a) and $10^9 < \tau$ (yr) $< 2 \cdot 10^9$ (Fig. 2b), where τ is the time of perihelion passage.

with $H_{10} < 10.9$. The average number $N_{OC} \approx 6.25 \cdot 10^{12}$ comets with $H_{10} < 10.9$, that is consistent with Weissman (1996). Francis (2005) found that the flux of comets through the inner solar system and the outer Oort cloud population are much lower than previously published estimates (Bailey & Stagg 1988; Weissman 1996).

Duncan *et al.* (1987) studied the formation of the Oort cloud and its subsequent evolution over a time span of $4.5 \cdot 10^9$ yr, and found that approximately 70% of Oort cloud comets lie in its inner part ($a < 10^4$ AU). According to them, the inner comet cloud ($a < 10^4$ AU) acts as a source, replenishing the outer cloud ($a > 10^4$ AU). Comets evolve from the inner into the outer Oort cloud, and can be further brought into the visible region with $a > 10^4$ AU. Thus, comets with $a > 10^4$ AU observed in near-Earth space can originate from either the outer or the inner clouds. The calculated frequency (5.1) of passages of comets with $a > 10^4$ AU from the outer Oort cloud within $q < 1$ AU is $9.1 \cdot 10^{-13}$ yr^{-1}, that is roughly one order of magnitude larger than that from the inner cloud ($5 \cdot 10^{-14}$ yr^{-1}). Nevertheless, the flux of comets with $a > 10^4$ AU in the region $q < 1$ AU (as in the region $q < 35$ AU) comprises both outer cloud comets and inner cloud comets that have replenished the OOC. The combined flux of comets per interval $(q, q+dq)$, can be calculated by

$$n(q) = n_{OOC}(q) + n_{IOC}(q) = N_{OOC} \cdot f_{OOC}(q) + N_{IOC} \cdot f_{IOC}(q), \qquad (5.3)$$

where $n_{OOC}(q)$ is the flux of comets from the outer Oort cloud, $f_{OOC}(q)$ is the frequency of passages of OOC comets in a given interval $(q, q+dq)$, N_{OOC} is the population of comets in the outer cloud, $N_{OOC} = N_{OC}/(k+1)$. Analogously, $n_{IOC}(q)$ is the flux of comets from the inner cloud, $f_{IOC}(q)$ is the frequency of passages of IOC comets in a given interval $(q, q+dq)$, N_{IOC} is the population of comets in the IOC, $N_{IOC} = N_{OC} \cdot k/(k+1)$.

Let $n(1) \approx 3$ comets with $a > 10^4$ AU and $H_{10} < 10.9$ per year, the ratio $k = 1$, $N_{OC} \approx 6.25 \cdot 10^{12}$ comets; then, the contribution from the inner cloud to this flux ($q < 1$ AU) is $n_{IOC}(1) = 0.156$ comets per year, i.e. about one comet every 6.5 years.

The relative population of the inner cloud ($N_{IOC}/N_{OOC} = k$) was determined with a large uncertainty (Hills 1981; Duncan *et al.* 1987; Dones *et al.* 2000). However, when comparing simulation results and observations, it must be considered that the flux of comets coming directly from the Oort cloud to near-Earth space is the less uncertain parameter. Using the ratio $N_{IOC}/N_{OOC} = k$, Eq. (5.3) can be written as:

$$n(q) = N_{OOC} \cdot (f_{OOC}(q) + f_{IOC}(q) \cdot k). \tag{5.4}$$

From (5.4), using the model data (Table 2), one can compute the flux of comets in every interval $(q, q + dq)$, where $q < 31.5$ AU, $dq = 1.5$ AU, depending on value of k.

Figure 2 shows the distributions $f_{a>10^4}(q)$ for $\tau < 10^9$ yr and $10^9 < \tau$ (yr) $< 2 \cdot 10^9$, where $f_{a>10^4}(q)$ is the frequency of passages of comets with $a > 10^4$ AU per year in the interval of perihelion distance $(q, q + dq)$, $q \leqslant 30$ AU, $dq = 1$ AU. The maxima in these distributions $f_{a>10^4}(q)$ are located in the outer part of the planetary region, near the orbit of Neptune (Fig. 2). Comets with $a > 10^4$ AU mainly come from the outer cloud.

Fernández (2002) pointed out that the flux of comets with $a > 1.5 \cdot 10^4$ AU increases with increasing perihelion distance. The flux of outer Oort cloud comets with $a > 10^4$ AU near the orbits of Jupiter, Saturn, Uranus and Neptune is 1.28, 2.89, 9.42, and 14.7 times larger than that with $q < 1$ AU. The corresponding ratios for the flux of comets with $a > 10^4$ AU originating from the inner cloud are 2.2, 10.8, 40 and 50.8; however the frequency of passages of these comets within the planetary region is much less than the frequency of passages of OOC comets with $a > 10^4$ AU (Fig. 2).

6. The ratio between comets with a $> 10^4$ AU and 'new' comets

Dybczynski (2001) proposed that only 50% comets, defined as 'new', are actually in their first revolution through the inner parts of the Solar system ($q < 15$ AU). However, he modified the Oort definition of dynamically 'new' comets, having changed the threshold value of inverse semimajor axis from 10^{-4} AU^{-1} to $2.5 \cdot 10^{-4}$ AU^{-1}. As mentioned above, in this paper are denoted as 'new' all the comets that make their first passage in the region $q < 35$ AU.

Figure 3 shows the ratio $\nu(q)$ between $\nu_{a>10^4}(q)$ and $\nu_{new}(q)$, where $\nu_{a>10^4}(q)$ is the number of perihelion passages of comets with $a > 10^4$ AU into the region $q < 31$ AU, $\nu_{new}(q)$ is the number of perihelion passages of 'new' comets.

6.1. *The ratio $\nu(q)$ for comets from the outer Oort cloud*

The computed frequency of passages of 'new' comets (f_{new}) from the outer Oort cloud is equal to $1.2 \cdot 10^{-12}$ yr^{-1} within $q < 1.5$ AU (Table 2). The frequency of passages of outer cloud comets with $a > 10^4$ AU within the region $q < 1.5$ AU is $f_{a>10^4}(1.5) = 1.41 \cdot 10^{-12}$ yr^{-1} (Table 2). The ratio $\nu(q)$ for OOC comets is equal to 1.17 in the region $q < 1.5$ AU.

The number of returns in the planetary region of comets with $a > 10^4$ AU depends on q (Fig. 3). For small q planetary perturbations are strong enough to remove most of comets from the outer Oort cloud, either to hyperbolic orbits or to smaller elliptical orbits. Outside the orbit of Jupiter, the number of comets with $a > 10^4$ AU returning to the planetary region increases rapidly with increasing q (Fig. 3). Our computations show that in the region $q < 15$ AU approximately 30% of comets from the outer Oort cloud with $a > 10^4$ AU are really on their first passage in the planetary system.

6.2. *The ratio $\nu(q)$ for comets from the inner cloud*

Figure 3 shows also the ratio $\nu(q)$ for comets originating from the inner cloud (dashed line). The obtained frequency of passages of new comets (f_{new}) from the inner Oort cloud within $q < 1.5$ AU is equal to $7 \cdot 10^{-14}$ yr^{-1} (Table 2) that is more than one order of

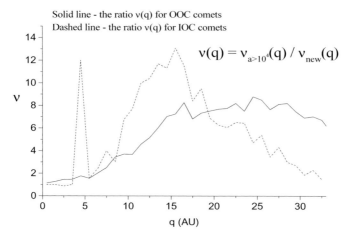

Figure 3. The ratio $\nu = \nu_{a>10^4}/\nu_{new}$ as a function of q, where $\nu_{a>10^4}$ is the number of perihelion passages of comets with $a > 10^4$ AU, ν_{new} is number of perihelion passages of 'new' comets.

magnitude less than the frequency of passages of new comets from the outer cloud. The ratio $\nu(q)$ has peaks near the orbit of Jupiter and in the Saturn-Uranus region (Fig. 3).

Comets from the inner cloud ($a < 10^4$ AU) that come to the region $q < 35$ AU may be removed to orbits with $a > 10^4$ AU by the planetary perturbations, and become subject by the stronger external perturbations. In fact, galactic and stellar perturbations can change the perihelion distances of these comets, bringing the perihelia of some of them into the visible region. The majority of outer Oort cloud comets evolve rapidly into smaller orbits, before their possible deflection by the action of giant planets back into the Oort region. Comets of the inner cloud evolve very slowly in the outside planetary region, and would be removed to orbits with $a > 10^4$ AU more likely than dynamically young or old comets originating from the OOC. The majority of IOC comets deflected due to the planetary perturbations into the outer cloud ($a > 10^4$ AU) come from orbits with semimajor axes in the range $5 \cdot 10^3 < a$ (AU) $< 10^4$.

Our computations show that in the region $q < 15$ AU only 5% of comets with $a > 10^4$ AU originating from the inner Oort cloud are on their first passage in the planetary system. Thus the majority of IOC comets move to orbits with $a > 10^4$ AU within the planetary region, mainly in the Saturn-Uranus region (Fig. 3).

7. Differences in the dynamical evolution of young comets from the inner and outer Oort cloud in the planetary region

Figure 4 shows the distributions $f_{young}(q)$ for $\tau < 10^9$ yr (Fig. 4a) and $10^9 < \tau$ yr $< 2 \cdot 10^9$ (Fig. 4b), where $f_{young}(q)$ is the frequency of passages of dynamically young comets with semimajor axes in the range $10^3 < a$ (AU) $< 10^4$ per year in the interval of perihelion distance $(q, q + dq)$, $q \leqslant 30$ AU, $dq = 1$ AU.

7.1. The dynamical evolution of young comets from the outer Oort cloud

In the Uranus-Neptune region ($q \approx 20 - 30$ AU) the evolution of cometary orbits under planetary perturbations is very slow, consequently galactic and stellar perturbations are important in this region. The perihelia of 'new' comets coming from the outer Oort cloud with $a > (1.5 - 2) \cdot 10^4$ AU in the region $q > 20$ AU are quickly transferred to

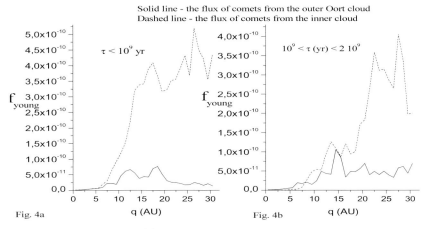

Figure 4. The frequency $f_{young}(q)$ of passages of comets with semimajor axes in the range $10^3 < a$ (AU) $< 10^4$ per year in the interval of perihelion distance $(q, q + dq)$, where $q \leqslant 30$ AU, $dq = 1$ AU, for $\tau < 10^9$ yr (Fig. 4a) and $10^9 < \tau$ (yr) $< 2 \cdot 10^9$ (Fig. 4b), τ is the time of perihelion passage.

$q < 20$ AU by the action of external perturbations. In a passage through the 'loss cone' ($q < 10$ AU) a comet undergoes a change in its reciprocal semimajor axes $\Delta\varepsilon$ of about $\Delta\varepsilon \approx (10^{-4} - 10^{-3})$ AU^{-1} (Fernández 1981), and is rapidly removed from the Oort cloud, either to smaller elliptical orbits or to hyperbolic orbits. In the region $q \approx (10 - 20)$ AU planetary perturbations are weaker, $\Delta\varepsilon \approx (10^{-5} - 10^{-4})$ AU^{-1}, therefore in this region the number of dynamically young comets returning to the planetary region increases. Thus, most of the perihelia of dynamically young comets with $a > 10^3$ AU from the outer Oort cloud are located in the Saturn-Jupiter region.

7.2. *The dynamical evolution of young comets from the inner cloud*

The maximum in the perihelion distribution of inner cloud comets on their first passage through the planetary region lies near Neptune's orbit, therefore IOC comets mainly come in the outer planetary region from the nearest regions ($q > 35$ AU). In the Uranus-Neptune region planetary perturbations are too weak ($\Delta\varepsilon \leqslant 10^{-5}$ AU^{-1} per perihelion passage) to remove comets with $a < 10^4$ AU. Comets from the inner cloud will drift in q very smoothly under the action of galactic perturbations in the Uranus-Neptune region.

Comets coming into the Jupiter-Saturn region will be affected by stronger planetary perturbations, and evolve into smaller orbits more rapidly. Therefore the flux of dynamically young comets from the inner cloud to the outer planetary region $q > 20$ AU is the largest (Fig. 4b).

Thus, the flux of dynamically young comets from the inner cloud is largest in the Uranus-Neptune region (Fig. 4), while the flux of dynamically young comets from the outer Oort cloud is largest in the Saturn-Uranus region (Fig. 4).

8. Oort cloud comets near the orbit of the Earth

'New' comets come directly from the Oort cloud into the inner solar system, and also in the near-Earth space ($q < 1.5$ AU). Those Oort cloud comets that come in the outer planetary region can evolve gradually to small perihelion distances. Thus there are two dynamical ways by which Oort cloud comets can reach near-Earth space.

The frequency of injection of outer Oort cloud comets directly from $q > 35$ AU to $q < 1.5$ AU is $1.2 \cdot 10^{-12}$ yr^{-1} (Table 2). This is the frequency of passages of 'new' comets. The frequency of supply of OOC comets evolving gradually to $q < 1.5$ AU from the outer planetary region is $5.6 \cdot 10^{-13}$ yr^{-1}. The frequency of injection of inner cloud comets directly from $q > 35$ AU into the region $q < 1.5$ AU is $7 \cdot 10^{-14}$ yr^{-1} (Table 2). Comets from the inner cloud can be injected from $q > 35$ AU directly to $q < 1.5$ AU only during comet showers due to a very close star passage (Hills 1981). From the outside planetary region IOC comets come gradually to $q < 1.5$ AU with frequency $1.5 \cdot 10^{-13}$ yr^{-1}. Thus, the majority of IOC comets reach the region $q < 1.5$ AU gradually evolving from the outer planetary region, while the majority of OOC comets come to $q < 1.5$ AU directly from $q > 35$ AU.

Considering comets that are coming gradually into the region $q < 1.5$ AU, let us examine their orbital elements on their previous perihelion passage, before entering to $q < 1.5$ AU region. The maximum in the distribution of perihelion distances of comets on their previous passage lies near the orbit of Saturn. The majority of the semimajor axes of IOC comets were located in the range $(0.5 - 1.5) \cdot 10^4$ AU. The semimajor axes of OOC comets before their entry to $q < 1.5$ AU were located mainly in the range $(0.5 - 3) \cdot 10^4$ AU, with the maximum in the interval $(2.5 - 3) \cdot 10^4$ AU. Therefore, dynamically young comets with semimajor axes of a few thousand AU ($a < 10^4$ AU) can also be perturbed from orbits with perihelia beyond the orbits of Saturn or Jupiter to perihelia inside the orbit of Jupiter. Near the orbit of Saturn these comets can be removed to orbits with $a > 10^4$ AU, become subject to strong external perturbations, and then come to the visible region on their next passage. The majority of new comets come into the outer planetary region. A fraction of ~ 0.015 new OOC comets and a fraction of ~ 0.011 new IOC comets can reach orbits with $q < 1.5$ AU from $15 < q < 35$ AU. The majority of the perihelia of IOC comets reaching near-Earth space ($q < 1.5$ AU), on their first passage in the planetary region were located in the interval $q = (30 - 35)$ AU.

The frequency of passages of long-period ($a > 34.2$ AU) comets from the inner cloud within $q < 1.5$ AU is $3.5 \cdot 10^{-13}$ yr^{-1}, that is roughly one order of magnitude less than the frequency of passages of LP comets from the outer Oort cloud ($5.28 \cdot 10^{-12}$ yr^{-1}). As mentioned above, the frequency of passages of new comets from the OOC within $q < 1.5$ AU is $1.2 \cdot 10^{-12}$ yr^{-1} (Table 2), and from the IOC is $7 \cdot 10^{-14}$ yr^{-1}. Thus the ratio of LP comets with $q < 1.5$ AU to new comets is ~ 5. Bailey & Stagg (1988) estimated that the flux of long-period comets with $H_{10} < 7$ near the Earth ($q < 1$ AU) is 1.0 ± 0.2 yr^{-1} AU^{-1}. Let us assume that the ratio $k = 1$, the Oort cloud contains $5.25 \cdot 10^{11}$ comets with $H_{10} < 7$ or $6.25 \cdot 10^{12}$ comets with $H_{10} < 10.9$, then from (5.4) the flux of long-period comets within $q < 1$ AU is equal to 0.8 yr^{-1} ($H_{10} < 7$), that is consistent with Bailey & Stagg (1988). Therefore, we can estimate the flux of comets in the vicinity of the Earth.

The contribution of OOC comets to the near-Earth flux ($q < 1.5$ AU) with $H_{10} < 7$ is ~ 1.4 LP comets per year, the contribution of IOC comets to this flux is ~ 0.1 LP comets per year. Thus, the total flux of LP comets from the Oort cloud with $q < 1.5$ AU is approximately 1.5 comets per year ($H_{10} < 7$). The total flux of LP comets within $q < 1.5$ AU with $H_{10} < 10.9$ is ~ 18 comets per year, of them ~ 1.1 LP comets originate from the inner cloud. The flux of new comets with $H_{10} < 7$ to $q < 1.5$ AU is ~ 0.34 comets per year, of them ~ 0.02 new comets come from the inner cloud (on average one comet per 50 years). The flux of new comets to $q < 1.5$ AU with $H_{10} < 10.9$ is ~ 4 comets per year, of them ~ 0.22 new comets from the inner cloud (on average one comet per 4.5 years). The frequency of passages of IOC comets that have been transferred to the OOC with $a > 10^4$ AU within $q < 1.5$ AU is $6 \cdot 10^{-14}$ yr^{-1}, that is more than one order of magnitude less than the frequency of passages of outer cloud comets with $a > 10^4$ AU

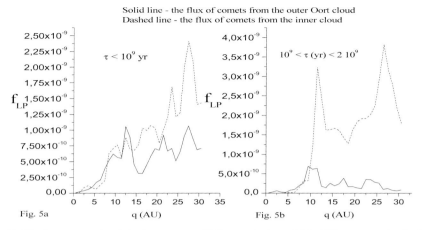

Figure 5. The frequency $f_{LP}(q)$ of passages of LP comets ($a > 34.2$ AU) per year in the interval of perihelion distance $(q, q + dq)$, where $q \leqslant 30$ AU, $dq = 1$ AU, for $\tau < 10^9$ yr (Fig. 5a) and $10^9 < \tau$ (yr) $< 2 \cdot 10^9$ (Fig. 5b), τ is the time of perihelion passage.

$(1.41 \cdot 10^{-12}$ yr$^{-1})$. The flux of OC comets with $a > 10^4$ AU within $q < 1.5$ AU is 4.6 comets with $H_{10} < 10.9$ per year, and is 0.4 comets per year with $H_{10} < 7$.

9. Dynamically old comets

Figure 5 shows the distributions $f_{LP}(q)$ for $\tau < 10^9$ yr (Fig. 5a) and $10^9 < \tau$ (yr) $< 2 \cdot 10^9$ (Fig. 5b), where $f_{LP}(q)$ is the frequency of passages of long-period (LP) comets with semimajor axes $a > 34.2$ AU per year in the interval of perihelion distance $(q, q+dq)$, $q \leqslant 30$ AU, $dq = 1$ AU.

The flux of comets with $a > 34.2$ AU originating from the inner Oort cloud strongly predominates in the region $q > 10$ AU (Fig. 5). The orbits of IOC comets evolve slowly to the outer planetary region. The flux of LP comets originating from the OOC is predominant within $q < 10$ AU. Comets of the outer cloud evolve into smaller orbits more rapidly than comets of the inner cloud. Inner cloud comets enter the planetary region with much larger perihelion distances, since they are only weakly affected by external perturbations. Comets with $a < (1.5 - 2) \cdot 10^4$ AU coming in the region $q > 20$ AU, besides, are weakly perturbed by the planets, and evolve slowly in the outside planetary system. Perturbations by Saturn act as a substantial barrier to the diffusion of cometary perihelia into the inner planetary system. The perihelia of comets, slowly drifting in q, will accumulate beyond the orbit of Saturn. The distributions of the frequency $f_{LP}(q)$ of passages of outer and inner cloud comets with $10^9 < \tau$ (yr) $< 2 \cdot 10^9$ have peaks slightly beyond the orbit of Saturn (Fig. 5).

10. Conclusions

The dynamical evolution of outer Oort cloud comets in the planetary region is faster than that of inner cloud comets. Outer cloud comets evolve into smaller orbits more rapidly. The flux of long-period comets originating from the inner cloud is substantially predominant in the region $q > 10$ AU, whilethe flux of LP comets from the outer cloud is dominant in the region $q < 10$ AU. The flux of dynamically young and old comets originating from the outer Oort cloud is largest in the Saturn-Uranus region. The flux of dynamically young comets originating from the inner cloud is largest in the Uranus-Neptune region.

Inner cloud comets ($a < 10^4$ AU) can replenish the outer Oort cloud ($a > 10^4$ AU). The frequency of passages of inner cloud comets that have replenished the outer cloud within $q < 1.5$ AU is $6 \cdot 10^{-14}$ yr^{-1}, that is more than one order of magnitude less than the frequency of passages of outer cloud comets with $a > 10^4$ AU ($1.41 \cdot 10^{-12}$ yr^{-1}).

The majority of new comets come into the outer planetary region. The flux of new comets with $15 < q < 31$ AU is higher than with $q < 15$ AU by a factor ~ 1.7 for comets from the outer cloud, and by a factor 7 for comets from the inner cloud. Inner cloud comets can reach the region $q < 1.5$ AU, mainly evolving gradually from the outer planetary region, while the majority of outer cloud comets come to $q < 1.5$ AU directly from $q > 35$ AU. The maximum in the distribution of perihelion distances of comets, that are coming gradually to near-Earth space, on their perihelion passage just before the entry into the region $q < 1.5$ AU, lies near the orbit of Saturn.

Near Saturn orbit, dynamically young comets with semimajor axes a few thousands astronomical units can be removed to orbits with $a > 10^4$ AU, become subject of strong external perturbations, and therefore may come to the visible region on their next passage.

The maximum in the distribution of perihelion distances of comets with $a > 10^4$ AU lies near Neptune orbit. Perturbations by Saturn act as a substantial barrier to the diffusion of cometary perihelia into the inner planetary system. The perihelia of comets evolving in the outer planetary region will accumulate beyond Saturn's orbit. The distribution of perihelion distances of LP comet has a peak slightly beyond the orbit of Saturn.

The frequency of passages of long-period ($a > 34.2$ AU) comets from the inner cloud within $q < 1.5$ AU is $3.5 \cdot 10^{-13}$ yr^{-1}, that is roughly one order of magnitude less than the frequency of passages of LP comets from the outer Oort cloud ($5.28 \cdot 10^{-12}$ yr^{-1}). The ratio of LP comets with $q < 1.5$ AU to new comets is ~ 5.

Acknowledgements

I thank the organizers for financial support enabling attendance at the meeting. This work also was supported by RFBR-Ural Grant 04-02-96042 and RFBR Grant 06-02-16512.

References

Bailey, M.E. & Stagg, C.R. 1988, *MNRAS* 235, 1

Dones, L., Levison, H.F., Duncan, M. & Weissman, P. 2000, *BAAS* 32, 1060

Duncan, M., Quinn, T. & Tremaine, S. 1987, *AJ* 94, 1330

Dybczynski, P.A. 2001, *Astron. Astrophys.* 375, 643

Emel'yanenko, V.V. 1999, *Evolution and source regions of asteroids and comets*, Proc. IAU Coll. 173, p. 339

Fernandez, J.A. 1981, *Astron. Astrophys.* 96, 26

Fernandez, J.A. 2002, *Earth, Moon & Planets* 89, 325

Francis, P.J. 2005, *AJ* 635, 1348

Heisler, J. & Tremaine, S. 1986, *Icarus* 65, 13

Heisler, J., Tremaine, S. & Alcock, C. 1987, *Icarus* 70, 269

Hills, J.G. 1981, *AJ* 86, 1730

Levison, H.F., Morbidelli, A., Dones, L., Jedicke, R., Wiegert, P.A. & Bottke, W.F. 2002, *Science* 296, 2212

Marsden, B.G., Sekanina, Z. & Everhart, E. 1978, *AJ* 83, 64

Marsden, B.G. & Williams, G.V. 2005, *Catalogue of Cometary Orbits*

Ogorodnikov, K.F. 1965, *Dynamics of stellar systems* (Oxford: Pergamon)

Oort, J.H. 1950, *Bull. Astron. Inst. Neth.* 11, 91

Weissman, P.R. 1993, *BAAS* 25, 1063

Weissman, P.R. 1996, *Completing the Inventory of the Solar System*, ASP Conf. Ser. 107, p. 265

Near Earth Objects, our Celestial Neighbors: Opportunity and Risk
Proceedings IAU Symposium No. 236, 2006
A. Milani, G.B. Valsecchi & D. Vokrouhlický, eds.
© 2007 International Astronomical Union
doi:10.1017/S1743921307003067

Migration of comets to the terrestrial planets

Sergei I. Ipatov[1,2] and John C. Mather[3]

[1]Department of Terrestrial Magnetism, Carnegie Institution of Washington
Washington DC, USA
email: siipatov@hotmail.com
http://www.dtm.ciw.edu/ipatov

[2]Space Research Institute, Moscow, Russia

[3]LASP, NASA/Goddard Space Flight Center, Greenbelt, USA
email: John.C.Mather@nasa.gov

Abstract. We studied the orbital evolution of objects with initial orbits close to those of Jupiter-family comets (JFCs), Halley-type comets (HTCs), and long-period comets, and the probabilities of their collisions with the planets. In our runs the probability of a collision of one object with the Earth could be greater than the sum of probabilities for thousands of other objects. Even without the contribution of such a few objects, the probability of a collision of a former JFC with the Earth during the dynamical lifetime of the comet was greater than 4×10^{-6}. This probability is enough for delivery of all the water to Earth's oceans during the formation of the giant planets. The ratios of probabilities of collisions of JFCs and HTCs with Venus and Mars to the mass of the planet usually were not smaller than that with Earth. Among 30 000 considered objects with initial orbits close to those of JFCs, a few objects got Earth-crossing orbits with semimajor axes $a < 2$ AU and aphelion distances $Q < 4.2$ AU, or even got inner-Earth ($Q < 0.983$ AU), Aten, or typical asteroidal orbits, and moved in such orbits for more than 1 Myr (up to tens or even hundreds of Myr). From a dynamical point of view, the fraction of extinct comets among near-Earth objects can exceed several tens of percent, but, probably, many extinct comets disintegrated into mini-comets and dust during a smaller part of their dynamical lifetimes.

Keywords. Transport of comets; transport of meteorites

---

1. Introduction

Farinella *et al.* (1993), Bottke *et al.* (2002), Binzel *et al.* (2002), Weissman *et al.* (2002) believe that asteroids are the main source of near-Earth objects (NEOs). Wetherill (1988) supposed that half of NEOs are former short-period comets. Trans-Neptunian objects (TNOs) can migrate to the near-Earth space. Duncan *et al.* (1995) and Kuchner *et al.* (2002) investigated the migration of TNOs to Neptune's orbit, and Levison & Duncan (1997) studied their migration from Neptune's orbit to Jupiter's orbit. Levison *et al.* (2006) studied the formation of Encke-type objects. More references on papers devoted to the migration of bodies from different regions of the solar system to the near-Earth space can be found in our previous publications on this problem (Ipatov 1995, 1999, 2000, 2001; Ipatov & Hahn 1999; Ipatov & Mather 2003, 2004a,b, 2006a). As the migration of TNOs to Jupiter's orbit was considered by several authors, Ipatov (2002) and Ipatov & Mather (2003, 2004a,b, 2006a) paid particular attention to the orbital evolution of Jupiter-crossing objects (JCOs), considering a larger number of JCOs than before.

In the present paper, we summarize our studies of migration of cometary objects into NEO orbits, paying particular attention to the probabilities of collision of cometary objects with the terrestrial planets. These studies are based on our previous runs and on some new runs. Earlier we did not consider the evolution of orbits of Halley-type comets and long-period comets and did not study the probabilities of collisions of different comets with the giant planets. Though some runs used here are the same as earlier, the discussion on migration of small bodies based on these runs is different.

2. Initial data

Ipatov & Mather (2003, 2004a,b, 2006a) integrated the orbital evolution of about 30 000 objects with initial orbits close to those of Jupiter-family comets (JFCs). We considered the gravitational influence of the planets, but omitted the influence of Mercury (except for Comet 2P/Encke) and Pluto. In about a half of the runs we used the method by Bulirsh-Stoer (1966) (BULSTO code), and in other runs we used a symplectic method (RMVS3 code). The integration package of Levison & Duncan (1994) was used. Usually we investigated the orbital evolution during the dynamical lifetimes of objects (until all the objects reached 2 000 AU from the Sun or collided with the Sun).

In the first series of runs (denoted as $n1$) we calculated the orbital evolution of 3 100 JCOs moving in initial orbits close to those of 20 real comets with period $5 < P_a < 9$ yr, and in the second series of runs (denoted as $n2$) we considered 13 500 initial orbits close to those of 10 real JFCs with period $5 < P_a < 15$ yr. We selected comets with the above periods among JFCs with numbers between 7 and 75 for $n1$ and with numbers between 77 and 113 for $n2$. In other series of runs, initial orbits were close to those of a single comet (2P/Encke, 9P/Tempel 1, 10P/Tempel 2, 22P/Kopff, 28P/Neujmin 1, 39P/Oterma, or 44P/Reinmuth 2). Comet 2P/Encke is the only comet with aphelion distance $Q < 4.2$ AU; comets 28P/Neujmin 1 and 44P/Reinmuth 2 are typical comets with semimajor axis $a \approx 7$ AU, and other four above comets are typical comets with $a \sim 3 - 4$ AU. In order to compare the orbital evolution of comets and asteroids, we also studied the orbital evolution of 1 300 test asteroids initially moving in the 3:1 and 5:2 resonances with Jupiter.

In our recent runs we also considered objects started from orbits of test long-period comets (initial eccentricity $e_o = 0.995$, $q_o = a_o (1 - e_o) = 0.9$ AU or $q_o = 0.1$ AU, initial inclination i_o between 0 and $180°$ in each run, objects started at perihelion; these runs are denoted as lpc runs) and test Halley-type comets ($a_o = 20$ AU, i_o between 0 and $180°$ in each run, objects started at perihelion; in some runs $e_o = 0.975$ and $q_o = 0.5$ AU, in other runs $e_o = 0.9$ and $q_o = 2$ AU; these runs are denoted as htc runs).

In our runs, planets were considered as material points, so literal collisions did not occur. However, using the algorithm suggested by Ipatov (1988, 2000) with the correction that takes into account a different velocity at different parts of the orbit (Ipatov & Mather 2003), and based on all orbital elements sampled with a 500 yr step, we calculated the mean probability P of collisions of migrating objects with a planet. We define P as P_Σ/N, where P_Σ is the probability for all N objects of a collision of an object with a planet. If it is not mentioned specially, all probabilities below are considered for dynamical lifetimes of objects. Note that our algorithm differed from the Öpik's scheme, and included calculations of a synodic period and the region where the distance between the "first" orbit and the projection of the "second" orbit onto the plane of the "first" orbit is less than the sphere of action (i.e., the Tisserand sphere). The code includes calculations of the probability that both objects are on the same line of sight from the Sun within the above region, of the difference in probabilities for planar and spatial models, and of the probability of a collision inside the sphere.

For BULSTO runs, the integration step error was less than ϵ, where ϵ varied between 10^{-13} and 10^{-8} (most of the runs were made for ϵ equal to 10^{-8} and 10^{-9}), and for a

RMVS3 runs an integration step d_s varied from 0.1 to 30 days (most runs were made for $d_s=10$ days). In a single run with N (usually $N = 250$) objects, ϵ or d_s was constant. Results obtained with the use of different methods of integration at different d_s or ϵ were similar (see Ipatov & Mather 2003 for details), except for probabilities of collisions with the Sun in such runs when this probability was large (for comets 2P/Encke and 96P/Machholz 1 from $n2$ series, and the 3:1 resonance with Jupiter). Probabilities of collisions of bodies with planets were close for different integrators even in the latter case because soon after close encounters with the Sun, bodies were ejected into hyperbolic orbits or moved in highly inclined orbits. In most "cometary" runs, the fraction P_{Sun} of comets collided with the Sun was less than 0.02; P_{Sun} exceeded 0.05 for some htc runs, and most of objects in 2P/Encke runs collided with the Sun.

Levison showed that it is difficult to detect solar collisions in any numerical integrator, so he removed objects with perihelion distance $q < q_{\min}$. Our runs were made for direct modeling of collisions with the Sun, but we studied what happens if we consider $q_{\min}$ equal to k_S radii r_S of the Sun. We obtained that the mean probabilities of collisions of bodies with planets, lifetimes of the objects that spent millions of years in Earth-crossing orbits, and other obtained results were practically the same if we consider that objects disappear when q becomes less than the radius r_S of the Sun or even several such radii (i.e., we checked $q < k_S \cdot r_S$, where k_S equals 1, 2, or another value). The only noticeable difference was for Comet 96P/Machholz 1 from the $n2$ series, and a smaller one was for Comet 2P/Encke, but the results of such runs were not included in our statistics. The eccentricity and inclination of Comet 96P/Machholz 1 are large, so usually even for these $n2$ runs, the collision probabilities of objects with the terrestrial planets were not differed by more than 15% at $k_S = 0$ and $k_S = 1$. Among more than a hundred considered runs, there were three runs, for each of which at $k_S = 0$ a body in an orbit close to that of Comet 96P/Machholz 1 was responsible for 70-75% of collision probabilities with the Earth, and for $k_S = 1$ a lifetime of such body was much less than for $k_S = 0$. Nevertheless, for all ($\sim 10^4$) objects from $n2$ series, at $k_S = 0$ the probabilities of collisions with the terrestrial planets were close to those at $k_S = 1$, even if we consider the above runs. The difference for times spent in Earth-crossing orbits is greater than that for the probabilities and is about 20%. For all runs of the 2P/Encke series, the difference in time spent in orbits with $Q < 4.7$ AU for $k_S = 0$ and for $k_S = 1$ was less than 4%. In the 2P/Encke series of runs (and also for the 3:1 resonance with Jupiter), at $k_S = 0$ we sometimes got orbits with $i > 90°$, but practically there were no such orbits at $k_S \geqslant 1$ (Ipatov & Mather 2004a,b). Among the objects with initial orbits close to that of Comet 96P/Machholz 1, we found one object which also got $i > 90°$ for 3 Myr. Inclinations of other such objects did not exceed $90°$.

3. Computer simulations of the migration of comets to near-Earth orbits

Some migrating JCOs got Earth-crossing orbits. Usually they spent in such orbits only a few thousands years, but a few objects moved in Earth-crossing orbits with aphelion distances $Q < 4.2$ AU for millions of years. The total times which 30 000 considered objects started from JFC orbits spent in Earth-crossing orbits with $a < 2$ AU were due to a few tens of objects, but mainly only to a few of them. In this section we consider only these few objects. With BULSTO at $10^{-9} \leqslant \epsilon \leqslant 10^{-8}$, six and nine objects, respectively from the 10P/Tempel 2 and 2P/Encke series, moved into Apollo orbits with $a < 2$ AU (*Al2* orbits) for at least 0.5 Myr each, and five of them remained in such orbits for more than 5 Myr each. Among the JFCs considered with BULSTO, only one and two JFCs reached inner-Earth orbits (IEO, $Q < 0.983$ AU) and Aten orbits, respectively. Only two objects in series $n2$ got $Al2$ orbits during more than 1 Myr. For the $n1$ series of runs, while moving in JCO orbits, objects had orbital periods $P_a < 20$ yr (JFCs) and $20 < P_a < 200$ yr (Halley-type comets) for 32% and 38% of the mean value T_J ($T_J = 0.12$ Myr) of the total time spent by one object in Jupiter-crossing orbits, respectively.

Four considered former JFCs even got IEO or Aten orbits for Myrs. Note that Ipatov (1995) obtained migration of JCOs into IEO orbits using the method of spheres to consider the gravitational influence of planets. In our BULSTO runs, one former JCO, which had an initial orbit close to that of 10P/Tempel 2, moved in Aten orbits for 3.45 Myr, and the probability of its collision with the Earth from such orbits was 0.344. It also moved for about 10 Myr in IEO orbits before its collision with Venus, and during this time the probability of its collision with Venus was $P_V = 0.655$. The above probabilities are greater than the total probabilities for 10^4 other JCOs. Another object (from the 2P BULSTO run) moved in highly eccentric Aten orbits for 83 Myr, and its lifetime before collision with the Sun was 352 Myr. Its probability of collisions with Earth, Venus, and Mars during its lifetime was 0.172, 0.224, and 0.065, respectively. With RMVS3 at $d_s \leqslant 10$ days for the 2P run, the probability of collisions with Earth for one object was greater by a factor of 30 than for 250 other objects. For series $n1$ with RMVS3, the probability of a collision with the Earth for one object with an initial orbit close to that of Comet 44P/Reinmuth 2 was 88% of the total probability for 1200 objects from this series, and the total probability for 1198 objects was only 4%. For series 44P/Reinmuth 2 with N=1500 there were no objects with $a < 2$ AU and $q < 1$ AU, though the 44P object in $n1$ run spent 11.7 Myr in such orbits. For series $n2$ with RMVS3, we obtained one object with an initial orbit close to that of Comet 113P/Spitaler with relatively large values of probabilities of collisions with Earth and Venus. This object is responsible for 10% of the total collision probability with Earth for all $n2$ objects, but most of the time spent by all these objects in orbits with $a < 2$ AU and $q < 1$ AU are due to this object. Though about a half of 30 000 considered objects belong to series $n2$, most of objects that spent a long time in Earth-crossing orbits with $Q < 4.2$ AU belong to other series of runs.

After 40 Myr one considered object with an initial orbit close to that of Comet 88P/Howell (from $n2$ RMVS3 runs) got $Q < 3.5$ AU, and it moved in orbits with $a = 2.60 - 2.61$ AU, $1.7 < q < 2.2$ AU, $3.1 < Q < 3.5$ AU, eccentricity $e = 0.2 - 0.3$, and inclination $i = 5 - 10°$ for 650 Myr. Another object (with an initial orbit close to that of Comet 94P/Russel 4) moved in orbits with $a = 1.95 - 2.1$ AU, $q > 1.4$ AU, $Q < 2.6$ AU, $e = 0.2 - 0.3$, and $i = 9 - 33°$ for 8 Myr (and it had $Q < 3$ AU for 100 Myr). So JFCs can very rarely get typical asteroid orbits and move in them for Myrs. In our opinion, it can be possible that Comet 133P/Elst-Pizarro moving in a typical asteroidal orbit (Hsieh & Jewitt 2006) was earlier a JFC and it circulated its orbit also due to non-gravitational forces. JFCs got typical asteroidal orbits less often than NEO orbits.

Levison et al. (2006) argued that our obtained orbits with $a \approx 1$ AU were due to the fact that collisions of objects with terrestrial planets were not taken into account in our runs and such orbits were caused by too close encounters of objects with planets which really result in collisions. Based on the orbital elements obtained in our runs, we can conclude that probabilities of collisions of migrating bodies with planets before bodies got orbits with $a < 2$ AU were very small and the reason of the transformations of orbits was not caused by such close encounters of objects with the terrestrial planets that really resulted in collisions with the planets. Some real probabilities of collisions of bodies moving in orbits with $a < 2$ AU with the terrestrial planets were only after bodies had already got such orbits and moved in them for tens or hundreds of Myr. Other scientists did not obtain the migration of JCOs into orbits with $a \approx 1$ AU because they considered other initial data. In series $n2$ with 13 500 objects, we also did not obtain orbits with $a \approx 1$ AU and obtained only two orbits with $a < 2$ AU (the latter orbits were also obtained by Levison *et al.* 2006). For other series of runs, we paid particular attention to those initial data for which migrating objects could spend a long time inside Jupiter's orbit.

4. Cometary objects in NEO orbits

Based on the results of migration of JFCs with initial orbits close to the orbit of Comet P/1996 R2 Lagerkvist obtained by Ipatov & Hahn (1999) (for these runs with about a hundred objects, there were no objects which spent a long time in Earth-crossing orbits), Ipatov (1999, 2001) found that $10 - 20\%$ or more of all 1-km Earth-crossers could have come from the Edgeworth-Kuiper belt into Jupiter-crossing orbits. Using our results of the orbital evolution of 30 000 JCOs and the results of migration of TNOs obtained by Duncan *et al.* (1995) and considering the total of 5×10^9 1-km TNOs with $30 < a < 50$ AU, Ipatov & Mather (2003, 2004a,b) estimated the number of 1-km former TNOs in NEO orbits. The results of their runs testify in favor of at least one of these conclusions: (1) the portion of 1-km former TNOs among NEOs can exceed several tens of percents, (2) the number of TNOs migrating inside the solar system could be smaller by a factor of several than it was earlier considered, (3) most of 1-km former TNOs that had got NEO orbits disintegrated into mini-comets and dust during a smaller part of their dynamical lifetimes if these lifetimes are not small. All these three scenarios could take place. We consider that the role of disintegration may be more valuable and most of former comets that could move inside Jupiter's orbit for millions of years really were disintegrated. As the number of TNOs, their rate of migration inside the solar system, and lifetimes of former comets before their disruption are not well known, the estimates of the fraction of former TNOs among NEOs are very approximate.

Disintegrated comets could produce a lot of mini-comets and dust. Therefore there could be a lot of cometary dust among zodiacal particles, some of them were produced by high eccentricity comets (such as Comet 2P/Encke). The same conclusion about cometary dust was made by Ipatov *et al.* (2006a,b) based on analysis of spectra of the zodiacal light. Frank *et al.* (1986a,b) concluded that there is a large influx of small comets into the Earth's upper atmosphere.

It is known (Merline *et al.* 2002; Noll 2006; Pravec *et al.* 2006) that about 15% of NEOs and 2-3% of main-belt asteroids are binaries. We can suppose that the fraction of NEO binaries is greater for those NEOs which are extinct comets than those for NEOs that came from the main asteroid belt. Comets more often split into smaller parts than asteroids, and probably there are former comets even among binary main-belt asteroids. Besides, if initial (before collisional destruction) small bodies were formed by compression of dust condensations, then the fraction of binary objects is greater for greater distances of the place of origin of bodies from the Sun (Ipatov 2004).

Comets are estimated to be active for $T_{act} \sim 10^3 - 10^4$ yr. T_{act} is smaller for closer encounters with the Sun (Weissman *et al.* 2002), so for Comet 2P/Encke it is smaller than for other JFCs. If considered as material points, some former comets can move for tens or even hundreds of Myr in NEO orbits, so the number of extinct comets can exceed the number of active comets by several orders of magnitude. The mean time spent by Encke-type objects in Earth-crossing orbits was $\geqslant 0.4$ Myr. This time corresponds to $\geqslant 40 - 400$ extinct comets of this type if we consider that Encke-type active comet is not an exceptional event in the history of the solar system. Note that the diameter of Comet

2P/Encke is about $5-10$ km, so the number of 1-km Earth-crossing extinct Encke-type comets can be greater by a factor of 25-100 than the above estimate for Encke-size comets and can exceed 1 000 for such estimates. The rate of a cometary object decoupling from the Jupiter vicinity and transferring to a NEO-like orbit can be increased by a factor of several due to nongravitational effects (Harris & Bailey 1998; Asher et al. 2001; Fernández & Gallardo 2002). The role of the Yarkovsky and YORP effects on dynamics of asteroids was summarized by Bottke et al. (2006).

Dynamical models of the NEO population considered by Bottke et al. (2002) allowed 6% of dead comets. From measured albedos, Fernández et al. (2001) concluded that the fraction of extinct comets among NEOs and unusual asteroids is significant (9%). Rickman et al. (2001) and Jewitt & Fernández (2001) considered that dark spectral classes that might include the ex-comets are severely under-represented and comets played an important and perhaps even dominant role among all km-size Earth impactors. Binzel & Lupishko (2006) studied albedos and spectra of NEOs and concluded that $15\pm5\%$ of the entire NEO population may be composed by extinct or dormant comets. Harris & Bailey (1998) concluded that the number of cometary asteroids becomes comparable to the number of bodies injected from the main asteroid belt if one considers non-gravitational effects. Typical comets have larger rotation periods than typical NEOs (Binzel et al. 1992; Lupishko & Lupishko 2001), but, while losing considerable portions of their masses, extinct comets can decrease these periods.

Our runs showed that if one observes former comets in NEO orbits, then most of them could have already moved in such orbits for millions (or at least hundreds of thousands) of years. Some former comets that have moved in typical NEO orbits for millions of years, and might have had multiple close encounters with the Sun, could have lost their mantles, which caused their low albedo, and so change their albedo (for most observed NEOs, the albedo is greater than that for comets (Fernández et al. 2001) and would look like typical asteroids.

5. Probabilities of collisions of comets with planets

The probability of a collision of one celestial body with a planet can be greater than the total probability for thousands of objects with almost the same initial orbit. All probabilities considered below were calculated for dynamical lifetimes of objects. A few JCOs (mentioned in Section 3) with the highest probabilities with planets were not included in the statistics presented below. For series $n1$, the probability P_E of a collision of an object with the Earth (during a dynamical lifetime of the object) was about 4.5×10^{-6} and 4.8×10^{-6} for BULSTO and RMVS3 runs, respectively (but for RMVS3 it is by an order of magnitude greater if we consider one more object with the highest probability). For series $n2$, the mean value of P_E was $\sim(10-15)\times10^{-6}$ for BULSTO and RMVS3 runs.

Probabilities of collisions of JFCs with planets were different for different comets. The probability of a collision of Comet 10P/Tempel 2 with the Earth was 36×10^{-6} and 22×10^{-6} for BULSTO and RMVS3 runs, respectively ($P_E=140\times10^{-6}$ if we include objects with high collision probabilities). For 2P/Encke runs, P_E was relatively large: $\approx(1-5)\times10^{-4}$. For most other considered JFCs, $10^{-6}\leqslant P_E\leqslant10^{-5}$. For Comets 22P/Kopff and 39P/Oterma, $P_E\approx(1-2)\times10^{-6}$, and for Comets 9P/Tempel 1, 28P/Neujmin 1, and 44P/Reinmuth 2, $P_E\approx(2-5)\times10^{-6}$. The Bulirsh-Stoer method of integration and a symplectic method gave similar results. Values of P_E were about $(0.5-2)\times10^{-6}$ for htc runs, with greater values for smaller q_o. For lpc runs, $P_E=0.6\times10^{-6}$ at $q_o=0.9$ AU and $P_E=0.25\times10^{-6}$ at $q_o=0.1$ AU. Dynamical lifetimes of some objects in htc and *lpc* runs exceeded several Myr. Note that we considered collision probabilities for objects starting from different types of orbits, but a type of orbit (e.g., JFCs, HTCs, and LPCs) can change during the orbital evolution of objects.

The fraction of asteroids migrated from the 3:1 resonance with Jupiter that collided with the Earth was greater by a factor of several than that for the 5:2 resonance ($P_E \sim 10^{-3}$ and $P_E \sim (1-3) \times 10^{-4}$, respectively). The probabilities of collisions with the Earth for resonant asteroids (per one object) were about two orders of magnitude greater than those for typical JFCs. The difference in P_E for the asteroids and TNOs is greater than that for the asteroids and typical JFCs, as only about 1/3 of TNOs that had left the trans-Neptunian belt reached Jupiter's orbit (Duncan *et al.* 1995). The present mass of the Edgeworth-Kuiper belt is considered to be about two orders of magnitude greater than that of the main asteroid belt. For dust particles started from comets and asteroids, P_E was maximum for diameters $d \sim 100$ μm (Ipatov *et al.* 2004; Ipatov & Mather 2006a,b). These maximum values of P_E were usually (exclusive for 2P/Encke runs) greater at least by an order of magnitude than the values for parent comets.

The probabilities P_V of collisions of JFCs and HTCs with Venus usually did not differ by more than a factor of 2 from those with Earth. For 2P/Encke runs, they were greater than those with Earth, but in most of other runs they were smaller. The probabilities P_M of collisions of JFCs and HTCs with Mars usually were smaller by a factor of 3-6 (10 for the 2P/Encke runs) than those with Earth, i.e., Mars accreted more cometary bodies than Earth per unit of mass of the planet. For *lpc* runs, the values of P_E and P_V can differ by a factor of 3, and $P_E/P_M \sim 7-10$.

For most our runs, the probability P_J of a collision of a JFC with Jupiter (during a dynamical lifetime of the comet) was ~ 0.01. Usually it was less than 0.03, though it can be up to 0.06 in a single run. Due to resonances, the actual values of P_J can be smaller than those in our runs. The mean time T_J spent by objects in Jupiter-crossing orbits was 0.12 Myr for *n1* runs. So the collision frequency of an object started from a JFC orbit and moving in a Jupiter-crossing orbit is about 10^{-7} yr^{-1}. Though T_J can be a little greater for 2P/Encke runs than for *n1* and *n2* runs, and it can exceed 1 Myr for *htc* runs, P_J was only about 5×10^{-4} for some 2P/Encke and *htc* runs. In other 2P/Encke runs, P_J can be greater or smaller by a factor of 20 than the above value. For *lpc* runs, P_J was smaller by an order of magnitude than that for *htc* runs though T_J did not differ much.

Probabilities P_S of collisions of objects from *n1* and *n2* runs with Saturn typically were smaller by an order of magnitude than those with Jupiter, and collision probabilities with Uranus and Neptune typically were smaller by three orders of magnitude than those with Jupiter. The ratio of probabilities of collisions of bodies with different giant planets, for a pair of planets can vary by more than an order of magnitude from run to run. As only a small fraction of comets collided with all planets during dynamical lifetimes of comets, the orbital evolution of comets for the considered model of material points was close to that for the model when comets collided with a planet are removed from integrations.

6. Delivery of water and volatiles to planets

Using $P_E = 4 \times 10^{-6}$ (this value is smaller than the mean value of P_E obtained in our runs for JFCs) and assuming that the total mass of planetesimals that ever crossed Jupiter's orbit is about $100 m_{\oplus}$ (Ipatov 1987, 1993), where $m_{\oplus}$ is the mass of the Earth, we obtain that the total mass of water delivered from the feeding zone of the giant planets to the Earth could be about the mass of water in Earth's oceans. We considered that the fraction k_w of water in planetesimals equaled 0.5. For present comets $k_w < 0.5$ (Jewitt 2004), but it is considered that k_w could exceed 0.5 for planetesimals. The fraction of the mass of the planet delivered by JFCs and HTCs can be greater for Mars and Venus than that for the Earth. This larger mass fraction would result in relatively large ancient oceans on Mars and Venus. The conclusion that planetesimals from the zone of the giant planets could deliver all the water to the terrestrial oceans was also made by Ipatov (2001) and Marov & Ipatov (2001) on the basis of runs by Ipatov & Hahn (1999).

The above estimate of water delivery by cometary bodies to the Earth is greater than those by Morbidelli et $al.$ (2000) and Levison et $al.$ (2001), but is in accordance with the results by Chyba (1989) and Rickman et $al.$ (2001). The larger value of P_E we have calculated compared to those argued by Morbidelli et $al.$ (2000) ($P_E \sim (1-3) \times 10^{-6}$) and Levison et $al.$ (2001) ($P_E = 4 \times 10^{-7}$) is caused by the fact that in our runs we considered other initial orbits and a larger number of JCOs. Levison et $al.$ (2001) did not take into account the influence of the terrestrial planets, so probably that is why his values of P_E are even smaller than those by Morbidelli et $al.$ (2000). The latter authors used results of integrations of objects initially located beyond Jupiter's orbit. For 39P/Oterma runs ($a_o = 7.25$ AU and $e_o = 0.25$), we obtained P_E equal to 1.2×10^{-6} and 2.5×10^{-6} for BULSTO and RMVS3 runs, respectively. These values are in accordance with the values of P_E obtained by Morbidelli et $al.$. Morbidelli et $al.$ (2000) considered reasonable that about $50 - 100 m_\oplus$ of planetesimals primordially existed in the Jupiter-Saturn region and about $20 - 30 m_\oplus$ of planetesimals in the Uranus-Neptune region. We think that they considerably underestimated the mass of planetesimals in the Uranus-Neptune region.

Lunine (2004, 2006) concluded that possible sources of water for Earth are diverse, and include Mars-sized hydrated bodies in the asteroid belt, smaller "asteroidal" bodies, water adsorbed into dry silicate grains in the nebula, and comets. Lunine et $al.$ (2003) considered most of the Earth's water as a product of collisions between the growing Earth and planet-sized "embryos" from the asteroid belt. Drake & Campins (2006) noted that the key argument against an asteroidal source of Earth's water is that the O's iso-topic composition of Earth's primitive upper mantle matches that of anhydrous ordinary chondrites, not hydrous carbonaceous chondrites. Kuchner et $al.$ (2004) investigated the possibility that the Earth's ocean water originated as ice grains formed in a cold nebula, delivered to the Earth by drag forces from co-orbital nebular gas. Dust particles could also deliver water to the Earth from the feeding zone of the giant planets. Ipatov & Mather (2006a,b) obtained that the probability of collisions of $10 - 100$ μm particles originated beyond Jupiter's orbit is about $(1 - 3) \times 10^{-4}$. Therefore the water in the terrestrial oceans ($2 \times 10^{-4} m_\oplus$) can be delivered by particles (for the model without sublimation) which had contained $\sim m_\oplus$ of water when they had been located beyond Jupiter. So dust particles could also play some role in the delivery of water to the terrestrial planets during planet formation.

There is the deuterium/hydrogen paradox of Earth's oceans (the D/H ratio is different for oceans and comets), but Pavlov et $al.$ (1999) suggested that solar wind-implanted hydrogen on interplanetary dust particles provided the necessary low-D/H component of Earth's water inventory, and Delsemme (1999) considered that most of the seawater was brought by the comets that originated in Jupiter's zone, where steam from the inner solar system condensed onto icy interstellar grains before they accreted into larger bodies. It is likely (Drake & Campins 2006) that the D/H and Ar/O ratios measured in cometary comas and tails are not truly representative of cometary interiors.

Small bodies which collided with planets could deliver volatiles and organic/prebiotic compounds needed for life origin. Marov & Ipatov (2005) concluded that dust parti-cles could be most efficient in the delivery of organic or even biogenic matter to the Earth, because they experience substantially weaker heating when passing through the atmosphere (an excess heat is radiated effectively due to high total surface-to-mass ratio for dust particles). They assumed that life forms drastically different from the terrestrial analogs are unlikely to be found elsewhere in the solar system (if any), e.g., either extinct or extant life on Mars.

7. Conclusions

Some Jupiter-family comets can reach typical NEO orbits and remain there for millions of years. From the dynamical point of view (if comets didn't disintegrate) there could be (not 'must be') many (up to tens of percent) extinct comets among the NEOs, but, probably, many extinct comets disintegrated into mini-comets and dust during a smaller part of their dynamical lifetimes if these lifetimes were large. Disintegration of comets can provide a considerable fraction of cometary dust among the zodiacal dust particles. The probability of a collision of one object moving for a long time in Earth-crossing orbits, with the Earth could be greater than the sum of probabilities for thousands of other objects, even having similar initial orbits. Even without a contribution of such a few bodies, the probability of a collision of a former JFC (during its dynamical lifetime) with the Earth was greater than 4×10^{-6}. This probability is enough for delivery of all the water to Earth's oceans during formation of the giant planets. The ratios of probabilities of collisions of JFCs and HTCs with Venus and Mars to the mass of the planet usually were not smaller than that for Earth.

References

Asher, D.J., Bailey, M.E. & Steel, D.I. 2001, in: M.Ya. Marov & H. Rickman (eds.), *Collisional Processes in the Solar System*, ASSL 261, 121

Binzel, R.P., Xu, S., Bus, S.J. & Bowell, E. 1992, *Science* 257, 779

Binzel, R.P., Lupishko, D. F., Di Martino, M., *et al.* 2002, in: W.F. Bottke Jr., A. Cellino, P. Paolicchi & R. P. Binzel (eds.), *Asteroids III* (Tucson: Univ. of Arizona), p. 255

Binzel, R.P. & Lupishko, D.F. 2006, in: D. Lazzaro, S. Ferraz-Mello & J.A. Fernández (eds.), *Asteroids, Comets, and Meteors*, IAU Symposium 229 (Cambridge Univ. Press), p. 207

Bottke, W.F., Morbidelli, A., Jedicke, R., Petit, J.-M., Levison, H.F., Michel, P. & Metcalfe, T.S. 2002, *Icarus* 156, 399

Bottke, W.F., Vokrouhlický, D., Rubincam, D.P. & Nesvorný, D. 2006, *Annu. Rev. Earth Planet. Sci.* 34, 157

Bulirsh, R. & Stoer, J. 1966, *Numer. Math.* 8, 1

Chyba, C.F. 1989, *Nature* 343, 129

Delsemme, A.H. 1999, *Planet. Space Sci.* 47, 125

Drake, M. & Campins, H. 2006, in: D. Lazzaro, S. Ferraz-Mello & J.A. Fernández (eds.), *Asteroids, Comets, and Meteors*, IAU Symposium 229 (Cambridge Univ. Press), p. 381

Duncan, M.J., Levison, H.F. & Budd, S.M. 1995, *Astron. J.* 110, 3073

Farinella, P., Gonczi, R., Froeschlé, Ch. & Froeschlé, C. 1993, *Icarus* 101, 174

Fernández, J.A. & Gallardo, T. 2002, *Icarus* 159, 358

Fernández, Y.R., Jewitt, D.C. & Sheppard, S.S. 2001, *ApJ* 553, L197

Frank, L.A., Sigwarth, J.B. & Graven, J.D. 1986a, *Geophys. Res. Lett.* 13, 303

Frank, L.A., Sigwarth, J.B. & Graven, J.D. 1986b, *Geophys. Res. Lett.* 13, 307

Harris, N.W. & Bailey, M.E. 1998, *Mon. Not. R. Astron. Soc.* 297, 1227

Hsieh, H.H. & Jewitt, D. 2006, in: D. Lazzaro, S. Ferraz-Mello & J.A. Fernández (eds.), *Asteroids, Comets, and Meteors*, IAU Symposium 229 (Cambridge Univ. Press), p. 425

Ipatov, S.I. 1987, *Earth, Moon, & Planets* 39, 101

Ipatov, S.I. 1988, *Soviet Astron.* 32, 560

Ipatov, S.I. 1993, *Solar Syst. Res.* 27, 65

Ipatov, S.I. 1995, *Solar Syst. Res.* 29, 261

Ipatov, S.I. 1999, *Celest. Mech. & Dyn. Astr.* 73, 107

Ipatov, S.I. 2000, *Migration of celestial bodies in the solar system*, (Moscow: Editorial URSS Publishing Company) (in Russian)

Ipatov, S.I. 2001, *Adv. Space Res.* 28, 1107

Ipatov, S.I. 2002, in: B. Warmbein (ed.), *Asteroids, comets, meteors, 2002*, ESA SP-500, p. 371

Ipatov, S.I. 2004, in: S.S. Holt & D. Deming (eds.), *The Search for Other Worlds*, AIP Conference Proceedings 713, p. 277

Ipatov, S.I. & Hahn, G.J. 1999, *Solar Syst. Res.* 33, 487

Ipatov S.I. & Mather J.C. 2003, *Earth, Moon, & Planets* 92, 89

Ipatov, S.I. & Mather, J.C. 2004a, in: E. Belbruno, D. Folta & P. Gurfil (eds.), *Astrodynamics, Space Missions, and Chaos*, Annals New York Acad. Sci. 1017, p. 46

Ipatov, S.I. & Mather, J.C. 2004b, *Adv. Space Res.* 33, 1524

Ipatov, S.I. & Mather, J.C. 2006a, *Adv. Space Res.* 37, 126

Ipatov, S.I. & Mather, J.C. 2006b, in: H. Kruger & A. Graps (eds.), *Dust in Planetary Systems*, in press, available from: http://arXiv.org/format/astro-ph/0606434

Ipatov, S.I., Mather, J.C. & Taylor, P.A. 2004, in: E. Belbruno, D. Folta, & P. Gurfil (eds.), *Astrodynamics, Space Missions, and Chaos*, Annals New York Acad. Sci. 1017, 66

Ipatov, S.I., Kutyrev, A.S., Madsen, G.J., Mather, J.C., Moseley, S.H. & Reynolds, R.J. 2006a, *37th LPSC*, #1471

Ipatov, S.I., Kutyrev, A.S., Madsen, G.J., Mather, J.C., Moseley, S.H. & Reynolds, R.J. 2006b, *AJ*, submitted, available from: http://arXiv.org/format/astro-ph/0608141

Jewitt, D. 2004, in: M.C. Festou, H.U. Keller & H.A. Weaver (eds.), *Comets II* (Tucson: Univ. Arizona Press), p. 659

Jewitt, D. & Fernández, Y. 2001, in: M.Ya. Marov & H. Rickman (eds.), *Collisional Processes in the Solar System*, ASSL 261, p. 143

Kuchner, M.J., Brown, M.E. & Holman, M. 2002, *Astron. J.* 124, 1221

Kuchner, M.J., Youdin, A. & Bate, M. 2004, *Proc. of the Second TPF/Darwin Int. Conf.*, http://planetquest1.jpl.nasa.gov/TPFDarwinConf/confProceedings.cfm

Levison, H.F. & Duncan, M.J. 1994, *Icarus* 108, 18

Levison, H.F. & Duncan, M.J. 1997, *Icarus* 127, 13

Levison, H.F., Dones, L., Chapman, C.R., *et al.* 2001, *Icarus* 151, 286

Levison, H.F., Terrel, D., Wiegert, P.A., Dones, L. & Duncan, M.J. 2006, *Icarus* 182, 161

Lunine, J.I. 2004, *Proc. of the Second TPF/Darwin International Conference*, http://planetquest1.jpl.nasa.gov/TPFDarwinConf/confProceedings.cfm

Lunine, J.I., Chambers, J., Morbidelli, A. & Leshin, L.A. 2003, *Icarus* 165, 1

Lunine, J.I. 2006, in: D.S. Lauretta & H.Y. McSween Jr. (eds.), *Meteorites and the Early Solar System II* (Tucson: Univ. Arizona Press), p. 309

Lupishko, D.F. & Lupishko, T.A. 2001, *Solar Syst. Res.* 35, 227

Marov, M.Ya. & Ipatov, S.I. 2001, in: M.Ya. Marov & H. Rickman (eds.), *Collisional Processes in the Solar System*, ASSL 261, p. 223

Marov, M.Ya. & Ipatov, S.I. 2005, *Solar Syst. Res.* 39, 374

Merline, W.J., Weidenschilling, S.J., Durda, D.D., Margot, J.-L., Pravec, P. & Storrs, A.D. 2002, in: W.F. Bottke Jr., A. Cellino, P. Paolicchi & R.P. Binzel (eds.), *Asteroids III* (Tucson: Univ. of Arizona), p. 289

Morbidelli, A., Chambers, J., Lunine, J.I., Petit, J.-M., Robert, F., Valsecchi, G.B. & Cyr, K.E. 2000, *Meteorit. Planet. Sci.* 35, 1309

Noll, K.S. 2006, in: D. Lazzaro, S. Ferraz-Mello & J.A. Fernández (eds.), *Asteroids, Comets, and Meteors*, IAU Symposium 229 (Cambridge Univ. Press), p. 301.

Pavlov, A.A., Pavlov, A.K. & Kasting, J.F. 1999, *J. Geophys. Res.* 104, 30 725

Pravec, P., Scheirich, P., Kusnirak, P., *et al.* 2006, *Icarus* 181, 63

Rickman, H., Fernández, J.A., Tancredi, G. & Licandro, J. 2001, in: M.Ya. Marov & H. Rickman (eds.), *Collisional Processes in the Solar System*, ASSL 261, p. 131

Weissman, P.R., Bottke, W.F. Jr. & Levison, H.F. 2002, in: W.F. Bottke Jr., A. Cellino, P. Paolicchi & R.P. Binzel (eds.), *Asteroids III* (Tucson: Univ. of Arizona), p. 669

Wetherill, G.W. 1988, *Icarus* 76, 1

Near-Earth Objects, our Celestial Neighbors: Opportunity and Risk
Proceedings IAU Symposium No. 236, 2006
A. Milani, G.B. Valsecchi & D. Vokrouhlický, eds.
© 2007 International Astronomical Union
doi:10.1017/S1743921307003079

Search for small trans-Neptunian objects by the TAOS project

W.P. Chen[1], C. Alcock[2], T. Axelrod[3], F.B. Bianco[2,6], Y.I. Byun[4],
Y.H. Chang[1], K.H. Cook[5], R. Dave[6], J. Giammarco[6], D.W. Kim[4],
S.K. King[7], T. Lee[7], M. Lehner[2], C.C. Lin[1], H.C. Lin[1], J.J. Lissauer[8],
S. Marshall[9], N. Meinshausen[10], S. Mondal[1], I. de Pater[10],
R. Porrata[10], J. Rice[10], M.E. Schwamb[6,11], A. Wang[7], S.Y. Wang[7],
C.Y. Wen[7] and Z.W. Zhang[1]

[1]Institute of Astronomy, National Central University, Taiwan,
email: wchen@astro.ncu.edu.tw

[2]Harvard-Smithsonian Center for Astrophysics, USA

[3]University of Arizona, USA

[4]Department of Astronomy, Yonsei University, Korea

[5]IGPP, Lawrence Livermore National Laboratory, USA

[6]Department of Physics & Astronomy, University of Pennsylvania, USA

[7]Institute of Astronomy and Astrophysics, Academia Sinica, Taiwan

[8]NASA Ames Research Center, USA

[9]Stanford Linear Accelerator Center, USA

[10]University of California, Berkeley, USA

[11]California Institute of Technology, Pasadena, CA

Abstract. The Taiwan-America Occultation Survey (TAOS) aims to determine the number of small icy bodies in the outer reach of the Solar System by means of stellar occultation. An array of 4 robotic small (D=0.5 m), wide-field (f/1.9) telescopes have been installed at Lulin Observatory in Taiwan to simultaneously monitor some thousand of stars for such rare occultation events. Because a typical occultation event by a TNO a few km across will last for only a fraction of a second, fast photometry is necessary. A special CCD readout scheme has been devised to allow for stellar photometry taken a few times per second. Effective analysis pipelines have been developed to process stellar light curves and to correlate any possible flux changes among all telescopes. A few billion photometric measurements have been collected since the routine survey began in early 2005. Our preliminary result of a very low detection rate suggests a deficit of small TNOs down to a few km size, consistent with the extrapolation of some recent studies of larger (30–100 km) TNOs.

Keywords. Photometry; occultations; Trans Neptunian objects; comets, discovery; surveys

1. The Taiwan-America Occultation Survey (TAOS) Project

As of September 2006, there are over 1000 Trans-Neptunian Objects (TNOs) known (www.boulder.swri.edu/ekonews). Because of their large distances and small sizes, only the brightest, hence the largest, TNOs could be detected by their reflected sunlight. The faintest objects detected so far correspond to a size of about a few tens of km. There seems a deficit of TNOs in this size range in comparison with larger ones, perhaps as a result of collisional disruption (Bernstein *et al.* 2004). To determine the number of even smaller TNOs, TAOS has implemented an array of 4 small (aperture 0.5 m), wide-field (f/1.9) telescopes at Lulin Observatory in Taiwan to monitor for chance stellar occultation by

Figure 1. The photometry aperture mask to process the TAOS data. Each star has a different aperture size for a maximum signal-to-noise light curve. The position of the aperture changes according to the centroid of the star to accommodate image motion.

TNOs (King *et al.* 2001; Alcock *et al.* 2003). TNOs as small as 1–2 km across should be detectable by TAOS. The occurrence rate of occultation events will provide constraints on the number and spatial distribution of TNOs.

Occultation by a TNO a few km across lasts for a fraction of a second, so rapid photometry is necessary. The 4 TAOS telescopes are located with a maximum separation of $\sim$ 100 m, too close to detect the shadow speed by timing difference at different telescopes, but allowing coincidence detection by multiple telescopes to reduce the false alarm rate. In addition to the TNO science, the huge TAOS database, some 100 GB per night, should be valuable in other studies such as stellar variability. Our robotic system is also responding to gamma-ray bursts.

2. Current Status of the TAOS Project

2.1. *Data Acquisition and Adaptive Aperture Photometry*

The TAOS CCD camera, instead of reading out the entire chip as in a regular imaging observation, continuously integrates (pause) and reads out a block of pixels (shift) one at a time, while the shutter remains open and the telescope tracks on the target field. This electronic "pause-and-shift" scheme in effect produces a sequence of snapshots and allows for stellar photometry at a rate up to a few hertz (Chen *et al.* 2003). The image thus obtained is crowded because all the stars in the field of view are "folded" into each readout block. The shift of electrons and the open shutter also cause a star to leave a streak, which interferes with the signal of a "nearby" star that appears to be adjacent in the readout block, though the two stars in fact may be widely separated in the sky. In addition, photons from either a neighboring star or a patch of sky are recorded at different times.

A photometry pipeline has been developed to deal with the crowdedness and temporal/spatial blending in the data. For each star, an optimal aperture size is used to minimize contamination from neighboring stars or streaks. An appropriate patch of sky is chosen for subtraction. An "aperture mask" is then applied to measure the fluxes of all the stars (Fig. 1). The position of the aperture keeps centered on, i.e., being adaptive to, the peak pixel of a star from one readout block to another, so as to compensate for image motion. This Adaptive Aperture Photometry routine uses square apertures so accuracy is compromised—not a serious problem because TAOS images with a 3″ pixel scale are undersampled—but it is very efficient, capable in real time to process about a thousand stars sampled at 5 Hz. Other analysis methods are being evaluated.

2.2. *Event Detection — The Rank Statistics*

Figure 2 shows the occultation of the star TYC 076200961 by the asteroid (286) Iclea observed on 6 February 2006. TAOS observed this predicted event successfully with 3 telescopes. Iclea is known to have a diameter of 97 km, and the TAOS system detected the event readily.

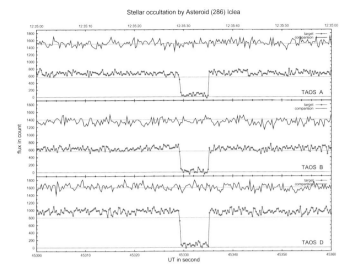

Figure 2. A showcase data of the star TYC 076200961 (m$_V$ ~ 11.83 mag) occulted by (286) Iclea (m$_V$ ~ 14.0 mag at the time) on 6 February 2006 observed by 3 TAOS telescopes. Data were taken at 4 Hz, and the duration of the event was estimated to be ≈ 5.75 s.

For short events, e.g., within one data point, we employ a nonparametric approach to identify simultaneous flux drops in stellar light curves. The flux at a given time is ranked among all data points. The rank statistic is then the product of the ranking orders of all telescopes. For example, if there are N data points from one telescope, the lowest flux has the rank $R = 1$ and the highest $R = N$. The rank statistic for data point i is then

$$Z_i = -\log\left(\prod_{k=1}^{N_{\text{tel}}} \frac{R_i^k}{N}\right) = \sum_{k=1}^{N_{\text{tel}}}\left(-\log\frac{R_i^k}{N}\right), \tag{2.1}$$

where N_{tel} is the total number of telescopes, and both the multiplication $\prod$ and summation $\sum$ are over N_{tel}. The quantity Z approximates to a Gamma distribution, if the noise between telescopes is independent. A probable occultation event would stand out as an "outlier" against an otherwise random distribution.

2.3. *Expected Event Rate*

The detected rate of TNO occultation events depends on the actual occurrence rate and the detectability. Relevant parameters include: (1) The surface number density (our goal) and angular size distribution of TNOs. (2) The surface number density and angular size distribution of background stars. Both these depend on the Galactic line of sight. With our current instrument setup, the 5 Hz observations reach about R=14 mag, which gives within the 3 deg^2 field of view typically several hundred to a few thousand stars in a target field. A dense field would be favorable for occultation but would create images too crowded for analysis. The angular size of a star can be estimated, for example, by its optical and infrared colors (van Belle 1999). Most stars have an angular size less than 0.1 milliarcsecond (mas). For reference, a TNO at 50 AU with an angular size of 0.1 mas has a physical diameter of ~ 4 km. Items (1) and (2) together specify the probability of area overlap (geometric consideration for occultation), plus the diffraction effect (King *et al.* 2006; Lehner *et al.* 2006), which tends to smooth out the flux drop and becomes important for small or distant TNOs. The extrapolated number density of small (> 1 km) TNOs varies widely, ranging from 10^2 to 10^6 per deg^2 (Bernstein *et al.* 2004). This amounts to about 1 event per few days to 1 event over a few years. (3) The shadow speed. Observing toward the opposition for example, as opposed to the quadrature, maximizes the chance of occultation but the event duration is short so difficult to detect. (4) CCD integration time and sampling rate, etc. These affect the limiting magnitude, hence the number of stars observable, and the capability to resolve a short flux drop.

Since 2005, TAOS with 3 telescopes has collected some billions of photometric measurements. Preliminary analysis shows a very low detection rate. A suspected event is shown in Fig. 3 where a relatively faint star (R~13.5 mag, RA= $13^h46^m26.7^s$ and Decl= $-10°50'31''$) was detected to have a 39%, 56%, and 50% flux drop, respectively, in 3 telescopes. The probability is, according to the Gamma approximation, 1.7×10^{-7} that the coincident flux drops were caused by chance. Given our high data rate, however, we cannot be highly confident that this was an actual occultation event. A fourth telescope, which is expected to be in service in early 2007, would increase greatly the credibility of a detection. Our result suggests a deficit down to km-sized TNOs. It is not clear at the moment how our low detection rate reconciles with the claim of numerous even smaller ($<$ 100 m) TNOs (Chang *et al.* 2006).

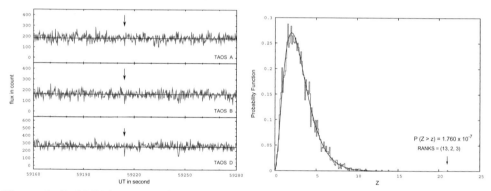

Figure 3. (Left) Light curves of a suspected TNO occultation event observed on 7 April 2005. (Right) Histogram of the rank statistics (2.1). This event ranks (13, 2, 3), respectively, in 3 telescopes out of 4728 data points sampled every 0.25 s. The smooth line is the Gamma distribution. The arrow marks the outlier event.

The NCU group acknowledges the NSC grant 95-2119-M-008-028. KHC's work was performed under the auspices of the US DOE, by the University of California, LLNL under contract No. W-7405-Eng-48.

References

Alcock, C., *et al.* 2003, *Earth, Moon & Planets* 92, 459

Bernstein, G.M., *et al.* 2004, *ApJ* 128, 1364

Chang, H.K., *et al.* 2006, *Nature* 442, 660

Chen, W.P., *et al.* 2003, *Baltic Ast.* 12, 568

King, S.K. 2001, in *Small-Telescope Astronomy on Global Scales*, eds. W.P. Chen, C. Lemme, & B. Paczynski, ASP Conf. Ser, 246, 253

King, S.K., *et al.* 2006, in: W. Ip & A. Bhardwaj (eds.), *Adv. in Geosci.*, Vol. 3 (World Sci.) p. 345

Lehner, M.J., *et al.* 2006, *Astron. Nach* 327, 814

Liang, C.L., *et al.* 2004, *Stat. Sci.* 19, 265

van Belle, G.T. 1999, *PASP* 111, 1515

Near Earth Objects, our Celestial Neighbors: Opportunity and Risk
Proceedings IAU Symposium No. 236, 2006
A. Milani, G.B. Valsecchi & D. Vokrouhlický, eds.
© 2007 International Astronomical Union
doi:10.1017/S1743921307003080

Some aspects of the statistics of Near-Earth Objects

J.R. Donnison

Astronomy Unit, School of Mathematical Sciences,
Queen Mary, University of London, Mile End Road

Abstract. The distribution of Near-Earth Objects, in particular Near-Earth asteroids is examined using maximum likelihood methods. These are analysed with respect magnitudes, taxonomic classes and to their orbital distances. Comparisons are made with the distributions of main-belt asteroids and short-period comets.

Keywords. Near-Earth objects, population

1. Introduction

The Near-Earth Object (NEO) population is defined as the group of small bodies with perihelion distance $q < 1.3$ AU and aphelion distance $Q > 0.983$ AU. These are composed of the 334 Atens that have semi-major axis $a < 1.0$ AU, the 1994 Apollos with perihelion $q < 1.0 AU$ and $a > 1.0$ AU, and the 1748 Amors with $1.0 < q < 1.3$ AU (as of August 2006). These NEOs with semi-major axes smaller than that of Jupiter are thought to be mainly asteroids that have escaped from the main asteroid belt, although as we shall see some may be extinct cometary nuclei.

In order to estimate the size of an NEO from its measured absolute magnitude its reflectivity (that is its geometric albedo) must be known. The measurements of albedo are only available for less than one percent of the NEOs, and the values span a wide of values from 0.023 to 0.63 (Binzel *et al.* 2002).

Attempts to debias the albedo and taxonomic distribution of NEOs have been made by Luu & Jewitt (1989) using Monte-Carlo simulations of the distribution. Binzel *et al.* (2002) conducted a similar study. They attempted to define a reasonable albedo distribution for each of the main belt source regions previously identified as sources of asteroidal and cometary material to the NEO population by Bottke *et al.* (2002). Stuart & Binzel (2004) utilize the direct observations of a taxonomically well determined subset of NEOs to determine the albedo distribution of the NEOs for which albedos are not available.

In this work we use maximum likelihood techniques to help determine the size distribution from the observed population with well determined absolute magnitudes. We draw heavily on the debiasing work particularly of Stuart & Binzel (2004). The absolute magnitude data used is taken from the MPC data sites at http://www.cfa.harvard.edu/iau/lists/Atens.html, http://cfa- www.harvard.edu/cfa/ps/lists/Amors.html, andhttp://cfa-www.harvard.edu/cfa/ps/lists/Apollos.html. In Fig. 1 we show a plot of the semi-major axis versus eccentricity of the Atens, Amors and Apollos used in this analysis.

2. Tisserand parameter

The Tisserand parameter is a dynamical quantity that is approximately conserved during an encounter between a planet and an interplanetary body. It therefore provides a

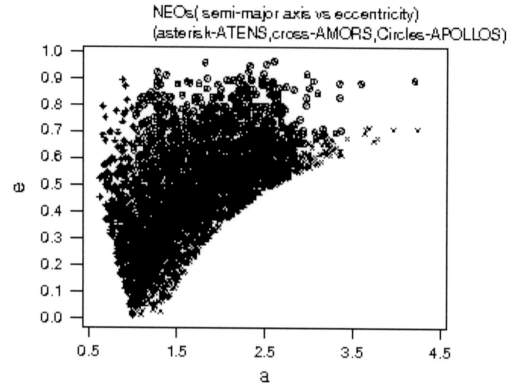

Figure 1. Plot of semi-major axis versus eccentricity of the Atens, Amors and Apollos used in this paper.

way to connect the post-encounter dynamical properties with the pre-encounter properties. The Tisserand parameter also provides a measure of the relative speed of an object when it crosses the orbit of a planet.

The Tisserand parameter T_J relative to Jupiter under the restricted circular three-body problem is given by

$$T_J = a_J/a + 2\sqrt{a/a_J (1 - e^2)} \cos i, \qquad (2.1)$$

where a_J is the semi-major axis of Jupiter, and a, e, and i are the semi-major axis, eccentricity, and inclination of the asteroid. Solar system bodies with $T_J \leqslant 3$ are dynamically coupled to Jupiter. A number of Near-Earth Asteroids have $2 < T_J < 3$. They tend to have low albedos and Fernandez *et al.* (2001) showed that there was a strong correlation between T_J and albedo which suggested that there is a significant cometary contribution to this asteroid population. The Jupiter family of comets have $2 \leqslant T_J \leqslant 3$ and the Halley and long-period comets tend to have $T_J \leqslant 2$. Bodies with $T_J > 3$ are generally decoupled from Jupiter and asteroids generally fall into this category. Binzel *et al.* (2004) did a similar analysis to Fernandez using taxonomic classes rather than albedo. They found a distinct taxonomic difference with respect to T_J, where C,D and X-type asteroids predominate for $T_J < 3$. This group is much more dominated by very dark objects than those with $T_J > 3$.

The Tisserand parameter will be incorporated here into the statistical determination of the distribution.

3. Magnitudes

The apparent magnitude m, of a near-Earth body is given by (Russell 1916; Jewitt 1999)

$$m = m_\odot - 2.5 \log_{10} \left(\frac{p_V R_A^2 \phi(\alpha)}{2.25 \times 10^{16} R^2 \Delta^2} \right), \tag{3.1}$$

where $m_\odot$ is the apparent magnitude of the Sun (-26.8 magnitude in the visible), p_V is the geometric albedo, R_A(km) is the radius of the asteroid, $R(AU)$ is the heliocentric distance, Δ(AU) is the geocentric distance, and $\phi(\alpha)$ is the phase function which at opposition with $\alpha = 0$ is $\phi(0) = 1$. This equation can be rewritten in terms of the absolute magnitude, H, as

$$m = H + 5 \log_{10} R + 5 \log_{10} \Delta - 2.5 \log_{10} \phi(\alpha), \tag{3.2}$$

where H is defined as the apparent magnitude that a near-Earth body would have if it was observed at 1 AU from the Sun, 1 AU from the Earth and at zero phase angle. In this analysis we consider the cumulative magnitude distribution using the absolute magnitudes H as listed in the Minor Planet Center list mentioned earlier. The H values are accurate to 0.05 magnitudes and clearly leading to errors in the estimation of the diameter. Combining equations (3.1) and (3.2), the diameter of the asteroid $D \, (= 2R_A)$ is related to the absolute magnitude and can be written in the form

$$H = C - 5 \log_{10} D - 2.5 \log_{10} p_V, \tag{3.3}$$

where $C = 15.618$ (Harris & Harris 1997; Stuart & Binzel 2004). The albedos of comet nuclei are typically $p_V \sim 0.04$, and those of the Centaurs, which probably originated in the Kuiper Belt, have a wide range of values from 0.04 to 0.17. The NEOs, as mentioned earlier, range from 0.023 to 0.63 (Binzel *et al.* 2002) giving a factor of 5 in possible diameter of an NEO for a given absolute magnitude. Stuart & Binzel (2004) found that for NEOs the correlation between albedo and absolute magnitude was not statistically significant and they assumed in their analysis that there was no correlation. We shall here make the same assumption. In order to assess the distribution more quantitatively, we consider the range of absolute magnitudes of the bodies. In Fig. 2 we have a plot of the cumulative number of the NEO population with absolute magnitude greater than H (as of August 2006). The bodies span the large magnitude range $9.45 \leqslant H \leqslant 30.01$.

The distribution shows a linear section for magnitudes up to $H \sim 20.0$ magnitudes. Above this magnitude the cumulative number no longer increases so steeply. For these higher magnitudes selection effects are presumed to be very important and many of the fainter and more distant bodies have yet to be discovered. Below this value the linear slope is not very sensitive to value of H chosen, though care must be taken in its assessment. In order to fit this linear section of the distribution we employ maximum likelihood estimation methods. The use of least-squares fits by many previous authors is not appropriate for cumulative plots as the data points are not independent of each other (Donnison & Sugden 1984; Gladman *et al.* 1998). We now proceed to the basic statistical model of this distribution.

4. Statistical model

We shall assume that the expected proportion of near-Earth bodies with diameters greater than D follows a Pareto power law of the form (that is a power law distribution

J.R. Donnison

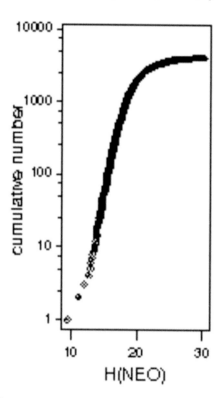

Figure 2. Plot of the cumulative number of the NEO population with absolute magnitude greater than H.

truncated at the lower end)

$$F(> D) = \left(\frac{D}{D_*}\right)^{-\alpha}. \tag{4.1}$$

Here α is the power law or cumulative size index and D_* is the lower limit in diameter that can be observed. Ideally we would like to work with the distribution of p_V but since the number of known albedos is very small we only have average values for taxonomic classes to work with (see Stuart & Binzel 2004). Therefore for a given p_V, in terms of absolute magnitude using Eq. (3.3), the expected proportion of bodies with magnitudes less than H is

$$F(< H) = \left(\frac{p_V}{p_{V_*}}\right)^{0.4\alpha} 10^{0.2\alpha(H - H_*)}, \tag{4.2}$$

where H_* is the critical upper limit in magnitude that is detectable corresponding to bodies with diameter D_*, and p_{V_*} is the corresponding albedo. The coefficient of the magnitudes is often denoted by β, so that $\alpha = 5\beta$.

An equivalent description of the diameter distribution is in terms of the probability density function given by

$$N(D) = \frac{\alpha}{D_*} \left(\frac{D}{D_*}\right)^{-(\alpha+1)}, \quad D > D_*, \tag{4.3}$$

where $N(D)\,dD$ is the expected proportion of bodies with diameters between D and $D+dD$. Eliminating the diameter using Eq. (3.3), we can write the distribution in terms of magnitude, so that we have $\Phi(H)$, the expected proportion of bodies with magnitudes between H and $H+dH$, for a given p_V is given by

$$\Phi(H) = \frac{\bar{C}p_{V_*}\alpha}{10^{-0.2H_*}}\left(\frac{p_V}{p_{V_*}}\right)^{\frac{1}{2}(\alpha+1)}10^{0.2(\alpha+1)(H-H_*)}, \qquad (4.4)$$

where $\bar{C}$ is a constant. To proceed further to determine α we use maximum likelihood estimation. The method of maximum likelihood is applicable to a random sample of observations taken from any given distribution. If we consider a set of n near-Earth bodies conditional on the number in the taxonomic class with absolute magnitudes $H_1, H_2, ...H_n$, then the corresponding likelihood function is given by the product of the joint probability density functions as

$$L(\alpha) = \prod_{i=1}^{n}\Phi_{c_i}(H_i), \qquad (4.5)$$

where c_i is the taxonomic class for i^{th} observation. It is easily seen that the maximum unconditional likelihood over α is the same as the maximum conditional likelihood given the numbers in the taxonomic classes. This can be handled more easily in log form so that

$$\ell(\alpha) = \log_e L(\alpha) = n\log_e \alpha + 0.2(\alpha+1)\sum_{i=1}^{n}(H_i - H_*)\log_e 10$$

$$+ 0.2nH_*\log_e 10 + \sum_{i=1}^{n}\log_e \bar{C}_i + \frac{1}{2}(\alpha+1)\sum_{i=1}^{n}\log_e\left(\frac{p_V c_i}{p_{V_*}}\right)$$

$$+ n\log_e p_{V*} \qquad (4.6)$$

Maximum likelihood estimation of the index now proceeds by maximizing $\ell(\alpha)$ as a function of α. The solution of the likelihood equation

$$\frac{\partial \ell}{\partial \alpha} = 0, \qquad (4.7)$$

is then given by

$$\frac{n}{\hat{\alpha}} + 0.2\log_e 10\sum_{i=1}^{n}(H_i - H_*) + \frac{1}{2}\sum_{i=1}^{n}\log_e\left(\frac{p_V c_i}{p_{V*}}\right) = 0, \qquad (4.8)$$

that is

$$\hat{\alpha} = \frac{n}{0.2\log_e 10\sum_{i=1}^{n}(H_* - H_i) - \frac{1}{2}\sum_{i=1}^{n}\log_e\left(\frac{p_V c_i}{p_{V*}}\right)}. \qquad (4.9)$$

The term involving the albedos can be written in terms of the various taxonomic orbital classes. That is

$$\sum_{i=1}^{n}\log_e\left(\frac{p_V c_i}{p_{V*}}\right) = \sum_{\text{taxonomic classes } c}\log_e\left(\frac{p_V}{p_{V_*}}\right)^{class_c}. \qquad (4.10)$$

Class_c is the observed number in class c. We can estimate this quantity from the fractional abundances and debiased albedos derived from the ten taxonomic complexes A, C, D, O, Q, R, S, U, V, X by Stuart & Binzel (2004). This takes into account that $T_J \leqslant 3$ for 30% of the NEOs and that $T_J > 3$ for the remaining 70%. Here we also extrapolate the abundances from the large sample size used by Stuart & Binzel (2004) to the current sample size used here. For the over simplified case where all the asteroids have the same albedo, that is $p_V = p_{V_*}$ then we have

$$\hat{\alpha} = \frac{n}{0.2 \log_e 10 \sum_{i=1}^{n} (H_* - H_i)}.$$ (4.11)

This gives an expression similar to that of Donnison (2006) used for trans-Neptunian bodies. The sampling variance of $\hat{\alpha}$ for large n for both the general case given by equation (4.9) and the simple case given by equation (4.11) is then approximately given by

$$\left\langle - \left(\frac{\partial^2 \ell}{\partial \alpha^2} \right)^{-1} \right\rangle = \frac{\alpha^2}{n},$$ (4.12)

so that the estimated standard error of $\hat{\alpha}$ in large samples is therefore

$$\frac{\hat{\alpha}}{\sqrt{n}}.$$ (4.13)

5. Results

From Fig. 2 the linear part ranges up to absolute magnitudes of around 20.0. The number of NEOs with magnitudes less than or equal to this value is 2079 (as of August 2006). These will form the data necessary for our determination. Before we proceed we can obtain a lower limit for the index if we assume the unrealistic situation that all the albedos are equal, that is $p_V = p_{V*}$, then $\hat{\alpha}$ given by Eq. (4.11) has a value of 1.16 ± 0.025. More realistically, estimating the albedo term using the fractional abundances and debiased albedos of Stuart & Binzel (2004) with a p_{V_*} of 0.05 (equivalent to D=0.6 km at magnitude 20.0) gives $\hat{\alpha}$ of 1.813 ± 0.040. This compares with the value of α of 1.95 found by Stuart & Binzel (2004) as the nearest power law fit and by Stuart (2001) who using a power law fit found a magnitude index β of 0.39. Bottke et $al.$ (2002) by modelling derived an α of 1.75 (based on a magnitude index β of 0.35). Rabinowitz et $al.$ (2000) also found a β of 0.35 using directly the debiased magnitude distribution observed by the NEAR survey. The result obtained does show some sensitivity to the value of p_{V_*} that is assumed.

6. Comparison with cometary distributions

The size distribution of cometary nuclei of long and short period comets has been investigated by a number of authors. Two approaches have been adopted. In the first approach, Donnison (1986, 1990, 1997, 1999), Hughes & Daniels (1980, 1982) and Hughes (2002) used the absolute magnitude, H_{10}, of the integrated dust and gas coma of $active$ $comets$ to estimate the magnitude, size and mass distributions of the cometary nuclei of both long and short period comets.

In the second approach the cometary diameters are measured directly. In the past only a few such size measurements were possible. However, the number of comet diameters accurately determined at large heliocentric distances has recently increased from ground based observations and the Hubble Space Telescope (HST) (Lamy et $al.$ 2000; Licandro et $al.$ 2000; Lowry et $al.$ 2001; Lowry & Fitzsimmons 2003 and Weissman & Lowry 2003). The cometary nuclei at these distances are not obscured by the surrounding coma and dust and are able to be measured directly. This has enabled the cometary index to be estimated for short period comets.

The distribution of masses of such bodies as comets asteroids and trans-Neptunian bodies is usually characterized by a power law index s defined through the mass distribution function $\zeta(m)$, such that the number of bodies with masses between m and $m + dm$ is given by

$$\zeta(m)dm = Am^{-s}dm, \tag{6.1}$$

where A is constant over a specific mass range. The s for near-Earth asteroids is related to α derived earlier by

$$s = \frac{\alpha}{3} + 1. \tag{6.2}$$

Currently using the active H_{10} short period comet data, the mass index s is about 1.6 (Donnison 1990; Hughes 2002). For the direct determination Fernandez *et al.* (1999), Tancredi *et al.* (2006) found for the Jupiter family of comets an s of 1.88, while Weissman & Lowry (2003) found an s of 1.53. Recently Meech *et al.* (2004) found 1.48 for the value of s. Since not all the data used has been published and is not is readily available a full explanation for the differences has not been found. The larger value of Fernandez *et al.* (1999) may however reflect the fact that their sample includes many comets observed at very small heliocentric distances where activity was possible and finding the index is complex. The present author is currently working on a new assessment of this index. These values for comets indicate that for the distribution of comets the majority of the mass lies in a few large bodies suggesting if considered in isolation that planetesimal accretion as opposed to collisional fragmentation is the most likely mode of formation. However, since they probably have their origin in the Kuiper Belt, their small sizes indicate that they may be partly collisional remnants of the larger bodies. For the near-Earth asteroids investigated in this paper we find s of around 1.65, suggesting that this is a distribution derived from larger bodies and that collisional fragmentation could be significant in their evolution. Yoshida *et al.* (2003), Yoshida & Nakamura (2004) have found that the slopes of the cumulative size distribution of Subaru-detected main belt asteroids in the magnitude range $16.5 < H < 18.5$, vary with semi-major axis with an α range from 1.11 ± 0.06 for outer belt asteroids to 1.91 ± 0.008 for those near the 4:1 resonance gap ($2.0AU < a < 2.2AU$). This corresponds to s values of 1.37 to 1.64. This supports the dynamical theory that the inner gaps can convey asteroids efficiently into the near-Earth region.

7. Conclusions

The size distribution index of NEOs has been estimated by using maximum likelihood methods and the fractional abundances and debiased albedos of Stuart & Binzel (2004). The value found is in line with previous estimates and with the main-belt asteroid size distribution and some estimates of the cometary distribution.

References

Binzel, R.P., Lupishko, D.F., Di Martino, M., Whiteley, R.J. & Hahn, G.J. 2002, in: W.F. Bottke, A. Cellino, P. Paolicchi & R.P. Binzel (eds.) *Asteroids III* (Univ. of Arizona Press), p. 251

Binzel, R.P., Rivkin, A.S., Stuart, J.S., Harris, A.W., Bus, S.J. & Burbine, T.H. 2004, *Icarus* 170, 259

Bottke, W.F., Morbidelli, A., Jedicke, R., Petit, J.-M, Levison, H.F., Michel, P. & Metcalfe, T.S. 2002, *Icarus* 156, 399

Donnison, J.R. 1986, *Astron. Astrophys.* 167, 359

Donnison, J.R. 1990, *Mon. Not. R. Astron. Soc.* 245, 658

Donnison, J.R. 2006, *Planet. Space Sci.* 54, 243

Donnison, J.R. & Pettit, L.P. 1997, *Planet. Space Sci.* 45, 841

Donnison, J.R. & Pettit, L.P. 1999, in: J.Svoreň, E.M. Pittich & H. Rickman (eds.) *Evolution and source Regions of Asteroids and Comets* , IAU Coll. 193, p. 289

Donnison, J.R. & Sugden, R.A. 1984, *Mon. Not. R. Astron. Soc.* 210, 673

Fernandez, Y.R., Jewitt, D.C. & Sheppard, S.S. 2001, *Astrophys. J.* 553, L197

Gladman, B., Kavelaars, J.J., Nicholson, P.D., Loredo, T.J. & Burns, J.A. 1998, *Astron. J.* 116, 2042

Harris, A.W. & Harris, A.W. 1997, *Icarus* 126, 450

Hughes, D.W. 2002, *Mon. Not. R. Astron. Soc.* 336, 363

Hughes, D.W. & Daniels, P.A. 1980, *Mon. Not. R. Astron. Soc.* 191, 511

Hughes, D.W. & Daniels, P.A. 1982, *Mon. Not. R. Astron. Soc.* 198, 573

Jewitt, D. 1999, *Annu. Rev. Earth Planet. Sci.* 27, 287

Lamy, P.L., Toth, I., Weaver, H.A., Delahodde, C., Jorda, L. & A'Hearn, M.F. 2000, *Bull. Am. Astron. Soc.* 32, 1061

Licandro, J., Tancredi G., Lindgren, M., Rickman, H. & Hutton R.G. 2000, *Icarus* 147, 161

Lowry, S.C. & Fitzsimmons, A. 2001, *Astron. Astrophys.* 397, 329

Lowry, S.C., Fitzsimmons, A. & Collander-Brown, S. 2003, *Astron. Astrophys.* 365, 204

Luu, J. & Jewitt, D. 1989, *Astron. J.* 98, 1905

Meech, K.J., Hainaut, O.R. & Marsden, B.G. 2004, *Icarus* 170, 463

Morbidelli, A., Jedicke, R., Bottke, W.F., Michel, P. & Tedesco, E.F. 2002, *Icarus* 158, 329

Rabinowitz, D., Helin, E., Lawrence, K. & Pravdo, S. 2000, *Nature* 403, 165

Russell, H.N. 1916, *Astrophys. J.* 43, 173

Stuart, J.S. 2001, *Science* 294, 1691

Stuart, J.S. & Binzel, R.P. 2004, *Icarus* 170, 295

Tancredi, G., Fernandez, J.A., Rickman, H. & Licandro, J. 2006, *Icarus* 182, 527

Weissman, P.R. & Lowry S.C. 2003, *Lunar & Planetary Sci.* XXXIV

Yoshida, F. & Nakamura T. 2004, abstracts/COSPAR04/03729/COSPAR04-A-03729

Yoshida, F., Nakamura, T., Watanabe, J.-I., Kinoshita, D., Yamamto, N. & Fuse, T. 2003, *PASJ* 55, 701

Near Earth Objects, our Celestial Neighbors: Opportunity and Risk
Proceedings IAU Symposium No. 236, 2006
A. Milani, G.B. Valsecchi, & D. Vokrouhlický, eds.
© 2007 International Astronomical Union
doi:10.1017/S1743921307003092

On secular resonances of small bodies in the planetary systems

Jianghui Ji[1,2]†, L. Liu[3] and G. Y. Li[1,2]

[1]Purple Mountain Observatory, Chinese Academy of Sciences, Nanjing 210008, China
email: jijh@pmo.ac.cn

[2]National Astronomical Observatory, Chinese Academy of Sciences, Beijing 100012, China
email: xhliao@nju.edu.cn

[3]Department of Astronomy, Nanjing University, Nanjing 210093, China

Abstract. We investigate the secular resonances for massless small bodies and Earth-like planets in several planetary systems. We further compare the results with those of Solar System. For example, in the GJ 876 planetary system, we show that the secular resonances ν_1 and ν_2 (respectively, resulting from the inner and outer giant planets) can excite the eccentricities of the Earth-like planets with orbits $0.21 \leqslant a < 0.50$ AU and eject them out of the system in a short timescale. However, in a dynamical sense, the potential zones for the existence of Earth-like planets are in the area $0.50 \leqslant a \leqslant 1.00$ AU, and there exist all stable orbits last up to 10^5 yr with low eccentricities. For other systems, e.g., 47 UMa, we also show that the Habitable Zones for Earth-like planets are related to both secular resonances and mean motion resonances in the systems.

Keywords. n-body simulations, planetary systems, habitable zone

1. Introduction

Since the discovery of the first Jupiter-mass planet orbiting the solar-type star 51 Peg (Mayor & Queloz 1995), it has been more than a decade. The breakthrough of scientific finding not only arouses great interests to search for other habitable planets or alien civilization worlds outside our own solar system, but also explicitly confirms that the planets can be at birth anywhere about their parent stars in the circumstellar disks, because these flat disks enshrouding young stars are considered to be a common feature of stellar evolution and of planetary system formation (Beichman *et al.* 2006). Primordial protoplanetary disks contain gas and dust and supply the raw ingredients from which the new planetary systems can form. To date, over 160 planetary systems (see also http://www.exoplanets.org and http://exoplanet.eu/) have been discovered by the measurements of Doppler radial velocity and other observational methods (Butler *et al.* 2006). More than 200 extrasolar planets have been detected about solar-type stars. Currently, amongst the detected systems, there are 20 multiple-planet systems, e.g., two-planet systems (e.g., 47 UMa etc.), three-planet systems (e.g., Upsilon And etc.) and a four-planet system, 55 Cancri. Then, the studies of the dynamics or formation of these systems are essential to understand how two (or more) planets originate from and evolve therein. In recent years, many authors have investigated the dynamical evolution of a planetary system and intended to reveal the possible mechanisms that stabilize a system, especially for those involved in the mean motion resonances (MMR), e.g., 2:1 MMR (GJ 876, HD 82943, HD 128311, HD 73651), and explored the secular interactions in the

† Present address: Department of Terrestrial Magnetism, Carnegie Institute of Washington, 5241 Broad Branch Road NW, Washington, DC 20015-1305

multiple systems (e.g., Hadjidemetriou 2002, 2006; Gozdziewski 2002, 2003; Ji *et al.* 2002, 2003; Lee & Peale 2002, 2003). Herein, we investigate the secular resonances for massless small bodies and Earth-like planets in several planetary systems, which is extremely important to make clear what dynamical structure of the newly-discovered systems may hold and how the secular resonances would have influence on the motions of the potential Earth-like planets and the location of the Habitable Zones (HZ). The HZs are generally believed as suitable locations where the biological evolution of life is able to develop on planetary surfaces in environment of liquid-water, subtle temperature and atmosphere components of CO_2, H_2O and N_2 (Kasting *et al.* 1993). In the meantime, the planetary habitability is also relevant to the stellar luminosity and the age of the star-planet system (Cuntz *et al.* 2003). However in Solar System, it is believed that the asteroids in the main belt can undergo secular resonances with respect to Jupiter (or Saturn), and their eccentricities can be greatly excited. The bodies can cross and approach Earth in million years, as a near-Earth object (NEO). Hence, the present study is mainly focusing on the issue that such secular resonances make a difference in other planetary systems.

Moreover, the scientific objectives of several space missions (e.g., SIM, TPF†) will be in part contributed to be hunting for the Earth-like planets, although this may come true after a significant improvement of precision of ground observations. Then, we also start such studies in the planetary systems advancing these projects. At first, we will quickly review the secular resonances taking place for the asteroids in the main belt.

2. Secular Resonances for Main Belt Asteroids in Solar System

It is well-known that the concentration or depletion for the asteroids in the main belt are associated with the mean motion resonances (MMR) with Jupiter's orbit and secular resonances (Williams 1969; Milani & Knežević 1992; Morbidelli & Moons 1993). The main belt asteroids are populated at the 3:2, 4:3 and 1:1 MMR with Jupiter, but rarely resided in the 2:1, 3:1, 5:2 and 7:3 resonant regions, which are called the Kirkwood gaps. Moreover, the secular resonances are responsible for the long-term dynamical evolution for small bodies. In general, secular resonances occur when the longitude of the perihelion or that of the ascending node of the small body shares the same precession rate as that of the massive giant planet (e.g. Jupiter and Saturn). There are three governing secular resonances in the asteroidal belt, known as the ν_5, ν_6 and ν_{16} resonances. The formed two are called apsidal secular resonances with respect to Jupiter and Saturn, respectively and can pump up the eccentricity of a small object; the latter one is the nodal secular resonance with respect to Saturn, which can enhance the inclination of the body. At present, it is believed that the NEOs are principally considered to be objects ejected from the main belt through a complicated dynamical process, where mean motion resonances as well as secular resonances play a vital role in their dynamical transportation (Morbidelli & Moons 1993; Froeschlé 1997; Morbidelli *et al.* 2002), indicating that the overlapping of mean motion resonances and secular resonances (Morbidelli & Moons 1993; Moons & Morbidelli 1995) can lead to large chaotic zones for the relevant asteroids.

For example, the small bodies trapped in a 3:1 orbital resonance with Jupiter (occupying the semi-major axes ~ 2.5 AU) are rarely distributed, involved in Kirkwood gaps. Wisdom (1983) pointed out that the chaotic motion for the asteroids in 3:1 MMR can increase the eccentricities and then make them approach and intersect the orbit of Mars (even Earth). Herein, Figure 1 shows that the orbital evolution for a massless test particle, however, over the time span of (0.65 Myr, 0.80 Myr) due to ν_5 and 3:1 resonance,

---

† http://planetquest.jpl.nasa.gov/SIM, and http://planetquest.jpl.nasa.gov/TPF

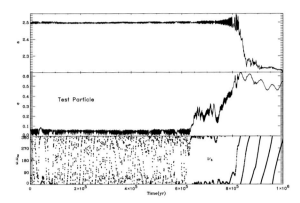

Figure 1. The time behavior of the semi-major axis a, the eccentricity e and $\varpi - \varpi_J$ for the test particle. a slightly oscillates about 2.50 AU within 0.6 Myr, over the time span of (0.65 Myr, 0.80 Myr) due to ν_5 resonance (see bottom and middle panels), and e is excited above 0.60, while a goes down to 2.20 AU. Eventually, the test body becomes a NEO candidate.

Planet	$M_{star}(M_\odot)$	$M \sin i (M_{Jup})$	Period P(d)	a(AU)	e
GJ 876 b	0.32	3.39	62.09	0.211	0.05
GJ 876 c	0.32	1.06	30.00	0.129	0.31
47 UMa b	1.03	2.86	1079.2	2.077	0.05
47 UMa c	1.03	1.09	2845.0	3.968	0.00

Table 1. Properties of 2 multiple planet systems (data adopted from Laughlin & Chambers 2001; Fischer *et al.* 2003)

the eccentricity e of the small body is excited up to 0.60 and meanwhile the semi-major axis a drops down to 2.20 AU, being an Earth-crossing body. In other numerical investigations for the dynamical evolution of the minor bodies over millions of years, we also find that several NEOs can be temporarily locked a 3:1 orbital resonance and also experience secular resonance ν_5 (or ν_6) with Jupiter (or Saturn), then confirm that the 3:1 orbital resonance and secular resonances play an important role in the origin for NEOs by previous studies (e.g., Morbidelli *et al.* 2002 and references therein).

3. Secular Resonances in Extrasolar Systems

In order to investigate the dynamical structure or Habitable Zones in the planetary systems, we also performed extensive numerical simulations for the planetary configurations of two giant planets with one fictitious low-mass body for several systems (e.g., 47 UMa, GJ 876, etc). We also show that the secular resonances can affect the motions of the small bodies in these systems, and shape the dynamical architecture in the debris disk as mean motion resonances. As for the methodology, we use a N-body code (Ji *et al.* 2002) of direct numerical simulations with the RKF7(8) (Fehlberg 1968) and symplectic integrators (Wisdom & Holman 1991). We always take the stellar mass and the minimum planetary masses from Table 1, while the mass of an assumed terrestrial planet is adopted to be in the range from 0.1 $M_\oplus$ to 10 $M_\oplus$. The used time stepsize is usually $\sim$ 1%-2.5% of the orbital period of the innermost planet. In addition, the numerical errors were effectively controlled over the integration timescale. The typical integration timescale for the simulation is 10^5 yr. The main results now follow.

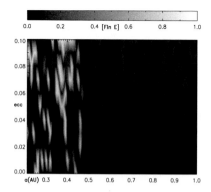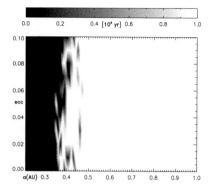

Figure 2. *Left panel*: Contour of the final eccentricities for the Earth-like planets in the GJ 876 system. Horizontal and vertical axes are the initial values of a and e. In the region $0.21 \leqslant a < 0.50$ AU, the eccentricities can be pumped up to high values ~ 1 or these bodies are directly ejected from the system due to the starting dynamical instability. Hence, in this region, the Earth-like planets are strongly chaotic and cannot survive in the system. *Right panel*: Surviving time for Earth-like planets in the system. The Earth-like planets evolve with short dynamical time before they end their destinies in the area $0.21 \leqslant a < 0.30$ AU, indicating that these orbits are completely unstable, for the initial conditions. The chaotic behaviors of the Earth-like planets in $0.21 \leqslant a < 0.50$ AU are, somewhat related to two secular resonances (ν_1 and ν_2).

3.1. *GJ 876*

The M dwarf main-sequence star GJ 876 with an estimated mass of 0.32 $M_\odot$ is the lowest mass star that hosts planets, and two Jupiter-like planets (Marcy *et al.* 2001) are revealed with minimum masses of $1.89 M_{Jup}$ and $0.56 M_{Jup}$ in this system. Moreover, the ratio of the orbital periods of two planets is close to a mean motion commensuration of 2:1. Being the first discovered 2:1 resonant planetary system, the GJ 876 has generated great interests and the long-term dynamics and planetary formation for two giant companions are extensively investigated (e.g., Hadjidemetriou 2002; Ji *et al.* 2002; Lee & Peale 2002; Beaugé & Michtchenko 2003; Kley *et al.* 2005; Laughlin *et al.* 2005, and references therein). However, the planetary formation theory (Lissauer 1993) suggests that even low-mass planets (e.g., Earth-like planets) may exist about the most abundant M dwarf stars with mass of $0.08 - 0.8 M_\odot$, which covers 75% of the total stellar population in the galaxy. For instance, Butler *et al.* (2004) announced the discovery of a Neptune-mass planet ($\sim 18 M_\oplus$) about M dwarf star GJ 436, implying the potential existence of the terrestrial or Neptunian planet in other systems.

Thus, we exhaustively investigated the case of two coplanar-configuration giant companions with one terrestrial planet. The initial orbital parameters were adopted as follows: the low-mass terrestrial bodies were placed at an equal interval of 0.01 AU from 0.21 AU to 1.0 AU in a, the eccentricities e were taken every 0.01 from 0 to 0.1, the inclinations are $0° < I < 5°$, and the other angles were randomly distributed between $0°$ and $360°$. Then each integration was carried out for 10^5 yr.

The numerical outcomes reveal that the two secular resonances ν_1 and ν_2 respectively arising from the inner and outer giant planets are responsible for the chaotic motions of the Earth-like planets. To understand the vital role of the secular resonances, we have carried out several computations. If a terrestrial planet has the mass of 1 $M_\oplus$, the location for ν_2 secular resonance is ~ 0.4550 AU, where the two eigenfrequencies for the terrestrial body and the outer giant planet are provided by Laplace-Lagrange secular theory (Murray & Dermott 1999) are, respectively, $1°.83$ yr^{-1} and $1°.90$ yr^{-1}. This is fairly in agreement with numerical results, where Figure 2 (*left panel*) shows the excitation of eccentricity of the Earth-like planets at ~ 0.45 AU. In addition, the relevant location for ν_1 secular resonance is ~ 0.2930 AU, in this case the terrestrial planet almost shares the same eigenfrequency as the inner giant planet, the values are $21°.30$ yr^{-1} and $20°.94$ yr^{-1}, respectively. The mutual Hill radius is $R_H = [(M_1 + M_2)/(3M_s)]^{1/3}(a_1 + a_2)/2$, where M_s, M_i are the masses of the host star and the planets (the subscript $i = 1, 2$, stands for the inner and outer planets, respectively), and a_1, a_2 the semi-major axes. Using the data in Table 1, we obtain $3R_H = 0.084$ AU, where $a_2 = 0.211$ AU, then we have an exterior influence boundary ~ 0.295 AU, which is almost equal to ν_1. Thus, the Earth-like planets with orbits ~ 0.30 AU are strongly affected by ν_1, and this also confirms our numerical explorations. On the other hand, plenty of mean motion resonances exist and overlap within Hill radius. The dynamical lifetime of the bodies will decrease drastically.

In a dynamical sense, for GJ 876, the potential existence of the Earth-like planets† concerns the region $0.50 \leqslant a \leqslant 1.00$ AU. Stable orbits exist, up to 10^5 yr with low eccentricities (see *right panel* of Figure 2) in the resulting evolution. This is because the initial orbits of Earth-like planets are a bit far away from the two secular resonances, free from secular perturbation. Moreover, the dynamical stability beyond 0.50 AU also suggests that outer belts for unaccreted planetesimals may exist in this system.

The formation of two giant planets in the GJ 876 system has been recently modelled by Lee & Peale (2002) and Kley *et al.* (2005), and the planets were likely captured into the 2:1 resonance by converging differential migrations in the protoplanetary disk. In this sense, the motions of the Earth-like planets or the planetesimals in the disk may be influenced by the orbital migration of the two giant planets, and they may be swept out directly or captured into the resonance with the two larger planets of GJ 876. This should be re-examined in future studies.

3.2. *47 UMa*

The main sequence star 47 UMa is of spectral type G0 V with a mass of $1.03M_\odot$. Butler & Marcy (1996) reported the discovery of the first planet in the 47 UMa system which has become one of the most amazing systems particularly after the subsequent release of an additional companion (Fischer *et al.* 2002, 2003). It is sometimes thought to be a close analog of our own solar system: for example, the mass ratio of the two giant companions in 47 UMa is ~ 2.62 (Table 1), as comparable to that of Jupiter-Saturn (JS) of 3.34; and the ratios of the two orbital periods are very similar. Hence, one may wonder whether there exists additional members in 47 UMa system (see Ji *et al.* 2005 for details).

Laughlin *et al.* (2002) and Gozdziewski (2002) studied the long-term stability of 47 UMa and pointed out that the secular apsidal resonance can help stabilize the two giant planets in an aligned configuration with the libration of their relative periapse longitudes (Ji *et al.* 2003). Then the eccentricities avoid larger oscillations due to this mechanism, as a result, this system can even survive for billion years (Barnes & Quinn 2004). Several pioneer works were concentrated on the structure of the system and presented a preliminary understanding of this issue. Jones, Sleep & Chambers (2001) investigated the existence of Earth-mass planets in the presence of one known giant planet, and subsequently Laughlin *et al.* (2002) and Asghari *et al.* (2004) further studied the stability of massless test particles about the so-called Habitable Zones (HZ) according to some earlier solutions (Fischer *et al.* 2002), where the dynamical model was treated as a restricted multi-body problem. Nevertheless, as the terrestrial planets possess significant masses, they can

---

† Rivera *et al.* (2005) reported a ~ 7.5 Earth-Mass planet about GJ 876, with the orbital period of 1.938 d, and they also indicated that additional planets may be revealed in this system with more observational data.

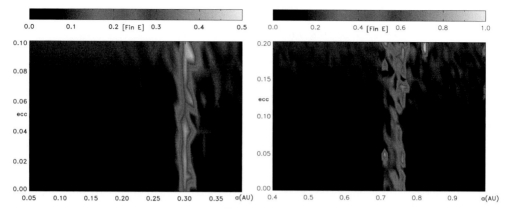

Figure 3. The contour of status of the final eccentricities for Earth-like planets, the vertical axis for the initial e. *Left panel*: $0.05 \leqslant a < 0.4$ AU for 1 Myr. Notice that the secular resonance at ~ 0.30 AU pumps up the eccentricities. *Right panel*: $0.4 \leqslant a < 1.0$ AU for 5 Myr. The eccentricity e of the orbits with $0.70 < a < 0.78$ AU is excited and in the 2:9 MMR at ~ 0.76 AU, e can reach ~ 0.90.

interact with the two giant planets by mutual gravitation, which may result in secular effects for the planetary system. Accordingly, we should take into account the masses of terrestrial bodies in the model when exploring the dynamical architecture. Herein, we performed extensive simulations to examine the dynamical architecture in both the HZ and the extended areas, for Earth-like planets (with masses from 0.1 $M_\oplus$ to 10 $M_\oplus$) of 47 UMa with stable coplanar planetary configuration, based on the best-fit orbital parameters given by Fischer *et al.* (2003). These new reliable orbital solutions are derived from additional follow-up observations, hence they can represent the actual motions of the system under study. On the other hand, the discovery of the close-in Neptune-mass planets (Butler *et al.* 2004; McArthur *et al.* 2004; Santos *et al.* 2004) demonstrates that it may be possible for less massive planets ($\sim M_\oplus$) to move close to the star. Therefore, we also explored low-mass planets in the region $0.05 \leqslant a < 0.4$ AU and we found that the secular resonance arising from the inner giant planet render the eccentricity excitations for the Earth-like planets.

For $0.05 \leqslant a < 0.4$ AU, we explored the secular evolution of hundreds of "hot Earths" or "hot Neptunes" over timescale of 1 Myr. All the simulations are dynamically stable for 10^6 yr, and 96% of the orbits posses $e_{final} < 0.20$. However, Figure 3 (*left panel*) shows that the eccentricities for the bodies at ~ 0.30 AU are excited to ~ 0.40, where the secular resonance ν_1 ($41''.11$ yr^{-1}) of the inner giant planet (similar to ν_5 for Jupiter) is responsible for excitation of the eccentricities. In addition, Malhotra (2004) also presented similar results, showing that the eccentricities of massless bodies are excited in the debris disk at ~ 0.30 AU, using nonlinear analytical theory.

In the region $0.4 \leqslant a < 1.0$ AU, more than one thousand of simulations were carried out for 5 Myr each (see *right panel* in Figure 3). Most of the Earth-like planets about 1:4 MMR at ~ 0.82 AU move stably in bounded motions with low-eccentricity trajectories, except for two cases where the eccentricities eventually grow to high values. The secular resonance ν_2 arising from the outer companion (similar to ν_6 for Saturn) can remove the test bodies. Herein ν_2 can also influence the Earth-like planets in this system. The terrestrial planets that all bear finite masses that may change the strength of this resonance; on the other hand, the location of the secular resonance is changed due to the orbital variation of the outer companion. For a terrestrial planet with a mass of $10M_\oplus$, the location for ν_2 secular resonance is about ~ 0.70 AU, where the two eigenfrequencies for the terrestrial body and outer giant planet given by the Laplace-Lagrange secular theory are, respectively, $211''.37$ yr^{-1} and $225''.48$ yr^{-1}. This indicates that both planets have almost the same secular apsidal precession rates in their motion. Hence, the ν_2 resonance, together with the mean motion resonance, can work at clearing up the planetesimals in the disk (see Fig. 3) by the excitation of the eccentricity (see also Nagasawa *et al.* 2005).

We point out that the most likely candidate for habitable environment is terrestrial planets with orbits in the ranges $0.8 \leqslant a < 1.0$ AU with low eccentricities (e.g., $0.0 \leqslant e \leqslant 0.1$). However, in our own solar system there are no terrestrial planets from the 1:4 MMR out to Jupiter, although there are stable orbits there. This may suggest that although some orbits are stable, the conditions are such that terrestrial planets cannot form so close to giant planets. Perhaps this is because runaway growth is suppressed due to the increased eccentricities from the perturbations of the giant planet. In 47 UMa, the corresponding region runs from 0.82 AU on out (see Fig. 3), almost completely covering the HZ. Hence, it would be reasonable to conclude that the only proper place to find habitable planets in this system would be at about 0.80 AU. But this should be carefully examined by forthcoming space measurements (e.g., SIM or GAIA) capable of detecting low-mass planets.

4. Summary and Discussion

In this work, we investigate the secular resonances for massless small bodies and Earth-like planets in several planetary systems with two giant planets (e.g., 47 UMa and GJ 876) by extensive numerical simulations, and further we have studied the potential existence of Earth-like planets in the related regions for these systems. In final, we summarize the following results:

(1) We can see that the 47 UMa planetary system may be a close analog of our solar system, and even it can also own several terrestrial members resembling the inner solar system (Ji *et al.* 2005). Besides, the two giant planets in the 47 UMa are similar to the Jupiter-Saturn pair in the solar system, and the corresponding secular resonances originating from them can stir the low-mass small bodies with low eccentricities in the initial "cold" disk. As to other systems, we also find that the Habitable Zones for Earth-like planets are related to both secular resonances and mean motion resonances in these systems, which may play an important role of shaping the asteroidal belts. A comparative study has been also performed in other planetary systems (see Érdi *et al.* 2004; Dvorak *et al.* 2004; Ji *et al.* 2006) with one or more giant planets to explore whether Earth-like planets can exist there, and further to locate less massive undetected planets or characterize the nature of the potential asteroidal structure in general planetary systems.

(2) The habitability for the development of biological evolution depends on many factors, such as the liquid water state, the temperature constraint, the atmosphere composition, the obliquity and rotation rate of a target terrestrial planet, the stellar luminosity, etc. However, in a dynamical viewpoint, it also requires that the habitable terrestrial planets have stable orbits in the HZ with low eccentricity at a nearly circular trajectory, herein we show that the secular resonances can excite some orbits residing in the HZ of a system, which may provide useful information or place some constraints on the observational strategy to discover such low-mass planets.

Acknowledgements

We are grateful to the referee, Anne Lemaitre for the valuable suggestions that helped to improve the contents. This work is financially supported by the National Natural

Science Foundations of China (Grants 10573040, 10673006, 10203005, 10233020) and the Foundation of Minor Planets of Purple Mountain Observatory.

References

Asghari, N., *et al.* 2004, *A&A* 426, 353

Barnes, R. & Quinn, T. 2004, *ApJ* 611, 494

Beauge, C. & Michtchenko, T.A. 2003, *MNRAS* 341, 760

Beichman, C.A., *et al.* 2006, AAS DPS Meeting, 38, 54.01

Butler, R.P. & Marcy, G.W. 1996, *ApJ* 464, L153

Butler, R.P., *et al.* 2004, *ApJ* 617, 580

Butler, R.P., *et al.* 2006, *ApJ* 646, 505

Cuntz, M., von Bloh, W., Bounama, C. & Franck, S. 2003, *Icarus* 162, 214

Dvorak, R., Pilat-Lohinger, E., Schwarz, R. & Freistetter, F. 2004, *A&A* 426, L37

Érdi, B., Dvorak, R., Sándor, Z., Pilat-Lohinger, E. & Funk, B. 2004, *MNRAS* 351, 1043

Fehlberg E. 1968, NASA TR R-287

Fischer, D., *et al.* 2002, *ApJ* 564, 1028

Fischer, D., *et al.* 2003, *ApJ* 586, 1394

Froeschlé, Ch. 1997, *Celest. Mech. & Dyn. Astron.* 65, 165

Gozdziewski, K. 2002, *A&A* 393, 997

Gozdziewski, K. 2003, *A&A* 398, 1151

Hadjidemetriou, J.D. 2002, *Celest. Mech. & Dyn. Astron.* 83, 141

Hadjidemetriou, J.D. 2006, *Celest. Mech. & Dyn. Astron.* 95, 225

Ji, J., Li, G. & Liu, L. 2002, *ApJ* 572, 1041

Ji, J., Liu, L. Kinoshita, H., Zhou, J.L., Nakai, H. & Li,G.Y. 2003, *ApJ* 591, L57

Ji, J., Liu, L., Kinoshita, H. & Li, G.Y. 2005, *ApJ* 631, 1191

Ji, J., Kinoshita, H., Liu, L. & Li, G.Y. 2006, *ApJ* accepted, [astroph/0611008]

Jones, B.W., Sleep, P.N. & Chambers, J.E. 2001, *A&A* 366, 254

Kasting, J.F., Whitmire, D.P. & Reynolds, R.T. 1993, *Icarus* 101, 108

Kley, W., Lee, M.H., Murray, N. & Peale, S.J. 2005, *A&A* 437, 727

Laughlin, G. & Chambers, J.E. 2001, *ApJ* 551, L109

Laughlin, G., Chambers, J.E. & Fischer D. 2002, *ApJ* 579, 455

Laughlin, G., Butler, R.P., Fischer, D.A., Marcy, G.W., Vogt, S.S. & Wolf, A.S. 2005, *ApJ* 622, 1182

Lee, M.H. & Peale, S.J. 2002, *ApJ* 567, 596

Lee, M.H. & Peale, S.J. 2003, *ApJ* 592, 1201

Lissauer, J.J. 1993, *ARAA* 31, 129

Malhotra, R. 2004, AAS DPS Meeting 36, 42.04

Marcy, G.W., *et al.* 2001, *ApJ* 556, 296

Mayor, M. & Queloz, D. 1995, *Nature* 378, 355

McArthur, B.E., *et al.* 2004, *ApJ* 614, L81

Milani, A. & Knežević, Z. 1992, *Icarus* 98, 211

Moons, M. & Morbidelli, A. 1995, *Icarus* 114, 33

Morbidelli, A. & Moons, M. 1993, *Icarus* 102, 316

Morbidelli, A., *et al.* 2002, in: W.F. Bottke Jr., A. Cellino, P. Paolicchi & R.P. Binzel (eds.) *Asteroids III* (Tucson: Univ. of Arizona Press), p. 409

Murray, C.D. & Dermott, S.F. 1999, *Solar System Dynamics* (New York: Cambridge Univ. Press)

Nagasawa, M., Lin, D.N.C. & Thommes, E. 2005, *ApJ* 635, 578

Rivera, E.J., *et al.* 2005, *ApJ* 634, 625

Santos, N.C., *et al.* 2004, *A&A* 426, L19

Williams, J.G. 1969, *Ph.D. Thesis*, Univ. of Califonia, Los Angeles

Wisdom, J. 1983, *Icarus* 56, 51

Wisdom, J. & Holman, M. 1991, *AJ* 102, 1528

Part 2

The Meteor/Asteroid Impact Transition

Near Earth Objects, our Celestial Neighbors: Opportunity and Risk
Proceedings IAU Symposium No. 236, 2006
A. Milani, G.B. Valsecchi & D. Vokrouhlický, eds.
© 2007 International Astronomical Union
doi:10.1017/S1743921307003110

(Mostly) dormant comets in the NEO population and the meteoroid streams that they crumble into

Peter Jenniskens

SETI Institute, 515 N. Whisman Road, Mountain View, CA 94043
email: pjenniskens@mail.arc.nasa.gov

Abstract. Many of our annual showers do not have active parent comets. In recent years, minor planets have been identified that move among the meteoroids streams. Some streams, such as the Quadrantids, Geminids, and Sextantids, are in such unusual orbits that the probability of a chance association is only of order 1 in 10^6. The streams identify those objects as dormant comet nuclei. Other streams, such as the Phoenicids and α-Capricornids are associated with minor planets that were found to be weakly active at their last perihelion passage. All the streams investigated so far are young, less than 2000 years old, and can not have been created in the classical sense of meteoroids being ejected from the comet nucleus by water vapor drag. Instead, these (mostly) dormant comets lost fragments at some point in the past, which crumbled into meteoroid streams. Scars of such events are now identified on the surface of active Jupiter family comets 9P/Tempel 1 and 81P/Wild 2. Thus, the meteor showers on Earth bear witness to what is the dominant mass-loss mechanism of comets in the inner solar system, a process that can account for much of the zodiacal cloud dust and the zodiacal dust bands.

Keywords. Meteors; meteoroids; meteor showers; interplanetary medium; comet mass loss; comet fragmentation

1. Introduction

When Whipple (1951) identified the drag of water vapor as a potent force to accelerating grains off the surface of a comet, his mathematical description of dust ejection by water vapor drag first described the formation of a meteoroid stream. After one orbit, the meteoroids return at a different time, spreading along the comet orbit into a comet dust trail. We now know that the meteor outbursts of some long-period comets are understood as being caused by the 1-revolution dust trails that wander occasionally in Earth's path. The 2-revolution trail is usually not detected because perturbations of the orbital period cause trails to fold and break and disperse in nodes. The orbital period of the dust of Halley-type comets is less changed, so that dust trails can persist for several revolutions, leading to multiple meteor storms when Earth travels through the debris field of a Halley-type comet. The dust of Jupiter-family comets is strongly perturbed when Jupiter meets the dust at aphelion, resulting in rapid distortions as a function of time of ejection. This changes the dust trail cross section and results in rapid spreading of the dust along the comet orbit. Some trail segments that avoid close encounters escape as intact trailets that can cause meteor outbursts on Earth (e.g., Jenniskens 2006a).

A problem with this scenario is that the majority of our meteor showers have no known parent body. As of January 2003, these included the major showers of Quadrantids, Geminids, Daytime Arietids, Taurids, and Sextantids, as well as the well-known α-Capricornids, δ-Aquariids, and Phoenicids. These are Jupiter-family comets and streams with a shorter orbital period that are now mostly decoupled from Jupiter. It was surmised, that the parent bodies of these streams had now completely disintegrated, or perhaps had evolved into orbits quite different from the meteoroids that we now see on Earth as a meteor shower. However, many of these streams, such as those responsible for the Quadrantids, the Geminids, the α-Capricornids and the Sextantid meteor shower, are so little dispersed that they can not be older than 2 000 years.

In the past three years, all of these streams were found to have associated minor planets. 2003 EH_1 moves in the highly inclined Quadrantid stream, 2002 EX_{12} moves among the α-Capricornids (and was weakly active when at perihelion), 2004 TG_{10} is a better match to the Northern Taurids than comet 2P/Encke, and 2005 UD moves among the small-q Daytime Sextantids (Jenniskens 2006a). Also, 2003 WY_{25} is a fragment of comet D/Blanpain responsible for the Phoenicid outburst of 1956, and the Daytime Arietids are in the same orbit as the Marsden Sungrazer group of comet fragments. This leaves no further doubt that (3200) Phaethon is also the parent of the Geminid shower.

These streams appear to originate in discrete formation events, presumably a form of fragmentation. In some cases, the single-year sighting of a comet may mark that event. In each case, the amount of mass in the stream is less or equal to the largest remaining fragment discovered so far. Instead of a catastrophic fragmentation, these streams appear to be the result of the spill-off of a comet fragment, which subsequently broke into meteoroids. The scars of this spill-off have now been identified on active short-period comet 9P/Tempel 1.

2. Individual meteor shower – parent body associations

2.1. The Geminids and (3200) Phaethon

Meteor showers were associated to minor planets ever since Whipple (1983) noticed that 1983 TB Phaethon moved among the very short-period ($P \approx 1.59$ yr) and eccentric Geminid meteoroids. However, in no other case was the agreement between meteoroid stream and proposed parent body as good. Even more, the Geminid-Phaethon association was long in doubt.

The Geminids have an unusually low perihelion distance ($q \approx 0.14$ AU), and both meteoroids and minor planet surface have been sintered in the heat of the Sun, which can warm a dust grain to above 700° K. Probably because of that, this minor planet resembles a B-type asteroid with a high 0.11 geometric albedo, much higher than that of known dormant comet nuclei, and the meteoroids penetrate deeper in the atmosphere than known cometary meteoroids (albeit not as deep as known asteroidal meteoroids). Many have assumed that Phaethon is an asteroid that had evolved independently in a similar orbit as the Geminids, despite the low probability for doing so (≈ 1 in 2×10^6, depending on the actual number of objects in this sparsely sampled population and the extent of the stream).

It has since been shown that the Geminids appear to have been created close to perihelion, where the probability of an asteroidal collision is low (Gustafson 1989).

2.2. The Quadrantids and 2003 EH_1

A second example was identified only twenty years later (Jenniskens 2003). In October 2003, the parent body of the Quadrantid shower was identified among newly discovered minor planets (Table 1). The mean orbit of the Quadrantids in Table 1 is that derived from photographic observations by the Dutch Meteor Society (Jenniskens et al. 1997). Due to the steep 72° inclination of the orbit, the probability of a chance association with unrelated minor planets is also only about 1 in 2×10^6. This object, 2003 EH_1, passed Earth's orbit in excess of 0.2 AU, and was therefore not immediately recognized as a near-Earth object. However, because the aphelion of the orbit is so close to that of Jupiter, the perihelion distance of the object (and of that of the Quadrantid meteoroids) is strongly perturbed, evolving in and out of Earth's orbit. It was shown that 2003 EH$_1$ was perturbed only very recently into an orbit on the outer parts of the stream (Jenniskens 2004).

Table 1. Comparison of orbital elements of Quadrantids and proposed parent bodies (angles referred to J2000).

Name	Epoch	a (AU)	q (AU)	i (°)	ω (°)	Ω (°)	ϖ (°)
Quadrantids	1995	3.14	0.979	71.2	172.0	283.3	95.3
2003 EH$_1$	2003	3.126	1.193	70.79	171.37	282.94	94.3
2003 EH$_1$	1490	3.10	0.580	65.7	163.7	286.5	90.2
C/1490 Y$_1$	1490	∞	0.761	73.4	164.9	280.2	85.1

Until that discovery, the Quadrantid shower was thought to originate from comet 96P/Machholz 1, now in a very different orbit, with the meteoroids having evolved at a different pace along a secular cycle than the parent comet (presumably in a slightly shorter orbital period). An age of the stream of about 4 000 years was implied. However, new precise photographic observations of Quadrantid meteoroids showed that the meteoroids moved in very similar orbits, passing just inside of Jupiter's orbit. That implied that the stream could not be older than about 500 years (Jenniskens *et al.* 1997). The Quadrantids are a massive stream, which can not have been created in normal Whipple-type ejection by water vapor drag in the short time available. A breakup is implied.

The moment of breakup may have been observed by Chinese astronomers in early 1491, when a comet was observed in a plane similar to that of the Quadrantids. Due to frequent close encounters with Jupiter, it is not easy to link 2003 EH$_1$ to the sightings of C/1490 Y$_1$. 2003 EH$_1$ was recovered at ESO/NTT on December 24, 2003 (E. Jehin, M. Billeres), resulting in a better defined orbit. Using the Orbfit program by Andrea Milani and coworkers, Giovanni B. Valsecchi, Jeremie Vaubaillon, and the author identified two possible solutions at less than 3 σ from the nominal orbital solution. A more precise orbit of 2003 EH$_1$ is needed before it can be established that this comet lost a fragment in or just before 1490 and at that time created the Quadrantid stream.

2.3. The Phoenicids and 2003 WY$_{25}$

A third case was identified a year later when minor planet 2003 WY$_{25}$ was discovered to be a fragment of comet D/1819 W$_1$ Blanpain (Jenniskens & Lyytinen 2005). That comet was observed only once in 1819 and the orbit is uncertain. After an initial similarity in angular elements of order $17°$, it was found that that similarity had improved to only $0.2°$ after the comet orbit was observed for a more extended period of time (Jenniskens & Lyytinen 2005). Subsequently, Jewitt (2006) discovered from past images taken on March 10, 2004, that 2003 WY$_{25}$ was at that time an active comet, the smallest known comet nucleus at this time, creating dust at a rate of only 0.01 kg/yr. At this rate, this comet fragment could not have generated the Phoenicid meteoroid stream. Blanpain was so active in 1819 that 2003 WY$_{25}$ is probably much smaller than Blanpain. This can be a fragment from a breakup that most likely occurred during or just before 1819.

An outburst of Phoenicids was observed in 1956. Jenniskens & Lyytinen (2005) investigated whether this breakup could have occurred at or just before 1819, with Blanpain being the manifestation of the immediate aftermath. After modeling the meteoroids released during a fragmentation event in 1819, we were able to demonstrate that the dust would have moved in Earth's path in 1951 and 1956, but not in other years since.

2.4. *The α-Capricornids and 2002 EX$_{12}$*

Next, Paul Wiegert (University of Western Ontario) and the author independently found that 2002 EX$_{12}$ moved in an orbit similar to that of the α-Capricornids, albeit slightly further evolved along the secular cycle. This object, too, was discovered to be a comet in July of 2005, and is now known as 169P/NEAT. The α-Capricornids have a very dispersed period of activity due to the low inclination of the orbit, but the radiant dispersion in perpendicular direction is very small.

2.5. *The DaytimeArietids and the Marsden Sunskirter comet fragments*

In the mean time, a similarity had been observed between the orbit of the Daytime Arietids and the comet fragments of the Marsden group of sungrazers by David Seargent (Kracht *et al.* 2002). The latter have a slightly smaller perihelion distance, but the orbit passes very close to Earth orbit. Due to the limited viewing area of the SOHO satellite, and the short activity curve of these small comet fragments, the orbital period of the Marsden sungrazer group could not be determined. In 2004, it was found that some fragments appeared to return after ~ 5.5 years (Marsden 2004), making this a short-period comet, making it more than likely that whatever was responsible for this large family of comet fragments is also responsible for the Daytime Arietids, even though this stream has a relatively short $P = 1.6 - 2.3$ yr orbital period.

The Marsden and Kracht group of sunskirters and comet Machholz, as well as the Daytime Arietids and δ-Aquariid showers are all part of a complex (the Machholz Complex) of objects (perhaps including the Quadrantid parent 2003 EH$_1$) that evolved along the same secular cycle. Recently, Sekanina & Chodas (2005) identified a mechanism for rapid dispersion along this cycle, by invoking an initial fragmentation of the progenitor body followed by a close encounter of the fragment stream with Jupiter in AD 1059. This puts the formation age of the Marsden group of fragments after 1059 AD.

The related δ-Aquariids may be associated with the Kracht group, or with a new yet undiscovered group of comet fragments.

2.6. *The Taurids*

Another such group of comet fragments is the Taurid Complex. It was originally proposed based on minor planets linked to Taurid meteoroid orbits by means of the *D*-criterion. In this manner a large number of associated bodies were found by Clube, Napier, and others, that implied a huge 43 km diameter parent broken up about 20 000 years ago (Clube & Napier 1984). Comet fragments hitting the Earth were proposed as the cause of climate and environmental changes on Earth. Since, it has been found that all of the originally proposed objects (outside of 2P/Encke) have asteroidal colors (S or O type taxonomy) and, therefore, in all likelihood are unrelated to the Taurid shower. All those objects have $q > 0.4$ AU (mostly with $q \approx 0.55$ AU), larger than the Taurid meteoroids with $q \approx 0.35$ AU.

Nevertheless, the large nodal extend of the Taurid stream (September – December) and the presence of southern and northern branches implies that the meteoroids evolved at least one full cycle of about 5 000 years, and possible multiple cycles. In contrast, clusters of orbits in the meteoroid stream and the general "fresh" appearance (numerous flares) of the meteors suggest that some Taurids were only fairly recently released.

Comet 2P/Encke itself was never a good fit to the Taurids observed on Earth. The comet has evolved further along the secular cycle, now with a relatively high inclination of 12° and a node at Mercury's orbit.

Only very recently, a parent body was identified with q and other orbital elements similar to that of the northern Taurids: 2004 TG$_{10}$ (Jenniskens 2006b). Next to comet 2P/Encke, this may be the second fragment of an original larger body. That body was perhaps as large as comet 1P/Halley, but certainly not as large as the 43 km sized parent body proposed by Clube and Napier.

2.7. *The Geminids and Sextantids*

It is now clear that also the Geminids originate from (3200) Phaethon. Phaethon and its stream are also part of a larger complex of objects. A second meteor shower is known with similar orbital elements, but at a different position along the secular cycle. That is the Daytime Sextantid shower. Again, it was unlikely that the Sextantids evolved from the same fragmentation event as that responsible for the Geminids. The Geminids are not old enough to account for the difference in orbital elements.

Recently, Ohtsuka (2005) identified 2005 UD as the likely body responsible for the Sextantid meteoroid stream in a separate fragmentation event. It is likely that other minor planets exist that are part of this same Phaethon Complex.

Table 2. Meteor showers and their parent comets. Orbital elements, Tisserand parameter, Nuclear magnitude, and Formation epoch.

Shower/comet	Epoch	a (AU)	q (AU)	i	ω	Ω	T_J	HN	Formation
Tau-Herculids				source of future showers					
73P/S.-W. 3	(1995)	3.06	0.933	11.4	198.8	69.9	+2.78	+17.7	AD 1995
Andromedids		2.90	0.777	7.5	242.7	225.5			
3D/Biela	(2004)	3.49	0.798	7.5	236.2	213.8	+0.78	> +7.1	AD 1840
Phoenicids		(3.05)	0.985	15.9	358.2	74.1			
2003 WY$_{25}$	(1956)	3.07	0.991	9.6	360.1	74.4	+0.51	> +8.5	≈ AD 1819
Quadrantids		3.14	0.979	72.0	172.0	283.3			
2003 EH$_1$	(2006)	3.13	0.979	70.8	171.4	283.0	+3.89	+17.7	≈ AD 1490
Daytime Arietids		1.75	0.094	27.9	29.5	78.7			
Marsden-group	(2004)	3.33	0.0483	26.8	23.2	81.5	≈ +1.8	> +18	⩾AD 1059
Geminids		1.37	0.141	24.0	324.4	261.5			
(3200) Phaethon	(2005)	1.27	0.140	24.2	325.3	262.5	+4.51	+14.6	≈ AD 1030
Northern Taurids		2.12	0.350	3.1	294.9	226.2			
2004 TG$_{10}$	(2005)	2.24	0.315	3.6	298.4	223.9	+2.99	+19.5	≈ AD 600
Capricornids		2.62	0.602	7.7	266.7	128.9			
2002 EX$_{12}$	(2005)	2.60	0.605	7.6	266.0	128.8	+2.89	+26.5	≈ AD 10

2.8. *Andromedids and τ-Herculids*

The idea that the fragmentation of comets is a source of meteoroids causing meteor showers on Earth was first proposed following the 1872 and 1885 Andromedid storms, which occurred after the breakup of lost comet 3D/Biela in 1840, and the continued fragmentation of the comet observed in the returns of 1846 and 1852. It is not certain, however, that the breakup added much to the normal dust ejection of this active comet.

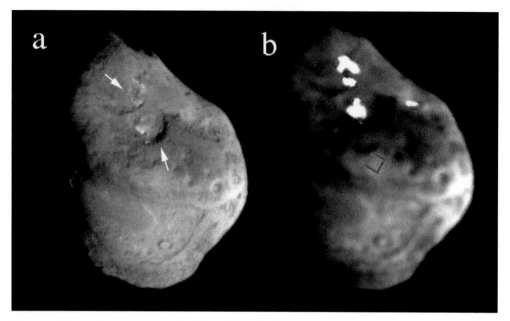

Figure 1. Deep Impact target 9P/Tempel 1. (a) Arrows mark areas that might be the scars from recent fragmentation. (b) White areas are where water ice was detected very near the surface. Photo: NASA/JPL/Deep Impact.

Although the meteor showers were spectacular, they were not more intense than the more recent Draconid storms of 1933 and 1946, thought to have been caused by comet 21P/Giacobini-Zinner through normal Whipple-type comet dust ejection through water vapor drag. Even though the Andromedid storms can be traced to dust trails in the Earth's path dating from 1846 and 1852, when the comet was in breakup, there were no dust trail encounters in other years that could have proven that normal activity of 3D/Biela wasn't capable of creating similar showers.

Another shower created during the breakup of an active comet is that of 73P/Schwassmann-Wachmann 3. This relatively young comet is still ice-laden and the fragmentation may prove helpful in understanding what underlies the breakup of other comets. The meteoroids from the 1995 breakup will be near Earth in 2022.

3. The scars of fragmentation

Active Jupiter Family Comets should show the scars of these fragmentation events. Ten days after the impact of Deep Impact on 9P/Tempel 1, I identified two circular features as the potential scars of cometesimal spill-off (from Jenniskens 2006b: Fig. 3a). Interestingly, it was subsequently reported that Deep Impact had detected water ice near the surface at these exact locations (A'Hearn *et al.* 2005). Water ice must be present relatively low below the surface, so that water vapor can seep to the surface and freeze there. It appears that relatively recently two (or three) 0.5-km sized fragments broke from the comet, creating a meteoroid stream in the process. These fragments must have been relatively pancake shaped, perhaps as a result of the accumulation process.

Indeed, fragmentation can explain morphological features on other comets as well. Pits and shallow depressions are common on 81P/Wild 2, all of which are more likely due to the loss of cometesimals than due to impacts. The morphology of those pits is very different than that of the impact craters on the Saturnian moon Phoebe, which is presumably a captured comet because of its retrograde motion.

4. Conclusions and interpretations

I conclude that our main showers are relatively young, less than 2 000 years old. The discrete fragmentation through the spill-off of boulders, loss of cometesimals, or catastrophic disruption is a common phenomenon among the population of dormant comet nuclei in the inner solar system. It is, in fact, their main mass-loss mechanism.

Many more ecliptic meteoroid streams have now associated (mostly) dormant comets that are still in very similar orbits. The ones listed in Table 2 are only the most well established associations. Many others are listed in Table 7 in the book "Meteor Showers and their Parent Comets" (Jenniskens 2006a). Most of these concern streams in low-inclination or less eccentric orbits, with much higher likelihood of random asteroids wandering among the meteoroids. Each of these cases needs further study.

From the mass of the streams and that of the remaining fragments, it appears that the streams typically represent an amount of mass similar to that of the mass of the remaining fragment. If each breakup adds about 1 000 billion kg of dust, then a steady state of about 300 000 fragmentation events over the course of 20 000 years could account for the current mass of dust in the zodiacal cloud (e.g., Hughes 1993). That amounts to 15 fragmentation events per year over the whole cloud of comets in all forms in the inner solar system. To produce our about 105 known antihelion streams (Jenniskens 2006a), this would demand that a fraction of $105/300\,000 = 0.04\%$ of those breakups evolved dust into Earth's path if the streams survive (at high enough dust density) for 20 000 years, or 0.4% if the streams survive for only 2 000 years. Those are reasonable numbers.

This young age of the zodiacal cloud argues against the asteroid families being the source of the zodiacal dust bands. In fact, the breakup of (mostly) dormant comets in the inner solar system is a much more likely mechanism. The argument that there are no known active Jupiter family comets in a suitable orbit does not longer apply. There is a wealth of potentially responsible minor bodies that are now dormant.

Acknowledgements

Jeremie Vaubaillon and Esko Lyytinen contributed to this work by numerical modeling the dynamics and evolution of meteoroid streams. PJ is supported by NASA's Planetary Astronomy and Planetary Atmospheres programs.

References

A'Hearn, M.F., Belton, M.J.S., Delamere, W.A., Kissel, J., Klaasen, K.P., McFadden, L.A., Meech, K.J., Melosh, H.J., Schultz, P.H., Sunshine, J.M., Thomas, P.C., Veverka, J., Yeomans, D.K., Baca, M.W., Busko, I., Crockett, C.J., Collins, S.M., Desnoyer, M., Eberhardy, C.A., Ernst, C.M., Farnham, T.L., Feaga, L., Groussin, O., Hampton, D., Ipatov, S.I., Li, J.-Y., Lindler, D., Lisse, C.M., Mastrodemos, N., Owen, W.M., Richardson, J.E., Wellnitz, D.D. & White, R.L. 2005, *Science* 310, 258

Clube, S.V.M. & Napier, W.M. 1984, *Mon. Not. R. Astron. Soc.* 211, 953

Gustafson, B.A.S. 1989, *Astron. Astrophys.* 225, 533

Hughes, D.W. 1993, in: J. Stohl & I.P. Williams (eds.), *Meteoroids and their Parent Bodies* (Bratislava: Astron. Inst. Slovak Acad. Sci.), p. 15

Jenniskens, P. 2003, *IAUC* 8252

Jenniskens, P., Betlem, H., de Lignie, M., Langbroek, M. & van Vliet, M. 1997, *Astron. Astrophys.* 327, 1242

Jenniskens, P. 2004, *Astron. J.* 127, 3018

Jenniskens, P. & Lyytinen, E. 2005, *Astron. J.* 130, 1286

Jenniskens, P. 2006a, *Meteor Showers and their Parent Comets* (Cambridge: Cambridge University Press)

Jenniskens, P. 2006b, in: A. Graps (ed.), *Proceedings of the "Dust in Planetary Systems" workshop, Kuaui, Hawaii*

Jewitt, D. 2006, *Astron. J.* 131, 2327

Kracht, R., Meyer, M., Hammer, D., Marsden, B.G. & Seargent, D.A.J. 2002, *MPEC 2002-E25*

Marsden, B.G. 2004, *MPEC 2004-X73*

Ohtsuka, K., 2005 *CBET* 283

Sekanina, Z. & Chodas, P.W. 2005, *Astron. J. Suppl. Ser.* 161, 551

Whipple, F.L. 1951, *Astrophys. J.* 113, 464

Whipple, F.L. 1983, *IAUC* 3881, 1

Near Earth Objects, our Celestial Neighbors: Opportunity and Risk
Proceedings IAU Symposium No. 236, 2006
A. Milani, G.B. Valsecchi & D. Vokrouhlický, eds.
© 2007 International Astronomical Union
doi:10.1017/S1743921307003122

NEO fireball diversity: energetics-based entry modeling and analysis techniques

Douglas O. ReVelle

Los Alamos National Laboratory, Los Alamos, New Mexico, USA
email: revelle@lanl.gov

Abstract. We have examined the behavior of a number of bolides in Earth's atmosphere from the standpoint of recent entry modeling techniques. The entry modeling has been carried out including a triggered progressive fragmentation model (TPFM) which maintains a maximum drag orientation for the fragments in either the collective or a non-collective wake limit during entry (ReVelle 2004). Specifically in this paper, we have proposed a new method of estimating the terminal bolide mass and have compared it against the corresponding single-body mass loss prediction. A new expression for the terminal mass is proposed that corrects the mass of the body for the changing mass to area ratio during the fragmentation process. As a result of this new work we have found two very interesting features that correspond very closely to those found from a direct analysis of the observational data. These include an instantaneous mass that closely resembles that directly observed and an ablation coefficient behavior that also strongly resembles meteor observations (such as those found recently by Ceplecha & ReVelle 2005). During fragmentation, the apparent ablation coefficient has now been shown to decrease dramatically approaching the intrinsic ablation coefficient proposed by Ceplecha & ReVelle (2005). In our modeling we have assumed a breakup into equal size fragments that are consistently and progressively multiples of two of the original unbroken leading piece. Had we assumed a multitude of many much smaller pieces that made up the totality of the original body, our predicted ablation coefficient would indeed have approached the very small intrinsic ablation parameter values predicted by Ceplecha and ReVelle. This is especially evident in the case of Sumava, but is also true in a number of other cases as well. The bolides whose properties have been modeled using our detailed entry code including a prediction of the panchromatic luminosity consist of the 1965 Revelstoke meteorite fall (Folinsbee 1967; Carr 1970; Shoemaker 1983), the 1974 Sumava fireball and the 1991 Benesov fireball as presented in Borovička & Spurný (1996) and in Borovička *et al.* (1998), the Tagish Lake meteorite fall of January 8, 2000 (Brown *et al.* 2002), the March 9, 2002 Park Forest meteorite fall (Brown *et al.* 2004), the June 6, 2002 Mediterranean (Crete) bolide as presented in Brown *et al.* (2002) and finally the September 4, 2004 Antarctic bolide respectively (Klekociuk *et al.* 2005). A self-consistent assessment of the detailed properties of each of the fireballs was made using all available information for each event. In the future, more reliable estimates of all of the necessary source parameters (including their overall degree of bulk porosity) will be made if all channels of information are reliably retrieved for bolide events (channels such as acoustic-gravity waves and specifically its infrasound emission, seismic waves, satellite optical and IR data, ground-based spectroscopy, ground-based photometry and radiometry, VLF radiation, meteorite fragment recovery, etc.).

Keywords. meteoroids, fireballs, bolides

---

1. Introduction and Overview

1.1. *Large Fireball Behavior in the Atmosphere*

During the hypersonic continuum flow entry of large meteoroids, the parameters of direct interest are the degree of bolide ablation, the drag and corresponding deceleration, the mass loss and the concomitant degree of fragmentation and the related behavior of the fragments in the near-wake region, the panchromatic light production as well as in other spectral bands of interest, an estimation of the total power budget, mechanical wave and radio frequency wave generation processes, possible impact and explosive cratering, etc. In this paper we will examine a number of fireballs, some of which have dropped meteorites and some of which have apparently disintegrated totally during their flight in the atmosphere. The degree to which we are able to understand these events through systematic analyses directly impacts on our ability to understand their possible origins and compositional variability which is very important to our understanding of our solar system and its evolutionary processes. To better understand this complete problem we demand not only entry modeling analyses, but also remote monitoring of the "optical" and acoustic/infrasonic and internal gravity wave production, i.e., acoustic-gravity waves and knowledge of their relationships to the structure and the composition of the meteoroids themselves.

The large range of expected phenomena have made the use of a wide range of observing techniques necessary for a proper understanding of these events and their source energies in order to properly calibrate the expected steady state influx rate of bolides (Brown *et al.* 2004). Readily available technology for the detection of bolides includes optical and infrared satellite data, ground-based radiometry (as has been developed by R.E. Spalding at Sandia National Laboratory in Albuquerque) and by the detection of acoustic-gravity (mechanical) and seismic waves by the IMS (International Monitoring System) network and by other available infrasonic and seismic sensors/arrays. Additional techniques include ground-based photometry (by the European Fireball Network, etc.), ground-based spectroscopy and detection by VLF radio receivers and of course information gained by direct meteorite recovery efforts, etc.

1.2. *Reliable Prediction of the Terminal Mass of Fireballs*

In previous treatments of entry dynamics (ReVelle 2004), the author has not focused his attention on precise predictions of the bolide's terminal mass. In this work we have provided two estimates of the terminal mass, one based strictly on the standard single-body theory (with no fragmentation assumed to be present throughout the entry) and a new estimate based upon the self-consistent correction of the mass based upon predicted changes in the mass to area ratio of the bolide during fragmentation. The latter approach reduces uniformly to the single-body predicted limit in the absence of fragmentation. Using this latter approach, significantly smaller values of the terminal mass are now generally predicted under a certain range of entry conditions.

These predictions are also based upon a new evaluation of the mass/area ratio in the non-collective wake behavioral limit. Previous non-collective wake solutions by the author had assumed that the multiplier, k_4 (see below), for p^*, the modified ballistic entry parameter (which is proportional to the mass to area ratio of the body) was exactly 1.0 based upon a mass reduction during each specified breaking interval with $k_3 = 0.50$ and a frontal cross-sectional area reduction $k_2 = 0.50$. If however we further demand a maximum drag orientation for the leading fragmented piece as we have also done for the collective wake limit, we now predict that $k_4 = 0.50$ which is identical to that of the collective wake solution (see below for further details). Intermediate solutions are certainly possible for the non-collective wake limit as well up to the limit for $k_4 = 1.0$, but this is in general an unknown detail without further extensive wake calculations. For a value of $k_4 = 0.50$ for both solutions, an identical end height for the two extreme wake behaviors is predicted to occur. For $k_4 = 0.75$ for example, we predict a slightly lower end height for the non-collective wake solution compared to the collective wake solution. For each of these solutions however, we obtain a different prediction of the breaking altitudes and the optical luminosity produced as well despite the fact that the end heights are almost identical. Full details of many of these proposed changes will be discussed below.

2. Entry Modeling Techniques

2.1. *Single-Body Mass Loss versus Mass Loss with Break-up*

We will start first from the fundamental differential equations of deceleration, mass loss, kinetic energy change, area change, shape factor change and the connection between geopotential altitude, z and time, t that describe the full behavior of the *single-body* meteor entry assuming negligibly small thermal conduction effects and negligible lift, namely (ReVelle 1979; Bronshten 1983; Ceplecha *et al.* 1998; ReVelle 2004):

$$\frac{dV}{dt} = -0.5 \cdot \rho(z) \cdot V^2(z) \cdot C_D(z) \cdot \left\{ \frac{A(z)}{m(z)} \right\}$$

$$\frac{dV}{dt} = -\frac{C_D(z) \cdot S_f(z) \cdot \rho(z) \cdot V^2(z)}{\rho_m^{2/3} \cdot m^{1/3}(z)} \tag{2.1}$$

$$\frac{dm(z)}{dt} = -0.5 \cdot \rho(z) \cdot V^3(z) \cdot C_D(z) \cdot \sigma(z) \cdot A(z)$$

$$\frac{dm(z)}{dt} = -\frac{C_D(z) \cdot S_f(z) \cdot \sigma(z) \cdot \rho(z) \cdot m^{2/3}(z) \cdot V^3(z)}{\rho_m^{2/3}} \tag{2.2}$$

$$\frac{dE_k(z)}{dt} = -\frac{I_{pan}(z)}{\tau_{pan}(\rho(z), m(z), V(z))} \tag{2.3}$$

$$\frac{dA(z)}{dt} = \mu \cdot \left\{ \frac{A(z)}{m(z)} \right\} \cdot \frac{dm(z)}{dt}; \qquad \mu = constant \tag{2.4}$$

$$\frac{dS_f(z)}{dt} = -\left[\frac{2}{3} - \mu \right] \cdot S_f(z) \cdot \left\{ \frac{1}{m(z)} \cdot \frac{dm(z)}{dt} \right\} \tag{2.5}$$

$$\frac{dz(t)}{dt} = -V(z) \cdot \sin \theta; \qquad \theta = constant \tag{2.6}$$

where:
- $m(z)$=instantaneous meteor mass;
- θ=horizontal entry angle (in the plane parallel earth approximation);
- C_D=coefficient of wave drag at hypersonic and supersonic speeds ($\mathcal{O}(1)$);
- E_k=kinetic energy of the body;
- I_{pan}=panchromatic radiation power emission;
- τ_{pan}=panchromatic differential luminous efficiency as given in ReVelle (2004);
- $A(z)$=frontal cross-sectional area of the meteor;
- $V(z)$=instantaneous meteor velocity;
- $\rho(z)$=atmospheric density as a function of altitude;
- $\sigma(z)$=ablation parameter as a function of altitude, z;
- $S_f(z) \equiv A/V_0^{2/3}$, $A(z)/A_\infty \equiv (m(z)/m_\infty)^\mu$, where V_0=meteor volume and μ=shape change parameter which is assumed constant;
- $S_f(z)$=shape factor;
- t=time of flight along the entry trajectory;
- z=geopotential altitude.

In equation (2.3) above, we can also write numerous similar expressions for the time rate of change of the kinetic energy in terms of the many other forms of kinetic energy deposited by the body into the atmosphere and their corresponding differential efficiencies.

These include acoustical energy deposition, heat deposition, ionization, dissociation, etc. (ReVelle 2004).

Next we will utilize the modified ballistic entry parameter, $p^*(z)$, as one integral part of the overall dynamical entry solution for the case of a constant σ. For this $p^*(z)$ parameter, which is a modified form of the object's mass/area ratio, we have the expression from single-body theory (ReVelle 1979; Bronshten 1983; ReVelle 2004):

$$p^*(z) = p^*_\infty \cdot e^{-\left[\sigma \cdot \frac{\mu-1}{2}\right] \cdot \left[V^2_\infty - V^2(z)\right]} \tag{2.7}$$

where
- $p^*(z) = m(z) \cdot g \cdot \sin\theta / (C_D \cdot A) =$ modified ballistic entry parameter;
- $g =$ acceleration due to gravity $(=$ constant$))$, so that changes in p^* are proportional to changes in m/A and which is used in equations (2.15)-(2.22) below.

Equation (2.7) may be immediately solved for the instantaneous mass in the form, if g, C_D and $\sin\theta$ are assumed constants of the hypersonic aerodynamic motion, i.e., the so-called simple ablation theory:

$$m(z) = m_\infty \cdot \left\{\frac{A(z)}{A_\infty}\right\} \cdot e^{\left[\sigma \cdot \frac{\mu-1}{2}\right] \cdot \left[V^2_\infty - V^2(z)\right]}. \tag{2.8}$$

Thus, if the cross-sectional area computed previously (A_∞) is larger than the current cross-sectional area (for collective wake behavior), the quantity in the curly brackets in equation (2.8) is < 1 with similar expectations for non-collective wake behavior, but with different values for the multiplying constants that are discussed below in Section 2.2. For the condition of *maximum drag* this area ratio is 0.5 for both the collective and the non-collective wake limits. Using the ancillary single-body cross-sectional area relationship in the form (ReVelle 1979):

$$A(z) = A_\infty \cdot e^{-\left[\sigma \cdot \frac{\mu}{2}\right] \cdot \left[V^2_\infty - V^2(z)\right]} \tag{2.9}$$

Equation (2.8) can now be simplified to the standard *single-body* mass loss form which is completely independent of the shape change parameter as expected (Bronshten 1983):

$$m(z) = m_\infty \cdot e^{-\frac{\sigma}{2} \cdot \left[V^2_\infty - V^2(z)\right]} \tag{2.10}$$

or. equivalently. changes in either the single-body radius, $r(z)$, in the frontal cross-sectional area, $A(z)$ or in the shape factor, S_f with altitude can be expressed in the form:

$$r(z) = \left(\frac{S_f(z)}{\pi}\right)^{1/2} \cdot \left(\frac{m(z)}{\rho_m}\right)^{1/3} \tag{2.11}$$

$$A(z) = S_f(z) \cdot \left(\frac{m(z)}{\rho_m}\right)^{2/3} \tag{2.12}$$

$$S_f(z) = S_{f\infty} \left(\frac{m_\infty}{m(z)}\right)^{2/3-\mu}. \tag{2.13}$$

These ancillary equations arise directly from the constant σ, ballistic entry, analytic solutions for the mass and area loss, etc. Since our entry dynamics solutions are entirely numerical, we must modify these expressions to their form expected during fragmentation as well as express them into a finite difference form for use in our entry dynamics solution scheme (which computes detailed entry characteristics for either a constant or a variable σ solution for either a hydrostatic, isothermal or non-isothermal atmosphere respectively). In all cases in our initial modeling, we have used $S_f \approx 1.209$ which is the value for a sphere.

2.2. *Mass Loss including Fragmentation Effects*

After fragmentation begins, we have modified the surface cross-sectional area and the corresponding heat transfer area (allowing in general for porous meteoroid entry ablation) as well as the mass and modified ballistic entry parameter as follows, depending on whether or not the wake behavior, which is modeled in terms of one of two possible extreme limits, i.e., either a collective or a non-collective wake behavior during entry (with A =drag cross-sectional area and A_H =heat transfer area). Thus, triggering of fragmentation is assumed to mechanically occur if:

$$p_{stag}(z) > S \tag{2.14}$$

where $p_{stag}(z) = 0.5 \cdot \rho(z) \cdot V^2(z) \cdot C_D(z)$ and S ="Breaking strength"; $S = S$(composition, bulk porosity, etc.) and mechanical breakup is allowed to occur as a cascade process (using our triggered progressive fragmentation model-TPFM)). The value of S is an assigned curve-fitted constant for each bolide type, etc. (ReVelle 2004).

The k_i multiplier parameters given below (for $k_i, i = 1, ...4$) have been written with the other parameters in a finite difference form so that the $z - 1$ refers to the previous altitude while z refers to the current altitude:

a) <u>Collective wake limit</u>: CWL as specified with a specific time delay after each break-up as a function of the velocity and angle of entry of the bolide, etc.

$$k_1 \equiv \frac{A(z)}{A(z-1)}; \qquad k_1 > 1(k_1 = 2.0 \text{ nominal}); \tag{2.15}$$

$$\text{applied at each specified breaking height}$$

$$k_2 \equiv \frac{A_H(z)}{A_H(z-1)}; \qquad k_2 > 1(k2 = 2.0 \text{ nominal}) \tag{2.16}$$

$$k_3 \equiv \frac{m(z)}{m(z-1)}; \qquad k_3 = 1.0 \tag{2.17}$$

$$k_4 \equiv \frac{p^*(z)}{p^*(z-1)}; \qquad k_4 = \frac{1}{2}. \tag{2.18}$$

In this CWL limit the broken mass quickly returns toward the leading surviving fragment in the form of "flying buckshot" in a *maximum drag* orientation while each piece continues to ablate so that the effective area has doubled while the effective mass that participates in the deceleration process has not changed. We have implictly assumed here that the drag coefficent of the swarm of fragments is the same for that of the original unbroken body. The mass is drawn forward because of the heavily decreased air density in the near wake.

b) <u>Non-collective wake limit</u>: NCWL

$$k_1 \equiv \frac{A(z)}{A(z-1)}; \qquad k_1 \leqslant 1(k_1 = 1.0 \text{ nominal}); \tag{2.19}$$

$$\text{applied at each specified breaking height}$$

$$k_2 \equiv \frac{A_H(z)}{A_H(z-1)}; \qquad k_2 \leqslant 1(k_1 = 1.0 \text{ nominal}) \tag{2.20}$$

$$k_3 \equiv \frac{m(z)}{m(z-1)}; \qquad k_3 = \frac{1}{2} \tag{2.21}$$

$$k_4 \equiv \frac{p^*(z)}{p^*(z-1)}; \qquad k_4 = \frac{1}{2}. \tag{2.22}$$

In this NCWL limit the broken mass is assumed to quickly be "lost" to the far wake and no longer interacts with the leading fragment which continues its flight in a *maximum drag* orientation while it continues to ablate. Thus, the effective area has remained the same while the effective mass that participates in the deceleration process has been reduced each time by a factor $k_3 = 1/2$.

Equations (2.18) and (2.22) imply that the effective m/A ratio has decreased following each of these fragmentation events. Specific values of the above constants determine the possible end height behavior. In the computations, the enhancement of the frontal area is only allowed right at each of the breakup locations which are assumed to be progressive once breakup begins. One physical reason for the continuation of the breakup process after equation (2.14) has been initially satisfied is that for sufficiently small fragments, heat conduction processes can produce thermal shock effects that will be sure to continue the fragmentation process. After the time step where each individual breakup occurs, the p^* ratio, etc. is maintained at the original value of the constants (all numerical k_i values are set to unity). We list below the key points to be noted for the above range of k_i values:

i) The CWL evaluated with $k_4 = 0.50$, has the same end height as for the NCWL evaluated with $k_4 = 0.50$ with all other factors the same. This occurs because the net mass to area ratio is the same for both situations for the case of *maximum drag* penetration.

ii) The CWL produces optically much brighter entries (due to the greatly increased frontal area of the body in this limit during fragmentation) than does the NCWL with all other factors the same. For Sumava for example, this brightness difference is more than 2 stellar magnitudes (at maximum luminosity the predicted change between these extremes is from ≈ -20 to -22).

The detailed bolide results shown below for these extreme wake behavior limits are determined by the specific set of values listed above even though they are only "turned on" at each of the specific breaking heights in the fragmentation cascade being modeled i.e., in our TPFM model. Each of the limiting wake behavior cases now occurs with nearly identical end heights (for $k_4 = 0.50$ specified for both extreme cases) until the predicted end height velocity was reached.

In addition to the above, there are specific limits to the modified ballistic parameter behavior for either the CWL or NCWL limits. For the former, mass does not change initially and the frontal area increases as the particles rearrange themselves into a *maximum drag orientation* profile. This causes p^* to be reduced to $1/2$ briefly at each specified fragmentation height compared to its pre-fragmentation value. In the NCWL, the mass is reduced by a factor of $1/2$ and for *maximum drag* conditions, the drag area remains the same so that the k_4 is identical to that for the CWL. This not only produces an identical end height for the numerical values specified for the two wake extremes, but also an identical end height velocity (see ReVelle 2004).

In addition, in a future publication the author will explore both the effects of fragmentation on the end height and the luminosity for the case in which the *maximum drag* orientation can not be achieved. In addition, the future specification of a fragment particle size distribution will also allow a more realistic modeling of the debris field of "flying buckshot" that is envisioned to form during the CWL.

2.3. *Finite Difference Form of the Resulting Equations*

Equation (2.8) may also be written in a finite difference, iterative entry dynamics solution format:

$$m(z) = m(z-1) \cdot \frac{A(z)}{A(z-1)} \cdot e^{-\frac{\sigma(z)}{2} \cdot [V^2(z-1) - V^2(z)]} \qquad (2.23)$$

where $z-1$ indicate greater heights than the current altitude, z. Since σ is lagged to previous heights, so are area and mass values. We have also demanded that

$$\frac{1}{\sigma(z)} \cdot \frac{\partial \sigma(z)}{\partial z} \cdot \delta z << 1$$

be satisfied so that any altitude changes are not too severe and only slowly varying.

Prior to fragmentation, Equation (2.10) is just the single-body model ablation equation originating from the drag/deceleration and the mass loss equation where $A(z)/A(z-1)$ is available from Equation (2.9).

i) **Prior to and after the termination of progressive fragmentation effects:** *single-body model*

$$m(z) = m(z-1) \cdot e^{-\frac{\sigma(z)}{2} \cdot [V^2(z-1) - V^2(z)]} \qquad (2.24)$$

ii) **After the onset of <u>progressive fragmentation</u> effects:**

$$m(z) = m(z-1) \cdot \frac{A(z)}{A(z-1)} \cdot e^{-\frac{\sigma(z)}{2} \cdot [V^2(z-1) - V^2(z)]}. \qquad (2.25)$$

Equation (2.25) is our proposed correction to the mass loss while accounting for the cross-sectional area changes generated by the fragmented bolide. These expressions *correct the mass for changes in the m/A ratio during the fragmentation process*. In the same way equation (2.10) *expresses changes in the single-body mass* prior to fragmentation. The advantage of using the m/A approach is that fragmentation influences on the mass loss can be directly incorporated in the modeling process unlike the conventional single-body mass loss equation which only accounts for a quasi-continuous ablation process for a single unbroken body.

Thus, the predicted final meteor masses during fragmentation are extreme <u>lower</u> limits to the "true" mass. As an indicator of the *upper* mass limit, we have also plotted the single-body model ablation results.

3. **Examples of NEO Fireball Diversity and its Modeling**

3.1. *Benesov/Sumava Fireball Dichotomy: Type I and IIIB Extremes*

In Table 1 we specify the key input parameters such as entry velocity, entry angle, etc. needed for a proper modeling of the drag, deceleration, the corresponding end heights (which in all cases agree quite well with the observations) and the single-body and fragmentation modeled mass loss (using the mass to area ratio approach developed in this paper) as well for modeling of the panchromatic luminosity output for the Benesov and Sumava bolides and for the other five very diverse bolide events presently under consideration. We also list in the same table some of the key outputs resulting from the modeling process such as initial mass, number of fragments produced, terminal velocity at the end height, the required wake behavior model, the necessary bolide type and/or the degree of bulk porosity, etc.

In Figure 1, we have plotted the altitude behavior of panchromatic luminosity, mass loss and σ for Benesov and Sumava (Borovivcka & Spurný 1996; Borovička *et al.* 1998).

Table 1. Summary of input modeling parameters assigned for the various bolides (all with $D = 4.605$ corresponding to 99% kinetic energy removal at the end height with $S_f = 1.209$ for a spherical shape).

Bolide	Observed V_∞ (km/s) Predicted V_t (km/s)	Predicted m_∞ (kg)	Observed θ (°): Radiant elevation angle	Assumed S_f; μ; Limiting wake model	Predicted bolide type and bulk porosity	Predicted number of fragments for light curve modeling
Sumava	26.90 25.70	3988.6	27.5	1.209; 2/3 CWL	IIIA/IIIB 86%	128
Benesov	21.18 12.55	1550.0	80.6	1.209; 2/3 CWL	I 20%	8
Revelstoke	13.0 3.25	$6.54 \cdot 10^5$	15.0	1.209; 2/3 CWL	II 20%	16
Tagish Lake	15.8 9.93	$1.405 \cdot 10^5$	18.0	1.209; 0.10 CWL	II/IIIA 60%	4
Park Forest	20.0 10.98	3405.0	61.0	1.209; 2/3 NCWL	I 20%	128
Crete bolide	13.5 3.38	$7.160 \cdot 10^5$	60.0	1.209; 0.05 CWL	I 5%	8
Antarctic bolide	13.0 4.99	$1.237 \cdot 10^6$	41.9	1.209; 2/3 NCWL	I/II 50%	4

Sumava has previously been shown to be very low in bulk density (group IIIB, with $m_\infty \approx 5000$ kg) while Benesov has been shown to be more like an ordinary chondrite, (group I with $m_\infty \approx 2000 - 4000$ kg). Svetsov (2000) also analyzed this bolide (assuming type Group II) and had determined that Sumava's behavior was inconsistent with the standard steady state, continuum theory, hydrodynamic flow solutions. In ReVelle (2004) the earlier group IIIB interpretation has been completely confirmed however.

Prompted by this extreme set of end heights, we reevaluated both Benesov and Sumava which are indicative of the two possible extreme compositions of observed fireballs. The mass loss prediction for Benesov in Figure 1 is extremely similar to that predicted by the approach of Ceplecha & ReVelle (2005) from analyses of the observational flight data (see their Figure 28 for the analysis of the mass loss for Benesov for example). This agreement has also motivated us to reevaluate Revelstoke (Type II) in Carr (1970), Tagish Lake (Type II/IIIA) in Brown *et al.* (2002), Park Forest (Type I) in Brown *et al.* (2004), the 6/06/02 (Crete) bolide (Type I) in Brown *et al.* (2002) and the 9/03/04 Antarctic bolide (Type I/II) in Klekociuk *et al.* (2005).

3.2. Additional Bolide Modeling Applications

First we consider the Revelstoke meteorite fall over western Canada: 3/31/1965: It was a type II-body (with numerous infrasound recordings summarized in Bayer & Jordan (1967) and in Shoemaker & Lowery (1967), but with no light curve data) as reported in Folinsbee (1967), in Carr (1970) and in Shoemaker (1983) with one or possibly two very small carbonaceous chondrite meteorites having been recovered. In Figure 2, we have plotted the panchromatic luminosity, mass loss and σ for Revelstoke. Shoemaker (1983) estimated the initial energy to be ≈ 20 kt from the analysis of the infrasonic waves, but estimates by Edwards *et al.* (2004, 2006) are ≈ 20 times smaller which is quite puzzling. Our Revelstoke entry modeling estimate resulted in an initial source energy ≈ 13.2 kt, but this value can't be constrained by a light curve since no data were available.

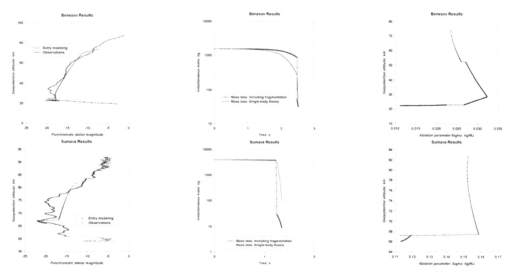

Figure 1. Top left: Benesov panchromatic light curve versus altitude: observed and modeled behavior; top center: Benesov mass loss: single-body and fragmentation modeling (see text); top right: Benesov ablation parameter (σ) versus geopotential altitude; bottom left: Sumava panchromatic light curve versus altitude: observed and modeled behavior; bottom center: Sumava mass loss versus altitude: single-body and fragmentation modeling (see text); bottom right: Sumava ablation parameter (σ) versus geopotential altitude.

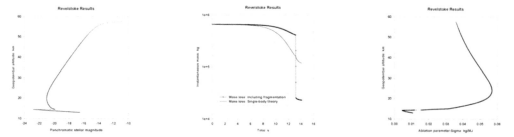

Figure 2. Left: Revelstoke panchromatic light curve versus altitude: observed and modeled behavior; center: Revelstoke mass loss versus altitude: single-body and fragmentation modeling (see text); right: Revelstoke ablation parameter (σ) versus geopotential altitude.

Next we consider the Tagish Lake Bolide: 1/18/2000 (in the southern Canadian Yukon) —type II/IIIA with very large porosity noted in the recovered meteorites (with no infrasonic data currently available as discussed in Edwards *et al.* 2004, 2006). In Figure 3, we have plotted panchromatic luminosity, mass loss and σ for Tagish Lake.

Next we modeled the Park Forest Meteorite Fall: 3/27/2003 (over Chicago)—type I with low porosity (with satellite detections, ground-based video and infrasonic detections as well as meteorites) given in Brown *et al.* (2004). In Figure 4, we have plotted the panchromatic luminosity, mass loss and σ for Park Forest.

Next we considered the Crete Bolide: 6/06/2002: type I (with satellite detections and infrasound), in Brown *et al.* 2002. In Figure 5, we have plotted the behavior of the panchromatic luminosity, mass loss and σ for the Crete bolide.

Finally, we considered the Antarctic bolide of 9/03/04—type I/II bolide (with satellite detections and infrasound data available). In Figure 6 we have plotted the behavior of this small Aten asteroid (Klekociuk *et al.* 2005) for the panchromatic luminosity, the mass loss and σ.

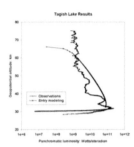 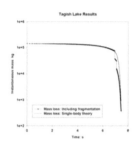 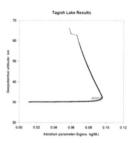

Figure 3. Left: Tagish Lake panchromatic light curve versus altitude: observed and modeled behavior; center: Tagish Lake mass loss versus altitude: single-body and fragmentation modeling (see text); right: Tagish Lake ablation parameter (σ) versus geopotential altitude.

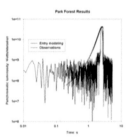 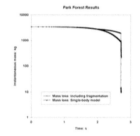 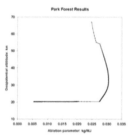

Figure 4. Left: Park Forest panchromatic light curve versus time: observed and modeled behavior; center: Park Forest mass loss versus time: single-body and fragmentation modeling (see text); right: Park Forest ablation parameter (σ) versus geopotential altitude.

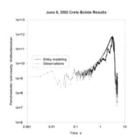 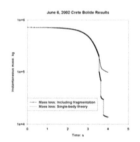 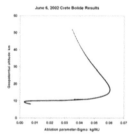

Figure 5. Left: Crete bolide panchromatic light curve versus time: observed and modeled behavior; center: Crete bolide mass loss versus time: single-body and fragmentation modeling (see text); right: Crete bolide ablation parameter (σ) versus geopotential altitude.

One unusual circumstance for this event, based upon our modeling is that the "best fit" for the light curve has an extremely low end height (see the right panel og Figure 6), comparable to that of Great Siberian Meteor of 1908.

Overall the success of the agreement of our entry modeling code predictions with fireball observational data and its analysis seems to be rapidly converging.

4. Summary and Conclusions

4.1. *New Entry Modeling Technique*

In this paper we have introduced a new modeling technique to determine the terminal mass of a meteoroid during extensive fragmentation. It is based upon a new method for the prediction of the mass to area ratio which can be used to calculate the subsequent mass loss during entry quite reliably (compared to the single-body model predictions of

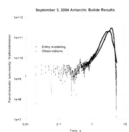
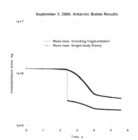
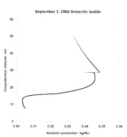

Figure 6. Left: Antarctic bolide: Power time curve (watts/m^2/sterad) versus time: observed and modeled behavior; center: Antarctic bolide mass loss versus time: single-body and fragmentation modeling (see text); right: Antarctic bolide ablation parameter (σ) versus geopotential altitude.

the past). We have compared the prediction of the single-body mass loss equation for each of the bolides against the new approach. In each case with extensive fragmentation, we have found that the single-body mass loss is an extreme upper limit to the "true" mass. Said another way, we have found that the mass loss calculated including fragmentation is generally significantly lower than that predicted using the single-body method. The predicted behavior is also very similar to the results presented in Ceplecha & ReVelle (2005).

In addition, we have also found that our ablation parameter, $\sigma(z)$, is predicted to strongly decrease during fragmentation events which is also in excellent agreement with the systematic analyses of observations in Ceplecha & ReVelle (2005). In fact, had we allowed fragmentation into a very large number of very small fragments instead of the simple factor of two reduction in mass allowed at each fragmentation point, the predicted σ would have become even much smaller, in excellent agreement with Ceplecha & ReVelle (2005). This strongly argues that many observed events have disintegrated into a large number of very small fragments so that the *intrinsic* σ is at least an order of magnitude or more less than the *apparent* σ.

There is still much interesting physics to be incorporated into the bolide modeling process. Although results exist to correct the shock wave radiation flux to the body as a function of the "bluntness" of the body, the author has not yet incorporated this behavior as a function of the μ parameter, etc. Additional physics yet to incorporate also includes interference heating (shock wave/gas cap viscous boundary layer interaction at larger Knudsen number indicative of the transitional flow regime or generally for smaller bodies), ablation products absorption of the shock wave radiation flux, precursor ionization or free stream absorption effects produced by shock wave ultraviolet radiation (a preheating of the ambient air ahead of the body that modifies the "static" atmospheric density structure), etc. Still other effects include thermal conduction and melting of small fragmented particles, rotational effects on shape change and on light production, and finally the evaluation of lift forces, all need much further work for a greater understanding of this extremely complicated natural and wondrous phenomenon.

4.2. *Improved Bolide Initial Mass Estimates*

In addition, we have applied this new entry technique to a number of fireballs to evaluate its overall predictability for a diverse group of NEOs. The fireballs tested include not only meteorite-dropping events, but also highly friable bolides like Sumava, etc. This friable behavior also includes the Tagish Lake bolide whose detailed behavior includes remarkably the recovery of meteorites from an extremely low density and highly porous and friable body. Using our new mass loss technique, we have been able to determine more reliable terminal mass estimates for each of the seven fireballs examined. Our analyses have determined that Benesov (type I), Park Forest (type I), the Crete bolide of June 6, 2002 (type I) and finally the Antarctic bolide of September 3, 2004 (type I/II) are all indicative of stronger and denser bodies indicating the relative likelihood of a significant and recoverable meteorite fall. It is extremely remarkable that Tagish Lake is both extremely friable and yet produced a well documented meteorite fall (having impacted on a frozen lake). The particles that did not arrive as ponderable bodies on the earth must have been extremely friable and of very low density or else they were completely pulverized into extremely small bodies during entry (again in agreement with the very much reduced *intrinsic* σ in comparison to the predicted *apparent* σ). The bulk porosity of Tagish Lake meteorite samples was about 45% (Brown *et al.* 2002) while our entry modeling results indicated the best agreement with the observed light curve at a porosity of $\approx 55 - 60\%$ for the original body (with respect to an ordinary chondrite bulk density). This strongly reminds us that the entering bodies are highly inhomogeneous even if they are not significantly porous, a fact that the current author has been recently emphasizing (ReVelle 2004).

Acknowledgements

We would like to thank the Los Alamos GNEM program and the US DOE HQ in NA-22 for continued financial support.

References

Bayer, K.C. & Jordan, J.N. 1967, *J. Acoust. Soc. of America* 41, 1580

Borovička, J. & Spurný, P. 1996, *Icarus* 121, 484

Borovička, J., Popova, O.P., Nemtchinov, I.V., Spurný, P. & Ceplecha, Z. 1998, *Astron. Astrophys.* 334, 713

Bronshten, V.A. 1983, *Physics of Meteoric Phenomena* (Dordrecht: Kluwer)

Brown, P.G., ReVelle, D.O., Tagliaferri E. & Hildebrand, A.R. 2002, *Meteorit. & Planet. Sci.* 37, 661

Brown, P.G., Spalding, R.E., ReVelle, D.O., Tagliaferri, E. & Worden, S.P. 2002, *Nature* 420, 314

Brown, P.G., Pack, D., Edwards, W.N., ReVelle, D.O., Yoo, B.B., Spalding, R.E. & Tagliaferri, E. 2004, *Meteorit. & Planet. Sci.* 39, 1781

Carr, M.H. 1970, *Geochim. Cosmochim. Acta* 34, 689

Ceplecha, Z., Borovička, J., Elford, W.G., ReVelle, D.O., Hawkes, R.L., Porubčan, V. & Šimek, M. 1998, *Space Sci. Rev.* 84, 327

Ceplecha, Z. & ReVelle, D.O. 2005, *Meteorit. & Planet. Sci.* 40, 35

Edwards, W.N., Brown, P.G. & ReVelle, D.O. 2004, *Earth, Moon & Planets* 95, 501

Edwards, W.N., Brown, P.G. & ReVelle, D.O. 2006, *J. Atmos. & Sol. Terrestr. Phys.* 68, 1136

Folinsbee, R.E., Douglas, J.A.V. & Maxwell, J.A. 1967, *Geochim. Cosmochim. Acta* 31, 1625

Klekociuk, A.R., Brown, P.G., Pack, D.W., ReVelle, D.O., Edwards, W.N., Spalding, R.E., Tagliaferri, E., Yoo, B.B. & Zagari, J. 2005, *Nature* 436, 1132

ReVelle, D.O. 1979, *J. Atmos. & Terrestr. Phys.* 41, 453

ReVelle, D.O. & Ceplecha, Z. 2001, in: B. Warmbein (ed.) ,*Meteoroids 2001*, ESA Special Publ. 495, p. 551

ReVelle, D.O. 2004, *Earth, Moon & Planets* 95, 441

Shoemaker, E.M. & Lowery, C.J. 1967, *J. Meteorit. Soc.* 3, 123

Shoemaker, E.M. 1983, *Annu. Rev. Earth & Planet. Sci.* 11, 461

Svetsov, V.V. 2000, *Sol. Syst. Research* 34, 302

Near Earth Objects, our Celestial Neighbors: Opportunity and Risk
Proceedings IAU Symposium No. 236, 2006
A. Milani, G.B. Valsecchi & D. Vokrouhlický, eds.
© 2007 International Astronomical Union
doi:10.1017/S1743921307003134

Properties of meteoroids from different classes of parent bodies

Jiří Borovička

Astronomical Institute of the Academy of Sciences, 251 65 Ondřejov, Czech Republic
email: borovic@asu.cas.cz

Abstract. Meteoroids observed to disintegrate in the terrestrial atmosphere can be directly linked to their parent bodies in case that they belong to certain meteor showers. We present a list of two dozens of parent bodies reliably associated with well recognized meteor showers. Among the parent bodies are long period comets, Halley-type comets, Jupiter family comets, comets of the inner solar system (such as 2P/Encke) and asteroids.

Physical and chemical properties of meteoroids coming from various parents are compared on the basis of meteor heights, decelerations, light curves and spectra. Jupiter family comets produce meteoroids with the lowest strength, namely porous aggregates of dust grains with bulk densities of about 0.3 g cm^{-3} or less. Halley type material is somewhat stronger and the material related to comet Encke is even stronger. In addition, small strong constituents, perhaps similar to carbonaceous chondrites, can be encountered within the normal cometary material. The strength of cometary material is also enhanced by long-term exposure to cosmic rays and by solar heating in the vicinity to the Sun ($r < 0.2$ AU). Both these processes lead to the loss of volatile sodium. Southern δ-Aquariids, Geminids and partly also Quadrantids were influenced by solar radiation. We argue that these showers, the asteroids associated with them ((3200) Phaethon and 2003 EH$_1$), and the whole interplanetary complexes they belong to are of cometary origin. The argument is supported by lower than chondritic Fe/Mg ratio found in Geminids as well as in Halley type comets. The typical property of stony meteoroids of asteroidal origin is the presence of internal cracks which cause that the incoming meteoroids are much weaker than the recovered meteorites.

Keywords. Meteors, meteoroids; comets: general; asteroids

---

1. Introduction

Meteoroids, by definition, are all bodies on heliocentric (or interstellar) orbits in the size range from several tens of microns to about 10 meters. Size is the only parameter which discriminates meteoroids from dust particles on one side and asteroids on the other side. The primary sources of meteoroids in the Solar System are comets and asteroids. Only a minority of meteoroids come from the solid surfaces of planets (e.g. Mars) and planetary satellites (e.g. Moon) or from the interstellar space. Since the dynamical lifetime of objects in the near-Earth space is of the order of 10 Myr only (Gladman *et al.* 2000; Foschini *et al.* 2000), no meteoroids could survive here from the beginning of the Solar System.

With the exception of the so called *cometary dust trails* (Davies *et al.* 1984; Sykes *et al.* 1986, 2004), meteoroids can be observed only as meteors in the planetary atmospheres. Earth-based meteor observations provide the orbits of an unbiased sample of meteoroids reaching the heliocentric distance of 1 AU. Physical properties (e.g. mechanical strength) and approximate composition of meteoroids can be also derived from meteor observations. This method provides an opportunity to study the properties of large variety of meteoroid parent bodies, in addition to the laboratory studies of meteorites, which is accessible only

Table 1. Long period and Halley-type comets with known meteoroid streams.

Comet	P [yr]	q [AU]	i [°]	Stream	Activity	Ref
	Long period					
C/1911 N1 Kiess	2500	0.684	148	Aurigids	O	[1]
C/1983 H1 IRAS-Araki-Alcock	963	0.991	73	η-Lyrids	AL	[2]
C/1861 G1 Thatcher	415	0.921	80	Lyrids	AM+O	[1]
C/1739 K1 (Zanotti)	unknown	0.674	124	Leonis Minorids	AL	[1]
	Halley-type					
C/1917 F1 Mellish	145	0.190	33	Dec. Monocerotids	AL	[1]
109P/Swift-Tuttle	133	0.960	113	Perseids	AH+O	[1]
1P/Halley	75.3	0.586	162	η-Aquariids	AH	[1]
				Orionids	AM+O	[1]
55P/Tempel-Tuttle	33.2	0.976	163	Leonids	AL+S	[1]
8P/Tuttle	13.5	0.997	55	Ursids	AL+O	[1,3]

for explanation see Table 3

for the strong part of meteoroid population, and still rare and expensive sample return missions. It is the main purpose of this review to discuss the physical and chemical properties of meteoroids of various parent bodies.

There are basically three processes which lead to the separation of meteoroids from their parent bodies. During the normal cometary activity near perihelion, the drag of vapors from evaporating ices takes away also the solid particles, dust and meteoroids (Whipple 1951). Secondly, a catastrophic disruption of comets produces not only secondary nuclei but also a large amount of dust and meteoroids (Jenniskens 2006). Finally, collisions among solar system bodies, in particular among asteroids, produces collisional fragments, including meteoroids (Nesvorný *et al.* 2003). In all cases, the separation velocities of meteoroids are much smaller than the orbital velocity. The orbits of young meteoroids are therefore very similar to the orbits of their parent bodies. If there is a range of orbital periods, then a meteoroid stream is formed in the next orbit. In this stage, it is relatively easy to link the meteoroid stream with its parent body. As time proceeds, various gravitational and non-gravitational forces lead to the dispersion of the stream and/or its separation from the parent body (see e.g. Vaubaillon *et al.* 2006).

2. Known parent bodies

It is likely that every active comet produces a meteoroid stream during perihelion passage. The common presence of cometary dust trails (Reach 2005) is an evidence of this process. Of course, the corresponding meteor shower can be observed only if at least a part of the stream intersects the orbit of the Earth. On the other hand, there are well recognized meteor showers with still unknown parent bodies (Vaubaillon *et al.* 2006). In Tables 1–3, we have compiled the known associations between meteoroid streams and their parent bodies. Only the associations which are certain or very likely are listed. Some other associations found in the literature (e.g. Hughes & Williams 2000) were omitted.

There are meteoroid streams related to almost all types of objects which cross the orbit of the Earth: long period comets ($P > 200$ years), Halley-type comets ($P < 200$ years, Tisserand parameter with respect to Jupiter, $T_J < 2$), Jupiter-family comets ($2 < T_J < 3$), Encke-type comets ($T_J > 3$), as well as objects classified according to their appearance as asteroids (which may also be dormant cometary nuclei). Tables 1–3 contain the current orbital period, P, perihelion distance, q, and inclination, i, of the parent bodies taken from

Table 2. Jupiter-family comets with known meteoroid streams.

Comet	P [yr]	q [AU]	i [°]	Stream	Activity	Ref
3D/Biela	6.65	0.879	13	Andromedids	S	[1]
7P/Pons-Winnecke	6.37	1.257	22	June Bootids	O	[1,4]
21P/Giacobini-Zinner	6.62	1.038	32	October Draconids	AL+S	[1]
26P/Grigg-Skjelerupp	5.11	0.997	21	π-Puppids	O	[5]
73P/Schwassmann-Wachmann 3	5.34	0.933	11	τ-Herculids	O	[1,6]
169P/NEAT	4.20	0.605	11	α-Capricornids	AL	[7]
Marsden group of comets	≈ 5.5	0.047	26	Daytime Arietids	AH	[7,8,9]
Kracht group of comets	≈ 5	0.045	13	S δ-Aquariids	AM	[9]
D/1819 W1 Blanpain = 2003 WY$_{25}$	5.10	0.892	9.1	December Phoenicids	AL+O	[1,10]

for explanation see Table 3

Table 3. Encke-type comets and asteroids with known meteoroid streams.

Object	P [yr]	q [AU]	i [°]	Stream	Activity	Ref
			Encke-type			
2P/Encke	3.3	0.339	12	Taurids	AM	[1]
				Daytime β-Taurids	AM	[1]
			Asteroids			
2003 EH$_1$	5.53	1.193	71	Quadrantids	AH	[11]
2005 UD	1.44	0.163	29	Daytime Sextantids	AM	[12]
(3200) Phaethon	1.43	0.140	22	Geminids	AH	[13,14]

Explanation to Tables 1–3: *Activity type:* AL – annual low (ZHR $\leqslant$ 10), AM – annual medium, AH – annual high (ZHR > 50), O – occasional outbursts (ZHR > 50), S – occasional storms (ZHR > 1000); ZHR is zenithal hourly rate of meteors;
References: [1] Cook (1973), [2] Lyytinen & Jenniskens (2003), [3] Jenniskens *et al.* (2002), [4] Asher & Emel'yanenko (2002), [5] Vaubaillon & Colas (2005), [6] Wiegert *et al.* (2005), [7] Jenniskens (2006), [8] Gorbanev & Knyaz'kova (2003). [9] Sekanina & Chodas (2005), [10] Jenniskens & Lyytinen (2005), [11] Jenniskens (2004), [12] Ohtsuka *et al.* (2006), [13] Whipple (1983) [14] Williams & Wu (1993)

the JPL Small-Body Database (http://ssd.jpl.nasa.gov). Note that the current orbits may differ from the situation at the time when the meteoroid stream was formed. For example, the perihelion of comet 7P/Pons-Winnecke has moved far outside the Earth's orbit since 1825, when the meteoroid swarm, which collided with the Earth in 1998, was ejected (Asher & Emel'yanenko 2002).

Not all bodies listed in Tables 1–3 are independent. Numerical simulations have shown that the orbits of some comets, asteroids and meteoroid streams are related. Such groups are called interplanetary complexes. Table 4 lists three complexes most discussed in the literature. It is a common view that interplanetary complexes are products of disintegration of a common progenitor (grand parent), presumably a comet. Note, however, that in neither case such a scenario has been proved by a detailed study.

The Machholz complex consists of objects on currently different orbits but showing the same orbital evolution. Of particular interest is the small perihelion distance reached during the orbital evolution and exhibited currently by the Marsden and Kracht groups of comets ($q \sim 0.05$ AU), Southern δ-Aquariids ($q \sim 0.07$ AU), and Daytime Arietids ($q \sim 0.09$ AU). The most detailed discussion of the complex was published by Sekanina & Chodas (2005). The Phaethon-Geminid complex was recently extended by the discovery

Table 4. Interplanetary complexes

Complex	Member Bodies	Member Streams	Ref
Machholz complex	96P/Machholz Marsden group of sunskirting comets Kracht group of sunskirting comets 2003 EH$_1$ C/1490 Y1	 Daytime Arietids S δ-Aquariids Quadrantids	[1,2]
Phaethon-Geminid complex	(3200) Phaethon 2005 UD	Geminids Daytime Sextantids	[3]
Taurid complex	2P/Encke possibly a number of asteroids, in particular 2004 TG$_{10}$	Taurids (N and S) Piscids χ-Orionids Daytime β-Taurids Daytime ζ-Perseids	[4,5]

References: [1] Sekanina & Chodas (2005), [2] Jenniskens (2006), [3] Ohtsuka *et al.* (2006), [4] Babadzhanov (2001) [5] Porubčan *et al.* (2006),

of the asteroid 2005 UD (Ohtsuka *et al.* 2006). This relatively compact complex does not contain any active comet. The members also have small perihelion distance ($q = 0.14 - 0.16$ AU) and the orbital period is very short (1.4 yr). The Taurid complex is an extensive complex of meteoroid streams with low inclination and perihelia between 0.2 and 0.5 AU. The center of the stream is clearly related to comet 2P/Encke. The relation of other parts of the complex to about two dozens of Apollo asteroids has been proposed (Babadzhanov 2001; Porubčan *et al.* 2006). It is, however, possible that some coincidences are random, since the orbits are similar to those of asteroids evolving from the ν_6 resonance (Valsecchi *et al.* 1995).

We do not list as complex the comet D/1819 W1 Blanpain, its probable fragment 2003 WY$_{25}$ (Jenniskens & Lyytinen 2005) and the December Phoenicid stream. The number of involved bodies and streams is not large enough to classify as a complex.

3. Asteroids as potential parent bodies

A number of near-Earth asteroids other than those listed in Table 3 have been proposed as parent bodies of observed meteors by various authors. They include asteroids (2101) Adonis (Babadzhanov 2003), (1620) Geographos (Ryabova 2002), and 2001 YB$_5$ (Meng *et al.* 2004) which have been associated with individual meteors found in meteor orbit databases or with rather doubtful minor showers. The problem for objects on typical asteroidal orbit is the generally low encounter velocity with the Earth. The potential meteors are faint, the shower radiant has a large area and unless the stream is very dense, it is difficult to distinguish individual meteors from sporadic background. At the present time, it is quite possible that the proposed associations are just chance coincidences.

Several links were proposed between meteorites of certain types and asteroids of certain taxonomic classes. These links are based on similar reflectance spectra and presumably similar mineralogical composition (see Burbine *et al.* 2002, for a detailed review). The direct orbital link is missing from two reasons. First, orbits of meteorites, except for nine cases (Trigo-Rodríguez *et al.* 2006), are unknown. Secondly, unlike the cometary streams which are only hundreds or thousand years old, the meteorites separated from their parent bodies millions of years ago. Their orbits were then modified chaotically under the influences of the Yarkovsky effect, orbital resonances and close encounters

Table 5. Orbital elements (J2000.0) of some fireballs of special interest

Fireball	a [AU]	q [AU]	i [°]	ω [°]	Ω [°]	Note	Ref
EN 041089	2.501 ±0.006	0.8310 ±0.0002	19.25 ±0.02	234.71 ±0.03	191.8631 ±0.0001	Probable diogenite	[1]
Karlštejn	3.49 ±0.09	1.0124 ±0.0001	137.90 ±0.05	174.60 ±0.07	71.5461 ±0.0001	Probable cometary crust	[2]
Příbram	2.401 ±0.002	0.78951 ±0.00006	10.482 ±0.004	241.750 ±0.013	17.79147 ±0.00001	Recovered H5 chondrite	[3]
Neuschwanstein	2.40 ±0.02	0.7929 ±0.0004	11.41 ±0.03	241.20 ±0.06	16.82664 ±0.00001	Recovered EL6 chondrite	[3]

References: [1] Borovička (1994), [2] Spurný & Borovička (1999b), [3] Spurný *et al.* (2003)

with terrestrial planets (Vokrouhlický & Farinella 2000). So, the preatmospheric orbit of an old meteoroid can be used to infer the character of its orbital evolution and the probable source region of the meteoroid but does not tell us the actual parent body.

The most firmly established asteroid-meteorite link is between the HED meteorites (howardites, eucrites, and diogenites) and the asteroid (4) Vesta and its family (Pieters *et al.* 2006). The fragments from the cratering on Vesta occupy the space up to the 3:1 resonance with Jupiter, which is the dynamical gateway to deliver bodies from the main belt to Earth-crossing orbits. No HED meteorite orbit is known, nevertheless, the spectrum of the fireball EN 041089 suggested diogenite composition (Borovička 1994). The orbit (see Table 5) indeed places the EN 041089 meteoroid exactly into the 3:1 resonance. The other proposed parent bodies include (8) Flora for L-chondrites, (6) Hebe for H-chondrites (Vokrouhlický & Farinella 2000), (3103) Eger for aubrites (Gaffey *et al.* 1992) and few others (Burbine *et al.* 2002).

4. Meteor data

Only the properties of meteorites can be studied in laboratory in detail. Since meteorites represent only the strongest part of the strongest meteoroids encountering the atmosphere, we would obviously like to know how do the meteoroids look before the atmospheric encounter and, even more desirably, how does the weaker cometary material look like. Meteor observations provide us at least a partial answer.

Physical studies of meteoroids are based on investigation of meteor heights, atmospheric deceleration, light curves, and spectra. The methods have been explained in detail in my previous review (Borovička 2006a), where also references to original work can be found. In the first approximation, physical properties of most meteoroids (excluding perhaps the iron meteoroids) can be well characterized by an one-dimensional parameter. The parameter expresses the degree of meteoroid fragility or, more exactly, mechanical strength. The classical classification of bolides into four types, I, II, IIIA, and IIIB, corresponds to increasing ability of meteoroids to disintegrate during the atmospheric entry by fragmentation. From I to IIIB, meteoroid mechanical strength and bulk density decreases, while porosity and fragility increases.

The signs which can be used to recognize and quantify meteoroid fragility are:

(*a*) In case of bright meteors caused by large meteoroids (> 1 cm),

- end height increases with increasing fragility (for a given velocity and mass);
- apparent ablation coefficient is large for more fragile bodies; and

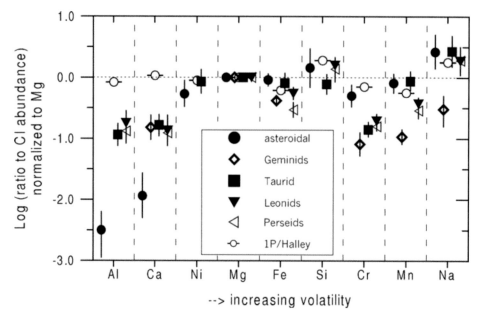

Figure 1. Abundances of eight elements relative to Mg in radiating plasma of selected fireballs as derived from photographic spectra. The abundances are shown as deviation from CI carbonaceous chondrite element-to-Mg ratio taken from Lodders (2003). The composition of Halley comet dust as determined by Jessberger *et al.* (1988) from the VEGA-1 spacecraft flyby is shown for comparison. The fireball data include the asteroidal fireball EN 270200 (Borovička 2006b), one Taurid fireball (this work), and an average of two Geminids (Borovička 2006c), three Leonids, and five Perseids (both from Borovička 2005).

- dynamic pressure causing meteoroid fragmentation in flight is smaller for more fragile bodies.

(*b*) In case of faint meteors,

- beginning height is larger (and increases with mass for a given shower) for more fragile bodies;
- light curves are symmetrical or skewed to the beginning for bodies which are subject to early disruption; and
- sodium may be released preferentially at the beginning of the luminous trajectory, if the body is disrupted very early.

Several authors attempted to compute meteoroid bulk densities from meteor data. Such task is very difficult and the results are model dependent. There is, nevertheless, an agreement that Geminids contain the meteoroids with highest density of all showers (Babadzhanov 2002; Bellot Rubio *et al.* 2002). Other meteoroids with relatively high density are δ-Aquariids and Quadrantids (Babadzhanov 2002).

Approximate chemical composition of meteoroids can be inferred from meteor spectra. In case of sufficiently rich spectra of bright meteors, the spectra can be modeled assuming thermal equilibrium and abundances of observable elements can be determined. In case of not-so-good but more numerous spectra of faint meteors, line intensities (mainly of Na, Mg, and Fe) can be mutually compared and the sample evaluated statistically.

Figure 1 shows relative elemental abundances in radiating plasma of several fireballs of different origin. In order to present an homogeneous dataset, only the data obtained recently by me with a uniform method have been included. Unfortunately, the plasma composition does not reflect directly the meteoroid composition. The low abundances of Al and Ca are caused by incomplete evaporation of meteoric matter during the atmospheric flight. The Al and Ca data for the asteroidal fireball are consistent with a $\sim 90\%$ evaporation of chondritic body (Borovička 2006b). The generally low abundances of Cr are not well understood and may be at least partly caused by another effect during ablation. The most important aspects in Fig. 1 are the differences in Fe abundances and the depletion of volatiles in Geminids. These facts will be discussed in the relevant sections. Figure 2 presents a comparison of the observed Mg, Na, and Fe line intensity ratios in

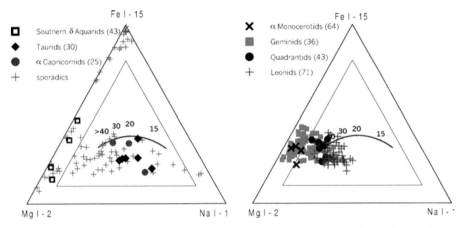

Figure 2. Comparison of brightness of three spectral lines – Mg I (multiplet 2), Na I (multiplet 1), and Fe I (multiplet 15) – in video spectra of faint meteors of seven showers and sporadic meteors. The thick curve shows the expected positions of meteors of chondritic composition and various velocities. The Na line is expected to be brighter in slower meteors (< 40 km s^{-1}), because of lower plasma temperature. The shower velocities in km s^{-1} are given in the legend in parenthesis. According to Borovička *et al.* (2005), Koten *et al.* (2006), and Borovička (2006c).

faint sporadic and shower meteors observed by video techniques. The thing to remember is that there is much larger diversity in composition among sporadic meteoroids in this size range (few millimeters) than among larger meteoroids. Shower meteoroids are better confined but there is real scatter even among meteoroids of the same shower. Differences between showers also exist. Southern δ-Aquariids do not contain Na and there is significant Na depletion in most Geminids and α-Monocerotids and partly also in Quadrantids. Most shower meteoroids have lower than chondritic Fe/Mg ratio.

5. Properties of meteoroids

In this section we will summarize the physical and chemical properties of meteoroids of various origin, without going into too much detail. At the moment we have only limited information on meteoroids from the long period comets in Table 1. Most of them are Lyrids. They do not show obvious differences from meteoroids of Halley-type comets, which is understandable because the 200 yr distinction period is quite arbitrary. Long period cometets are therefore not discussed here separately.

5.1. Halley-type comets

Halley-type comets are represented by well-studied meteor showers Perseids and Leonids (see Table 1). The data show that typical Halley-type cometary material has very low strength (< 0.05 MPa), low density ($\leqslant 1$ g cm^{-3}) and high porosity. Individual meteoroids are most likely irregular aggregates of dust grains. Small meteoroids often disintegrate into the constituents grains early during the atmospheric entry, producing meteors with nearly symmetrical light curves (Beech & Murray 2003; Koten *et al.* 2004).

Meteor beginning heights increase with meteoroid mass in Leonid and Perseid showers, indicating that the ablation (mostly by sputtering) starts very early (Koten *et al.* 2004). Early ablation is supported by volatility of some meteoroid components. Large meteoroids do not break-up at the beginning but often disintegrate catastrophically at lower heights, leading to quick evaporation of the grains and to spectacular meteor flares (Spurný *et al.* 2000a; Borovička & Jenniskens 2000). There are also evidences of spontaneous fragmentation of meteoroids in the interplanetary space (Watanabe *et al.* 2003).

The diversity of light curve shapes (Beech & Murray 2003; Koten *et al.* 2004) and the various degree of early evaporation of sodium in faint Leonid meteors (Borovička *et al.* 1999) indicate that there are differences in shape, structure and possibly also composition of mm-sized meteoroids even within one meteor shower. In addition, Perseids seem to be somewhat stronger on average than Leonids. This indication follows from the average light curve shape of faint meteors (Koten *et al.* 2004), from the comparison of end heights of Perseids (Spurný 1995) and Leonids (Spurný *et al.* 2000a), and from the meteoroid density estimates (Babadzhanov 2002).

Chemically, the Halley-type material is depleted in Fe relative to Mg in comparison with the composition of CI chondrites (Borovička 2005; Borovička *et al.* 2005, see also Figs. 1 and 2). Also the abundances of Cr and Mn seem to be lower, while Si and Na are enhanced. The same results were obtained by the *in situ* mass spectroscopy of comet 1P/Halley dust onboard the VEGA-1 spacecraft (Jessberger *et al.* 1988). Although the differences from the CI composition in the VEGA-1 experiment have been originally attributed to inaccurately known ion yields in the mass spectrometers, meteor data show that they are probably real.

5.1.1. *Strong constituents*

There are evidences that quite strong constituents are embedded within the generally weak Halley-type material. At least in two cases, small Leonid fragments separated from much larger original bodies and were observed to continue well below the position of the main flare (Spurný *et al.* 2000b; Borovička & Jenniskens 2000). In the latter case, a mm-sized particle penetrated to the height of 56 km before being ablated, and survived dynamic pressure up to 2 MPa without fragmentation. The spectrum did not suggest any significant difference in chemical composition from the original meteoroid. The particle was therefore not a calcium-aluminum rich inclusion, for example. Possibly, it may have been material similar to carbonaceous chondrites. Note that Gounelle *et al.* (2006) argued for cometary origin of the CI Orgueil meteorite, based on the entry speed of the fireball.

5.1.2. *Probable cometary crust material*

Most sporadic meteoroids on Halley-type orbits show similar properties to Leonids and Perseids. However, a quite different material was also found on such orbits. The material is characterized by significantly larger strength (~ 1 MPa) and the lack of sodium within it. A typical example is the Karlštejn fireball (Spurný & Borovička 1999a, b; see Table 5 for orbital elements). Such material is quite rare among cm-sized bodies but was found to comprise about 10% of mm-sized sporadic meteoroids (Borovička *et al.* 2005). This type of material is not present within the showers of active comets. However, the α-Monocerotids – meteoroids of a long period shower with unknown parent body – have somewhat similar properties. They are stronger than Perseids (Jenniskens *et al.* 1997) and have lower content of sodium (Fig. 2).

It is likely that this material represents fragments of primordial cometary crust. The compaction and loss of volatile Na were produced by cosmic ray irradiation during long time residence of comets in the Oort cloud. We can speculate that α-Monocerotids were ejected from near surface layers of a long period comet during one of its first approaches to the Sun.

5.2. *Jupiter family comets*

The meteoroids from Jupiter family comets are even softer then those of Halley-type comets. The October Draconids are a prototype of the most fragile material entering the Earth's atmosphere. The likely bulk density is about 0.3 g cm^{-3} (see the discussion in Borovička 2006a). Recent observations at the Ondřejov Observatory (Borovička *et al.*, in preparation) have shown that small Draconids disintegrate under the dynamic pressures less than 1 kPa. Their strength is therefore comparable to the strength of fresh snow. The results of the Deep Impact experiment yielded even lower strength (< 65 Pa) of the surface layers of comet 9P/Tempel 1 (A'Hearn *et al.* 2005). The Fe/Mg ratio in the Jupiter family meteoroids was not firmly established yet.

5.3. *Taurid complex*

Taurid meteoroids are of medium strength on average, they are stronger than Halley-type and Jupiter family cometary material. Nevertheless, very weak material can also be encountered within the complex – for example the Šumava fireball (Borovička & Spurný 1996) was a χ-Orionid. The Fe/Mg ratio seems to be nearly chondritic, at least in some meteoroids (Fig. 1). On the other hand, some faint Taurids show low Fe/Mg (Fig. 2). The Taurid complex is therefore not homogenous. In any case, Taurid meteoroids suggest that comet Encke is physically distinct from Jupiter family comets and may have either different origin or different evolutionary history. The difference was, however, not due by solar heating, since Taurids do not show any hint of Na loss.

5.4. *Phaethon-Geminid complex*

Geminid meteoroids are known to be the strongest of all meteor showers (e.g. Spurný 1993). Their bulk density is about 3 g cm^{-3} (Ceplecha & McCrosky 1992; Babadzhanov 2002). They are severely depleted in Na and also in other volatiles like Mn (Figs. 1 and 2; Borovička 2006c). Unlike other meteor showers, Geminid meteor beginning height does not increase with meteoroid mass (Koten *et al.* 2004). All these distinct properties of Geminids are likely a consequence of their low perihelion distance of 0.14 AU. At 0.14 AU, the meteoroids are heated by solar radiation to about 700 K. Borovička *et al.* (2005) have found all small sporadic meteoroids with perihelia less than 0.2 AU completely depleted in Na and compacted. The Na depletion in Geminids is not complete and varies from meteoroid to meteoroid. Geminids therefore seem to be younger than most sporadic meteoroids. The variations in Na content suggest that the meteoroids are of various ages, i.e. they were released from Phaethon at various times in the past (Borovička 2006c). Alternatively, they may come from various depths inside Phaethon.

In our interpretation, the high strength of Geminids is not an evidence of their asteroidal origin. The Fe/Mg ratio is similar to Halley type cometary material (Fig. 1) and suggests cometary origin of the Geminids, Phaethon, and other members of the complex.

5.5. *Machholz complex*

Quadrantids and Southern δ-Aquariids, the two night time showers of the Machholz complex, have currently quite different orbits but show the same orbital evolution and may have evolved from a common progenitor (Sekanina & Chodas 2005). Southern δ-Aquariids have extremely low perihelion distance of 0.07 AU. They are completely Na-free and compacted (Fig. 2; Borovička *et al.* 2005), in accordance with the low perihelion.

At 0.07 AU, the meteoroid temperature reaches 1000 K. Southern δ-Aquariids are old enough to have lost all sodium by thermal desorption.

Quadrantids, with perihelion distance of 0.98 AU, do not go any close to the Sun. Orbital simulations, however, show that the perihelion distance was low (~ 0.1 AU) 1000–2000 years ago (Porubčan & Kornoš 2005). The narrow width of the core of the stream and the relatively recent appearance of the shower on terrestrial skies, nevertheless, indicate that the core is only 200–500 years old (Jenniskens 2004; Wiegert & Brown 2005). Physical and chemical properties of Quadrantid meteoroids – the strength and Na content – were found to be intermediate between Leonids and Geminids (Koten et al. 2006, see also Fig. 2). There is therefore evidence on moderate influence by solar radiation at low perihelia. Since the stream is so young, the material alternation must have occurred near the surface of the parent body, most likely 2003 EH$_1$.

In this interpretation, asteroid 2003 EH$_1$ was active few hundred years ago when it produced the core of the Quadrantid stream. The meteoroids were released from near surface layers. A cometary splitting, which would produce meteoroids from deep interior of the body, is not consistent with the solar alternation of the Quadrantid material. On the other hand, 2003 EH$_1$ itself is likely a product of earlier fragmentation of the Machholz complex parent comet.

Outer portions of the Quadrantid stream are several thousands years old (Wiegert & Brown 2005). That material can be expected to be more influenced by solar radiation, unfortunately, we do not have physical data on meteors from that part of the stream.

5.6. Asteroidal streams

The search for meteoroid streams associated with asteroids is an evergreen topic in meteor science. The major streams Geminids and Quadrantids are associated with asteroids that very likely exhibited cometary activity in the past. Regular non-cometary asteroids may form meteoroid streams by their mutual collisions. Since asteroid collisions in near-Earth region are rare events, such streams, if any, are expected to be old, disperse and hard to detect. An interesting observation in this respect is the close orbital correspondence (see Table 5) of Příbram and Neuschwanstein meteorites, two of nine meteorites with known orbit (Spurný et al. 2003). The meteorites are, however, of different types – Příbram is an ordinary chondrite (H5) and Neuschwanstein is enstatite chondrite (EL6). If they are part of a larger meteoroid stream, they challenge our understanding of asteroid structure, stream formation and evolution. Pauls & Gladman (2005), nevertheless, concluded on statistical basis that the orbital closeness of both meteorites can be explained by chance coincidence.

5.7. Sporadic meteoroids

Actual parent bodies of sporadic meteors are not known. Nevertheless, we can compare physical and orbital properties of sporadic bodies and still get some information about source regions of different types of material. Figure 3 shows the distribution of semimajor axes and eccentricities of different types of fireballs (Shrbený 2005). The graph gives an idea of orbits of sporadic meteoroids in the size range from several cm to about a meter, which intersect Earth's orbit. There is no significant difference between the orbits of fireballs of types I and II, i.e. stony and carbonaceous bodies. Both groups have a peak at $a \sim 2.3$ AU and $e \sim 0.7$, indicating the origin in the inner asteroid belt. The fragile meteoroids of types IIIA and IIIB have flatter distribution of semimajor axes (axes larger than 5 AU are not shown in Fig 3). The most interesting aspects is the bimodal distributions of eccentricities of the soft cometary material IIIB. While the IIIA meteoroids tend to have large eccentricities, IIIB's have a secondary peak at $e \sim 0.7$. This

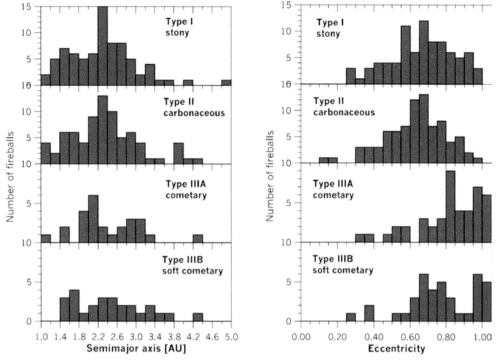

Figure 3. Distribution of semimajor axes (left) and eccentricities (right) of sporadic fireballs of four strength categories. Fireballs with $a > 5$ AU or $a < 1$ AU are not shown in the left graph. According to Shrbený (2005).

peak is clearly related to Jupiter family comets. The observation of sporadic meteoroids therefore confirm the softness of Jupiter family comets.

We should note that the picture would be different for smaller meteoroids. As noted already by Ceplecha (1988), stony bodies are almost absent among mm-sized meteoroids. The spectroscopic survey (Borovička *et al.* 2005) surprisingly revealed that iron type material prevails among the few small meteoroids on typically asteroidal orbits. Since large stony meteoroids are delivered form the main belt to the Earth with the help of the Yarkovsky effect (Vokrouhlický & Farinella 2000), which is inefficient for mm-sized bodies, the absence of small stones near the Earth can be understood. The reason for the presence of irons is, however, not clear.

5.8. *Properties of stony meteoroids*

Meteoroids of type I are related to stony meteorites. Stony meteorites are compact objects of high strength. Incoming meteoroids, on the other hand, easily fragment in the atmosphere under dynamic pressures two orders of magnitude smaller than the strength of meteorites (Petrovic 2001, Popova *et al.*, in preparation). This is valid both for cm-sized (Ceplecha *et al.* 1993) and meter sized bodies (Borovička *et al.* 1998; Borovička & Kalenda 2003). The typical property of stony meteoroids is therefore presence of various internal cracks, which cause structural weakness of the body. Recovered meteorites then represent the most compact part of incoming bodies. Of course, there are large differences between individual bodies but only rarely the incoming meteoroid does not break up under the loading pressure of a few MPa.

6. Summary

We have demonstrated that the analysis of meteor data can reveal differences in physical structure and chemical composition of various Small Bodies in the Solar System (comets and asteroids). The results presented here should be, nevertheless, considered in some sense as preliminary. The analysis of more data in the future is expected to provide larger statistical sample and more firmly established results.

The emerging picture shows that the cometary dust is fragile, porous, and irregular conglomerate of individual dust grains. The fragility varies among different types of comets. Jupiter family comets are composed from the most porous and fragile material. The bulk density is about 0.3 g cm^{-3} (or even less) and the strength of small meteoroids does not exceed 1 kPa. Although this type of material is present also in Halley type comets, typical Halley type material is markedly stronger with probable densities of about 0.8 g cm^{-3}. The material related to comet Encke is even stronger on average. Comet Encke has therefore either different origin or different evolutionary history than Jupiter family comets.

Comets, however, contain also much stronger constituents than is the typical cometary material. Millimeter sized particles of the strength of at least 2 MPa were found to be embedded in larger (~ 10 cm) Leonid meteoroids. The strength of these particles may be comparable to the strength of carbonaceous chondrites. In any case, cometary material is inhomogeneous, certainly on mm-scale, and perhaps on larger scales as well.

Meteor spectroscopy suggests that the Fe/Mg ratio in comets, at least in Halley type comets, is lower than chondritic. This conclusion is in conflict with widespread belief that cometary abundances of non-volatile elements are identical to solar photosphere and CI chondrites. There are, however, other suggestions that this is not the case: the mass spectroscopy of comet Halley dust and the infrared spectroscopy of comet Hale-Bopp (Jessberger *et al.* 1988; Hanner *et al.* 1999). If confirmed, the low Fe/Mg ratio in cometary dust will deserve further studies.

Cometary material can be significantly reprocessed by space environment. The thermal solar radiation in the vicinity to the Sun (< 0.2 AU) and long term exposure to cosmic rays in the Oort cloud have similar effects. They lead to the depletion of more volatile elements, namely Na, and to general compaction of the dust. In consequence, two populations of strong and Na-free meteoroids exist in the Solar System: the Sun approaching meteoroids and the remnants of cometary primordial crust.

The solar heating is responsible for the peculiar properties of Geminid meteoroids. Despite of their strength, we consider Geminids to be of cometary origin, based on their Fe/Mg ratio and also on their presence within an interplanetary complex. The situation is similar with Quadrantids. Both (3200) Phaethon and 2003 EH$_1$ are products of cometary fragmentation, exhibited cometary activity in the past and now are extinct or dormant.

Asteroidal fragments of centimeter to meter size contain many internal cracks. Their bulk strength is much lower than the strength of recovered meteorites. The meteoroid strength lies typically in the range 0.1 – 5 MPa. There is no clear dependence of the strength on meteoroid mass. It can be expected that small asteroids, even if they are not rubble piles, have low bulk strength.

Acknowledgements

This work was supported by grant no. 205/05/0543 from GAČR. The institutional research plan number is AV0Z10030501.

References

A'Hearn, M.F., Belton, M.J.S., Delamere, W.A. *et al.* 2005, *Science* 310, 258

Asher, D.J. & Emel'yanenko, V.V. 2002, *MNRAS* 331, 126

Babadzhanov, P.B. 2001, *A&A* 373, 329

Babadzhanov, P.B. 2002, *A&A* 384, 317

Babadzhanov, P.B. 2003, *A&A* 397, 319

Beech, M. & Murray, I.S. 2003, *Mon. Not. R. Astron. Soc.* 345, 696

Bellot Rubio, L.R., Martínez González, M.J., Ruiz Herrera, L., Licandro, J., Martínez-Delgado, D., Rodríguez-Gil, P. & Serra-Ricart, M. 2002, *A&A* 389, 680

Borovička, J. 1994, in: Y. Kozai *et al.* (eds.) *Seventy-Five Years of Hirayama Asteroid Families*, Astron. Soc. Pacific Conf. Ser. 63, p. 186

Borovička, J. 2005, *Earth, Moon & Planets* 95, 245

Borovička, J. 2006a, in: D. Lazzaro, S. Ferraz-Mello & J.A. Fernández (eds.) *Asteroids, Comets, Meteors, IAU Symp.* 229, p. 249

Borovička, J. 2006b, *Earth, Moon & Planets* (in press)

Borovička, J. 2006c, *Meteorit. Planet. Sci. Suppl.* 41, A25 (abstract)

Borovička, J. & Jenniskens, P. 2000, *Earth, Moon & Planets* 82-83, 399

Borovička, J. & Kalenda, P. 2003, *Meteorit. Planet. Sci.* 38, 1023

Borovička, J. & Spurný, P. 1996, *Icarus* 121, 484

Borovicka, J., Popova, O.P., Nemtchinov, I.V., Spurný, P. & Ceplecha, Z. 1998, *A&A* 334, 713

Borovicka, J., Stork, R. & Bocek, J. 1999, *Meteorit. Planet. Sci.* 34, 987

Borovička, J., Koten, P., Spurný, P., Boček, J. & Štork, R. 2005, *Icarus* 174, 15

Burbine, T.H., McCoy, T.J., Meibom, A., Gladman, B. & Keil, K. 2002, in: Bottke Jr., W.F. *et al.* (eds.), *Asteroids III* (Univ. Arizona Press), p. 653

Ceplecha, Z. 1988, *Bull. Astron. Inst. Czech.* 39, 221

Ceplecha, Z. & McCrosky, R.E. 1992, in: A. W. Harris, E. Bowell (eds.), *Asteroids, Comets, Meteors 1991* (Houston: Lunar Planet. Inst.), p. 109

Ceplecha, Z., Spurný, P., Borovička, J. & Keclíková, J. 1993, *A&A* 279, 615

Cook, A.F. 1973, in: C.L. Hemenway *et al.* (eds.) *Evolutionary and Physical Properties of Meteoroids*, NASA-SP 319, p. 183

Davies, J.K., Green, F.S., Stewart, B.C., Meadows, A.J. & Aumann, H.H. 1984, *Nature* 309, 315

Foschini, L., Farinella, P., Froeschlé, C., Gonzi, R., Jopek, T.J. & Michel, P. 2000, *A&A* 353, 797

Gaffey, M.J., Reed, K.L. & Kelley, M.S. 1992, *Icarus* 100, 95

Gladman, B.J., Michel, P. & Froeschlé, C. 2000, *Icarus* 146, 176

Gorbanev, Yu.M. & Knyaz'kova, E.F. 2003, *Sol. Syst. Res.* 37, 506 (*Astron. Vestnik* 37, 555)

Gounelle, M., Spurný, P. & Bland, P.A. 2006, *Meteorit. Planet. Sci.* 41, 135

Hanner, M.S., Gehrz, R.D., Harker, D.E., Hayward, T.L., Lynch, D.K., Mason, C.C., Russell, R.W., Williams, D.M., Wooden, D.H. & Woodward, C.E. 1999, *Earth, Moon & Planets* 79, 247

Hughes, D.W. & Williams, I.P. 2000, *MNRAS* 315, 629

Jenniskens, P. 2004, *AJ* 127, 3018

Jenniskens, P. 2006, *Meteor Showers and their Parent Comets* (Cambridge Univ. Press)

Jenniskens, P. & Lyytinen, E. 2005, *AJ* 130, 1286

Jenniskens, P., Betlem, H., de Lignie M. & Langbroek, M. 1997 *ApJ* 479, 441

Jenniskens, P., Lyytinen, E., de Lignie M.C., Johannink, C., Jobse, K., Schievink, R., Langbroek, M., Koop, M., Gural, P., Wilson, M.A., Yrjölä, I., Suzuki, K., Ogawa, H. & de Groote, P. 2002, *Icarus* 159, 197

Jessberger, E.K., Christoforidis, A. & Kissel, J. 1988, *Nature* 332, 691

Koten, P., Borovička, J., Spurný, P., Betlem, H. & Evans, S. 2004, *A&A* 428, 683

Koten, P., Borovička, J., Spurný, P., Evans, S., Štork, R. & Elliott, A. 2006, *MNRAS* 366, 1367

Lodders, K. 2003, *Astrophys. J.* 591, 1220

Lyytinen, E. & Jenniskens, P. 2003, *Icarus* 162, 443

Meng, H., Zhu, J., Gong, X., Li, Y., Yang, B., Gao, J., Guan, M., Fan, Y. & Xia, D. 2004 *Icarus* 169, 385

Nesvorný, D., Bottke, W.F., Levison, H.F. & Dones, L. 2003, *ApJ* 591, 486

Ohtsuka, K., Sekiguchi, T., Kinoshita, D., Watanabe, J.-I., Ito, T., Arakida, H. & Kasuga, T. 2006, *A&A* 450, L25

Pauls, A. & Gladman, B. 2005, *Meteorit. Planet. Sci.* 40, 1241

Petrovic, J.J. 2001, *J. Materials Sci.* 36, 1579

Pieters, C.M., Binzel, R.P., Bogard, D., Hiroi, T., Mittlefehldt, D.W., Nyquist, L., Rivkin, A. & Takeda, H. 2006, in: D. Lazzaro, S. Ferraz-Mello & J.A. Fernández (eds.) *Asteroids, Comets, Meteors, IAU Symp.* 229, p. 273

Porubčan, V. & Kornoš, L. 2005, *Contrib. Astron. Obs. Skalnaté Pleso* 35, 5

Porubčan, V., Kornoš, L. & Williams, I.P. 2006, *Contrib. Astron. Obs. Skalnaté Pleso* 36, 103

Reach, W.T. 2005, *BAAS* 37, 471

Ryabova, G.O. 2002, *Sol. Syst. Res.* 36, 234 (*Astron. Vestnik* 36, 254)

Sekanina, Z. & Chodas, P.W. 2005, *ApJS* 161, 551

Shrbený, L. 2005, diploma thesis, Charles Univ. Prague

Spurný, P. 1993, in: J. Štohl, I.P. Williams (eds.), *Meteoroids and Their Parent Bodies* (Bratislava: Astron. Inst. Slovak Acad. Sci.), p. 193

Spurný, P. 1995, *Earth, Moon & Planets* 68, 529

Spurný, P. & Borovička, J. 1999a, in: W.J. Baggaley & V. Porubčan (eds.), *Meteoroids 1998* (Bratislava: Astron. Inst. Slovak Acad. Sci.), p. 143

Spurný, P. & Borovička, J. 1999b, in: J. Svoreň *et al.* (eds.), *Evolution and Source Regions of Asteroids and Comets* (Tatranská Lomnica: Astron. Inst. Slovak Acad. Sci.), p. 163

Spurný, P., Betlem, H., van't Leven, J. & Jenniskenns, P. 2000a, *Meteorit. Planet. Sci.* 35, 243

Spurný, P., Betlem, H., Jobse, K., Koten, P. & van't Leven, J. 2000b, *Meteorit. Planet. Sci.* 35, 1109

Spurný, P., Oberst, J. & Heinlein, D. 2003, *Nature* 423, 151

Sykes, M.V., Lebofsky, L.A., Hunten, D.M. & Low, F. 1986, *Science* 232, 1115

Sykes, M.V., Grün, E., Reach, W.T. & Jenniskens, P. 2004, in: M.V. Festou, H.U. Keller & H.A. Weaver (eds.) *Comets II* (Univ. Arizona Press), p. 677

Trigo-Rodríguez, J.M., Borovička, J., Spurný, P., Ortiz, J.L., Docobo, J.A., Castro-Tirado, A.J. & Llorca, J. 2006, *Meteoritics & Planet. Sci.* 41, 505

Valsecchi, G.B., Morbidelli, A., Gonczi, R., Farinella, P., Froeschlé, Ch. & Froeschlé, C. 1995 *Icarus* 118, 169

Vaubaillon, J. & Colas, F. 2005, *A&A* 431, 1139

Vaubaillon, J., Lamy, P. & Jorda, L. 2006, *MNRAS* 370, 1841

Vokrouhlický, D. & Farinella, P. 2000, *Nature* 407, 606

Watanabe, J., Tabe, I., Hasegawa, H., Hashimoto, T., Fuse, T., Yoshikawa, M., Abe, S. & Suzuki, B. 2003, *PASJ* 55, L23

Whipple, F.L. 1951, *ApJ* 113, 464

Whipple, F.L. 1983, *IAU Circ.* no. 3881

Wiegert, P.A. & Brown, P.G. 2005, *Icarus* 179, 139

Wiegert, P.A., Brown, P.G., Vaubaillon, J. & Schijns, H. 2005, *MNRAS* 361, 638

Williams, I.P. & Wu, Z. 1993, *MNRAS* 262, 231

Near Earth Objects, our Celestial Neighbors: Opportunity and Risk
Proceedings IAU Symposium No. 236, 2006
A. Milani, G.B. Valsecchi & D. Vokrouhlický, eds.
© 2007 International Astronomical Union
doi:10.1017/S1743921307003146

Automation of the Czech part of the European fireball network: equipment, methods and first results

Pavel Spurný, Jiří Borovička and Lukáš Shrbený

Astronomical Institute of the Academy of Sciences, 251 65 Ondřejov, Czech Republic
email: spurny@asu.cas.cz

Abstract. In the last several years the manually operated fish-eye cameras in the Czech part of the European fireball Network (EN) have been gradually replaced with new generation cameras, the modern and sophisticated completely autonomous fireball observatories (AFO), which were recently developed in the Czech Republic. The main motivation for construction of this new observing system was to continue in regular fireball observations and to make these observations more complex and efficient. In this paper we briefly describe basic design and work of this new instrument and its deployment at the Czech stations of the EN. The current dislocation of the individual stations and their equipment is also discussed. Along with this new modern instrument we developed also new software for measurement of photographic negatives which makes this time consuming work more efficient and easier. The AFOs provide us with data on fireballs far richer and more interesting than those we were able to get in the past. This is illustrated by the cases of two recently observed fireballs which were recorded by the AFOs. We describe the high precision of all the measureded values as well as the very detailed information about light curves in both cases.

Keywords. Meteors; meteoroids; instrumentation

1. Introduction

The Earth steadily interacts with meteoroids – small interplanetary bodies of various dimensions, masses, composition and structure. A part of meteoroids are clearly linked to comets as shower meteors and another part is linked to asteroids and represents the densest part of the interplanetary matter. All meteorites with known pre-atmospheric orbits have the origin in the main belt of asteroids. The penetration of larger meteoroids through the atmosphere gives rise to spectacular luminous events – fireballs. Their photographic recordings provide excellent means to examine physical properties as well as the temporal and spatial distribution of extraterrestrial matter in the near-Earth space. The most efficient tools for registration of these very scarce events are the fireball networks, systems covering large areas of the Earth's surface, with multiple camera stations designed to image a large fraction of the night sky. The first such camera network has been established in the Czech Republic and without interruption has been in regular operation until now.

2. Basic instrument

In the last several years the manually operated fish-eye cameras in the Czech part of the European Fireball Network (EN) have been gradually replaced with new generation cameras, modern and sophisticated completely autonomous fireball observatories. The

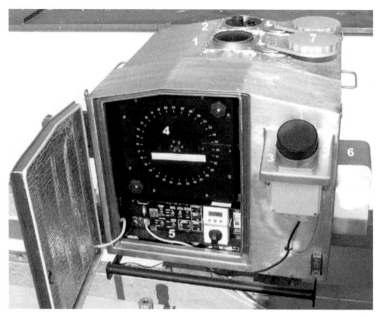

Figure 1. The Autonomous Fireball Observatory. 1- fish-eye lens; 2 – battery of sensors (for cloudiness, sky brightness and sound); 3 – precipitation detector; 4 – rotating cartridge with films; 5 – control panel; 6 – fans for cooling; 7 – covers.

main motivation for the construction of this new observing system was to continue in regular fireball observations and to make these observations more complex and efficient.

2.1. *Autonomous Fireball Observatory (AFO) – description of design and work*

The main function of the Autonomous Fireball Observatory is the same as for manually operated cameras: to take photographic images of the whole sky with good resolution. The AFOs main lens is the very precise Zeiss Distagon 3.5/30 mm fish-eye lens, and we use large format panchromatic sheet films (9 × 12 cm, Ilford FP4), the same used in the old manually operated cameras. The diameter of the sky on the image is 8 cm and an angular resolution of 1 arc minute over the whole sky is reached on the average. We initially thought of the use of CCD sensors, however due to specific requirements for fireball registrations, such as all-night continuous recording, very wide dynamic range of recorded events (∼ 15 stellar magnitudes) and high angular resolution, we found that the photographic emulsion is still the most suitable for this purpose.

The observatory is shown in Fig. 1. The mass of the whole unit is about 120 kg and the dimensions are 80 × 60 × 60 cm. The observatory is weather resistant and sealed against dust, guaranteeing reliable operations also in remote areas with hostile climatic conditions. Lens and camera interior are maintained at constant temperature. The camera is designed to use minimal electrical power and short outages of the electricity distribution network are taken care of by a back-up power supply.

The back-up supply ensures that the camera is safely put into sleep mode, where it stands by for reinstatement of the electricity supply. The camera takes exposure automatically according to a prescribed schedule corrected by actual weather conditions, which are recognized directly by the camera. Each camera unit contains a magazine of 32 films. Human intervention is thus limited to the replacement of the magazine every 5–7 weeks, depending on the number of clear nights. The camera contains a simple weather station that automatically evaluates climatic conditions before and during the

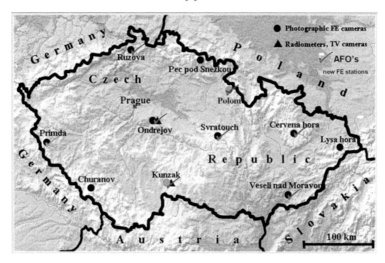

Figure 2. Map of the Czech Fireball Network.

exposure. The exposure is started and proceeds under favorable conditions only, that is when no precipitation is detected and the sky is sufficiently clear. Cloudiness is detected by a very sensitive CCD camera. This information is digitally processed and continuously evaluated. The result of this evaluation is the number of stars in the observed section of sky, which is compared with a specified limit. The film is held exactly flat by under-pressure from a vacuum pump. A two-blade rotating shutter with remotely selectable rotation speed is located just above the film. This allows the speed of the fireball to be determined. The unit receives absolute time/date signals from a public internet server. The times of the beginning and end of the exposure are written to a hard disk, and an exposure identification number printed on the film. If a film is exposed it is exchanged at the end of the night and protective covers are closed over the objective and the sensors. Because only one film is used each clear night we need another method to record the time of a fireball. The camera therefore includes a photoelectric brightness sensor which monitors the total brightness of the sky continuously during the night when no precipitation is detected. The brightness level and time are recorded with a frequency of 500 measurements per second. Therefore, along with precise time information we obtain also very detailed light curves. After long tests we know that these detectors are so sensitive that we have such information for all fireballs bright enough to be recorded photographically. This radiometer is accompanied by a sound detector, which is activated by sufficiently bright and long light pulses and records the sound for up to 10 minutes. In the case of a bright bolide passage, the camera distinguishes it and automatically informs us by email or SMS message. This fact very nicely documents the high technical level of this very modern instrument, which is completely a Czech product. The cameras are networked with internet connection. Each night, the camera logs every operation and these files along with exposure diaries and event reporting are transferred to the central server at Ondřejov.

The AFOs provide us with more complex data on fireballs than what we got before with older equipment. This instrument was designed and developed by the Czech company Space Devices Ltd. with close cooperation with the Astronomical Institute of the Czech Academy of Sciences which has funded this entire project. The Space Devices company is able to produce large numbers of AFOs.

Table 1. Positions of the stations of the Czech Fireball Network.

Number and Name		WGS84 System		
		$\lambda(^\circ E)$	$\varphi(^\circ E)$	$h(m)$
2	Kunžak	15.200930	49.107290	656.1
3	Růžová	14.286526	50.834114	348.5
4	Churáňov	13.614948	49.068430	1119.2
9	Svratouch	16.033917	49.735111	736.0
10	Polom	16.322249	50.350154	748.2
11	Přimda	12.677897	49.669358	745.0
12	Veselí nad Moravou	17.369622	48.954122	176.2
14	Červená hora	17.541962	49.777261	749.7
16	Lysá hora	18.447643	49.546408	1324.4
17	Pec pod Sněžkou	15.729276	50.692143	823.1
20	Ondřejov	14.780042	49.910128	525.0

2.2. Current status of the Czech Fireball Network and AFOs deployment in the Czech Republic

The Czech Fireball Network, the founding part of the European Fireball Network, started regular operation in the fall of 1963, and has remained in full operation until the present time. This network has undergone important changes in the past four decades in terms of geographical coverage, camera equipment and management. The last important change and improvement of the network operation is the replacement of manual all-sky cameras with the new automated fireball observatories described in the previous subsection. The big advantage of the new AFO consists in its completely autonomous work. It does not need every-night human intervention and service which was the most important limiting factor for reliable work of each particular station. In terms of service, the AFO needs only a site with electric power and internet connection. Other requirements for good location of fireball station, i.e. sufficiently dark sky and good horizon remain the same. Most of our stations satisfy these needs and have stayed at the same locations, but some of them were replaced. Since December 2003 we have built two completely new stations – 10 Polom and 2 Kunžak, and replaced old manual all-sky cameras with the AFOs at stations 12 Veselí nad Moravou, 9 Svratouch, 3 Růžová, 14 Červená hora and 11 Přimda. One AFO is also in regular operation at the central station of the Czech Fireball Network, at Ondřejov Observatory along with other instruments for fireball observations which are still operated manually. The current dislocation of the fireball stations over the Czech Republic is shown in Figure 2. At present time we operate 11 stations and 8 of them are equipped with the new AFOs. The remaining two mountain stations Churáňov and Lysá hora will be rebuilt for automated operation in 2007 and the last station Pec pod Sněžkou, which is close to a quickly developing tourist center, will be gradually closed because of unsatisfactory observing conditions. The list of present stations with their geographic coordinates is given in Table 1.

For completeness, we still operate two additional all-sky cameras (with the same lens) in a driven regime at the Ondřejov Observatory and Churáňov station. The main purpose of operation of these two guided cameras was to obtain the time of the fireball passage from simultaneous fixed and guided photographic records. However, the AFOs provide us this information more precisely and therefore the operation of guided cameras will be reduced in near future.

Figure 3. Parts of the all-sky photographs. The left one displays the EN280506 fireball as recorded at Ondřejov observatory, the right one the EN300706 fireball as recorded at Polom.

Along with the all-sky cameras, we operate also a battery of six long focus horizontal cameras at Ondřejov Observatory with objectives Tessar (1/3.5, $f = 360$ mm). These cameras cover a part of the sky where we have very good horizon (covering about 230 degrees in azimuth and 30 degrees in elevation) to get information on precise position of either deep penetrating fireballs close to Ondřejov or very distant events.

To obtain more complex information on some suitably located and bright events, the photographs of fireballs in the direct light are accompanied by the spectral records from a battery of another six cameras at Ondřejov observatory equipped with similar type of lenses as the horizontal ones.

3. Methods

After the photographic films are developed, they are searched visually for meteors. Negatives containing meteors recorded at least at two stations are measured. The classical way of measurement makes use of the *Ascorecord* measuring device. This provides excellent positional accuracy but is time consuming. The operator has to identify the star trails and meteor points and adjust their positions manually. In order to compute meteor brightness, the widths of star trails and the meteor trail are to be measured. To make the process quicker and easier, we have developed a new software, called *Fishscan*, which enables the measurement to be done on PC using scanned negatives. After entering the coordinates of the station and the exposure times, star trails are identified semi-automatically. Positional measurement is done simply by clicking on the beginnings or ends of selected star trails, on some points along the meteor trail, and on the positions of shutter breaks along the meteor trail (for the velocity measurement). Brightness measurements are realized by making photometric scans across the star trails and across the meteor trail (between the shutter breaks) and by integrating the signal along the scans.

We have made extensive tests of the accuracy of the new method. The positional accuracy critically depends on the precision of the scanner used to convert the photographic negative into digital form. No commercial scanner was found to give a precision comparable to the manual *Ascorecord* measurement. Systematic positional errors were present across the negative in all cases. Only a professional photogrammetric scanner gave results equally good as the *Ascorecord* method. On the other hand, even cheap scanners yielded

Table 2. Atmospheric trajectory data of the EN280506 and EN300706 fireballs.

Fireball No.		EN280506	EN300706
Date		28.5.2006	30.7.2006
Time	(UT)	$23^h 16^m 42.0^s \pm 0.1^s$	$20^h 23^m 21.3^s \pm 0.1^s$
Maximum abs. magnitude		-11.4	-7.3
Initial mass	(kg)	65.	2.3
Beginning height	(km)	88.04 ± 0.02	83.424 ± 0.014
Beginning longitude	(deg E)	15.8335 ± 0.0003	16.6433 ± 0.0003
Beginning latitude	(deg N)	50.43274 ± 0.00011	50.09692 ± 0.00012
Terminal height	(km)	30.50 ± 0.02	37.020 ± 0.007
Terminal longitude	(deg E)	16.0474 ± 0.0003	16.73842 ± 0.00013
Terminal latitude	(deg N)	51.25030 ± 0.00011	50.54122 ± 0.00006
Length of trajectory	(km)	109.41	68.48
Duration	(s)	6.79	5.01
Slope	(deg)	31.32	42.44
PE/Type		-4.59/I	-4.58/I
Stations used for computation		$10, 3, 20, 14$	$10, 14, 9, 20$

Table 3. Radiants and orbital elements (J2000.0) of the EN280506 and EN300706 fireballs.

Fireball No.		EN280506	EN300706
RA of apparent radiant	(deg)	243.324 ± 0.013	265.034 ± 0.014
DE of apparent radiant	(deg)	-6.99 ± 0.02	3.235 ± 0.019
Initial velocity	(km/s)	17.576 ± 0.005	15.964 ± 0.002
RA of geocentric radiant	(deg)	241.369 ± 0.015	263.063 ± 0.016
DE of geocentric radiant	(deg)	-14.99 ± 0.02	-4.95 ± 0.02
Geocentric velocity	(km/s)	13.650 ± 0.007	11.480 ± 0.003
Heliocentric velocity	(km/s)	33.377 ± 0.004	38.028 ± 0.003
Semimajor axis	(AU)	1.3936 ± 0.0006	2.942 ± 0.002
Eccentricity		0.4691 ± 0.0003	0.6737 ± 0.0002
Perihelion distance	(AU)	0.73983 ± 0.00015	0.95979 ± 0.00004
Aphelion distance	(AU)	2.0474 ± 0.0013	4.924 ± 0.004
Argument of perihelion	(deg)	261.10 ± 0.03	210.173 ± 0.013
Ascending node	(deg)	67.41320 ± 0.00007	127.45038 ± 0.00001
Inclination	(deg)	2.550 ± 0.010	5.566 ± 0.007

meteor photometry more accurate than the *Ascorecord* method based on trail widths (see Section 4.2).

Computation of meteor trajectories, orbits and other data analysis is done by the procedures developed at the Ondřejov Observatory in the last decades. Positional reduction of all-sky images is performed by the REDSKY procedure of Borovička *et al.* (1995). Meteor trajectories and orbits are computed by the methods described by Ceplecha (1987). The method of Borovička (1990) is used to check the consistency of results. The approach of Ceplecha *et al.* (1993) is applied to study meteor dynamics, ablation and fragmentation. If a meteorite fall is suspected, the meteorite landing area is computed by the method described in Ceplecha (1987). The camera system achieves usually an absolute positional accuracy of 10–20 m for each point on a fireball's luminous trajectory (for a body at distances up to 150 km from the camera), and a dynamic precision of the order of 10 m/s. The precision decreases with increasing distance but we can detect fireballs and reliably determine all important values up to distances of 300–400 km from the stations. In some particular cases we can get reliable results from distances even larger than 500 km. This means that we can obtain excellent orbital data on fireballs as well as the

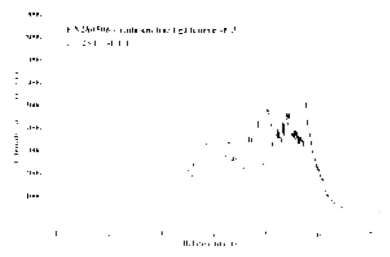

Figure 4. The light curve of the EN280506 as detected by sky brightness sensor installed in the AFO at Růžová station.

precise positions necessary among others for rapid recovery of meteorites not only from the territory of the Czech Republic but also from surrounding countries.

4. Results

We present here complete data on two recent fireballs as an example of the capabilities of our observing system based on AFOs. We show the high precision of all determined values as well as very detailed information about the light curves of both the cases presented.

4.1. The EN280506 fireball

This bright and long fireball with a peak absolute magnitude −11.4 was photographed on 28th May 2006 at $23^h16^m42.1^s$ UT (the time of the beginning of the photographed trajectory). Photographic records of this fireball were taken on 4 stations of the Czech Fireball Network (see Fig. 3 for the image of this fireball). Autonomous Fireball Observatories recorded this event on stations 10 Polom, 3 Růžová and also at the Ondřejov Observatory, where also fixed and guided manually operated fish-eye cameras were in operation. At station 14 Červená hora the manually operated camera was still in use. It was one of the cases, where the AFOs immediately announced detection of a large event, which enabled us to have first very qualified estimates about the atmospheric trajectory already the second day after the event. Such prompt information is especially important in the cases which could produce a meteorite fall. After the reduction of all available all-sky images, all geophysical, orbital and dynamical data have been computed. The fireball started its almost 110 km long luminous trajectory at an altitude of 88 km over the Czech highest mountains Krkonoše with the low entry velocity of 17.6 km/s and terminated its flight after 6.8 seconds at a height of 30.5 km near the Polish town Legnica. The average slope of the atmospheric trajectory to the Earth's surface was 31.3 degrees. Basic results are presented along with the EN300706 fireball in Tables 2 and 3. All values were determined with high precision and well document the capabilities of this observing system.

From its behavior during the atmospheric flight and from its orbit in the Solar system it was a typical asteroidal meteoroid which could produce a significant meteorite if it

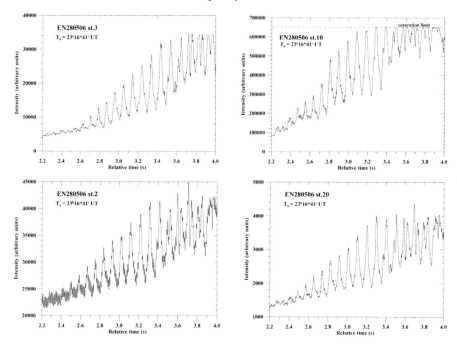

Figure 5. The light curve of the EN280506 fireball as independently seen from four different stations of the Czech Fireball Network. Periodical behavior of the brightness is evident.

would have larger initial mass. In this case the total terminal mass was less than 100 grams which is too low for any systematic search.

The main exceptionality of this fireball consists in its unusual light curve, especially in its beginning part.

The light curve was recorded by the brightness sensors at all AFOs which were that night in the operation. The whole light curve from station 3 Růžová, which was close to the event but not so close like station 10 Polom where the signal was saturated in the maximum, is shown on the Figure 4. The beginning part of the fireball is enlarged in Figure 5 along with another three stations. Figures 4 and 5 nicely show the high resolution of these records (500 samples/s). From the comparison of independent records from different stations, high fidelity of these records is evident. Even the tiniest details of the light curves are well reproduced. The differences in intensity and noise level are caused only by different distances of individual AFOs from the fireball path. All these details are visible only on these radiometric records and cannot be distinguishable from the photographs where we can observe only some small irregular changes of brightness.

A very interesting and unusual feature is observed in the first half of the light curve. As it is shown in Figure 5, significant periodic fluctuations in brightness are observed shortly after the beginning. There is an easily distinguishable main periodicity, at an almost constant frequency of 10.61 Hz from the very beginning with increasing amplitude. This effect is not very common. We already observed similar behavior for only one fireball detected by another kind of radiometers with still higher time resolution but lower sensitivity, which is described in Spurný & Borovička (2001). Such significant periodic variations in brightness could be possibly explained by rotation of highly elongated body before its fragmentation. These records hold unique information about luminous processes during meteoroids atmospheric flight which we have had never before from our observations and therefore the evaluation of these light curves is still not finished

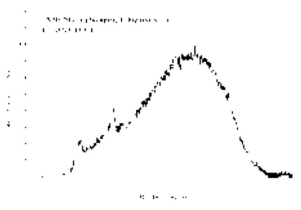

Figure 6. The light curve of the EN300706 as detected by radiometer at Polom station.

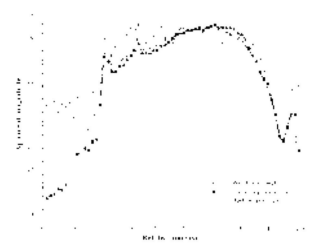

Figure 7. The apparent light curve of the EN 300706 fireball at station 10 Polom as determined by three different methods.

and will be described in more detail in a separate paper. It is presented here mainly for illustration of the capabilities of our new observing instruments.

4.2. *The EN300706 fireball*

The second example of the work of our fireball network is the fireball EN300706. This fireball was not so bright and long as the previous case but it well documents the work of AFOs and also the consistence in results from our new measuring methods, especially the photometry described in Section 3. This fireball was recorded at almost all stations of our network, but the four closest stations equipped only by AFOs were taken for computations. The fireball image taken by the AFO at station 10 Polom is shown in Figure 3. All main results are collected in Tables 2 and 3. This slow moving fireball of −7.3 maximum absolute magnitude started its 68 km long luminous flight at a height of 83 km and disappeared after 5 second at a height of 37 km. Like the previous case, this fireball belongs to the type I fireballs and its asteroidal origin is the most likely.

Figure 7 shows a comparison of the fireball light curve at station 10 Polom obtained by three different methods: photographic photometry using trail width measurements by the *Ascorecord* measuring device, photographic photometry done the scanned image with the *Fishscan* software (described in Sect. 3), and the signal from the photoelectric sky brightness sensor. The sensor data (shown in original form in Fig. 6) have been smoothed to match the temporal resolution of the photograph and transformed to logarithm scale. Since the radiometric signal is not calibrated in magnitudes, the data were shifted vertically to nearly correspond to the photographic light curves at the relative time 2.5 s, near the fireball maximum.

The comparison demonstrates the almost excellent correspondence of the light curve shape between the radiometric and the *Fishscan* method. Marked differences between both curves can be seen only near the fireball beginning, where, nevertheless, the errors of both methods exceed one magnitude because of the fireball brightness. The sensitivity limit of the radiometer is about magnitude -3. The *Ascorecord* curve shows larger scatter and deviates significantly near the fireball beginning and end. The absolute values of the fireball brightness near the fireball maximum, nevertheless, agree well from both photographic methods.

Acknowledgements

We thank Mrs. J. Keclíková who measured most of the photographic records, and Mr. J. Boček for his maintenance of the fireball stations. This work has been supported by project AV0Z10030501.

References

Borovička, J. 1990, *Bull. Astron. Inst. Czech.* 41, 391

Borovička, J., Spurný, P. & Keclíková, J. 1995, *A&AS* 112, 173

Ceplecha, Z. 1987, *Bull. Astron. Inst. Czech.* 38, 222

Ceplecha, Z., Spurný, P., Borovička, J. & Keclíková, J. 1993, *A&A* 279, 615

Spurný, P. & Borovička, J. 2001, in: B. Warmbein (ed.) *Proc. Meteoroids 2001 Conf.*, ESA-SP 495, p. 519

Near Earth Objects, our Celestial Neighbors: Opportunity and Risk
Proceedings IAU Symposium No. 236, 2006
A. Milani, G.B. Valsecchi & D. Vokrouhlický, eds.
© 2007 International Astronomical Union
doi:10.1017/S1743921307003158

Ondřejov's double-station observation of faint meteors using a TV technique (image intensifiers Dedal): 1998–2005

Rostislav Štork

Astronomical Institute, Academy of Sciences of the Czech Republic
Fričova 298, CZ–251 65 Ondřejov, Czech Republic
email: stork@asu.cas.cz

Abstract. TV cameras equipped with image intensifiers are suitable and often used instruments for observing faint meteors. Our regular double-station observations with image intensifiers Dedal (type Dedal 41) started in 1998. Since 2006 we perform observations using new Mullard XX1332 intensifiers. In this paper we present an overview of the eight years of Dedal era. The amount of observing hours, number of meteors, statistics how many meteors were recognized using the automatic system MetRec or found manually, etc. are presented for each year. We also briefly overview the science results from these observations.

Keywords. Meteors; instrumentation

---

1. Instrumentation

The left panel of Fig. 1 shows image intensifiers Dedal 41 mounted in front of video-cameras (S-VHS Panasonic). Dedal 41 is a second generation image intensifier with micro channel plate (MPC), which provides high gain and image quality. The positions of our stations (together with Prague) are marked on the map of the Czech Republic (right panel on Fig. 1). The distance between Ondřejov and Kunžak is 92.5 km.

We use the objectives Arsat 1.4/50 (field of view 28°), Flektogon 2.4/35 (field of view 40°) and Zenitar 2.8/16 (field of view 85°) for wide-field camera. The videosignal is recorded on the external S-VHS videorecorders. The time is inserted into the videoframes (Fig. 2). Originally we used a DCF inserter Cuno but now we prefer a GPS inserter AstroLab because the DCF signal is not stable enough in our area and the AstroLab digits do not cover part of the field of view as the Cuno did.

2. Overview of observations

Our regular double-station observation started in 1998. We usually observe major meteor showers and minor shower whenever their higher activity is expected. Since 2001 we have also been using a third camera, the wide-field one, which is located in Kunžak and covers a larger field of view. Only part of a meteor is often recorded in the field of view of normal (narrow-field) cameras. In these cases a wide-field camera allows to record the whole visible trajectory of bright meteors.

Figure 3 shows the number of observing hours per year and the number of recorded meteors. The large drop of observation in 2002 is clearly visible. This was due to very bad weather during our observing campaigns.

The number of single-station (dark column) and double-station (light column) meteors is displayed in the histogram of Fig. 3. The higher number of meteors recorded by one

131

Figure 1. Left: Image intenfiers Dedal 41 mounted in front of videocameras (S-VHS Panasonic); right: the map of the Czech Republic — the positions of our stations are marked (Prague is also shown).

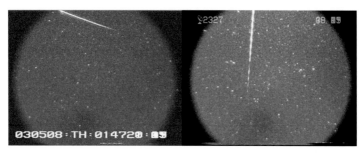

Figure 2. Examples of recorded meteors (sum images of about ten videoframes): left: with older DCF time inserter (Cuno); right: with new GPS inserter (AstroLab).

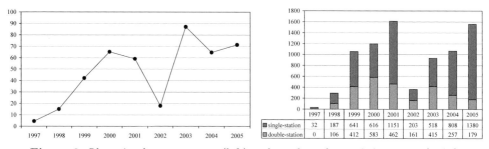

Figure 3. Observing hours per year (left) and number of recorded meteors (right).

station only is mainly caused by the use of the third camera and in 2005 also by the fact that not all the records are thoroughly processed yet.

2.1. *Meteors and individual meteor records*

We distinguish between meteors (phenomena) and meteor records (records on videotape). It means we have one record for single-station meteor and two records for double-station meteor.

The left panel on Fig. 4 shows number of meteors and individual meteor records. The total number of meteor observations is 8 111, numbers per year are shown by the solid line. The total number of individual meteor records is 11 105, numbers per year are shown by dashed line.

For every single-station meteor we check the other station videotape. The dotted line in the left graph (Fig. 4) shows how many meteors were not found on the other station record although the check was done. The number of meteors which were not found despite of the additional check of videotape is 2 911.

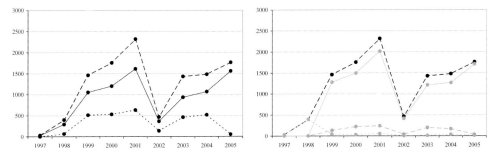

Figure 4. Left part: Number of meteors (solid) and number of individual meteor records (dashed); dotted line shows the number of meteors not found on the other station (although the tapes were additionally manually checked). Right part: the analysis of the ways we found meteors on videotapes. The majority of meteors have been found using the automatic system MetRec – solid blue (grey on b/w print) line. The dotted blue (grey) line represents number of meteors found visually (watching TV screen). The meteors found by checking a tape at the time when the meteor was detected on the other station are displayed by the dashed blue (grey) line. The dashed black line shows the total number of individual meteor records (the same as on the left graph).

The reason for not observing the meteor at one of the stations may be either worse observing conditions at that time or the fact that the meteor was outside of the field of view. We configure our cameras so that the meteors are at similar ranges from both stations and meteor apparent brightness is therefore nearly the same.

2.2. *Methods of finding meteors on videotapes*

The right panel on Fig. 4 shows the statistics of the different methods with which we used to find meteors on videotapes. Since 1999 the majority of meteors have been found using the automatic system MetRec (Molau 1998, www.metrec.org) – solid blue (grey on b/w print) line (total number 9 409); the dotted blue (grey) line represents number of meteors found visually (watching TV screen) – the total number is 633. The meteors found by checking a tape at the time where meteor was detected on the second station are displayed by the dashed blue (grey) line (total number 1 063). Nearly one in four checked meteors was found. The dashed black line shows the total number of individual meteor records, i.e. the sum of all (solid, dashed, and dotted) blue (grey) lines.

3. Results

The data from this observing program have been used in several publications and other papers are in preparation. Here we present a short list of them.

Štork *et al.* (2003) analysed sporadic meteors using Ceplecha's K_B parameter (Ceplecha 1988) and found that no fragile cometary meteoroids (i.e. meteoroids with low value of K_B) are present on orbits with low perihelion distance.

Koten *et al.* (2003) presented the catalogue of the heliocentric orbits of 817 meteors recorded within our double-station program in 1998–2001. The catalogue is also available in electronic version.

Koten *et al.* (2004) studied light curves and atmospheric trajectories of almost 500 meteors of five major meteor showers. Although the parameter F describing the shape of the light curve lays in a broad range, they succeeded to determine the typical value of this parameter for all the studied showers. Investigation of height data showed significant differences between Geminids and other showers. The Geminids proved to be the most compact meteoroids.

Borovička *et al.* (2005) studied 97 spectra of mainly sporadic meteors. For the majority of them the heliocentric orbits were also known from our double-station observations. Three distinct populations of Na-free meteoroids were identified. The first population are meteoroids on asteroidal orbits containing only Fe lines in their spectra and possibly related to iron-nickel meteorites. The second population are meteoroids on orbits with small perihelia ($q \leqslant 0.2$ AU), where Na was lost by thermal desorption. The third population of Na-free meteoroids resides on Halley type cometary orbits.

Koten *et al.* (2006) studied 51 Quadrantid meteors. The data analysis showed that the Quadrantids are similar in some aspects to meteor showers of cometary origin, but in other aspects they are closer to the Geminid meteors. Quadrantid meteoroids have partially lost volatile component, but are not depleted to the same extent as the Geminid meteoroids. These results lead to the conclusion that the parent body is a dormant comet.

In 2005 we succeeded to observe enhanced activity of October Draconids (Koten *et al.*, in preparation). For the first time we detected also the atmospheric deceleration of faint meteors, which can be used to describe meteoroid structure and fragmentation (Borovička *et al.*, in preparation).

4. Conclusions

This paper presents a summary of data collected using Dedal image intensifiers during 1998–2005. The recorded meteors have been catalogued, digitized, analyzed and meteor trajectories are being computed.

The era of Dedal lasted for eight years and is over. We are now using new Mullard XX1332 image intensifiers. It has been tested at one station already in autumn 2005. This intensifier has similar sensitivity but lower level of noise. The result is that the stellar limiting magnitude is about 8 for a field of view $55°$, i.e. the same value as for Dedal with field of view $28°$. The number of recorded meteors is several times higher.

Acknowledgements

The author would like to thank to the team members Pavel Koten, Pavel Spurný and Jiří Borovička, who participate in the observations, computation of meteors trajectories and data analysis. I am also indebted for their valuable comments in preparation of this paper. Last but not least many thanks to the referee Tadeusz Jopek and the editor David Vokrouhlický for their suggestions. This work was supported by the Academy of Sciences of the Czech Republic scientific project AV0Z10030501.

References

Borovička, J., Koten, P., Spurný, P., Boček, J. & Štork R. 2005, *Icarus* 174, 15

Ceplecha, Z. 1988, *Bull. Astron. Inst. Czechosl.* 39, 221

Koten, P., Spurný, P., Borovička, J. & Štork, R. 2003, *Publ. Astron. Inst. ASCR*, No. 91, 1

Koten, P., Borovička, J., Spurný, P., Betlem, H. & Evans, S. 2004, *A&A* 428, 683

Koten, P., Borovička, J., Spurný, P., Evans, S., Štork, R. & Elliot, A. 2006, *MNRAS* 366, 1367

Molau, S. 1998, in: R. Arlt & A. Knoefel (eds.), *Proceedings of the International Meteor Conference*, Stará Lesná 20–23 August 1998, p. 9

Štork, R., Koten, P., Borovička, J. & Spurný, P. 2003, in: B. Warmbein (ed.), *Proceedings of the Asteroids, Comets, Meteors (ACM 2002) conference*, ESA SP-500, p. 189

Near Earth Objects, our Celestial Neighbors: Opportunity and Risk
Proceedings IAU Symposium No. 236, 2006
A. Milani, G.B. Valsecchi & D. Vokrouhlický, eds.
© 2007 International Astronomical Union
doi:10.1017/S174392130700316X

Is the near-Earth asteroid 2000 PG_3 an extinct comet?

Pulat B. Babadzhanov[1] and Iwan P. Williams[2]

[1]Institute of Astrophysics, Dushanbe 734042, Tajikistan
email:P.B.Babadzhanov@mail.ru

[2]Queen Mary University of London, E1 4NS, UK
email:I.P.Willliams@qmul.ac.uk

Abstract. The existence of an observed meteor shower associated with some Near-Earth Asteroid (NEA) is one of the few useful criteria that can be used to indicate that such an object could be a candidate for being regarded as an extinct or dormant cometary nucleus. In order to identify possible new NEA-meteor showers associations, the secular variations of the orbital elements of the NEA 2000 PG_3, with comet-like albedo (0.02), and moving on a comet-like orbit, was investigated under the gravitational action of the Sun and six planets (Mercury to Saturn) over one cycle of variation of the argument of perihelion. The theoretical geocentric radiants and velocities of four possible meteor showers associated with this object are determined. Using published data, the theoretically predicted showers were identified with the night-time September Northern and Southern δ-Piscids fireball showers and several fireballs, and with the day-time meteor associations γ-Arietids and α-Piscids. The character of the orbit and low albedo of 2000 PG_3, and the existence of observed meteor showers associated with 2000 PG_3 provide evidence supporting the conjecture that this object may be of cometary nature.

Keywords. Near-Earth Asteroids, meteoroid streams, meteor showers

1. Introduction

The real distinction between comet and asteroid is in terms of composition, the first is dominated by ices and the second by rock and metals, which is determined by their location at the formation stage. Direct determination of the internal composition is however nearly impossible for most bodies and so other indirect methods have been used. In the past, the distinction between the orbits of comets and those of asteroids were regarded as an obvious discriminant, asteroids moved on near circular orbits located somewhere between Mars and Jupiter while comets moved on highly elliptical orbits possibly with high inclination and larger semimajor axis. Over recent years this distinction has become more blurred as it was realized that the dynamical lifetime of many orbits was much shorter than the age of the solar system. New observational discoveries, particularly of the Near-Earth Object (NEO) population, with the possibility that many asteroids, particularly in this NEO population are in reality dormant or dead cometary nuclei (see for example Williams 1997), re-enforced this conclusion.

An additional discriminant that has been used is the albedo. The albedo of comets generally lie in the range 0.02 to 0.12 (Jewitt 1992) while the albedos of asteroids are much higher. Of course, this is not an infallible test, the surface of a comet nucleus can become devoid of ices and thus take on an asteroidal appearance. Conversely collisions can cause resurfacing which exposes ices on asteroids that could have been buried for millenia.

Another indication of the cometary nature of an NEA is the existence of a related meteoroid stream produced during the period of cometary activity. At present about 1700 minor meteor showers and associations have been detected either optically or by radar. In the overwhelming majority of cases the parent comets of these showers have not been identified. There are two obvious reasons why the identification has not been made, either the parent and stream have experienced very different orbital evolution as Williams & Wu (1993) suggested for the comet of 1491 and the Quadrantids, or by the transformation of the parent comets into an asteroid-like body following the cessation of outgassing.

A meteor shower can only be produced from a meteoroid stream that intersects the Earth's orbit. Hence, the search for dead or dormant comet that is associated with a meteoroid stream can only be meaningful when conducted within the NEA population. Currently, there are several thousand known NEOs and the number is increasing very rapidly. Up to now only a dozen NEOs have been shown to have associated meteor showers, (3200) Phaethon and the Geminid shower (Fox et al., 1984), the Taurid NEO complex and about 40 observable meteor showers (Babadzhanov 2001), 2003 EH$_1$ and the Quadrantid meteor shower (Jenniskens 2004; Williams et al. 2004), and 9 asteroid-fireball stream association (Porubčan et al. 2004).

From our general understanding of meteoroid stream formation (Babadzhanov & Obrubov 1992; Babadzhanov 1998, 2001; Williams 2002), the number of meteor showers produced by a meteoroid stream corresponds to the Earth-crossing class of the parent body orbit. During a year's orbiting around the Sun, the Earth collides with those stream meteoroids which have orbital nodes at a heliocentric distance close to 1 AU, i.e. satisfying the expression:

$$\omega = \pm \arccos\{[a(1 - e^2) - 1]/e\}. \tag{1.1}$$

For a given a and e the Earth's orbit may be intersected at four values of ω. As a result, one meteoroid stream may produce two night-time showers at the pre-perihelion intersections and two day-time showers at the post-perihelion intersections with the Earth. For example, asteroid (3200) Phaethon is a quadruple crosser of the Earth's orbit and the meteoroids of the stream that separated from Phaethon, having various values of the argument of perihelion, can form four meteor showers: the pre-perihelion Geminids and Canis Minorids, post-perihelion Daytime Sextantids and δ-Leonids (Babadzhanov & Obrubov 1992).

In accordance with these concept, to investigate possible genetic relationships between NEOs and meteor showers we need to include the following steps (Babadzhanov & Obrubov 1992; Babadzhanov 1998, 2001):

1) The calculation of the orbital evolutions of a near-Earth object for a time interval covering one cycle of the variation in the argument of perihelion.

2) The determination of the number of crossings of the Earth orbit during one cycle of variation of the perihelion argument. The number of crossings may be from one to eight.

3) The calculation of the theoretical geocentric radiants and velocities for the Earth-crossing orbits.

4) The search for theoretically predicted radiants in catalogues of observed meteor showers and of individual meteors.

In this paper we are concerned with asteroid 2000 PG$_3$.

2. Asteroid 2000 PG$_3$

The Near-Earth Asteroid 2000PG$_3$, discovered on August 1, 2000, has the following orbital elements (equinox 2000.0):

Semi-major axis	$a = 2.83$ AU
Eccentricity	$e = 0.859$
Perihelion distance	$q = 0.400$ AU
Inclination	$i = 20.5°$
Longitude of ascending node	$\Omega = 326.8°$
Argument of perihelion	$\omega = 138.5°$
Longitude of perihelion	$(\pi = \Omega + \omega)\,\pi = 105.3°$

By any definition this would be regarded as more of a comet-like than asteroid-like orbit with high eccentricity and inclination. Fernandez, Jewitt & Shepard (2001) have determined the visual geometric albedo and radiometric effective radius of 2000PG$_3$. They give an effective radius R and geometrical albedo p in the range of $R = 3.08 - 3.49$ km and $p = 0.021 - 0.015$. Hence, based on both its albedo and orbit, the 2000 PG$_3$ appears as a good candidate for an extinct or dormant cometary nucleus.

We calculated the secular variations of the orbital elements of 2000 PG$_3$ using the Halphen-Goryachev integration method (Goryachev 1937). Gravitational perturbations from the six planets (Mercury-Saturn) were taken into account. The perturbations by other planets are very small, and are neglected. Results of calculations show that during the time interval embracing one cycle of variations of the argument of perihelion (~ 5.000 yrs) 2000 PG$_3$ intersects the Earth's orbit four times.

Figure 1 shows the secular variations of the heliocentric distance to the ascending node, R_a, and descending node, R_d, of the orbit of 2000 PG$_3$ plotted against the argument of perihelion ω. As can be seen, one or the other of R_a and R_d has a value of unity, so that the orbit of 2000 PG$_3$ crosses the Earth's orbit, at the values of ω equal to 69°, 111°, 246° and 294°. It is therefore possible that any meteoroid stream associated with 2000 PG$_3$ might produce four meteor showers. The theoretical orbital elements and the theoretical geocentric coordinates of radiant (right ascension α and declination δ) and geocentric velocity V_g (km/s), the dates of activity (and the solar longitudes L. corresponding to these dates) of the meteor showers associated with asteroid 2000 PG$_3$ (denoted as A, B, C, D) are given in Table 1.

We undertook a search for the predicted showers in published catalogues of observed meteor showers: Cook (1973) (C), Kashcheev, Lebedinets & Lagutin (1967) (K), Lebedinets, Korpusov & Sosnova (1972) (L), Sekanina (1973), Sekanina (1976) (S1,S2), Terentyeva (1989) (T), Cannon (2001) (C1) and Halliday, Griffin & Blackwell (1996)(H) (note that we use Halliday's notation in Table 1). In this search we required the positions of the predicted and the observed radiant to be closer than $\pm 10°$ in both right ascension an declination, the difference in geocentric velocity $\Delta V_g \leqslant 5$ km/s and the period of activity to be within ± 15 days of each other. We also calculated D_{SH}, the Southworth & Hawkins (1963) criterion, which serves as a measure of similarity of two orbits, and required $D_{SH} \leqslant 0.3$.

All four theoretically predicted showers associated with the asteroid 2000 PG$_3$ were identified. Two were identified with the real night-time September Southern and Northern δ-Piscids together with fireballs from the three fireball networks European (EN), Prairie (PN) (McCrosky *et al.* 1978) and MORP (Halliday *et al.* 1996). In the records of the IAU Meteor Orbit Data Center we find 11 orbits that are very close to the theoretical shower C and 15 orbits very close to the theoretical shower D. The mean orbital elements of these two meteor groups, which we will call the Daytime γ-Arietids and Daytime α-Piscids, because of the position of their mean radiant, are as follows:

Daytime γ-Arietids: $q = 0.355 \pm 0.016, e = 0.830 \pm 0.018, i = 12.1 \pm 1.4, \Omega = 40.5 \pm 1.9, \omega = 64.3 \pm 1.9, \alpha = 21.6 \pm 2.2, \delta = 19.2 \pm 1.3, V_g = 30.3 \pm 0.7,$

P. B. Babadzhanov & I. P. Williams

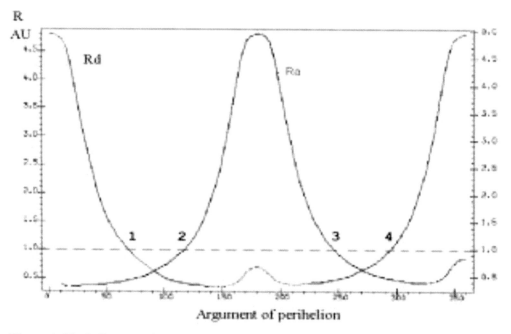

Figure 1. The heliocentric distance of the ascending R_a and descending R_d nodes of the asteroid 2000 PG$_3$ plotted against the argument of perihelion. Crossing 1 correspond to the association with the Daytime γ-Arietids; crossing 2 to that with the September Southern δ-Piscids; crossing 3 to that with the Daytime α-Piscids; crossing 4 to that with the September Northern δ-Piscids.

Daytime α-Piscids: $q = 0.372 \pm 0.008, e = 0.805 \pm 0.013, i = 7.2 \pm 1.1, \Omega = 222.3 \pm 1.6, \omega = 245.2 \pm 1.3, \alpha = 30.0 \pm 1.4, \delta = 5.6 \pm 1.0, V_g = 28.9 \pm 0.5$,

Period of activity of both these meteor association: April 25 − May 10.

Table 1 lists the observed (O) orbital elements, solar longitudes and corresponding dates of maximum activity, the geocentric coordinates of the radiants and velocities for all four showers. The values of D_{SH} given in Table 1 show satisfactory agreement between the theoretically predicted and the observed showers, i.e. all four possible meteor showers associated with 2000 PG$_3$ are active to date. The character of the orbit and low albedo of 2000 PG$_3$, and the existence of the meteor showers associated with 2000 PG$_3$ provide evidence supporting the conjecture that this asteroid may be of a cometary nature.

Porubčan, Kornoš & Williams (2004) noted that if a relation between NEAs and meteoroid streams exist, it will be best recognized for fireball streams represented by larger meteoroids. As we see, the Southern and Northern δ-Piscides meteor showers, seen in September, described by Terentyeva (1989) from the data of the Prairie and MORP fireball networks, consist of fireballs brighter than −15 magnitude produced by bodies of decameter sizes. It was also shown (Babadzhanov 2003) that meteoroid streams may be populated as well by large bodies of several tens of meters in diameter. Porubčan, Kornoš & Williams (2004) found 26 asteroids associated with 20 different fireball streams. Therefore, it may be useful to search for small extinct comets along the orbits of meteoroid streams during periods of meteor shower activity. This inference is confirmed by the detection of 17 objects of some meters to some tens of meters, which passed within a few million km of the Earth. They were observed by Barabanov, Zenkovich, Mikisha, et al. (2001) during the period of activity of the Capricornids, Perseids, Leonids, and Coma Berenicids meteor showers, near the radiant positions of these showers, using 60-cm and 1-m telescopes with CCD-cameras ST-6, at the Zvenig-

Table 1.The theoretical (T) and observed (O) orbital elements, geocentric radiants and velocities of the meteor showers associated with the Near-Earth Asteroid 2000 PG$_3$. In the first column, EN stands for EN140977a, MORP for MORP841001, and PN for PN680912. For the type of shower (T), N denotes a night-time shower, and D a day-time shower; the catalogues (C) are described in the text, and in that column, M stands for IAU Meteor Orbit Data Center. Units: q is in AU, all the angles are in degrees, V_g is in km/s.

Shower	q	e	i	Ω	ω	π	$L_\odot$	Date	α	δ	V_g	D_{SH}	T	C
T "A"	0.340	0.880	14.2	353.6	111.0	104.6	173.6	9/17	10.7	−7.2	31.0			
O δ-Psc^S	0.436	0.781	4.6	354.8	107.0	101.8	174.8	9/18	4.0	−3.0	25.6	0.22	N	T
O S Psc	0.420	0.820	2.0	357.0	107.0	104.0	177.0	9/20	6.0	0.0	26.3	0.23	N	C
O Psc	0.400	0.792	3.3	357.0	111.5	108.5	177.0	9/20	8.0	0.0	26.0	0.22	N	C1
O EN	0.388	0.830	2.4	351.5	112.2	103.7	171.5	9/14	2.7	−1.2	27.8	0.22	N	EN
T "B"	0.342	0.879	11.6	170.9	293.7	104.6	170.9	9/14	359.4	9.5	30.5			
O δ-Psc^N	0.446	0.775	2.7	179.6	285.8	105.4	179.6	9/23	6.0	5.0	25.4	0.21	N	T
O Psc	0.344	0.816	3.8	167.6	298.5	106.1	167.6	9/10	0.2	3.6	27.4	0.15	N	S2
O 508	0.412	0.810	10.4	174.3	287.1	101.4	174.3	9/17	359.1	10.7	26.3	0.11	N	L
O MORP	0.435	0.855	11.3	187.8	283.1	110.9	187.8	10/01	9.7	15.9	27.8	0.15	N	H
O PN	0.270	0.860	5.1	169.0	305.3	114.3	169.0	9/12	4.1	5.6	30.5	0.20	N	PN
T "C"	0.364	0.871	13.2	36.6	68.7	105.3	36.6	4/26	19.1	19.7	29.9			
O γ-Ari	0.355	0.830	12.1	40.5	64.3	104.8	40.5	5/01	21.6	19.2	30.3	0.05	D	M
T "D"	0.342	0.879	10.6	219.0	246.3	105.3	39.0	4/29	27.9	2.6	30.4			
O α-Psc	0.372	0.805	7.2	222.3	245.2	107.5	42.3	5/03	30.0	5.6	28.9	0.10	D	M

orod (Moscow district) and Simeiz (Crimea) observatories of the Institute of Astronomy, Russian Academy of Sciences.

3. Conclusions

Investigation of the orbital evolution of 2000 PG$_3$ shows that this object is a quadruple Earth-crosser and, therefore, its hypothetical meteoroid stream might produce four meteor showers observable from Earth in April and September. A search of the theoretically predicted radiants in the catalogues of observed meteor showers show that all these showers, namely, the night-time September Southern and Northern δ-Piscides meteor showers, and the day-time meteor associations γ-Arietid and α-Piscids are active at present. The existence of the meteor showers associated with 2000 PG$_3$ provides evidence supporting the conjecture that this asteroid may be of a cometary nature.

Acknowledgements

We would like to express our gratitude to the referee Dr. Vladimír Porubčan for useful comments. This work was supported by the International Science and Technology Center Project T-1086.

References

Babadzhanov, P.B. 1998, *Celest. Mech. & Dynam. Astron.* 69, 221
Babadzhanov, P.B. 2001, *A&A* 373, 329
Babadzhanov, P.B. 2003, *A&A* 397, 319
Babadzhanov, P.B. & Obrubov, Yu.V. 1987, in: Z. Ceplecha, P. Pecina (eds.) *Interplanetary matter*, Proc. 10th European Regional Astronomy Meeting of the IAU 2, p. 141
Babadzhanov, P.B. & Obrubov, Yu.V. 1992, *Celest. Mech. & Dynam. Astron.* 54, 111

Barabanov, S.I., Zenkovich, A.D., Mikisha, A.M., Smirnov, M.A 2001, in: Near-Earth Astronomy of the XXI Century, Proceedings of Conference, Zvenigorod 2001, May 21-25, Moscow, GEOS, p. 158

Cannon, E. 2001, *Visual Meteor Showers, http://web.austin.utexas.edu/edcannon/aka-date.htm*

Cook, A.F. 1973, in: Hemenway C.L., Millman P.M., Cook A.F. (eds.), *Evolutionary and Physical Properties of Meteoroids*, NASA SP-319, Washington, D.C., p. 183

Fernandez, Y.R., Jewitt, D.C. & Shepard, S.S. 2001, *Ap J* 553: L197

Fox, K., Williams, I.P. & Hughes, D.W. 1984, *MNRAS* 2108, 11P

Goryachev, N.N. 1937, Halphen's Method for Calculation of Planetary Secular Perturbations and its Application to Ceres. Krasnoe znamya, Tomsk

Halliday, I., Griffin, A.A. & Blackwell, A.T. 1971, MORP network fireball data (1971-84). The IAU Meteor Data Center in Lund, Sweden

Halliday, I., Griffin, A.A. & Blackwell, A.T. 1996, *Meteor & Planet. Sci.* 31, 185

Jenniskens, P. 2004, *WGN, The Journal of the IMO* 32:1, 7

Jewitt, D.C. 1992, in: R.L. Newburn *et al.* (eds.), Comets in the Post-Halley Era (Dordrecht: Kluwer), p. 19

Kashcheev, B.L., Lebedinets, V.N. & Lagutin, M.F. 1967, Meteoric Phenomena in the Earth atmosphere, Nauka. Moscow

Lebedinets, V.N., Korpusov, V.N. & Sosnova, A.K. 1972, *Trudy Inst. Eksper. Meteorol.* No 1 (34), 88

McCrosky, R.E., Shao, C.Y. & Posen, A. 1978, Prairie network fireball data (1963-75). *Meteoritika (Russian)* 37, 44

Porubčan, V., Kornoš, L. & Williams, I.P. 2004, *Earth, Moon & Planets* 95, 697

Sekanina, Z. 1973, *Icarus* 18, 253

Sekanina, Z. 1976, *Icarus* 27, 265

Southworth, R.B. & Hawkins, G.S. 1963, *Smit. Contrib. Astrophys.* 7, 261

Terentyeva, A.K. 1989, in: C.-I. Lagerkvist, H. Rickman, B.A. Lindblad, M. Lindgren (eds.) Asteroids, Comets, Meteors III, (Uppsala Universitet, Reprocentralen HSC), Uppsala, p. 579

Williams, I.P. 1997, *Astronomy & Geophysics* 38, 23

Williams, I.P. 2002, The Evolution of Meteoroid Streams, in: E. Murad and I.P. Williams (eds.), Meteors in the Earth's Atmosphere, Cambridge University Press, p. 13

Williams, I.P., Ryabova, G.O., Baturin, A.P. & Chernitsov, A.M. 2004, *MNRAS* 355, p. 1171

Williams, I.P. & Wu, Z. 1993, *MNRAS* 264, 659

Part 3

Rotation, Shapes and Binaries

Near Earth Objects, our Celestial Neighbors: Opportunity and Risk
Proceedings IAU Symposium No. 236, 2006
A. Milani, G.B. Valsecchi & D. Vokrouhlický, eds.
© 2007 International Astronomical Union
doi:10.1017/S1743921307003183

Radar reconnaissance of near-Earth asteroids

Steven J. Ostro, Jon D. Giorgini and Lance A. M. Benner

Jet Propulsion Laboratory, California Institute of Technology Pasadena, California, USA
email: ostro@reason.jpl.nasa.gov

Abstract. Radar is a uniquely powerful source of information about near-Earth asteroid (NEA) physical properties and orbits. This review consists largely of edited excerpts from Ostro and Giorgini (2004).

Keywords. Minor planets, asteroids; techniques: radar astronomy

1. Introduction

Radar is a uniquely powerful source of information about near-Earth asteroid (NEA) physical properties and orbits. Measurements of the distribution of echo power in time delay (range) and Doppler frequency (radial velocity) constitute two-dimensional images that can provide spatial resolution finer than a decameter if the echoes are strong enough.

The best radar images reveal geologic detail, including craters and blocks. Radar wavelengths at Arecibo (13 cm) or Goldstone (3.5 cm), the world's only continuously active NEA radars (Ostro 2006a), are sensitive to bulk density (a joint function of mineralogy and porosity) and the decimeter-scale structural complexity of approximately the uppermost meter of the target's surface.

Radar can determine the masses of binary NEAs via Kepler's third law and of solitary NEAs via measurement of the Yarkovsky acceleration. With adequate orientational coverage, a sequence of images can be used to construct a three-dimensional model, to define the rotation state, to determine the distribution of radar surface properties, and to constrain the internal density distribution.

As of August 2006, radar has detected echoes from 195 NEAs, of which 110 are designated Potentially Hazardous Asteroids (Ostro 2006b). Radar has revealed both stony and metallic objects, principal-axis and non-principal-axis rotators, smooth and extremely rough surfaces, objects that appear to be monolithic fragments and objects likely to be nearly strengthless gravitational aggregates, spheroids and highly elongated shapes, contact-binary shapes, and binary systems.

Delay-Doppler positional measurements often have a fractional precision finer than 1/10,000,000, comparable to sub-milliarcsecond optical astrometry. Radar can add centuries to the interval over which close Earth approaches can accurately be predicted, significantly refining collision probability estimates compared to those based on optical astrometry alone.

If a small body is on course for a collision with Earth in this century, delay-Doppler radar could almost immediately let us recognize this by distinguishing between an impact trajectory and a near miss, and would dramatically reduce the difficulty and cost of any effort to prevent the collision.

2. Post-discovery astrometric follow-up

Once an asteroid is discovered, its orbital motion must be followed well enough to permit reliable prediction and recovery at the next favorable apparition. A single radar detection of a newly discovered NEA shrinks the instantaneous positional uncertainty at the object's next close approach by orders of magnitude with respect to an optical-only orbit, thereby preventing "loss" of the object.

Comparison of radar+optical with optical-only positional predictions for recoveries of NEAs during the past decade shows that radar-based predictions have had pointing errors that average about 310 times smaller than their optical-only counterparts, dramatically facilitating recovery. Furthermore, radar astrometry (Yeomans *et al.* 1987; Ostro *et al.* 1991; see Giorgini 2006 for a tabulation) can significantly reduce ephemeris uncertainties even for an object whose optical astrometry spans many decades.

A goal of optical searches is to provide as much warning as possible of any possibly dangerous approach of NEAs as large as 140 m. However, since an orbit estimate is based on a least-squares fit to measurements of an asteroid's position over a small portion of its orbit, knowledge of the future trajectory generally is limited by statistical uncertainties that increase with the length of time from the interval spanned by astrometry.

Trajectory uncertainties are greatest and grow most rapidly during close planetary encounters, as the steeper gravity field gradient differentially affects the volume of space centered on the nominal orbit solution within which the asteroid is statistically located. Eventually the uncertainty region grows so large, generally within the orbit plane and along the direction of motion, that the prediction becomes meaningless.

Current ground-based optical astrometric measurements typically have angular uncertainties of between 0.2 and 1.0 arcsec (a standard deviation of 0.5 to 0.8 arcsec is common), corresponding to tens or hundreds or thousands of km of uncertainty for any given measurement, depending on the asteroid's distance. Radar can provide astrometry with uncertainties as small as $\simeq 10min$ range and $\simeq 1mm/s$ in range rate. Since radar measurements are orthogonal to plane-of-sky angular measurements and have relatively high fractional precision, they offer substantial leverage on an orbit solution and normally extend NEO trajectory predictability intervals far beyond what is possible with optical data alone.

3. Radar and collision probability prediction

For newly discovered NEOs, a collision probability is now routinely estimated (Milani *et al.* 2002) for close Earth approaches, and is combined with the object's estimated diameter and the time until the approach to rate the hazard using the Palermo Technical Scale (Chesley *et al.* 2002). The JPL Sentry program's risk page (Chesley 2003) lists objects found to have a potential for impact within the next 100 years.

However, for newly discovered objects, the limited initial astrometry typically does not permit accurate trajectory prediction. When an object's optical astrometric arc is only days or weeks long, the orbit is so uncertain that a potentially hazardous close approach cannot be distinguished from a harmless one or even a non-existent one. The object is placed on the Sentry page, then typically removed later, when additional optical astrometry is obtained and the span of observations is extended. It is extremely rare for a radar-observed object to be on the Sentry page.

4. Negative predictions, positive predictions, and warning time

To a great extent, the dominance of NEA trajectory uncertainties is a temporary artifact of the current discovery phase. Predictions are made for single-apparition objects having a few days or weeks of measurements. The uncertainty region in such cases can encompass a large portion of the inner solar system, thereby generating small but finite impact probabilities that change rapidly as the data arc lengthens, or if high-precision radar delay and Doppler measurements can be made. Impact probabilities in such cases are effectively a statement that the motion of the asteroid is so poorly known that the Earth cannot avoid passing through the asteroid's large uncertainty region – hence the apparent impact "risk". As optical measurements are made, the region shrinks. The resulting change in impact probability, up or down, is effectively a statement about where the asteroid won't be – a "negative prediction" – rather than a "positive prediction" of where it will be. This is due to the modest positional precision of optical measurements.

In contrast, radar measurements provide strong constraints on the motion and hence "positive predictions" about where an asteroid will be decades and often centuries into the future. Thus radar substantially opens the time-window of positive predictability.

5. 99942 Apophis

The several-hundred-meter asteroid (99942) Apophis, formerly known as 2004 MN$_4$, was discovered in June 2004 and lost until it was rediscovered in December 2004. Integration of the orbit calculated from the half-year-long set of optical astrometry revealed an extremely close approach to Earth on April 13, 2029, and possibly hazardous subsequent approaches. Arecibo delay-Doppler radar astrometry obtained during late January 2005 showed the object to be several hundred kilometers closer than had been predicted by the optical measurements (Benner *et al.* 2005, IAU Circ. 8477). Radar observations in August 2005 (Giorgini *et al.* 2005, IAU Circ. 8593) and May 2006 (Benner *et al.* 2006, IAU Circ. 8711) further refined the orbit, moving the predicted 2036 Earth encounter to a lower-probability region within the distribution of possible orbits. The current radar+optical collision 2036 collision probability is about one-third of the optical-only value (S. R. Chesley, personal communication)

6. 29075 (1950 DA)

Integrations of the radar-refined orbit of (29075) 1950 DA (Giorgini *et al.* 2002) revealed that in 2880 there could be a hazardous approach not indicated in the half-century arc of pre-radar optical data. During the observations, a radar time-delay measurement corrected the optical ephemeris's prediction by 7.9 km, changing an optical-only prediction of a nominal miss distance of 20 Earth radii in 2880 into a radar-refined prediction of a 0.9-Earth-radius approach. The uncertainty in the collision probability (which could be as low as zero or as high as 1/300) is dominated by the Yarkovsky acceleration, which is due to the thermal reradiation of absorbed sunlight and depends on the object's mass, size, shape, spin state, and global distribution of optical and thermal properties. This example epitomizes the fundamental inseparability of NEA physical properties and long-term prediction of their trajectories.

7. Images and physical models

With adequate orientational coverage, delay-Doppler images can be used to construct three-dimensional models (e.g., Hudson *et al.* 2000, 2003), to define the rotation state, and to constrain the internal density distribution. Even a single echo spectrum jointly constrains the target's size, rotation period, and sub-radar latitude. A series of Doppler-only echo spectra as a function of rotation phase can constrain the location of the center of mass with respect to a pole-on projection of the asteroid's convex envelope (e.g., Benner *et al.* 1999). For objects in a non-principal-axis spin state, the hypothesis of uniform internal density can be tested directly (Hudson & Ostro 1995). Given a radar-derived model and the associated constraints on an object's internal density distribution, one can use a shape model to estimate the object's gravity field and hence its dynamical environment, as well as the distribution of gravitational slopes on the surface, which can constrain regolith depth and interior configuration.

For most NEAs, radar is the only Earth-based technique that can make images with useful spatial resolution. Therefore, although a sufficiently long, multi-apparition optical astrometric time base might provide about as much advance warning of a possibly dangerous close approach as a radar+optical data set, the only way to compensate for a lack of radar images is with a space mission.

8. Extreme diversity

As reviewed by Ostro *et al.* (2002), NEA radar has revealed both stony and metallic objects, principal-axis and complex rotators, very smooth and extraordinarily rough surfaces, objects that must be monolithic and objects that almost certainly are not, spheroids and highly elongated shapes, objects with complex topography and convex objects virtually devoid of topography. It is meaningless to talk about the physical characteristics of a "typical" NEA.

9. Surface roughness and bulk density

Porous, low-strength materials are very effective at absorbing energy (Asphaug *et al.* 1998). The apparently considerable macroporosity of many asteroids (Britt *et al.* 2002) has led Holsapple (2004) to claim that explosive deflection methods may be ineffective, even for a non-porous asteroid if it has a low-porosity regolith only a few cm deep.

The severity of surface roughness would be of concern to any reconnaissance mission designed to land or gather samples. The wavelengths used for NEAs at Arecibo (13 cm) and Goldstone (3.5 cm), along with the observer's control of the transmitted and received polarizations, make radar experiments sensitive to the surface's bulk density and to its roughness at cm-to-m scales (e.g., Magri *et al.* 2001). Bulk density is a function of regolith porosity and grain density, so if an asteroid can confidently be associated with a meteorite type, then the average porosity of the surface can be estimated. Values of porosity estimated by Magri *et al.* (2001) for nine NEAs range from 0.28 to 0.78, with a mean and standard deviation of 0.53 + 0.15. The current results suggest that most NEAs are covered by at least several centimeters of porous regolith, so the above warning by Holsapple may be valid for virtually any object likely to threaten collision with Earth.

The fact that NEAs' circular polarization ratios (SC/OC) range from near zero to near unity means that the cm-to-m structure on these objects ranges from negligible to much more complex than any seen by the spacecraft that have landed on Eros (whose SC/OC is about 0.3, near the NEA average), the Moon, Venus, or Mars.

10. Binary NEAs: mass and density

The most basic physical properties of an asteroid are its mass, its size and shape, its spin state, and whether it is one object or two (or more; Shepard *et al.* 2006). The current Arecibo and Goldstone systems are uniquely able to identify binary NEAs and at this writing have observed 20, most of which are designated PHAs (see the chronological history table in Ostro 2006b). Current detection statistics, including evidence from optical lightcurves (Pravec 2003) suggest that between 10% and 20% of PHAs are binary systems.

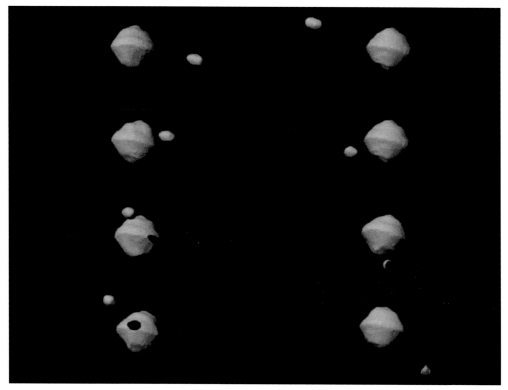

Figure 1. The asteroid (66391) 1999 KW$_4$ is a binary for which there are excellent radar data, allowing for a detailed shape model of both primary and secondary.

Analysis of echoes from these binaries is yielding our first measurements of PHA densities. Delay-Doppler images of 2000 DP$_{107}$ (Margot *et al.* 2002) reveal a 800-m primary and a 300-m secondary. The orbital period of 1.767 d and semimajor axis of 2620 + 160 m yield a bulk density of 1.7 + 1.1 g cm-3 for the primary. DP107 and other radar binaries have spheroidal primaries spinning near the breakup point for strengthless bodies. Whether binaries' components were mutually captured following a highly dispersive impact into a much larger body (Richardson *et al.* 2002 and references therein) or formed by tidal disruption of an object passing too close to an inner planet (Margot *et al.* 2002), it seems likely that most of the primaries are unconsolidated, gravitationally bound aggregates, so Holsapple's warning applies to them.

11. Mission design and spacecraft navigation

Whether a PHA is single or binary, mitigation will involve spacecraft operations close to the object. Maneuvering near a small object is a nontrivial challenge, because of the weakness and complexity of the gravitational environment (Scheeres *et al.* 2000). Maneuvering close to either component of a binary system would be especially harrowing.

The instability of close orbits looms as such a serious unknown that unless we have detailed information about the object's shape and spin state, it would be virtually impossible to design a mission capable of autonomous navigation close to the object. Control of a spacecraft operating close to an asteroid requires knowledge of the asteroid's location, spin state, gravity field, size, shape and mass, as well as knowledge of any satellite bodies that could pose a risk to the spacecraft. Radar can provide information on all these parameters. Knowledge of the target's spin state as well as its shape (and hence nominal gravity harmonics under the assumption of uniform density; Miller *et al.* 1999) would permit design of stable orbits immune to escape or unintended surface impact. (Upon its arrival at (433) Eros, the NEAR Shoemaker spacecraft required almost two months to refine its estimate of the gravity field enough to ensure reliable close-approach operations.)

If it ever turns out to be necessary to have a sequence of missions beginning with physical reconnaissance and ending with a deflection, then a radar-derived physical model would speed up this process, reduce its cost, decrease complexity in the design and construction of the spacecraft, and improve the odds of successful mitigation. [Radar-derived shape models of small NEAs have made it possible to explore the evolution and stability of close orbits (e.g., Scheeres *et al.* 1996, 1998). This radar imaging results for Itokawa (Ostro *et al.* 2001, 2005) were used by the Japanese Institute of Space and Astronautical Science in planning Hayabusa's encounter.] A reduced need for contingency fuel could be significant enough to allow a smaller launch vehicle for the mission. For example, the result might save $100 million via a switch from a Titan III launch vehicle to a Titan IIS, or $200 million for a switch from a Titan IV to a Titan III. The ability of prior radar reconnaissance to reduce mission cost, complexity and risk was embraced by the Department of Defense in their proposed Clementine II multiple-flyby mission (Hope *et al.* 1997), all of whose candidate targets either had already been observed with radar (Toutatis, Golevka) or were radar observable prior to encounter (1987 OA, 1989 UR).

12. Uniqueness of radar opportunities

How much effort should be made to make radar observations of NEAs? For newly discovered objects, it is desirable to guarantee recovery to ensure accurate prediction of close approaches at least throughout this century. Moreover, a target's discovery apparition often provides the most favorable radar opportunity for decades and hence a unique chance for physical characterization that otherwise would require a space mission. Similarly, even for NEAs that have already been detected, any opportunity offering a significant increment in echo strength and hence imaging resolution should be exploited. Binaries and non-principal-axis rotators, for which determination of dynamical and geophysical properties requires a long, preferably multi-apparition time base, should be observed extensively during any radar opportunity.

Acknowledgements

This research was conducted at the Jet Propulsion Laboratory, California Institute of Technology, under contract with the National Aeronautics and Space Administration

(NASA). This material is based in part upon work supported by NASA under the Science Mission Directorate Research and Analysis Programs.

References

Ahrens, T.J. & Harris, A.W. 1992, *Nature* 360, 429

Asphaug E., Ostro, S.J., Hudson, R.S., Scheeres, D.J. & Benz, W. 1998, *Nature* 393, 437

Benner, L.A.M., Ostro, S.J., Rosema, K.D., Giorgini, J.D., Choate, D., Jurgens, R.F., Rose, R., Slade, M.A., Thomas, M.L., Winkler, R. & Yeomans, D.K. 1999, *Icarus* 137, 247

Britt, D.T., Yeomans, D., Housen, K. & Consolmagno, G. 2002 in: W.F. Bottke, A. Cellino, P. Paolicchi & R.P. Binzel (eds.), Asteroids III (Tucson: The University of Arizona Press), p. 485

Chesley, S.R. 2003, `http://neo.jpl.nasa.gov/risks`

Chesley, S.R., Chodas, P.W., Milani, A., Valsecchi, G.B. & Yeomans, D.K. 2002, *Icarus* 159, 423

de Pater, I., Palmer, P., Mitchell, D.L., Ostro, S.J., Yeomans, D.K. & Snyder, L.E. 1994, *Icarus* 111, 489

Giorgini, J.D. 2006, http://ssd.jpl.nasa.gov/radar_data.html

Giorgini, J.D., Ostro, S.J., Benner, L.A.M., Chodas, P.W., Chesley, S.R., Hudson, R.S., Nolan, M.C., Klemola, A.R., Standish, E.M., Jurgens, R.F., Rose, R., Chamberlin, A.B., Yeomans, D.K. & Margot, J.-L. 2002, *Science* 296, 132

Holsapple, K.A. 2004, in: M.J.S. Belton, D.K. Yeomans & T.H. Morgan (eds.), Mitigation of Hazardous Comets and Asteroids (Cambridge), p. 113

Hope, A.S., Kaufman, B., Dasenbrock, R. & Bakeris, D. 1997, in: I.M. Wytrzyszczak, J.H. Lieske & R.A. Feldman, (eds.), Dynamics and Astrometry of Natural and Artificial Celestial Bodies, Proceedings of IAU Colloquium 165 (Dordrecht: Kluwer), p. 183

Hudson, R.S. & Ostro, S.J. 1995, *Science* 270, 84

Hudson, R.S., *et al.* 2000, *Icarus* 148, 37

Hudson, R.S., Ostro, S.J. & Scheeres, D.J. 2003, *Icarus* 161, 348

Magri, C., Consolmagno, G.J., Ostro, S.J., Benner, L.A.M. & Beeney, B.R. 2001, *Meteoritics Planet. Sci.* 36, 1697

Margot, J.L., Nolan, M.C., Benner, L.A.M., Ostro, S.J., Jurgens, R.F., Giorgini, J.D., Slade, M.A. & Campbell, D.B. 2002, *Science* 296, 1445

Milani, A., Chesley, S.R., Chodas, P.W. & Valsecchi, G.B. 2002, in: W.F. Bottke, A. Cellino, P. Paolicchi & R.P. Binzel (eds.), Asteroids III (Tucson: The University of Arizona Press), p. 55

Miller, J.K., Antreasian., P.J., Gaskell, R.W., Giorgini, J.D., Helfrich, C.E., Owen, W.M., Jr, Williams, B.G. & Yeomans, D.K. 1999, Amer. Astron. Soc. Paper # 99-463, Girdwood, Alaska.

Ostro, S.J. 1997, in: J. Remo (ed.), Near-Earth Objects: The United Nations International Conference, Annals of the New York Academy of Sciences 822, p. 118

Ostro, S.J. 2006a, `http://echo.jpl.nasa.gov/~ostro/snr`

Ostro S.J. 2006b, `http://echo.jpl.nasa.gov/asteroids/index.html`

Ostro, S.J. & Giorgini, J.D. 2004, in: M.J.S. Belton, D.K. Yeomans & T.H. Morgan (eds.), Mitigation of Hazardous Comets and Asteroids (Cambridge), p. 38

Ostro, S.J., Campbell, D.B., Chandler, J.F., Shapiro, I.I., Hine, A.A., Velez, R., Jurgens, R.F., Rosema, K.D., Winkler, R. & Yeomans, D.K. 1991, *Astron. J.* 102, 1490

Ostro, S.J., *et al.* 1996, *Icarus* 121, 44

Ostro, S.J., *et al.* 2004, *Meteoritics Planet. Sci.* 39, 407

Ostro, S.J., *et al.* 2005, *Meteoritics Planet. Sci.* 40, 1563

Ostro, S.J., Hudson, R.S., Benner, L.A.M., Giorgini, J.D., Magri, C., Margot, J.-L. & Nolan, M.C. 2002, in: W.F. Bottke, A. Cellino, P. Paolicchi & R.P. Binzel (eds.), Asteroids III (Tucson: The University of Arizona Press), p. 151

Pravec, P. 2003, `http://www.asu.cas.cz/~asteroid/binneas.htm`

Richardson, D.C., Leinhardt, Z.M., Melosh, H.J., Bottke, W.F. Jr. & Asphaug, E. 2002, in: W. Bottke, A. Cellino, P. Paolicchi & R. Binzel (eds.), Asteroids III (Tucson: The University of Arizona Press), p. 501

Scheeres, D.J., Ostro, S.J., Hudson, R.S. & Werner, R.A. 1996, *Icarus* 121, 67

Scheeres, D.J., Ostro, S.J., Hudson, R.S., Suzuki, S. & de Jong, E. 1998, *Icarus* 132, 53

Scheeres, D.J., Williams, B.G. & Miller, J.K. 2000, *J. Guidance, Control and Dynamics* 23, 466

Shepard, M.K., Schlieder, J., Estes, B., Magri, C., Nolan, M.C., Margot, J.-L., Bus, S.J., Volquardsen, E.L., Rivkin, A., Benner, L.A.M., Giorgini, J.D., Ostro, S.J. & Busch, M.W. 2006, *Icarus* 184, 198

Yeomans, D.K., Ostro, S.J. & Chodas, P.W. 1987, *Astron. J.* 94, 189

Near Earth Objects, our Celestial Neighbors: Opportunity and Risk
Proceedings IAU Symposium No. 236, 2006
A. Milani, G.B. Valsecchi & D. Vokrouhlický, eds.

© 2007 International Astronomical Union
doi:10.1017/S1743921307003195

Inverse problems of NEO photometry: Imaging the NEO population

Mikko Kaasalainen[1] and Josef Ďurech[1,2]

[1]Department of Mathematics and Statistics, P.O.Box 68,
FI-00014 University of Helsinki, Finland
email: mikko.kaasalainen@helsinki.fi

[2]Astronomical Institute, Charles University in Prague, V Holešovičkách 2,
18000, Prague, Czech Republic
email: durech@sirrah.troja.mff.cuni.cz

Abstract. The physical properties of NEOs and other asteroids are mostly obtained with photometry. The resulting models describe the shapes, spin states, scattering properties and surface structure of the targets, and are the solutions of inverse problems involving comprehensive mathematical analysis. We review what can and cannot be obtained from photometric (and complementary) data, and how all this is done in practice. The role of photometry will become completely dominating with the advent of large-scale surveys capable of producing calibrated brightness data. Due to their quickly changing geometries with respect to the Earth, NEOs are the population that can be mapped the fastest.

Keywords. data analysis; photometry; infrared; scattering; asteroid, surfaces

1. Introduction

Photometry is the main source of information on the NEO population and the other asteroids as a whole. Traditional dense lightcurves have been obtained for a large number of targets over several decades, and this mode of observation will continue in the future. However, a real paradigm change will be brought about by surveys such as Pan-STARRS and LSST that will produce colossal photometric databases that, though sparse in time, can readily be used for obtaining the physical characteristics of asteroids everywhere (Kaasalainen 2004; Ďurech *et al.* 2006, 2007). This is a unique opportunity to map NEOs (and other asteroids) both as individual targets and as a population: no other observing mode can reach such a vast number of targets. The survey datasets are efficiently enriched by any additional dense photometric or other observations.

The analysis of sparse photometric datasets will very soon become an automated industry, resulting in tens of thousands of asteroid models, a large portion of them NEOs. The computational effort in this is considerable in both computer and human time, which means that most of the targets will not be analyzed with close scrutiny: we will have to trust the computer. This, again, means that we have to have a good understanding of the reliability of our models, and this is impossible without a thorough understanding of the mathematical nature of the inverse problem(s) involved. Indeed, the inversion of photometric data involves some profound mathematical truths, and the effect of these theorems is quite visible in all parts of the inversion process. Very important concepts are the uniqueness and stability of the solution, the parameter spaces, the so-called inverse crimes in simulations and error prediction, and the domination of systematic errors over random ones.

This paper is organized as follows. In §2 we summarize and discuss some main facts about photometric inversion from both theoretical and practical points of view, and investigate the role of systematic errors. Some particular properties of NEO (and other asteroid) observations and modelling are discussed in §3, where we also introduce a convenient way of producing synthetic asteroids and a useful method for initial period scanning of sparse datasets. In §4 we discuss the sparse observation mode in combination with dense lightcurves and/or other data such as thermal infrared or radar. We sum up and discuss future observations in §5.

2. Inversion of photometric data

Asteroid lightcurve observations and their analysis have perhaps three main components. In historical order these are:

I. Period analysis for almost 2000 asteroids, including the detection of many binary systems.

II. Full physical (spin, shape and scattering) modelling from combined datasets, also with data other than photometric (including, e.g., radar, stellar occultations, thermal infrared, and adaptive optics).

III. A vast quantity of physical models using accurately calibrated photometry from large surveys (Pan-STARRS, LSST, etc.) as the main database.

Item I has resulted in statistically important catalogues by Pravec, Harris and others (Pravec *et al.* 2007; Harris, private communication): despite the inevitable observational selection effects, we are beginning to have some idea of the period distribution of asteroids. Item II has produced the first reasonably large (more than 100 objects) catalogue giving us some idea what asteroids are really like: how their spin axes are distributed in space, what kinds of shapes and irregularities they exhibit, what their actual (spin/shape corrected) solar phase curves are like, what we can say about their surface properties, etc. We now have several ground truths from space missions, laboratory studies, etc. from which we know that photometric modelling gives a good global portrait of the target. For example, the Keck adaptive optics images of several asteroids coincide, within uncertainties of the two methods, with the predicted plane-of-sky images from photometry-based models determined earlier (Marchis *et al.* 2006; the same level of correspondence is also obtained between the Hubble Space Telescope images of Storrs *et al.* (1999) and the photometry-based predictions). Similar shape and pole agreement is also obtained with the laboratory model of Kaasalainen *et al.* (2005), or the radar and fly-by target models in Kaasalainen *et al.* (2001). Indeed, we can say that a good set of photometric data essentially enables us to image an asteroid crudely.

Similarly, we know that combining thermal infrared observations with these models yields, e.g., accurate estimates of surface regolith properties (Mueller *et al.* 2005). The models can even be used for getting a colour map of the surface using data at different wavelengths and thus gain some insight on mineral distributions (Nathues *et al.* 2005). The spin properties can reveal evolutionary surprises, in particular in connection with the YORP effect from thermal radiation (Vokrouhlický *et al.* 2004; Bottke *et al.* 2006 and references therein). While the level of detail from groundbased observations cannot reach that of in situ space missions, the latter are going to remain few in number. Photometry alone has the chance to give us a well representative and statistically significant coverage of the physical properties of asteroids and asteroid populations such as NEOs. Photometric inversion is now a routine process, so obtaining the models is straightforward. We should also like to note that, from the mathematical point of view, lightcurve inversion is of considerable interest as it is one of the rather few difficult inverse problems that now have rigorous uniqueness and stability results and a robust, well-converging practical numerical procedure.

2.1. *Fundamentals of the problem*

There are a few fundamental theorems on lightcurve inversion that we review here in a nutshell as they are the key to understanding the potential and limitations of photometric inversion. For details, see Kaasalainen *et al.* (2001, 2003) and Kaasalainen & Lamberg (2006) and references therein. With ω and ω_0 we denote the viewing and illumination directions on S^2 (unit sphere), and L is the measured brightness (intensity).

THEOREM 2.1 (UNIQUENESS THEOREM). *$L(\omega_0, \omega)$-data on $S^2 \times S^2$ uniquely determine the curvature of a convex surface. In other words, different viewing and illumination geometries break the Russell degeneracy $\omega = \omega_0$, i.e., the ambiguity in modelling a target from the areas of its simple projections.*

In fact, many asteroid observations such as photometric, radar, interferometric, or adaptive optics ones, are mathematically close cousins: they are essentially representations of generalized projection operators. This is why the inverse problems of these data modes are closely related and can be studied with a common approach (Kaasalainen & Lamberg 2006). Uniqueness theorems can also be shown for, e.g., radar analysis.

It can also be shown that the curvature function of a convex surface $\mathcal{B}$ uniquely determines its shape $\mathbf{x}_{\mathcal{B}}$ (up to a translation of $\mathbf{x}$), and that the shape construction converges (the so-called Minkowski problem, see Lamberg & Kaasalainen 2001).

COROLLARY 1. *$S^2 \times S^2$-data uniquely determine the shape of a convex surface: the mapping $L(\omega_0, \omega) \to \mathbf{x}_{\mathcal{B}}$ is unique.*

THEOREM 2.2 (STABILITY THEOREM). *The mapping $L(\omega_0, \omega) \to \mathbf{x}_{\mathcal{B}}$ is continuous for convex bodies in usual topologies (the inverse problem is conditionally well-posed in the sense of Tikhonov).*

We call this *Minkowski stability*; its role is now understood to be important in protecting the solution not only from random noise, but also from systematic errors in both data and modelling.

It should be noted here that the above theorems assume that the intrinsic darkness (albedo) of the surface is uniform. So far, probe data from asteroids have shown no striking albedo contrasts over large areas. Also, noticeable violation of the convexity condition can be used as an indicator of albedo asymmetry in inversion, and so far only a couple of asteroids have displayed moderate asymmetry in this sense (Kaasalainen *et al.* 2003). Significant albedo variegation without asymmetry would be physically rather implausible, so albedo uniformity is apparently quite well satisfied on global resolution level. Furthermore, Minkowski stability applies to *shape* determination, not to attributing brightness variations to albedo variegation on the surface, so the former is much safer than the latter. In other words, visible instability in the inferred albedo map is replaced by minor changes in the inferred shape (one can imagine a polyhedron modification where the facet areas are changed here and there, but the overall shape changes little due to the basic nature of the Minkowski problem).

There is a fundamental difference between the stability and reliability of convex and nonconvex modelling, and between the parameter spaces the two employ. The convex modelling is performed in the parameter space describing the Gaussian image of a shape, and this image is then transformed into shape information in radius space (Kaasalainen & Lamberg 2006 and references therein). Nonconvex inversion is performed in the radius space ($\mathbb{R}^3$) directly, which makes the whole process much more vulnerable and ambiguous. On conjectural level, it appears that the photometric $L(\omega_0, \omega)$ data of at least a starlike nonconvex body (at least when observed at solar phase angles α sufficiently high) determine its shape as robustly as with a convex body, but *only if* the scattering function is known accurately and the noise level is low (Kaasalainen *et al.* 2001; Ďurech & Kaasalainen 2003). In all such simulations we have carried out, a nonconvex model efficiently converges towards the correct minimum of χ^2 (with suitable regularization of radius variation), faithfully displaying the same features as the target even with model resolution (discretization of the problem) much lower than that of the simulation model. This seems to be a mathematically interesting extension of the uniqueness theorem for convex bodies. However, accurate photometry of real targets of known nonconvex shapes but with inadequately known (and modelled) scattering functions does not yield similar convergence (and the fit remains worse than with a convex model). This is an example of how taking the numerical conjecture directly as a model of reality would be an "inverse crime" (Kaipio & Somersalo 2005) leading to overoptimistic results: nonconvex photometric inversion is sensitive to (systematic) errors and the insufficiency of the scattering model, and in reality the scattering behaviour is never known well enough for this purpose. On the other hand, due to Minkowski stability, *convex* inversion is quite stable against the incorrectness of the scattering model (including slight albedo variegation) or other systematic errors.

We discuss the pole longitude ambiguity theorem with some detail here as the concepts are necessary in later discussion in the paper. Let $\mathbf{x} = (x, y, z)$ denote a vector in a coordinate system fixed to the target (i.e., rotating with it, z-axis aligned with the rotation axis), and $\mathbf{x}'$ a vector in a nonrotating system (denoted by primes) where the rotation vector points at the direction given by the spherical coordinates (θ', φ') (e.g., ecliptic or equatorial coordinates; rotation is in the positive direction around this vector, with period P). Then $\mathbf{x}'$ and $\mathbf{x}$ are related by

$$\mathbf{x} = \mathsf{R}\mathbf{x}', \tag{2.1}$$

where

$$\mathsf{R} = \mathsf{R}_z(\phi_0 + \Omega(t - t_0))\mathsf{R}_y(\theta')\mathsf{R}_z(\varphi'), \tag{2.2}$$

where t is the time, $\Omega = 2\pi/P$, ϕ_0 and the epoch t_0 are some initial values, and $\mathsf{R}_i(\phi)$ is the rotation matrix corresponding to the rotation of the coordinate frame through angle ϕ counterclockwise about the positive i-axis. In particular, $\mathsf{R}_z(\phi)$ is

$$\mathsf{R}_z(\phi) = \begin{pmatrix} \cos\phi & \sin\phi & 0 \\ -\sin\phi & \cos\phi & 0 \\ 0 & 0 & 1 \end{pmatrix}. \tag{2.3}$$

THEOREM 2.3 (AMBIGUITY THEOREM). *If the viewing and illumination directions ω' and ω_0' in a nonrotating frame remain in the same plane at all times of observation t_i, the infinite-distance observations of a body $\mathcal{B}$, with surface points $\mathbf{b} = (x, y, z)$ and rotation vector $\beta'(\theta', \varphi')$ given in a coordinate system whose $x'y'$-plane is the invariant plane, are indistinguishable from those of a body $\hat{\mathcal{B}}$ with $\hat{\mathbf{b}} = (x, y, -z)$ and $\hat{\beta}' = \beta'(\theta', \varphi' + \pi)$. We call this ecliptic degeneracy.*

Proof: We choose the invariant plane defined by ω' and ω_0' as the $x'y'$-plane of the nonrotating system, so the z'-coordinates of the viewer and the illumination source are zero. From (2.3) we have

$$\mathsf{R}_z(\varphi' + \pi)\mathbf{x}'|_{z'=0} = -\mathsf{R}_z(\varphi')\mathbf{x}'|_{z'=0}, \tag{2.4}$$

so (2.2) yields

$$\mathbf{x}(\varphi' + \pi, \mathbf{x}'|_{z'=0}) = -\mathbf{x}(\varphi', \mathbf{x}'|_{z'=0}). \tag{2.5}$$

Since ϕ_0 is arbitrary, we can set $\hat{\phi}_0 = \phi_0 + \pi$ (as $(x, y) \to (-x, -y)$ corresponds to a trivial shape rotation of π in the xy-plane). Therefore a vertical mirror-image shape $\hat{\mathcal{B}}$ ($z \to -z$) with a rotation direction changed by $\varphi' \to \varphi' + \pi$ has the same viewing and illumination directions with respect to the body shape as $\mathcal{B}$ and thus yields exactly the same observations as those of $\mathcal{B}$. □

This ambiguity property affects all data that are not two-dimensionally resolved in a plane projection, i.e., in this sense equivalent to observations made at infinity. Thus it appears also with radar data in addition to photometric observations. The coplanarity of ω' and ω_0' is often the approximate case as many asteroids move close to the plane of the ecliptic. For such targets, only observations with resolved plane-of-sky projections can properly remove the spin direction ambiguity.

Both Russell and ecliptic degeneracies break fast: asymmetric shapes can easily produce asymmetric lightcurves at solar phases less than $10°$ (i.e., asymmetry or nonzero first Fourier harmonic of the lightcurve does *not* imply albedo variegation), and ecliptic observation latitudes higher than $10 - 15°$ can already be sufficient for a unique pole solution (depending on the dataset).

2.2. Surface scattering properties

The main signature of surface scattering properties is the asteroid's phase curve, particularly its behaviour near zero phase angle. Proper definition of a phase curve necessarily includes pole and shape modelling, as then we can plot both the disk-integrated phase curve for the whole body (with equatorial viewing geometry for standard reference) and the phase curve for the surface material alone as if it were measured from a sandbox (Kaasalainen *et al.* 2001, 2004). If the size of the target is known, we obtain a scale factor for the latter curve, automatically defining the intrinsic darkness of the material (rather than use less well-defined concepts such as geometric albedo). Proper modelling also automatically produces, e.g., the "amplitude-phase relationship" which is thus actually a redundant concept. Often the phase curves are not very well described by the H-G system; usually it is better to take the phase curves as such (as they are readily reproduced from a limited set of parameters) and derive from them any particular values needed.

In this context, we should discuss the choice of the scattering model in some detail, rather than take Hapke's or other models for granted. From the general modelling point of view, we actually should not use a ready-made scattering model: instead, we should let it be constrained only by basic principles. Thus, the scattering model S should be of the general form $S(\mu, \mu_0, \alpha)$ (with the natural assumption that scattering from the surface material is invariant in the rotation of the surface patches in the tangent plane) where μ, μ_0 are the customary normal cosines of the emergent and incident rays. (Using α here is essentially equivalent to using the azimuthal angle between the two rays.) With some physical guidance, a general form for such a model is

$$S = f(\alpha)S_0 = f(\alpha)\frac{\mu\mu_0}{\mu + \mu_0}\sum_{ijk} b_{ijk}\mu^i \mu_0^j \cos(k\alpha), \tag{2.6}$$

with $b_{00k} \equiv 0$, $b_{ijk} \equiv b_{jik}$ due to ray reciprocality, and $\cos(k\alpha)$ due to symmetry.

The Lommel-Seeliger and Lambert combination typically used in lightcurve inversion (Kaasalainen *et al.* 2001) is thus equivalent to using only the lowest-order coefficients b_{000} (set to unity) and $b_{100} = b_{010}$. It does not have to have any particular physical justification: we just use a suitable truncated series of the general scattering form. As it happens, it already mimicks "physical" scattering models very well, as discussed in Kaasalainen *et al.* (2001). In particular, the separation of a phase function $f(\alpha)$ is physically consistent as by far the most of α-dependent variation of scattering is in this part. From the inversion point of view, this is enhanced by Minkowski stability: the result is not very dependent on the scattering model. Thus there is no practical necessity to carry the scattering series expansion further; indeed, many further coefficients would cause instability as the disk-integrated data do not contain proper information on them. However, it is interesting to note that just by including the next two coefficients, b_{001} and $b_{101} = b_{011}$, i.e.,

$$S_0 = \mu\mu_0 \left[\frac{1 + c_1 \cos\alpha}{\mu + \mu_0} + c(1 + c_2 \cos\alpha) \right], \tag{2.7}$$

we can already get a virtually exact match with the four-parameter Hapke model, and by adding the next few orders we get the same for the five-parameter rough-surface Hapke model (Helfenstein & Veverka 1989). This underlines the fact that any scattering behaviour can be expected to be modellable with a short scattering series, even when physical scattering models are not sufficient or are ambiguous. Disk-resolved data and other additional information can be used in determining the coefficients. Also, solving for b_{ijk} is essentially a linear problem, which is easier than, e.g., the determination of Hapke parameters.

2.3. *Systematic errors*

The scattering behaviour of asteroid surfaces can be studied accurately with in situ measurements; so far, only (433) Eros and possibly Itokawa have been studied extensively enough for this purpose. Li *et al.* (2004) derived a set of Hapke parameters for Eros from disk-resolved data. It is natural to ask whether the exactly known shape model (Konopliv *et al.* 2002), combined with the determined Hapke model, can reproduce the observed lightcurves (Lagerkvist *et al.* 2001) – in other words, can the whole of Eros' surface be described by a uniform scattering model? The answer is a conditional yes. Most lightcurves are fitted perfectly; however, some show clear discrepancies. Upon closer inspection, the discrepancies are due to systematic data errors. An example of this is shown in Fig. 1. The first and last lightcurves are fitted exactly, whereas the middle ones, essentially in the same geometries, are not. Thus the fit deviation is caused by persistent systematic errors in the data, even though the random noise level is low. This is a good example of the domination of systematic errors over random ones.

Another example of systematic (modelling) errors in inversion is presented by Ida and Gaspra. As discussed in Simonelli (1995, 1996), no single scattering model over the surface can be found that would fit the data with the known shape models (down to the noise level and without any systematic offset etc. effects). Scattering apparently varies over the surface. On the other hand, we get an excellent fit with our low-order scattering model and convex shape. This underlines both the role of Minkowski stability and the fact that local (nonconvex) shape details cannot be obtained from disk-integrated data. In short, systematic errors in both data and the model set a resolution limit to our modelling.

How much can we expect our result to be off due to insufficiently well modelled (or indeed not accurately modellable) scattering? While the effect of random noise is easy to study, unknown systematic effects are much harder to predict. Thus, it is important to search for ground truth whenever possible. In this respect, the laboratory study of Kaasalainen *et al.* (2005) was particularly useful as there we modelled a target with a decidedly incorrect scattering law. Even so, the shape obtained was surprisingly accurate, and the worst pole estimate differed by less than ten degrees of arc from the correct pole

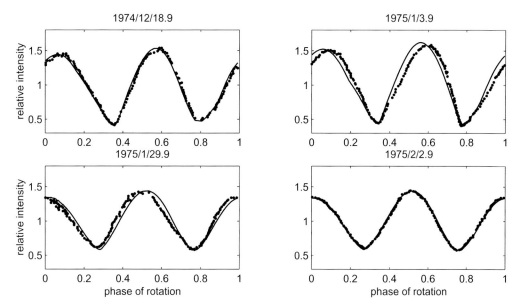

Figure 1. Examples of simulated lightcurves of the probe-based Eros model against observations: two lightcurves have systematic errors.

direction, while the best one was only a few degrees away. This seems to set an upper "worst case" limit – again Minkowski stability comes to our rescue. Accuracy from the ground truth by space probes is consistently good: the poles in Kaasalainen *et al.* (2001) were only a few degrees away from the in-situ-determined ones, while the pole for Itokawa (Kaasalainen *et al.* 2006) was only one degree away from the Hayabusa-determined one (Demura *et al.* 2006). Thus we can usually safely set a standard error of five or ten degrees of arc to pole estimates from lightcurve inversion.

Systematic errors can also muddle the details of asteroid models even with data sources that are expected to be more informative than photometry. For example, delay-Doppler radar techniques ostensibly provide an excellent chance of obtaining detailed shape information. However, ground truth from Itokawa shows that the shape inversion of the apparently detailed high signal-to-noise radar plots did not yield the most obvious global nonconvex feature ("otter's head") of the asteroid – an indication of the feature was visible, but its actual nature was not revealed (compare the figures in Ostro *et al.* 2005 and Demura *et al.* 2006). In this case this was mostly due to the very limited observing geometries, but nevertheless Minkowski stability in convex inversion is apparently a rather unique (and lucky) phenomenon as a guard against systematic errors in data and the model. With partially disk-resolving data sources such as radar, there is the danger of producing overdetailed models with features that are not actually there.

3. NEO observations and modelling

The foremost characteristic of NEOs is the quickly changing and wide-ranging observing geometry. The spin solution seldom has the ecliptic degeneracy as the observational ecliptic latitudes are often sufficiently high; just one or two apparitions may already offer suitably varying geometries for modelling, and the solar phase angle α reaches large values. Typical NEO photometric observations and their modelling are presented in Kaasalainen *et al.* (2004).

3.1. *The effect of solar phase angle*

The larger the phase angle α, the further away the target is from the Russell degeneracy $S^2 \times S^2 \to S^2$, i.e., the larger the shading effects, the more information there is on the shape (and spin). In principle, photometry at large phase angles obviously carries information on nonconvex shape features. In practice, this information is seldom obtainable as the phase angle should be unrealistically high, especially since the scattering properties are not modellable accurately enough.

It is possible to express a rough relationship between the size of nonconvex features of a body and the minimum solar phase angle α_{min} needed for the photometric detection of those features (Ďurech & Kaasalainen 2003). A convenient measure of the nonconvexity degree of a body is the dimensionless quantity of *nonconvexity measure* $\mathcal{V}_{nc}$, $0 \leqslant \mathcal{V}_{nc} < 1$ defined as

$$\mathcal{V}_{nc} = 1 - \frac{V}{V_{ch}}, \tag{3.1}$$

where V is the volume of the body and V_{ch} the volume of its convex hull. For example, for Gaspra $\mathcal{V}_{nc} = 0.05$, while for Castalia (radar), Eros, Ida, and Kleopatra (radar) the corresponding values are, respectively, 0.08, 0.15, 0.18, and 0.36. For two ellipsoids in contact with each other, $\mathcal{V}_{nc} = 0.2$, while for two separated ellipsoids with an ellipsoid-size gap $\mathcal{V}_{nc} = 0.5$.

If a set of lightcurves can be explained with a convex shape down to the noise level, this set obviously contains no information on nonconvexities; α_{min} is defined as the phase angle at which a convex model no longer fits the data as well as a nonconvex one. A large set of both synthetic and real shape models displays a clear correlation between α_{min} and $\mathcal{V}_{nc}$ (Ďurech & Kaasalainen 2003); we roughly have

$$\alpha_{min} = 120° - 220°\mathcal{V}_{nc}. \tag{3.2}$$

Usually only small asteroids have $\mathcal{V}_{nc} > 0.2$, so it is clear that very few MBAs can show any photometric information on nonconvexities. Even strongly nonconvex NEOs seldom have $\mathcal{V}_{nc} > 0.3$, so photometry cannot detect nonconvexities on most NEOs either.

Surface undulation gives more pronounced shadowing effects at larger phase angles, as is well known from, e.g., the role of the surface roughness parameter in the Hapke model. However, simple crater simulations (Kaasalainen *et al.* 2004) show that statistical cratering mostly yields the same correlation effect as α_{min} for separate nonconvex features. Only datasets of very densely cratered surfaces at high solar phase angles ($\alpha > 90°$) cannot be explained with a locally smooth surface (i.e., a matte-like surface containing only small-scale roughness included in the scattering law).

3.2. *Modelling considerations*

More complex dynamical behaviour than constant-period principal-axis rotation is found in some NEOs due to their small size. Any well-modellable dynamics is straightforward to analyze in the lightcurve inversion scheme: we just modify the rotation equation (2.2) suitably. Precessing motion (and its photometric distinguishability from binary motion) is discussed in Kaasalainen (2001, *et al.* 2003) and Pravec *et al.* (2005, 2007). YORP effect causes an essentially linear change in the angular speed Ω (for some range of time) due to thermal torque (Vokrouhlický *et al.* 2004, Bottke *et al.* 2006):

$$\Omega(t) = \Omega(t_0) + \upsilon\Delta t, \tag{3.3}$$

where the effect parameter is υ, and $\Delta t = t - t_0$. Thus the period change rate is

$$\dot{P} = -\frac{\upsilon}{2\pi}P^2 \qquad [\Rightarrow P(t) = (\frac{\upsilon}{2\pi}\Delta t + P(t_0)^{-1})^{-1}]. \tag{3.4}$$

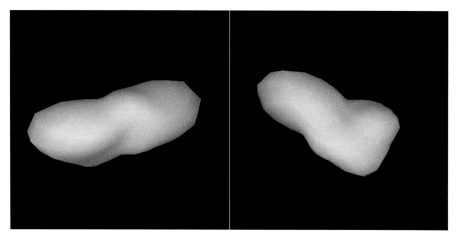

Figure 2. Offspring of Golevka and Eros (left), and Ida and Gaspra (right).

This causes a phase lag quadratic in time, described by just the one additional parameter υ that can be directly included in inversion. To be detected, the phase lag $\delta\phi$ due to YORP must be larger than the linear phase shift due to period uncertainty ΔP (estimated at an epoch earlier than the comparison time Δt) and the difference between YORP base period $P(t_0)$ and the best-fit constant period P_c:

$$\delta\phi = \frac{1}{2}|\upsilon|(\Delta t)^2 > \Delta t\left[\frac{2\pi}{P_c^2}\Delta P + \left|\Omega(t_0) - \frac{2\pi}{P_c}\right|\right]. \qquad (3.5)$$

The latter contribution to the linear shift occurs since the constant-period fit tries to compensate for YORP effect by finding a period slightly different from $P(t_0)$ such that the phase shifts are minimized over the timeline used for period determination. The phase lag as such is basically just a visual effect. The ultimate test and measurement of YORP effect is the inclusion of υ in (2.2) in inversion: if the resulting χ^2 is clearly better with a distinctly nonzero υ, the change in rotation is properly measurable.

As demonstrated by the colour map of Eunomia in Nathues *et al.* (2005), the use of various filters can provide information on the global variegation of surface material. Surveys such as Pan-STARRS will obtain data through a number of filters, so even sparse datasets should be sufficient for at least indicating colour variegation.

Since even very small NEOs are observable, the target shapes can vary considerably, as can be seen from the variety of models in Kaasalainen *et al.* (2004). In modelling synthetic datasets, this should be taken into account. Arbitrary synthetic shapes can be produced in many ways; here we introduce a rather natural means of obtaining quasiasteroidal shapes. Rather than use random-generated realizations of statistical or fractal measures for describing surface undulation and global shape variation, we can draw from the supply of existing real shape models and modify these. The results look usually more realistic than synthetic random models. For example, we can combine the shapes of two asteroids in a genetic fashion; with such asteroid breeding from an initial population, we can obtain as large and as varying an offspring population of models as desired. In Fig. 2 we show two such examples from pairing Eros and Golevka (Konopliv *et al.* 2002; Hudson *et al.* 2000) or Ida and Gaspra (Thomas *et al.* 1994, 1996). These were generated by expressing the original shapes (with the same volume and centroids at origin) as spherical harmonics series, and then picking coefficients for a new series by randomly switching between the two series. Coefficients can also be "mutated" randomly. Bodies generated in this manner present typical-looking features – indeed, one can breed a series of varying shapes with

some common main characteristics, if wanted. Morphological characteristics, such as the
"banana" shape of Eros, are easy to preserve in this way.

3.3. *Sparse photometry and surveys as future basis*

It is at first sight almost counterintuitive that sparse sequences are sufficient for asteroid
modelling as shown in Kaasalainen (2004) and Ďurech *et al.* (2006, 2007). After all, the
sampling interval is for most targets much longer than the sidereal period, so ordinary
time-series methods such as Fourier or power spectrum analysis (used for initial period
estimation of traditional lightcurves) are completely useless. The reason for the sufficiency
is that the underlying well-defined mathematical model is highly constraining: only a
certain type of an object can create a given sparse sequence as long as there are enough
points at various observing geometries. In fact, sparse sequences are handled just like the
ordinary lightcurve inversion problem: the mathematical model takes care of filling the
gaps. Now we just have to scan a wide range of potential periods as the rotation period
is not apparent in the data before the actual modelling.

The YORP effect can easily be included in the inversion just like for dense lightcurves.
For the latter, one can directly see the phase lag in plots; for sparse sets, there have
to be enough data points in different apparitions so that the lagged phases affect the
fit χ^2 properly, i.e., the inclusion of the YORP parameter υ significantly improves the
fit. One may also see an indication of YORP if the constant-period fit is good for one
time interval while the deviations grow larger further away from it. For certain targets,
υ should possibly be included at the outset as otherwise the effect might go unnoticed
(there could still be a clearly best period for the dataset, and the increased deviation
would be taken for noise).

A typical requirement for a sparse dataset is some 100 well-distributed data points over
five years for main-belters, while for NEAs even less is sufficient (Kaasalainen 2004). The
calibration accuracy should be at least around 0.05 mag, so the new surveys can indeed
meet the requirements. The surveys can make use of data at smaller solar elongations
than those typical for ordinary lightcurves since only one point is needed at a time;
thus the geometry coverage is wider (i.e., the observational gaps between apparitions are
narrower). This is why the observation geometry range of groundbased surveys is not
really very much smaller than that of satellite-based ones.

3.3.1. *Fast initial period scanning*

The sparse datasets have to be scanned for all potential period ranges; in certain cases
there are some indications of the range (fast or slow rotators; see Ďurech *et al.* 2007),
but basically the sparse data do not show any period signatures prior to modelling.
Doing full modelling for each possible period "slot" roughly separated by the interval
$\Delta P \approx (1/2)P^2/T$, where P is the trial period and T the timeline of the dataset, is
somewhat time-consuming. The period ranges can effectively be initially scanned by
using a simplified model to detect the most likely values or value ranges. Indeed, in
many cases even using just a brick to model the asteroid will highlight the best few
period locations. We have found that a practical model for initial scanning is simply
a geometrically scattering ellipsoid (Ostro and Connelly 1984) with linear-exponential
phase function $f(\alpha)$ (Kaasalainen *et al.* 2001). Thus the model reads (with an arbitrary
scaling factor absorbed in $f(\alpha)$ if desired)

$$L = f(\alpha)\left[\sqrt{\mathbf{e}^T\mathsf{M}\mathbf{e}} + \frac{\mathbf{e}^T\mathsf{M}\mathbf{s}}{\sqrt{\mathbf{s}^T\mathsf{M}\mathbf{s}}}\right], \tag{3.6}$$

where

$$f(\alpha) = A_0 \exp\left(-\frac{\alpha}{D}\right) + k\alpha + 1,$$ (3.7)

with A_0, D, k as free parameters, and

$$\mathsf{M} = \begin{pmatrix} 1/a^2 & 0 & 0 \\ 0 & 1/b^2 & 0 \\ 0 & 0 & 1/c^2 \end{pmatrix},$$ (3.8)

and

$$\mathbf{e} = \mathsf{R}\mathbf{e}', \quad \mathbf{s} = \mathsf{R}\mathbf{s}',$$ (3.9)

where $\mathbf{e}'$ and $\mathbf{s}'$ are, respectively, the unit vectors of the Earth's and the Sun's positions in an inertial coordinate system (equatorial or ecliptic), and R is given by (2.2). The model has only nine parameters: three for $f(\alpha)$, two for semiaxes a, b (setting $c = 1$), two for the pole direction, and P, ϕ_0 for period and initial rotational phase (of a-axis). These parameters are optimized very fast by the Levenberg-Marquardt algorithm (Press *et al.* 1990) as both the model and the gradients w.r.t. parameters are very simple to evaluate in the fully analytical form above. (Using a more realistic scattering model for the ellipsoid would be useless as then the numerical computation of the surface integral would take just as long as for the proper full model.) To ensure positive values for a, b as well as realistic shapes $(a, b > c = 1)$ it is useful to write the semiaxes in the form

$$a = 1 + \exp(p), \quad b = 1 + \exp(q),$$ (3.10)

where p, q are the parameters to be optimized. It is interesting to note that though the best fit from this initial model is, of course, much worse than that obtained with a full model, the fit for incorrect periods is also considerably worse than with a full model. In other words, the level of the typical chi-square "thicket", below which the best period must stand out, increases, making it possible for the good period(s) to show even if their actual fits (and the rest of the simplified model) are bad. Of course this does not apply to all targets, but the majority can be initially scanned in this manner (Ďurech *et al.* 2007).

4. Combined datasets

4.1. *Sparse and dense photometry*

Of course one cannot expect to get too much out of a handful of data points, especially if a number of these are noisier than expected (there are bound to be several outliers no matter what the engineers say). The models from sparse photometry are rough and in many cases (mildly) nonunique. This is where follow-up observations and observer networks come in. Even just one additional dense lightcurve would be of great help in at least the following cases:

1. *There are more than one possible periods fitting the sparse data.* This happens if the number of sparse data points is subcritical. The number is a complicated function of survey strategy and technical choices and can thus vary a lot. This may also happen with faint targets for which data are noisy, or with very spheroidal targets. Also, since the number of objects is so large, each target will only be given some standard computer time for analysis. Most of this time is spent in period scanning, and some targets will run out of the allocated pipeline time for period search as there will be both very fast and very slow rotators. Such targets will thus be flagged with "period not found" and saved

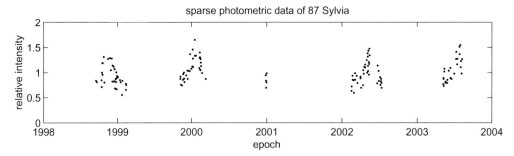

Figure 3. Sparse USNO dataset of (87) Sylvia from five apparitions (143 epochs). The observed intensity is reduced to unit distance from the Earth and the Sun.

for later analysis. Additional lightcurves will help to determine what the actual period region is.

2. *Even if the period is known, there may be more than one independent pole solutions.* This, again, happens with slightly too small sparse datasets. Here one should note that for objects moving close to the plane of the ecliptic, the ecliptic degeneracy cannot be removed by photometric means, regardless of the method.

3. *There are sparse data points that just cannot be fitted.* This usually means that the target is a binary (mutual events affect some data) or a tumbler (or otherwise somehow strange and thus interesting), or the points are just outliers. A dense lightcurve can help in clearing the matter. We expect several targets to be flagged for follow-up observations in this manner.

4. *Quality and reliability check.* Even if everything seems to be fine with the sparse data analysis and we get a full model, we must do random checks to make sure that essentially the same model pops out from the combined sparse and dense datasets. If the models are different, we have overlooked something.

5. *More detail needed.* If the object seems to be strangely shaped, we need more data points to get additional details.

The following example well portrays the power of combined databases, even when the survey data are very noisy. Even a limited additional dataset thus greatly boosts the value of the survey set. In this case, it actually makes the inversion possible in the first place as the sparse set alone is insufficient for modelling due to noise. The sparse set of asteroid 87 Sylvia (Fig. 3) was extracted from the astrometric database available at the Asteroids Dynamic Site (AstDys); the data were obtained at the U.S. Naval Observatory, Flagstaff. The dense lightcurves of the same asteroid in Fig. 4 essentially correspond to just one observation geometry, simulating a typical additional dataset for a target. These were chosen from the Uppsala Asteroid Photometric Catalogue (Lagerkvist *et al.* 2001). Lightcurve inversion of the full set of all available 40 lightcurves of Sylvia leads to the shape model shown in Fig. 5, left. The model derived from only the four lightcurves and sparse data (Fig. 5, right) is somewhat different as expected as the USNO data are obviously very noisy, more than 0.1 mag on average. Nevertheless, the model portrays many similar global characteristics. Above all, the pole directions of the two models are only seven degrees of arc apart and the rotation periods are the same. The inversion would not have been possible with either dataset alone.

There are currently almost 2000 asteroids for which one or more lightcurves have been observed – thus we can expect that many of these can be modelled after only a few years of survey photometry. For targets with limited datasets it will probably be useful to run cross-check analyses by using the dataset geometries and a synthetic model; this will give

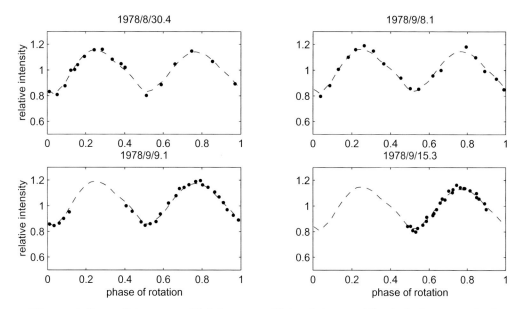

Figure 4. Dense lightcurves of Sylvia as an additional dataset. The dashed curve is the fit from the shape model.

Figure 5. Left: shape model (pole-on view) of 87 Sylvia from all dense lightcurves (Uppsala catalogue), right: the model from the noisy sparse+dense data. The spin solutions are essentially the same.

an indication of how much the result of the analysis might be off (even when the result appears to be unique and well-behaving).

4.2. *Multidatainversion with generalized projection operators*

Multidatainversion is usually based on photometry combined with any other data sources. Typically, photometry (sparse or dense) already provides a good estimate of the parameters, and the complementary source is employed to yield a detailed solution. Most often the improvement lies in revealing nonconvex features of the body or in removing the ecliptic degeneracy.

The general principle of multidatainversion is to form a joint goodness-of-fit χ^2_{tot} that combines the χ^2 from the main source with the χ^2_i of the complementary sources i, multiplied by suitable weights λ_i. Thus we have

$$\chi^2_{\text{tot}} = \chi^2 + \sum_i \lambda_i \chi^2_i. \qquad (4.1)$$

The weights are adjusted in minimizing χ^2_{tot} with the condition that each χ^2_i as well as χ^2 be acceptable. This condition usually leads to a certain degree of nonuniqueness in the solution as there may be several feasible sets of weights that fulfill the condition and

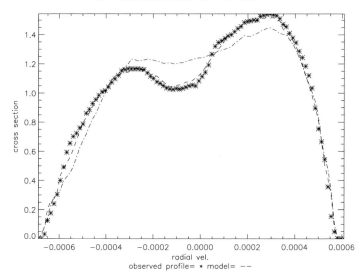

Figure 6. Doppler spectrum of the target (asterisks), with fits from convex model based on photometry alone (dot-dash) and the nonconvex model based on combined photometry and radar (dashed line).

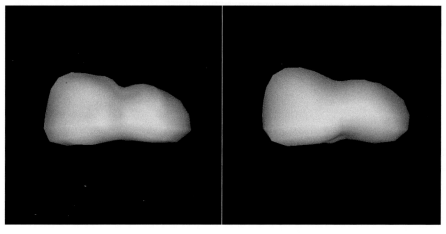

Figure 7. View of the target (left) and the nonconvex model (right) in the same direction in which the Doppler spectrum was computed.

lead to virtually equal values of $\chi^2_{\rm tot}$. Also, the exact values of the maximal allowed χ^2_i are usually not well defined in practice: these depend not only on the noise levels of the individual sources, but also on systematic effects such as the expected reliability of a source. Furthermore, insufficiency of the model affects fits for separate data sources differently. In practice, the multidatainversion results from generalized projection operators appear to be quite stable; one reason for this is the stability of the solution from the main source, photometry.

Here we demonstrate the important role of even small datasets of complementary information in multidatainversion. In Fig. 6 we show the simulated Doppler radar spectrum (in arbitrary units) of the target on the left in Fig. 7. The viewing direction ω is the same in both cases (note that the sign convention of the Doppler velocity makes the spectrum look like a "mirror image"). The convex inversion of simulated photometric data produces a model close to the convex hull of the target, giving an excellent fit to the data. Thus the data do not contain proper information on the sizable nonconvex features of the target. However, the convex model gives a poor fit to the observed Doppler spectrum (dot-dash line), so even one spectrum already contains significant information when added to other data. Using a spherical harmonics series for the radius of the (starlike) model and Levenberg-Marquardt algorithm to minimize $\chi^2_{\rm tot}$, we obtained the nonconvex model displayed on the right in Fig. 7. The generalized projection integrals were computed with a densely tessellated polyhedral surface representation. While still giving an excellent fit to photometric data, the model also fits the Doppler spectrum very well (dashed line), and well reproduces the main features of the target (with lower resolution: largest degree and order l, m for the function series are both 6).

As shown in Kaasalainen & Lamberg (2006), similar multidatainversion can be performed with, e.g., interferometric data. For NEOs, radar is the most important additional data source. With survey photometry analysis, we can expect to detect interesting targets (binaries, tumblers) that need radar (as well as dense lightcurve) observations for complete analysis. Such synergy will make all observations much more time-efficient than now. Another important additional data source is thermal infrared. Combining photometry-based models with IR, we can get very good estimates of the sizes and regolith properties of NEOs, as Hayabusa ground truth from Itokawa well confirms (Mueller *et al.* 2005, 2007).

5. Conclusions

Photometric datasets at well-covered observing geometries essentially allow us to image asteroids and to obtain their spin states (including tumbling, YORP effect, etc.) and scattering characteristics. Observations of NEOs offer widely ranging and quickly changing geometries, so data sufficient for modelling are usually obtained faster than for other asteroids. The future of photometric observations and analysis lies in sparse datasets obtained in surveys, and often also in their combination with other datasets. Systematic errors dominate over random noise, which should be remembered in the analysis and error estimation. The most time-consuming part of sparse data analysis is the period determination, but it can often be greatly speeded up with a simplified model for initial period scanning.

We will obtain thousands of good NEO models from surveys such as Pan-STARRS and LSST. With such numbers, the selection effects and biases are smaller than hitherto as the surveys simply record everything. Obtaining additional data (radar, dense lightcurves, etc.) will also be more time-efficient as we know what to observe and when, and even small amounts of additional data will improve sparse photometry considerably. This makes observer networks very important in coordinating databases and follow-up work.

Software packages, manuals and links for lightcurve inversion (and links for downloading) can be found at `www.rni.helsinki.fi/~mjk`. While the programs are straightforward to use, the user must have some knowledge of what they require and produce, and how to interpret the result. This paper and some of the references here should be sufficient for the purpose, but experience is the only good teacher. We hope that the open software will be developed further by various users.

Acknowledgements

We thank Alan Harris, Robert Jedicke, Anna Marciniak and Brian Warner for useful comments and discussions. This work was supported by the Academy of Finland.

References

Bottke, W.F., Vokrouhlický, D., Rubincam, D.P. & Nesvorný, D. 2006, *Annu. Rev. Earth Planet. Sci.* 34, 157

Demura, H. & 19 colleagues 2006, *Science* 312, 1347

Ďurech, J. & Kaasalainen, M. 2003, *A&A* 404, 709

Ďurech, J., Grav, T., Jedicke, R., Kaasalainen, M. & Denneau, L. 2006, *Earth, Moon & Planets*, in press

Ďurech, J., Scheirich, P., Kaasalainen, M., Grav, T., Jedicke R. & Denneau, L. 2007, this volume

Helfenstein, P. & Veverka, J 1989, in: R.P. Binzel *et al.* (eds.), *Asteroids II* (Tucson: University of Arizona Press), 557

Hudson, R. & 26 colleagues 2000, *Icarus* 148, 37

Kaasalainen, M., Torppa, J. & Muinonen, K. 2001, *Icarus* 153, 37

Kaasalainen, M. 2001, *A&A* 376, 302

Kaasalainen, M., Mottola, S. & Fulchignoni, M. 2003, in: W.F. Bottke, A. Cellino, P. Paolicchi & R.P. Binzel (eds.), *Asteroids III* (Tucson: University of Arizona Press), p. 139

Kaasalainen, M. & 21 colleagues 2004, *Icarus* 167, 178

Kaasalainen, M. 2004, *A&A* 422, L39

Kaasalainen, M. & Lamberg, L. 2006, *Inverse Problems* 22, 749

Kaasalainen, M. & 20 colleagues 2006, Proceedings of the 1st Hayabusa Symposium, in press

Kaasalainen, S., Kaasalainen, M. & Piironen, J. 2005, *A&A* 440, 1177

Kaipio, J. & Somersalo, E. 2005, *Statistical and computational inverse problems*, Springer, New York

Konopliv, A., Miller, J., Owen, W., Yeomans, D., Giorgini, J., Garnier, R. & Barriot, J. 2002, *Icarus* 160,289

Lagerkvist, C.-I., Piironen, J. & Erikson, A. 2001, Asteroid Photometric Catalogue, 5th update. Uppsala Univ. Press, Uppsala.

Lamberg, L. & Kaasalainen, M. 2001, *J. Comp. Appl. Math* 137, 213

Li, J., A'Hearn, M. & McFadden, L. 2004, *Icarus* 172, 415

Marchis, F., Kaasalainen, M., Hom, E., Berthier, J., Enriquez, J., Hestroffer, D., Le Mignant, D. & de Pater, I. 2006, *Icarus* 185, 39

Mueller, T., Sekiguchi, T., Kaasalainen, M., Abe, M. & Hasegawa, S. 2005, *A&A*, 443, 347

Mueller, T., Sekiguchi, T., Kaasalainen, M., Abe, M. & Hasegawa, S. 2007, this volume

Nathues, A., Mottola, S., Kaasalainen, M. & Neukum, G. 2005, *Icarus* 175, 452

Ostro, S.J. & Connelly, D. 1984, *Icarus* 57, 443

Ostro, S.J. & 12 colleagues 2005, *Meteor. Planet. Sci.* 40, 1563

Pravec, P., Harris, A., Scheirich, P. *et al.* 2005, *Icarus* 173, 108

Pravec, P., Harris, A. & Warner, B. 2007, this volume

Press, W., Flannery, B., Teukolsky, S. & Vetterling, W. 1990, *Numerical Recipes*, Cambridge University Press

Simonelli, D., Veverka, P., Thomas, P. & Helfenstein P. 1995, *Icarus* 114, 387

Simonelli, D., Veverka, P., Thomas, P., Helfenstein, P. & Carcich B. 1996, *Icarus* 120, 38

Storrs, A., Vilas, F., Landis, R., Wells, E., Woods, C., Zellner, B. & Gaffey, M. 1999, *Icarus* 137, 260.

Thomas, P., Veverka, J., Simonelli, D., Helfenstein, P.,Carcich, B., Belton, M., Davies, M. & Chapman, C. 1994, *Icarus* 107, 23

Thomas, P., Belton, M., Carcich, B, Chapman, C., Davies, M., Sullivan, R. & Veverka, J. 1996, *Icarus* 120, 20

Vokrouhlický, D., Čapek, D., Kaasalainen, M. & Ostro, S.J. 2004, *A&A* 414, L21

Near Earth Objects, our Celestial Neighbors: Opportunity and Risk
Proceedings IAU Symposium No. 236, 2006
A. Milani, G.B. Valsecchi & D. Vokrouhlický, eds.
© 2007 International Astronomical Union
doi:10.1017/S1743921307003201

NEA rotations and binaries

Petr Pravec[1], A. W. Harris[2] and B. D. Warner[3]

[1]Astronomical Institute, Academy of Sciences of the Czech Republic, Fričova 1, CZ-25165
Ondřejov, Czech Republic

[2]Space Science Institute, 4603 Orange Knoll Ave., La Canada, CA 91011, USA

[3]Palmer Divide Observatory, 17995 Bakers Farm Rd. Colorado Springs, CO 80908, USA

Abstract. Of the nearly 3900 near-Earth asteroids (NEAs) known as of June 2006, 325 have estimated rotation periods, with most of those determined by lightcurve analysis led by a few dedicated programs. NEAs with diameters down to 10 meters have been sampled. Observed spin distribution shows a major changing point around diameter of 200 meters. Larger NEAs show a barrier against spins faster than 11 d^{-1} (period about 2.2 h) that shifts to slower rates (longer periods) with increasing lightcurve amplitude (i.e., with increasing equatorial elongation). The spin barrier is interpreted as a critical spin rate for bodies in a gravity regime; NEAs larger than 200 meters are predominantly bodies with tensile strength too low to withstand a centrifugal acceleration for rotation faster than the critical spin rate. The cohesionless spin barrier disappears at sizes less than 200 meters where most objects rotate too fast to be held together by self-gravitation only, so a cohesion is implied in the smaller NEAs.

The distribution of NEA spin rates in the cohesionless size range ($D > 0.2$ km) is highly non-Maxwellian, suggesting that mechanisms other than just collisions have been at work. There is a pile up just in front of the barrier, at periods 2–3 h. It may be related to a spin up mechanism crowding asteroids to the barrier. An excess of slow rotators is observed at periods longer than 30 hours. A spin-down mechanism has no obvious lower limit on spin rate; periods as long as tens of days have been observed.

Most NEAs appear to be in their basic spin states with rotation around principal axis with maximum moment of inertia. Tumbling objects (i.e., bodies in excited, non-principal axis rotation) are present and actually predominate among slow rotators with estimated damping timescales longer than the age of the solar system. A few tumblers observed among fast rotating coherent objects appear to be either more rigid or younger than the larger (cohesionless) tumblers.

An abundant population of binary systems has been found among NEAs. The fraction of binaries among NEAs larger than 0.3 km has been estimated to be $15 \pm 4\%$. Primaries of binary systems concentrate at fast spin rates (periods 2–3 h) and low amplitudes, i.e., they lie just below the cohesionless spin barrier. The total angular momentum content in binary systems suggests that they formed from parent bodies spinning at the critical rate. The fact that a very similar population of binaries has been found among small main belt asteroids suggests a binary formation mechanism that may not be related to close encounters with the terrestrial planets.

Keywords. asteroid, rotation; binary asteroids; tumbling; asteroid, lightcurves

---

1. Introduction

During the last dozen years our data set on rotations of near-Earth asteroids (NEAs) has increased enormously. Most of the data have been obtained by a few dedicated programs (see, e.g., Pravec *et al.* 1998; Mottola *et al.* 1995a; Krugly *et al.* 2002) that placed a high priority within their observational strategies on suppressing selection effects against slow rotators as well as low amplitude objects, and on resolving complex lightcurves of tumblers and binaries among NEAs. Radar observations contributed to the rotation data as well, and they resolved more than half of the NEA binaries known to date (see Ostro *et al.* 2006).

Of the nearly 3900 near-Earth asteroids (NEAs) known as of June 2006, 325 have estimated rotation periods, 14 tumblers have been identified, and 30 binary systems have been found. In this paper, we present an overview of a few of the things we learned from the data.

2. Data Set

The principal method of asteroid period estimation is rotational lightcurve photometry. By using the harmonic series analysis proposed by Harris *et al.* (1989), period estimation from dense lightcurve data is mostly straightforward. There are selection effects against low amplitude and long period objects with the lightcurve technique, but they have been largely suppressed by the observational strategies of the dedicated NEA photometry programs, which allocated telescope time when and as needed to resolve more difficult cases. This led not only to suppressing the bias against low amplitude/long period NEAs, but also to resolving complex lightcurves of tumbling asteroids and binary systems, which show more than a single period.

Though this paper deals with near-Earth asteroids, we point out that so far there has not been found any significant difference between parameters of near-Earth asteroids and those of more distant asteroids (main belt, Mars-crossers) other than that would be attributable to a size dependence in a given parameter. It should be noted that the sample of spin rates of main-belt/Mars-crossing (MB/MC) asteroids is abundant only at sizes larger than 3 km, so there is actually little overlap between the NEA and the MB/MC samples in size; spin rates data above 3 km basically refer to MB/MC asteroids, while those below 3 km are mostly of NEAs. Extending the sample of MB/MC asteroid spin rates to km-sized bodies will be needed to study possible differences between them and the NEA population.†

3. Spin barrier

Asteroids with sizes from a few hundred meters up to about 10 km show a barrier against spins faster than f about 11 d^{-1} (period about 2.2 h), see Fig. 1. The limit shifts to slower rates (longer periods) with increasing lightcurve amplitude (i.e., with increasing equatorial elongation). The dependence of the spin limit on equatorial elongation is shown in Fig. 2, where limiting curves for cohesionless elastic-plastic solid bodies with the angle of friction $\phi = 90°$ and with bulk densities 1, 2, 3, 4, 5 g/cm^3 are plotted. The angle of friction in real asteroids is unknown, but it is expected to be on an order of 40° (Richardson *et al.* 2005). Considering that Holsapple (2001, 2004) calculated that the critical spin frequency for $\phi = 40°$ is about 10% lower than that for $\phi = 90°$ and that amplitudes of a few asteroids close to the spin barrier were measured at higher solar phases so they probably need to be corrected to lower values to represent the equatorial axes ratio, we get that 99% of measured NEAs larger than 0.2 km rotate slower than

---

† Dermawan (2004) has made a first attempt to obtain a sample of spin rates of main belt asteroids with sizes comparable to NEAs, using the Subaru telescope with a wide field imaging system. He reports a significant fraction of super-fast rotators (periods under 2 hours) among MBAs extending up to sizes larger than 1 km. If true, this would mark a provocative departure from the properties of NEAs. However, upon examining the lightcurves presented in the Dermawan thesis, we find the results questionable due to insufficient observational coverage, and having to press too close to the intrinsic noise level of the observations seeking periods.

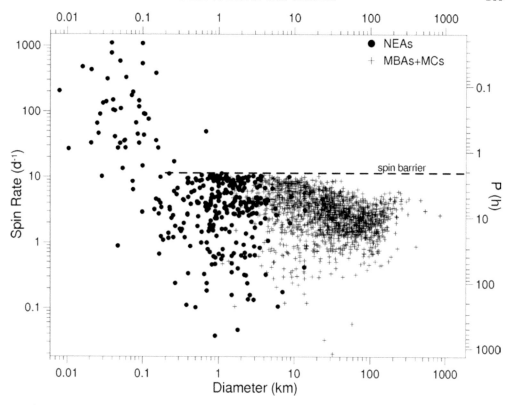

Figure 1. The spin barrier in spin rate (f) vs diameter (D) apparent at sizes from a few hundred meters to about 10 km.

the limit for bulk density of 3 g/cm^3 (data compiled by Harris *et al.* 2006). See Harris (1996) and Pravec & Harris (2000) for earlier data on the spin limit.

The spin barrier is interpreted as a critical spin limit for bodies in a gravity regime; NEAs larger than 0.2 km are predominantly bodies with tensile strength too low to withstand a centrifugal acceleration for rotation faster than the critical spin rate.

Above $D = 3$ km, an upper limit on the tensile strength given by the spin barrier is higher than a scaled tensile strength of cracked but coherent rocks, so the existence of the spin barrier does not constrain whether asteroids in the size range 3–10 km are strengthless objects or just cracked but coherent bodies. Below $D = 3$ km, the maximum possible tensile strength allowed by the spin barrier for a majority of asteroids in the size range is too low for them to be cracked but coherent bodies; this implies that a cohesionless structure is predominant among asteroids with $D = 0.2$ to 3 km (Holsapple 2006).

The cohesionless spin barrier disappears at sizes less than 200 meters where most objects rotate too fast to be held together by self-gravitation only, so a cohesion is implied in the smaller NEAs.

The distribution of NEA spin rates in the cohesionless size range ($D > 0.2$ km) is highly non-Maxwellian, see Fig. 3. It suggests that mechanisms other than just collisions were involved. There is a pile up just in front of the barrier, at spin rates 9–10 d^{-1}(periods 2–3 h). It may be related to a spin up mechanism crowding asteroids to the barrier. An excess of slow rotators is observed at periods longer than 30 hours. A spin-down mechanism has no obvious lower limit on spin rate; periods as long as tens of days have

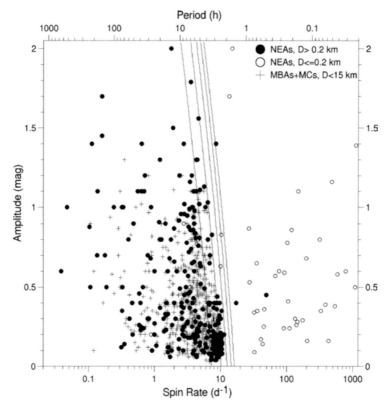

Figure 2. The spin barrier in amplitude (A) vs spin rate (f). The curves are limits for cohesionless elastic-plastic solid bodies with the angle of friction $\phi = 90°$ and with bulk densities 1, 2, 3, 4, 5 g/cm^3, from left to right.

been observed. The YORP effect appears to be a qualitatively consistent explanation (see Bottke *et al.* 2002, 2006).

4. Non-principal axis rotators

The first detection of an asteroid in non-principal axis rotation state, near-Earth asteroid (4179) Toutatis, was made with radar (Hudson & Ostro 1995). Since then, a couple more NPA rotators have been found also by using radar (see Ostro *et al.* 2006).

The lightcurve photometry technique has provided data on more NPA rotators (tumblers) among near-Earth asteroids. Several NEAs showed deviations from single periodicity attributable to NPA rotation (Pravec *et al.* 2005; a few latest detections pre-published on Pravec's web page†). In a few cases where abundant data have been obtained, a fit with 2-dimensional Fourier series indicates two basic periods plus their linear combinations (Pravec *et al.* 2005; see also Kaasalainen 2001).

In Fig. 4, the f–D data with tumblers highlighted are plotted. All but the largest are near-Earth asteroids. (The largest, at $D = 58$ km, is the main belt asteroid (253) Mathilde, which was found to be in NPA rotation state by Mottola *et al.* (1995b).) From the plot, it is apparent that tumblers larger than a few hundred meters are generally slow rotators, while three fast rotating tumblers were found in the size range from 10 to a few hundred meters.

† http://www.asu.cas.cz/~ppravec/newres.htm

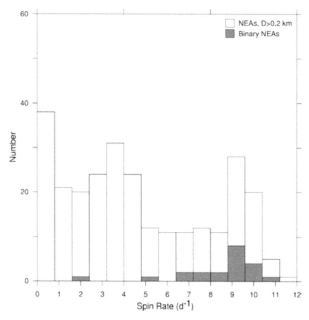

Figure 3. Excess at spin rates $f < 0.8$ d^{-1} (slow rotators) and $f = 9$–10 d^{-1} (pile up below the spin barrier). Binary primaries concentrate in the pile up.

An intepretation of the distribution of tumblers in the f–D parameter space uses estimated damping time scales based on the theory by Burns & Safronov (1973) and with "rubble pile" parameters estimated by Harris (1994). Tumblers are predominant among asteroids larger than a few hundred meters and with a damping time scale longer than the age of the solar system; most asteroids in the range for which abundant data have been obtained show NPA rotation. The small fast rotating tumblers found among coherent objects (that lie above the spin barrier) are more rigid or younger than the larger (cohesionless) tumblers (Pravec *et al.* 2005).

5. Binaries

As of mid-2006, 30 binary systems had been found among near-Earth asteroids. Twenty of them were resolved with radar observations (see Ostro *et al.* 2006), and 15 were resolved with the photometric technique (see Pravec *et al.* 2006; and two new ones by Reddy *et al.* 2005, 2006a,b); five were detected by both techniques.

The photometric technique of asynchronous binary detection, described in Pravec *et al.* (2006), is based on deconvolution of a lightcurve of the binary asteroid where (at least) one of its components rotates with a period different from orbital period. For a full, regular detection of the binary system, it has to show mutual events –occultations and/or eclipses– among the components of the binary system. From such data, the rotation period of the primary as well as orbital period together with a size ratio, or its lower limit in a case of partial events, are directly derived. A unique resolution of whether a rotational lightcurve component belongs to the primary is routinely done using the fact that the primary's rotational variation does not go away during mutual events while the secondary's variation, detected in cases where the amplitude is apparent even if diluted by the light of the primary, disappears when the smaller body is fully hidden behind the larger body.

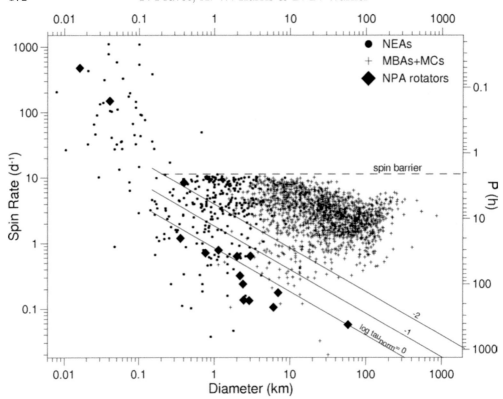

Figure 4. Tumbling asteroids in spin rate (f) vs diameter (D). Tumblers predominate below the line of constant damping times scale of 4.5 byr ($\log \tau_{norm} = 0$) at sizes larger than a few hundred meters; see text.

Pravec *et al.* (2006) simulated their binary NEA photometric survey and they estimated that $15 \pm 4\%$ of near-Earth asteroids larger than 0.3 km are binary systems with a secondary-to-primary mean diameter ratio $D_s/D_p \geqslant 0.18$. They found that the concentration of binaries with $D_s/D_p \geqslant 0.18$ is particularly high among NEAs smaller than 2 km in diameter, and that the abundance of such binaries decreases significantly among larger NEAs. Secondaries show an upper size limit of $D_s = 0.5$–1 km. Systems with $D_s/D_p < 0.5$ are abundant, but larger satellites are significantly less common.

The primaries of NEA binaries are mostly fast rotators with low equatorial elongations, most of them lying in the pile up in front of the spin barrier (see Figs. 3, 5, 6). The distribution of their rotation periods is concentrated between 2.2 and 2.8 h and has a tail up to ~ 4 h. Orbital periods show an apparent cut-off at $P_{orb} \sim 11$ h; closer systems with shorter orbital periods have not been observed, which is consistent with the Roche limit for strengthless bodies. Secondaries are more elongated on average than primaries. Most, but not all, of their rotations appear to be synchronized with the orbital motion; non-synchronous secondary rotations may occur especially among wider systems with $P_{orb} > 20$ h.

A population of asynchronous binary asteroids among main belt asteroids (MBAs) smaller than 10 km was found recently (see Pravec *et al.* 2006, and references therein; examples of recently detected ones see, e.g., Warner *et al.* 2005, 2006; Pray *et al.* 2006a,b; Higgins *et al.* 2006a,b; Jakubík *et al.* 2005; Cooney *et al.* 2006). The asynchronous MBA

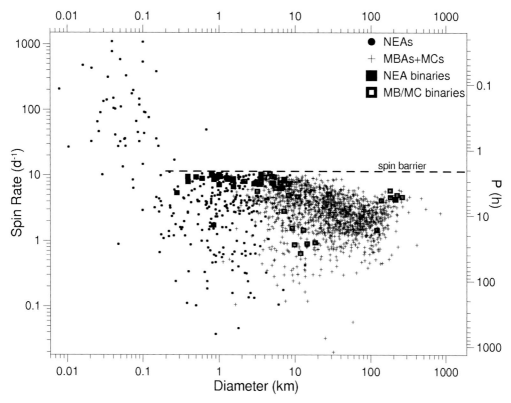

Figure 5. Primaries of binary NEAs in spin rate (f) vs diameter (D). Primaries of binary systems among MB/MC asteroids are also plotted, for comparison.

binaries are similar to the NEA binaries in most characteristics. The only prominent difference is that, unlike NEA binaries which concentrate at sizes $D_p < 2$ km ($D_s < 1$ km), the asynchronous main belt binaries extend up to nearly 10 km in D_p (their satellites are up to 3 km in D_s). Some smaller differences in other parameters appear to be a size dependence only (see below).

In addition to the asynchronous binaries population among NEAs as well as small MBAs, there is also a smaller population of fully synchronous, nearly equal sized binaries (D_s/D_p nearly 1). Such systems appear to be infrequent among NEAs, with only one such system having been found so far, (69230) Hermes (Margot *et al.* 2006; see also Pravec *et al.* 2006, and reference therein). Among small MBAs, a few such systems with sizes around 10 km have been found by Behrend *et al.* (2006) and Kryszczynska *et al.* (2005). The abundance (fraction) of fully synchronous, nearly equal-sized binaries among small MBAs has not yet been precisely estimated. It seems, however, that they are abundant only in a narrow size range just around the diameter of 10 km; see the "tail" of the distribution of primary spins to frequencies around 1 d^{-1} around $D = 10$ km in Fig. 5.

Recently we began a study of overall characteristics of the few known populations of small binaries (both synchronous and asynchronous) among NEAs as well as MBAs. Among the main underlying questions of the study are what all the systems have in common, and whether some apparent differences might be due only to a size dependence of formation/evolution mechanisms.

The first thing that we examined (Pravec & Harris, in preparation) is the angular momentum content in binary systems. We found that all the small binaries (both

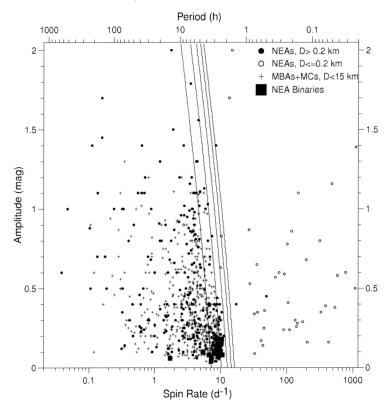

Figure 6. Primaries of binary NEAs in amplitude (A) vs spin rate (f). See also caption to Fig. 2.

synchronous and asynchronous, from NEAs to MBAs) have a total angular momentum very close to, but not generally exceeding, the critical limit for a single body in a gravity regime. It suggests that asteroid binaries with D_p about 10 km and smaller formed from parent bodies spinning at the critical rate (at the gravity spin limit for asteroids in the size range). Some small differences between characteristics of MBA and NEA binaries may be due to a size dependence of formation/evolution mechanisms. A suggested explanation of the apparent tendency to slower primary rotations and longer orbital periods with increasing size is that larger systems may be more tidally evolved. Over-all, known binaries among NEAs to main belt asteroids have characteristics so similar when corrected for effects of size depedence that they may be a part of a common binary population in which the same mechanism is related to the critical spins of their parent bodies.

6. Conclusions

Rotations and binary properties suggest that NEAs larger than 200 m are predominantly cohesionless structures held together by self-gravitation. Superfast rotations of most smaller asteroids indicate that they are held together by some cohesive forces.

Binary properties suggest that they originated from critically spinning cohesionless bodies. A similar binary population has been observed among small MB/MC asteroids;

it suggests that a binary formation mechanism may not be related to encounters with the major planets.

Acknowledgements

The work at Ondřejov was supported by the Grant Agency of the Czech Republic, Grant 205/05/0604. The work at Space Science Institute was supported by grant NAG5-13244 from the NASA Planetary Geology-Geophysics Program.

References

Behrend, R., Bernasconi, L., Roy, R., Klotz, A., Colas, F., Antonini, P., Aoun, R., Augustesen, K., Barbotin, E., Berger, N., Berrouachdi, H., Brochard, E., Cazenave, A., Cavadore, C., Coloma, J., Cotrez, V., Deconihout, S., Demeautis, C., Dorseuil, J., Dubos, G., Durkee, R., Frappa, E., Hormuth, F., Itkonen, T., Jacques, C., Kurtze, L., Laffont, A., Lavayssière, M., Lecacheux, J., Leroy, A., Manzini, F., Masi, G., Matter, D., Michelsen, R., Nomen, J., Oksanen, A., Pääkkönen, P., Peyrot, A., Pimentel, E., Pray, D., Rinner, C., Sanchez, S., Sonnenberg, K., Sposetti, S., Starkey, D., Stoss, R., Teng, J.-P., Vignand, M. & Waelchli, N. 2006, *Astron. Astrophys.* 446, 1177

Bottke, W.F. Jr., Vokrouhlický, D., Rubincam, D.P. & Brož, M. 2002, in: W.F.Bottke Jr., A. Cellino, P. Paolicchi & R.P. Binzel (eds.), *Asteroids III*, (Tucson: University of Arizona Press), p. 395

Bottke, W.F. Jr., Vokrouhlický, D., Rubincam, D.P. & Nesvorný, D. 2006, *Annu. Rev. Earth Planet. Sci.* 34, 157

Burns, J.A. & Safronov, V.S. 1973, *Mon. Not. Roy. Astron. Soc.* 165, 403

Cooney, W., Gross, J., Terrell, D., Pravec, P., Kusnirak, P., Pray, D., Krugly, Yu., Kornos, L., Vilagi, J., Gajdos, S., Galad, A., Reddy, V., Dyvig, R., Nudds, S., Kanuchova, Z., Pikler, M. & Husarik, M. 2006, *CBET* 504

Dermawan, B. 2004, PhD Thesis, School of Science, Univ. of Tokyo

Harris, A.W. 1994, *Icarus* 107, 209

Harris, A.W. 1996, *Proc. Lunar Planet Sci. Conf.* 27, 493

Harris, A.W., Young, J.W., Bowell, E., Martin, L.J., Millis, R.L., Poutanen, M., Scaltriti, F., Zappalà, V., Schober, H.J., Debehogne, H. & Zeigler, K.W. 1989, *Icarus* 77, 171

Higgins, D., Pravec, P., Kušnirák, P., Šarounová, L., Gajdoš, Š., Galád, A. & Világi, J. 2006a, *CBET* 389

Higgins, D., Pravec, P., Kušnirák, P., Cooney, W., Gross, J., Terrell, D. & Stephens, R. 2006b, *CBET* 507

Holsapple, K.A. 2001, *Icarus* 154, 432

Holsapple, K.A. 2004, *Icarus* 172, 272

Holsapple, K.A. 2006, *Icarus*, in press

Hudson, R.S. & Ostro, S.J. 1995, *Science* 270, 84

Kryszczynska, A., Kwiatkowski, T., Hirsch, R., Polinska, M., Kaminski, K. & Marciniak, A. 2005, *CBET* 239

Jakubík, M., Husárik, M., Világi, J., Gajdoš, Š., Galád, A., Pravec, P., Kušnirak, P., Cooney, W., Gross, J., Terrell, D., Pray, D. & Stephens, R. 2005, *CBET* 270

Kaasalainen, M. 2001, *Astron. Astrophys.* 376, 302

Krugly, Yu.N., Belskaya I.N., Shevchenko V.G., Chiorny V.G., Velichko F.P., Erikson A., Mottola S., Hahn G., Nathues A., Neukum G., Gaftonyuk N.M. & Dotto E. 2002, *Icarus* 158, 294

Margot, J.-L., *et al.* 2006, *This proceedings*

Mottola, S., de Angelis, G., di Martino, M., Erikson, A., Hahn, G. & Neukum, G. 1995a, *Icarus* 117, 62

Mottola, S., Sears, W.D., Erikson, A., Harris, A.W., Young, J.W., Hahn, G., Dahlgren, M., Mueller, B.E.A., Owen, B., Gil-Hutton, R., Licandro, J., Barucci, M.A., Angeli, C., Neukum, G., Lagerkvist, C.-I. & Lahulla, J.F. 1995b, *Planet. Space Sci.* 43, 1609

Ostro, S.J., Giorgini, J. & Benner, L.A.M. 2006, *This proceedings*

Pravec, P. & Harris, A.W. 2000, *Icarus* 148, 12

Pravec, P., Wolf, M. & Šarounová, L. 1998, *Icarus* 136, 124

Pravec, P., Harris, A.W., Scheirich, P., Kušnirák, P., Šarounová, L., Hergenrother, C.W., Mottola, S., Hicks, M.D., Masi, G., Krugly, Yu.N., Shevchenko, V.G., Nolan, M.C., Howell, E.S., Kaasalainen, M., Galád, A., Brown, P., Degraff, D.R., Lambert, J.V., Cooney, W.R. & Foglia, S. 2005, *Icarus* 173, 108

Pravec, P., Scheirich, P., Kušnirák, P., Šarounová, L., Mottola, S., Hahn, G., Brown, P., Esquerdo, G., Kaiser, N., Krzeminski, Z., Pray, D.P., Warner, B.D., Harris, A.W., Nolan, M.C., Howell, E.S., Benner, L.A.M., Margot, J.-L., Galád, A., Holliday, W., Hicks, M.D., Krugly, Yu.N., Tholen, D., Whiteley, R., Marchis, F., Degraff, D.R., Grauer, A., Larson, S., Velichko, F.P., Cooney, W.R., Stephens, R., Zhu, J., Kirsch, K., Dyvig, R., Snyder, L., Reddy, V., Moore, S., Gajdoš, Š., Világi, J., Masi, G., Higgins, D., Funkhouser, G., Knight, B., Slivan, S., Behrend, R., Grenon, M., Burki, G., Roy, R., Demeautis, C., Matter, D., Waelchli, N., Revaz, Y., Klotz, A., Rieugné, M., Thierry, P., Cotrez, V., Brunetto, L. & Kober, G. 2006, *Icarus* 181, 63

Pray, D., Pravec, P., Kušnirák, P., Cooney, W., Gross, J., Terrell, D., Galád, A., Gajdoš, Š., Világi, J. & Durkee, R. 2006a, *CBET* 353

Pray, D., Pravec, P., Pikler, M., Husárik, M., Stephens, R., Masi, G., Durkee, R. & Goncalves, R. 2006b, *CBET* 617

Reddy, V., Dyvig, R., Pravec. P. & Kušnirák, P. 2005, *IAU Circ.* 8483

Reddy, V., Dyvig, R., Pravec. P., Kušnirák, P., Gajdoš Š., Galád A. & Kornoš L. 2006a, *CBET* 384

Reddy, V., Dyvig, R.R., Pravec, P., Kušnirák, P., Kornoš, L., Világi, J., Galád, A., Gajdoš, Š., Pray, D.P., Benner, L.A.M., Nolan, M.C., Giorgini, J.D., Ostro, S.J. & Abell, P.A. 2006b, *37nd Annual Lunar and Planetary Science Conference* March 13-17, 2006, League City, Texas, abstract no. 1755

Richardson, D.C., Elankumaran, P. & Sanderson, R.E. 2005, *Icarus* 173, 349

Warner, B.D., Pravec, P., Harris, A.W., Galad, A., Kušnirák, P., Pray, D.P., Brown, P. Krzeminski, Z., Cooney Jr., W.R., Higgins, D., Masi, G., Gross, J., Terrell, D., Reddy, V., Dyvig, R., Behrend, R., Strajnic, J., Manzini, F., Revaz, Y., Ravonel, M. & Hoffmann, T. 2005, *In: Asteroids, Comets, Meteors 2005, IAU Symp. 229* Abstract No. 10.14

Warner, B.D., Pray, D.P., Pravec, P., Kušnirák, P., Cooney, W., Jr., Gross, J. & Terrell, D. 2006, *Minor Planet Bull.* 33, 57

Near Earth Objects, our Celestial Neighbors: Opportunity and Risk
Proceedings IAU Symposium No. 236, 2006
A. Milani, G.B. Valsecchi & D. Vokrouhlický, eds.
© 2007 International Astronomical Union
doi:10.1017/S1743921307003213

The Dynamics of NEO Binary Asteroids

D.J. Scheeres

Dept. of Aerospace Engineering, The University of Michigan, Ann Arbor,
MI 48109-2140, USA, email: scheeres@umich.edu

Abstract. The dynamics of binary Near-Earth objects (NEO) are discussed and a simple model for the study of their dynamics is introduced. Main results on the motion and stability of binary asteroids are reviewed. The effect of perturbations external to the binary system, including solar gravity, solar radiation pressure, and planetary gravity, are considered.

Keywords. asteroid, rotation; dynamics, rotation; binary asteroids

1. Introduction

Binary NEO exhibit a rich set of dynamics and are exposed to many external and internal perturbations. These include coupling of orbital and rotational angular momentum and energy, effect of non-spheroidal mass distributions on dynamical evolution, solar gravitational perturbations for bodies close to the sun, planetary tides during close approaches, and Solar irradiation effects.

The general dynamical problem of binary asteroids, or binary bodies in orbit about each other and subject to external perturbations, has received considerable study over the years. The relevant studies include investigations of point mass dynamics about non-spherical bodies Chauvineau *et al.* (1993), Scheeres (1994), Scheeres *et al.* (1996), Scheeres *et al.* 1998, point mass dynamics about point bodies incorporating solar gravitational and radiation perturbations Hamilton & Burns (1991), Scheeres & Marzari (2002), effect of solar radiation on finite bodies Ćuk & Burns (2005), motion of two massive bodies about each other Kinoshita (1972), Maciejewski (1995), Scheeres (2002a), Scheeres (2002b), Breiter *et al.* (2005), and motion of particles about binary asteroids Scheeres & Bellerose (2005). Recently, a detailed model and associated dynamics of the NEO binary asteroid (66391) 1999 KW$_4$ was studied in detail in Ostro *et al.* (2006), Scheeres *et al.* (2006). Despite these many studies the general problem still has many challenges that must be addressed, ranging from better constraints and understanding of the dynamical evolution of these systems to basic questions on what the most important physics for the evolution and energy dissipation of these systems are. The goal of the current paper is simply to define the basic dynamical problem of binary asteroids, introduce an ideal model for binary asteroids and present basic results on its dynamics, and identify the most relevant known perturbations acting on these systems and their characteristics.

2. The General Model

We first state the most general form of the binary asteroid dynamics problem. These have been derived in an alternate form in Maciejewski (1995) and provide the equations of motion for the relative translational motion between the two components of the binary asteroid and each body's rotational dynamics. We give the current statement as they are in a particularly compact form, as given in Scheeres *et al.* (2006):

$$\frac{M_1 M_2}{M_1 + M_2}\ddot{r}_i = U_{r_i} + F_i(r, T^1, T^2, t) \tag{2.1}$$

$$\dot{H}_i^I = -T_{ij}^I \epsilon_{jkl} T_{mk}^I U_{T_{ml}^I} + M_i^I(r, T^I, t) \tag{2.2}$$

$$\Omega_i^I = (I_{ji}^I)^{-1} T_{kj}^I H_k^I \tag{2.3}$$

$$\dot{T}_{ij}^I = T_{ik}^I \Omega_l^I \epsilon_{klj} \tag{2.4}$$

$$U(r, T^1, T^2) = \mathcal{G} \int_{B^1} \int_{B^2} \frac{dm^1 dm^2}{|r + T^1 \rho^1 - T^2 \rho^2|} \tag{2.5}$$

where $i = 1, 2, 3$ denote coordinates and $I = 1, 2$ denotes the two bodies. Here r is the relative position vector between the two centers of mass expressed in an inertial frame, M_I denotes the mass of body I, H^I is the inertial frame angular momentum vector of the Ith body, I^I is the inertia tensor of body I in its body-fixed frame, Ω^I is its angular velocity vector, T_{ij}^I is the attitude matrix of body I mapping its body-fixed frame to the inertial frame, ϵ_{ijk} is the skew-symmetric 3-tensor (with $\epsilon_{123} = 1$) that defines the cross product, $\mathcal{G}$ is the universal constant of gravitation, B^I signifies the mass distribution of the body with differential mass element dm^I, ρ^I is the location of that mass element in the Ith body frame, and U is the mutual gravitational potential between the bodies. The quantities F_i and M_i^I represent the external force and moment, respectively, acting on these systems. These external perturbations generally arise from the gravitational attraction of a planet or the sun, and from radiation induced forces acting on the bodies. Dots over a variable denote time derivatives, subscripts on all variables except U denote vector, matrix or tensor elements, and we assume the Einstein summation convention. A subscript on U denotes partial differentiation. The most difficult item to compute in the above equations of motion is the mutual force potential between the two bodies. Werner & Scheeres (2005) summarize the literature on this problem and provides a novel and efficient method for evaluation of the mutual potential given standard polyhedral shapes of the bodies.

In the absence of external perturbations, under their self-dynamics, the binary system will exhibit the classical constraints of conservation of energy and conservation of total angular momentum.

For the total energy we have the following scalar quantity

$$E = \frac{1}{2}\left(I_{ij}^1 \Omega_i^1 \Omega_j^1 + I_{ij}^2 \Omega_i^2 \Omega_j^2\right) + \frac{1}{2}\frac{M_1 M_2}{M_1 + M_2}\dot{r}_i \dot{r}_i - U(r, T^1, T^2) \tag{2.6}$$

To properly consider evolutionary behavior of a binary system we must also consider the self-potentials of each body, accounting for the ability of a system to absorb energy into changes in the mass distribution, or shape, of each of the bodies Scheeres (2004), such distortions are also associated with the dissipation of energy. Thus, conservation of energy assumes that the shapes of each body remain fixed, or rigid.

The total angular momentum of the system can be denoted as the following vector

$$K_i = \sum_{I=1}^{2} T_{ij}^I I_{jk}^I \Omega_k + \frac{M_1 M_2}{M_1 + M_2} r_j \dot{r}_k \epsilon_{ijk} \tag{2.7}$$

and is conserved even in the presence of body deformations. In that case the inertia tensors of the bodies may shift as well.

The conservation of these quantities plays an important role in defining and determining the dynamics of a binary asteroid without external perturbation. External perturbations can provide changes in the total conserved quantities, allowing the system to evolve over time.

3. An Ideal Model

The above relations and definitions can be applied to arbitrary binary systems. However, such detailed models are only in existence for very few binary systems, and most binaries are only known by the grossest properties such as their elongation, apparent diameter, and light-curve derived spin rates (Pravec *et al.* (2006)). Given this, and given the measured properties of known asteroids, it makes sense to define a simplified model for a binary system. Our basic model will be comprised of two ellipsoidal bodies, with the mutual potential defined by an expansion up to second order only. If one of these bodies is a sphere the problem has been called the Sphere Restricted Full 2-body problem in Scheeres (2002b) or more recently the name "Kinoshita Problem" has been proposed in Breiter *et al.* (2005) due to the first studies of this problem being made in Kinoshita (1972).

3.1. *Ideal Physical Model of Binaries*

The morphology of most binaries can be "fit" by a relatively few simple observations on their geometry and configuration. We should note that these observations are due, in part, to a dearth of high resolution models of most binaries, and that there are special cases known which may violate each of these assumptions.

- Primary shape is oblate
- Primary rotation rate is rapid and at or near the surface disruption limit
- Secondary shape is elongate
- Secondary rotation is synchronous with orbit
- Mutual orbit is near-circular with inclination near zero

Although there are notable exceptions, the above geometrical features appear to be representative of most NEO binary systems, and are what should be explored generically first. Indeed, the KW4 binary system fits with this ideal model very well.

3.2. *Mutual Potential*

The general mutual potential between two bodies can be stated in a simplified form if we apply MacCullagh's formula to each body. Following Maciejewski (1995) we can develop an explicit formula for the mutual potential between two mass distributions at the second order:

$$U = \frac{\mathcal{G}M_1 M_2}{\sqrt{r_k r_k}} \left[1 + \frac{1}{2 r_k r_k} \left(I_{ii}^1 + I_{ii}^2 \right) - \frac{3}{2(r_k r_k)^2} \left(I_{ij}^1 + I_{ij}^2 \right) r_i r_j \right] \tag{3.1}$$

We will use this basic formula for the mutual potential in our following discussions.

3.3. *Mechanics of the Ideal Binary Model*

We can separate the internal motion of a binary system, the motion only attributable to the mutual interaction between the two bodies, into two portions: relative orbit and angular evolution. The relative orbit evolution involves the planar orbit of the system and can be described by the classical orbit elements of semi-major axis, eccentricity, argument of periapsis and mean motion. The angular evolution couples orbital and rotational dynamics together and considers the rotational angular momentum of the primary and secondary and the angular momentum vector of the relative orbit, relating to the inclination and longitude of ascending node.

For each case we can identify a minimum energy configuration for the system, and the dynamics for deviations from these configurations. In the following we define each of these minimum energy configurations and describe the motion in their vicinity.

3.3.1. *Relative Orbit Dynamics*

For the ideal model the minimum energy configuration of the binary system has the secondary in a constant synchronous rotation and the orbit and primary angular momentum vectors aligned. This model fits the basic observational constraints for binary systems. The relative orbit describes a circular path, but the system is not in an osculating circular orbit in general. This is a somewhat subtle point, but can be understood by considering the osculating orbit elements of a simple binary system comprised of a central sphere and a smaller ellipsoidal body. Let us align the ellipsoid in a relative equilibrium with its longest axis pointing towards the sphere and rotating at the necessary rate to balance gravitational attractions. Due to the non-point mass mutual attraction of the two bodies the rotation rate differs from a Keplerian orbit, inducing a semi-major axis not strictly equal to the radius of the orbit. This in turn implies that the orbit has a non-zero eccentricity, with the result that the system has a constant true anomaly equal to 0 or 180 degrees, and a precessing argument of periapsis with period equal to the orbit period.

For the relaxed model the motion is indistinguishable from a classical circular orbit, however when the system is perturbed from this relaxed configuration the resulting dynamics will differ from a slightly eccentric Keplerian orbit. This can be shown in Fig. 1 which shows the eccentricity vector for a binary system comprised of a spherical primary and a non-symmetric secondary body nominally in a synchronous orbit. The relative equilibrium and periapsis libration curves correspond to the above situation, where the true anomaly librates about 0 degrees and the argument of periapsis circulates with a period equal to the orbit period. The circulating periapsis curve is sufficiently excited to break out of this configuration and has its true anomaly circulating and its argument of periapsis moving in a more traditional manner with a secular drift. This is an interesting issue that still requires study, and is relevant as for a general system the primary is not rotationally symmetric, and the orbit and angular evolution of the bodies will have small scale fluctuations about the synchronous state. If the relative equilibrium is energetically stable, these will only lead to small scale deviations from the minimum energy state.

3.3.2. *Angular Momentum Dynamics*

The dynamics of the orbit and primary rotational angular momentum vectors are dominated by conservation of angular momentum. Assuming a mild rotational asymmetry for the primary, the precession of the primary rotation pole and the orbit plane are locked to each other (Kinoshita (1972)). In general the primary and the orbit will lock up the vast majority of the system's angular momentum with the secondary's rotational angular momentum contributing a negligible component. Let the magnitude of the total primary rotational angular momentum equal $H = I_3^1 \Omega_3$ and the magnitude of the orbital angular momentum equal $G = \frac{M_1 M_2}{M_1 + M_2} \sqrt{\mathcal{G}(M_1 + M_2)a}$. Then due to the secondary's assumed on-average synchronous motion and the primary's assumed modest equatorial ellipticity, the respective magnitudes of the primary rotational angular momentum, H, and the orbit angular momentum, G, are constant on average. Denote the total angular momentum vector to be aligned with the inertial z-axis with a magnitude K. Let the angle between the primary angular momentum vector and the z-axis be δ. Let the angle between the orbit angular momentum and the z-axis be ι. Let the sum of the two angles be the mutual obliquity: $\Delta = \delta + \iota$. The inclination and obliquity have only small fluctuations from

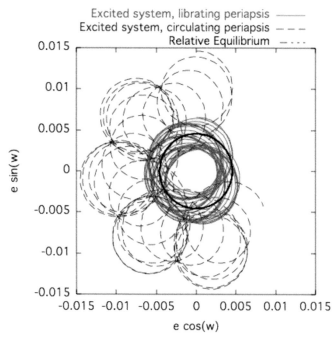

Figure 1. Plot of the eccentricity vector of a binary system at various levels of excitement.

their initial values, so the angular momentum vectors trace out cones in inertial space (Fig. 2).

The following constraints from conservation of total angular momentum then apply, assuming we can ignore the contribution of the secondary rotational angular momentum to the system.

$$K = H\cos\delta + G\cos i \qquad (3.2)$$

$$H\sin\delta = G\sin i \qquad (3.3)$$

$$\Delta \sim \text{Constant} \qquad (3.4)$$

These constraints enforce the following geometry on the system, to within the magnitude of the secondary angular momentum (Scheeres *et al.* (2006)):

• The primary, orbit and total angular momentum lie in a plane with the total angular momentum lying in between the others.

• The precession period and direction of the orbit plane must equal the precession period of the primary rotation pole.

• The orbit pole traces out a cone with half angle: $\sim (G/K)\Delta$

• The primary pole traces out a cone with half angle: $\sim (H/K)\Delta$

• The minimum energy configuration is $\Delta = 0$, as this forces the two spin rates associated with H and G to be minimized, thus minimizing kinetic energy.

By definition, the secondary orientation is locked into a Cassini state, driven by the difference between the orbit and primary angular momentum. If the system has no fluctuations due to non-rotational symmetry, then the attitude of the secondary will be locked in a constant orientation. Due to asymmetries in the system the secondary attitude will in general fluctuate about a mean Cassini state for that system Colombo (1966).

These secular dynamics can also be predicted from basic analysis of the mean orbit plane precession formula and the mean primary spin-axis precession formula. When carried out in detail each of these analyses predict the same mean precession rate, equal to:

$$\dot{\psi} = -\frac{3\sqrt{\mathcal{G}(M_1 + M_2)}}{\sqrt{a^3}} \frac{I_a - I_t}{a^2} \cos \Delta \tag{3.5}$$

where I_a and I_t are the axial and transverse moments of inertia from the assumed rotationally symmetric primary.

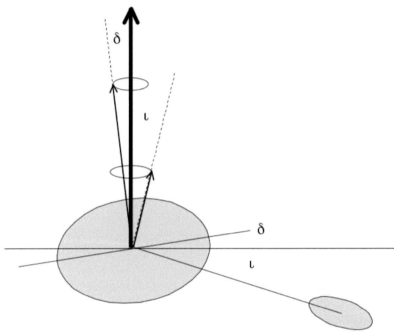

Figure 2. The diagram shows the path followed by the evolving angular momentum vectors. The large vertical arrow represents the total angular momentum, the smallest arrow represents the orbit angular momentum, which has an angle i with the total angular momentum, and the other arrow represents the primary's angular momentum, which has an angle δ with the total angular momentum. The secondary's rotational angular momentum is too small to show in general.

3.4. *Internal Stability*

An interesting and important topic for the ideal binary model is the stability of the proposed relative equilibrium configuration. When the formal stability of these configurations are determined, only those binary systems which match the generally observed binary morphology are stable, thus potentially explaining why other binary morphologies are not found. In the following we define internal stability as the stability of the system without external perturbation. Clearly, if a system is not internally stable it is not likely to be found in nature, independent of the external perturbations acting on it.

There are at least three different relevant definitions of stability for a binary asteroid:
- Stability against escape: Can or will the two components escape each other?
- Stability against impact: Can or will the two components impact each other?
- Orbit/Configuration stability: Can the current orbit or configuration of the system persist?

Each of these stability types can be evaluated using well-defined mathematical techniques and evaluations Scheeres (2002b), Scheeres (2006). These conditions can be reduced to a few basic observations.

First, configuration or orbit stability reduces to the energetic stability of the relative equilibrium. An equilibrium configuration is energetically stable if it inhabits the minimum energy configuration possible for that system at a given system angular momentum. We note that there are a number of possible relative equilibrium for a binary system that may be spectrally stable, but there is only one that ever has, in addition, energetic stability. These invariably have the minimum moment of inertia principal axis directed towards the primary, although for non-symmetric bodies the axis does not point precisely at the body Scheeres (2006). This generalizes to the common gravity gradient orientation of Earth and natural satellites. Its important to note that for mass distributions between the two binary bodies that are not dominated by the primary, that these configurations can be unstable. The classical example being that a particle placed in a relative equilibrium along the minimum moment of inertia axis of a massive ellipsoid is always unstable. Thus, there is a transition in these configurations from unstable to stable as the primary body mass becomes dominant. In Figs. 3 and 4 these stability curves are presented for two different ellipsoid-sphere relative equilibrium configurations. Each point on these diagrams defines a relative equilibrium between the sphere and ellipsoid, with the minimum moment of inertia axis of the ellipsoid pointing at the sphere and the system rotating about the maximum moment of inertia. Points above the stability limit line are energetically stable, and points below are not. The equal density distance is defined as the distance between the mass center of the ellipsoid and a sphere assuming an equal density for each body. Thus, this distance is only a function of the mass ratio between the bodies. Note that the distance has been normalized by the long axis of the ellipsoid. We note that even though the two ellipsoids have significantly different shapes, the qualitative features of their stability limit lines are similar. This indicates that sphericity of the primary is an essential ingredient for the overall stability of the given configuration.

Stability against mutual escape is ensured by the total energy being negative, $E < 0$, while if $E > 0$ it is possible for the system to escape. Note that most binaries technically violate this due to the extremely rapid rotation rate of the primary. However, if we note that the mass distribution of a spinning spheroidal body cannot, in general, interact with the orbital system we should remove this energy from consideration, which in general will reduce the energy of the system considerably. This reduced or "free" energy is what is shown in Figs. 3 and 4. We note that all binary systems whose relative equilibrium has a positive free energy are also unstable, this result holds across all parameter values for this ideal system. Having a positive free energy is only a necessary condition for the system to have a mutual escape, and it is possible for a positive energy system to also suffer impact.

Stability against impact is ensured by the relative equilibrium of the system being stable, or by the total angular momentum being high enough. Note that most binaries satisfy this formally, again due to the rapid rotation of the primary. In Scheeres (2002b) a specific condition for impact stability of a system is derived. A simpler observation can be made with the aid of the stability diagrams above. Here we note that whenever a relative equilibrium configuration is unstable, then there will exist unstable manifolds that travel from that relative equilibrium configuration, one of which increases the distance between the bodies and the other decreases the distance, this latter generally intersecting with the surface of the body. Thus, configuration instability implies that a pathway for the bodies to impact also exists. While the ideal situation has the system initially placed in

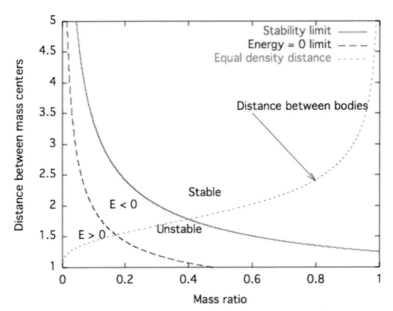

Figure 3. Stability diagram for a sphere-ellipsoid system in relative equilibrium, ellipsoid semi-major axes of 1:0.9:0.8. A mass fraction of 0 represents a particle about a massive ellipsoid, while a mass fraction of 1 represents an ellipsoid with negligible mass about a massive sphere.

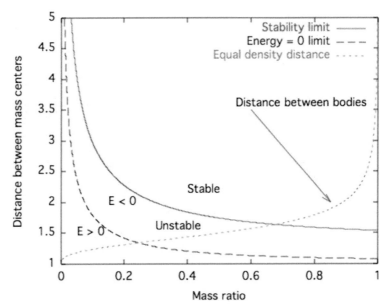

Figure 4. Stability diagram for a sphere-ellipsoid system in relative equilibrium, ellipsoid semi-major axes of 1:0.5:0.25.

a relative equilibrium, for a realistic system the "initial conditions" will be more general, but the existence of these pathways to impact will still hold and make it more likely that the system can transition into an impact, perhaps forming a contact binary. As the relative equilibrium approaches the stability limit, these unstable manifolds may no longer lead to mutual impacts, the limit for when this occurs has yet to be studied in detail, however.

From these stability results we can infer a number of general results on what sort of binary systems can survive naturally. We note that systems with elongate primaries are generally unstable at smaller separation distances and hence should either lead to mutual escape or impact. Systems with spheroidal primaries are more prone to have stable configurations for their orbiting member and should persist, which is in agreement for the ideal model of binaries inferred from observations. We also note that almost any binary system with sufficient separation can be stable, including elongate primaries. The apparent lack of such binary systems implies that, at some point of their formation or evolution, binary asteroids may reside in a close configuration which would then destroy such systems.

4. External Perturbations

External perturbations to an NEO binary system can arise from several sources, each of which will have a different signature. The perturbations we will consider, and the ones most relevant for these systems, are due to planetary flybys, solar gravity effects, and solar radiation effects.

4.1. *Gravitational Perturbations*

While the effect of planetary flybys and perihelion passages are similar in that in both cases the perturbations arise from a combination of gravity and centripetal accelerations, we find that the net effect of these flybys are quite different. Still, we can begin our analysis with a common investigation and later separate the effect of these interactions. Following from Marchal (1990) and Scheeres & Marzari (2002) we can define a modified form of the elliptic/hyperbolic Hill 3-body problem that only incorporates the effect of gravity. The main effect of the gravitational perturbation during a periapsis passage is characterized by the radial location of the libration point as measured from the binary center of mass towards or away from the sun, equal to $x_P = \left(\frac{\mu}{3\mu_P}\right)^{1/3} q$, where μ and μ_P are the gravitational parameters of the binary system and the perturbing body, respectively, and q is the periapsis radius, or distance between the binary center of mass and the perturbing body at closest approach. The ratio between the binary orbit semi-major axis and the libration point distance x_P controls the strength of the perturbation, and whether or not the perturbation can disrupt the binary during closest approach. A basic result is that if the relationship $a/x_P < 1/3$ holds, the binary cannot be directly disrupted due to the flyby Marchal (1990), although it can be significantly perturbed and may impact following closest approach.

Another important consideration is the angular rate of the flyby, specifically the comparison between the angular rate of the binary relative to the perturbing body and the angular rate of the binary orbit itself. During perihelion passage the binary system will be traveling at a rate $\dot{\nu} = \sqrt{\mu_P(1+e)/q^3}$, where e is the eccentricity of the flyby orbit and will be greater than one for a planetary flyby (i.e., be a hyperbolic orbit) and less than one for a solar close approach. The angular orbital rate of the binary is approximately equal to $n = \sqrt{\mu/a^3}$. The ratio between these two is important as it controls whether the perturbation is "impulsive", i.e. if it acts fast relative to the orbit rate of the binary system, or whether it must be averaged over several binary orbit periods. As we will demonstrate with some simple examples later these lead to qualitatively different effects on the orbit. Taking this ratio we see that the result is a function of the libration point at periapsis, x_P, defined earlier:

$$\frac{\dot{\nu}}{n} = \sqrt{1+e}\left(\frac{a}{x_P}\right)^{3/2} \tag{4.1}$$

When this ratio is small, the effect of the perturbing gravity will be averaged over more rotation angle of the binary, while when it is large the binary orientation will be approximately fixed during the fly-by and the perturbing gravity will act impulsively.

4.1.1. Planetary flybys

For planetary flybys the eccentricity equals $e = 1 + qV_\infty^2/\mu_P$ and is greater than 1. This, combined with the possibility of very close flyby distances which makes the ratio a/x_P relatively larger, will lead to an impulsive nature for these interactions. Such a flyby can inject energy and angular momentum into the binary system nearly instantaneously, leading to escape, impact, or major perturbation Farinella (1992). The expected signature of a close approach that doesn't destroy a binary will be a highly randomized system until relaxation effects come into play. We note that this process may also be more efficient at disrupting binaries than at creating binaries, as the disruption distance is greater than the creation distanceRichardson & Walsh (2007). In Fig. 5 we show some example perturbations to a binary system subject to a planetary flyby. The binary model used in these simulations is the Keplerian binary so that the effect of the external perturbation can be clearly seen without any other perturbations. It is important to note that even if the flyby does not directly destroy the binary, it is possible for the perturbed binary to run afoul of the solar perturbation due to an increase in semi-major axis, and become subject to such solar perturbations.

4.1.2. Solar gravity effects

Solar gravity effects are only active for asteroids with perihelia low enough such that the ratio a/x_P grows larger. For these cases the solar perturbation can be a significant source of excitation, however. As the eccentricity of the orbit is less than one by definition, and as the ratio a/x_P tends to be small, $\dot{\nu}/n \ll 1$ and the gravitational interaction generally occurs over a period of many binary revolutions, and hence the net effect is much more subdued than a planetary flyby. Since the effect is active for an extended period of time, however, it can create significant excitation of a system without disrupting it. Applying the above relations we find that for a binary system with $\mu = 10^{-7}$ km^3/s^2 the semi-major axis relation becomes $a < 32q$ where q is measured in astronomical units (AU). Thus, for a typical separation of less than 5 kilometers, perihelion must come down to 0.15 AU before the system can be completely disrupted by the sun. Prior to this, however, the system is subject to perturbations during each perihelion passage. The nature of these perturbations are much different than planetary flybys as the significant perturbation lasts over many binary orbit revolutions. Thus there is a large degree of averaging that occurs during perihelion passage which changes the effect of the perturbation, seen in Fig. 6. We note that the perihelion passage is able to measurably shift the system inclination. Also, the excitation of the eccentricity can stimulate internal motions that leave the binary in an excited state, even if the eccentricity reduces back to a small value as is the case for the simulations shown here (due to the nominal binary orbit being modeled as a Keplerian 2-body problem).

4.2. Solar radiation effects

Incident sunlight carries momentum that is transfered to the system and reemitted by reflection and thermal re-emission. This momentum flux can cause changes in the rotation

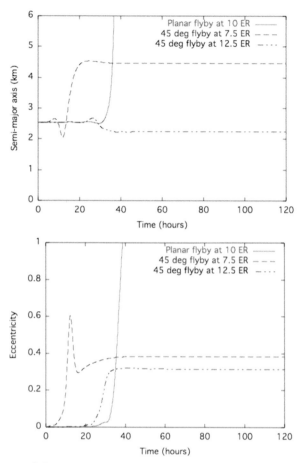

Figure 5. Evolution of the semi-major axis, eccentricity and inclination of a binary system during an Earth flyby

state and mutual orbit of the system. The radiation acts independently on each body, but can be coupled by the dynamical reaction of each body. There are two main effects, a force and a torque acting on each component of the system. The study and understanding of the effects of these forces and torques on binary systems is one of the most pressing issues for this field of study.

4.2.1. *Solar radiation pressure force*

The force acting on the primary will mainly affect the heliocentric trajectory of the system. In particular, the thermal characteristics of the primary become important as it is only the reemission of radiation transverse to the orbit that will have a measurable effect on the heliocentric trajectory, called the Yarkovsky effect, a topic that has been studied extensively in Bottke *et al.* (2002) and analyzed specifically for binary asteroids in Vokrouhlický *et al.* (2005). The force acting on the secondary will affect the mutual orbit of the binary system as this will act as a small non-gravitational force acting on the system. Any asymmetry in the net force acting on the body can cause a small but finite net torque to be delivered to the orbit and cause the orbit to grow or shrink in time, what is called the Binary YORP (BYORP) effect Ćuk & Burns (2005). To date the only detailed analysis of the long-term implications of this effect is in Ćuk & Burns

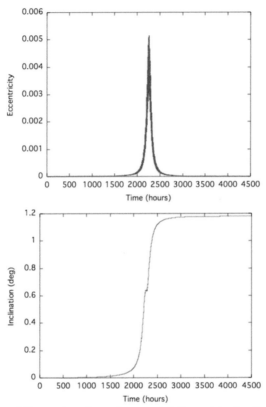

Figure 6. Evolution of the semi-major axis, eccentricity and inclination of a binary system during a perihelion passage

(2005) where they provide a first-order averaging analysis of this effect on binary orbits and predict that extremely rapid migration of these orbits may occur, with disruption or impact on the order of thousands of years. The analysis in that paper does make some simplifying assumptions, mainly that the net differential force acting on the secondary can be modeled as an averaged, constant force over long periods of time. This may be a reasonable assumption but the extremely short lifetimes are difficult to reconcile with the number of binary objects that have been observed in the NEO population, and imply an extremely fast production rate of binaries. The analysis by Vokrouhlický et al. (2005) investigates the effect of the Yarkovsky force on binary asteroids over shorter time spans, and also provide an analysis of the Yarkovsky-Schach effect, which arises due to the primary eclipsing the secondary. A deeper understanding of these dynamics over long time spans is a pressing topic in this area.

4.2.2. *Solar radiation pressure torques*

The effect of solar insolation on the rotation states of the primary and secondary can also be significant for the evolution of the system. Asymmetries in either of the bodies can lead to a net torque acting on them, with the rotational acceleration and obliquity of the bodies modified Rubincam (2000), this effect has been termed the YORP effect in the literature.

YORP can also cause the primary to either increase its spin rate or decrease it over time, pumping angular momentum into or out of the system. If it increases, it is possible for material to be spun off the surface of the primary, which would then transfer angular momentum to the orbit and cause the system to expand. If it decreases, it could eventually lead to a mutually synchronous system. YORP acceleration acting on the secondary will bias its attitude relative to the secondary-primary line, allowing for the transfer of angular momentum to the orbit. For a positive torque, this would lead to a growth in the orbit and, ultimately, in a decrease in the spin rate of the secondary. A negative YORP torque on the secondary should lead to a shrinking in the orbit and a speed up in the spin rate of the secondary. A possible characteristic of systems subject to YORP may be the rapid spin of the primary at or near the disruption limit. None of the other external perturbations can, as easily, lead to this state.

5. Conclusions

This paper reviews current literature on the dynamics of binary asteroids and presents a summary of what is currently known and hypothesized on these systems. The general governing equations of motion and constraints are stated. A simplified, but non-trivial, model for the study of binary asteroids is proposed and discussed. A discussion on the stability of binary asteroids is given, and we note that binary systems that should be unstable at close separation distances are not found in nature, implying that binary asteroids may go through a close configuration at some point in their lifetime. Finally, we discuss the main features and effects of the main external perturbations that act on binary asteroids in the near-Earth population.

Acknowledgements

This research was supported by a grant from NASA's Planetary Geology and Geophysics Program.

References

Bottke, W.F., Vokrouhlický, D., Rubincam, D.P. & Brož, M. 2002, in W.F. Bottke, A. Cellino. P. Paolicchi & R. Binzel (eds.), *Asteroids III* (Arizona University Press, Tucson), p. 395
Breiter, S., Melendo, B., Bartczak, P. & Wytrzyszczak, I. 2005, *Astron. Astrophys.* 437, 753
Chauvineau, B., Farinella, P. & Mignard, F. 1993, *Icarus* 105, 370
Colombo, G. 1966, *Astron. J.* 71, 891
Ćuk, M. & Burns, J.A. 2005, *Icarus* 176, 418
Farinella, P. 1992, *Icarus* 96, 284
Hamilton, D.P. & Burns, J.A. 1991, *Icarus* 92, 118
Kinoshita, H. 1972, *Publ. Astron. Soc. Japan* 24, 423
Maciejewski, A.J. 1995, *Celest. Mech. Dyn. Astron.* 63, 1
Marchal, C. 1990, The Three-Body Problem, Elsevier
Ostro, S.J., Margot, J.-L., Benner, L.A.M., Giorgini, J.D., Scheeres, D.J., Fahnestock, E.G., Broschart, S.B., Bellerose, J., Nolan, M.C., Magri, C., Pravec, P., Scheirich, P., Rose, R., Jurgens, R.F., Suzuki, S. & DeJong, E.M. 2006, Radar Imaging of Binary Near-Earth Asteroid (66391) 1999 KW4, *Science*, in press
Pravec, P., Scheirich, P., Kusnirák, P. & Sarounová, L. 2006, *Icarus* 181, 63
Richardson, D. & Walsh, K. 2007, this volume
Rubincam, D.P. 2000, *Icarus* 148, 2
Scheeres, D.J. 1994, *Icarus* 110, 225
Scheeres, D.J., Ostro, S.J., Hudson, R.S. & Werner, R.A. 1996, *Icarus* 121, 67
Scheeres, D.J., Ostro, S.J., Hudson, R.S., DeJong, E.M. & Suzuki, S. 1998, *Icarus* 132, 53
Scheeres, D.J. & Marzari, F. 2002, *J. Astronautical Sciences* 50, 35
Scheeres, D.J. 2002, *Icarus* 159, 271
Scheeres, D.J. 2002, *Celest. Mech. Dynam. Astr.* 83, 155

Scheeres, D.J. 2004, *Celest. Mech. Dynam. Astr.* 89, 127

Scheeres, D.J. & Bellerose, J. 2005, *Dynamical Systems: An International Journal* 20, 23

Scheeres, D.J. 2006, *Celest. Mech. Dynam. Astr.* 94, 317

Scheeres, D.J., Fahnestock, E.G., Ostro, S.J., Margot, J.-L., Benner, L.A.M., Broschart, S.B., Bellerose, J., Giorgini, J.D., Nolan, M.C., Magri, C., Pravec, P., Scheirich, P., Rose, R., Jurgens, R.F., Suzuki, S. & DeJong, E.M. 2006, Dynamical Configuration of Binary Near-Earth Asteroid (66391) 1999 KW4, *Science*, in press

Vokrouhlický, D., Čapek, D., Chesley, S.R. & Ostro, S.J. 2005, *Icarus* 179, 128

Werner, R.A. & Scheeres, D.J. 2005, *Celest. Mech. Dynam. Astr.* 91, 337

Near Earth Objects, our Celestial Neighbors: Opportunity and Risk
Proceedings IAU Symposium No. 236, 2006
A. Milani, G.B. Valsecchi & D. Vokrouhlický, eds.
© 2007 International Astronomical Union
doi:10.1017/S1743921307003225

Physical models of asteroids from sparse photometric data

Josef Ďurech[1,3], Petr Scheirich[2], Mikko Kaasalainen[3], Tommy Grav[4], Robert Jedicke[4] and Larry Denneau[4]

[1]Astronomical Institute, Charles University in Prague,
V Holešovičkách 2, 18000, Prague, Czech Republic,
email: durech@sirrah.troja.mff.cuni.cz

[2]Astronomical Institute, Academy of Sciences of the Czech Republic,
Fričova 1, 25165 Ondřejov, Czech Republic

[3]Department of Mathematics and Statistics, Rolf Nevanlinna Institute, University of Helsinki,
P.O. Box 68, 00014 Helsinki, Finland

[4]Institute for Astronomy, University of Hawaii,
2680 Woodlawn Drive, Honolulu, Hawaii, 96822, USA

Abstract. We present an overview of our work on shape and spin state determination of asteroids from photometric data sparse in time. Our results are based on simulations that were performed using realistic shape and light-scattering models and time sequences that will be provided by Pan-STARRS (Panoramic Survey Telescope and Rapid Response System). We show some typical examples of physical model reconstruction of main belt and near-Earth asteroids and discuss the lightcurve inversion of slow and fast rotators, binary asteroids and tumbling asteroids. We emphasize the scientific potential of sparse photometric data to produce models of a large number of asteroids within the next few years.

Keywords. survey; photometry; asteroid, lightcurves; asteroid, shape

1. Introduction

Astronomers have observed asteroid lightcurves – brightness variations as a function of time – for many decades. Since a typical asteroid rotates with a period of hours, a substantial part of its whole rotation cycle can be covered during one night. A set of lightcurves from several apparitions with different observing geometries contains so much information about the spin state and the shape of an asteroid, that those physical parameters can be unambiguously derived using lightcurve inversion. The lightcurve inversion method developed by Kaasalainen & Torppa (2001) and Kaasalainen *et al.* (2001) has led to about a hundred asteroid models (Kaasalainen *et al.* 2002b, 2004; Torppa *et al.* 2003) and its reliability has been proven by ground truths from, e.g., Kaasalainen *et al.* (2001, 2005) and Marchis *et al.* (2006).

However, it is not necessary to have lightcurves that densely cover the rotational phase for deriving a plausible physical model of an asteroid. As has been shown by Kaasalainen (2004), calibrated brightness measurements from several years that are sparse in time are fully sufficient. This approach is much more time efficient than the standard approach of observing dense lightcurves of selected targets. Future asteroid surveys will provide us with a huge amount of accurate sparse photometric data and we will have the opportunity to reconstruct physical models of a substantial part of the asteroid population. Lightcurve inversion of sparse photometric data will be an essential method for asteroid shape and spin state modelling in the near future.

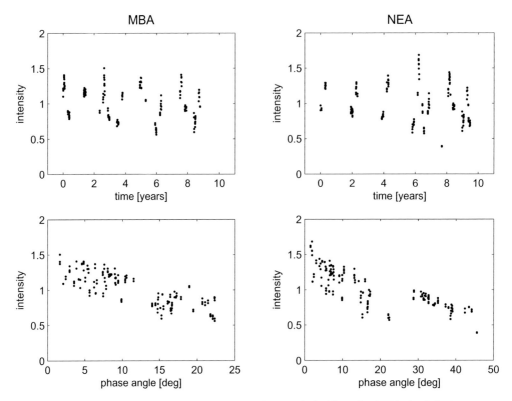

Figure 1. Typical Pan-STARRS observations of a MBA (left) and a NEA (right) shown as a time sequence covering ten years (top) and as a brightness vs. solar phase angle plot (bottom). The brightness is reduced to unit distances from the Earth and the Sun.

In the following text, we often refer to results presented by Kaasalainen & Ďurech, *Inverse problems of NEO photometry: Imaging the NEO population*, in this proceedings book (hereafter KD). This paper gives a general overview about methods for modelling asteroid physical parameters and about the lightcurve inversion in particular.

So far, sparse data have been available only as a by-product of astrometric observations. Such data usually suffer from large systematic errors and noise and cannot be used alone (see KD, Sect. 4.1). Future photometric surveys will reach much better photometric accuracy and the lightcurve inversion applied to such accurate data will lead to unambiguous and reliable physical models.

We summarize the results of sparse data analysis in this paper that were obtained using simulations based on Pan-STARRS (Panoramic Survey Telescope and Rapid Response System) cadences. Pan-STARRS is an ongoing project at the Institute for Astronomy of the University of Hawaii, the first light is scheduled for early 2007. It is the first project that will provide sparse photometry with sufficiently low calibration errors.

Another source of accurate photometric data sparse in time will be the Gaia space mission. Although its lifetime will be shorter than for ground-based surveys, it will observe at small solar elongations, thus covering a wider range of ecliptic longitudes in shorter time.

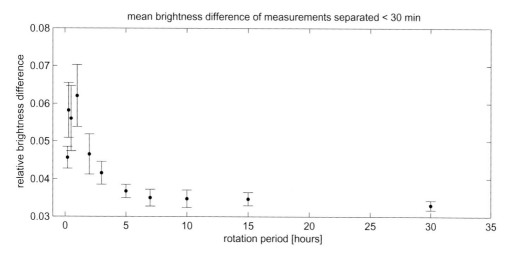

Figure 2. The mean relative brightness difference between observations separated by less than 30 min as a function of the rotation period.

2. Photometry sparse in time

Contrary to a set of individual lightcurves that densely cover a substantial part of the full rotation cycle, sparse data sets consist of individual brightness measurements, typically only one or a few points per night. Sparse photometry has to be calibrated in the absolute scale – we need to directly compare brightness measurements with each other on the time scale of many years. Examples of synthetic sparse data sets are shown in Fig. 1. The plots were generated using the Pan-STARRS observing scheduler, a realistic shape model and Hapke's light-scattering model. The brightness was reduced to unit distances from the Earth and the Sun. The main belt asteroid (MBA) data consist of 171 points and the near-Earth asteroid (NEA) data consist of 134 points. Both sets cover ten years of observation. The scatter in the phase angle plots is caused by asteroid's rotation and (to some extent) by a Gaussian noise of 3% that was added to the synthetic data. There are two groups of points at the phase angle plot of the MBA separated by a gap at $\alpha \sim 13°$ corresponding to observations near the opposition and to sweet spots† observations at $\alpha \gtrsim 15°$. Naturally, NEA observations reach much larger solar phase angles.

An important feature of the Pan-STARRS observing strategy is that each field of sky will be observed twice each night after about 15 min. This fact enables us to make some preliminary estimations about asteroid's rotation rate before actual modelling. If an asteroid rotates slowly, than its lightcurve variations at the time scale of ~ 15 min are very small (except when the object is extremely elongated or there are very sharp features in its lightcurve). On the other hand, when the rotation period is comparable with the time baseline of the subsequent measurements, the typical dispersion of such measurements is as large as the lightcurve amplitude. An example of this behaviour is shown in Fig. 2. We generated many lightcurves for various combinations of cadence/shape/pole/period and then computed the mean difference between all subsequent measurements separated by less than 30 min. The plot clearly shows that the brightness difference between close pairs is only slightly larger than the noise of the data (which was 3%) for rotation periods above ~ 5 hr. It means that if the mean brightness difference between close observations

---

† Sweet spots are sky regions near quadrature with $|\beta| < 10°$ and $60° < |l| < 90°$, where β is the ecliptic latitude and l is the solar elongation.

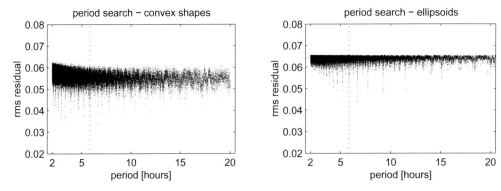

Figure 3. Period search results using convex shapes (left) and triaxial geometrically scattering ellipsoids (right). The correct period is marked with the dotted vertical line and it gives the best fit in both cases.

is significantly larger than the expected noise, the rotation period is not likely to be longer than 10–20 times the mean separation between those observations.

3. Lightcurve inversion of sparse data

The process of lightcurve inversion of sparse data sets is basically the same as the lightcurve inversion of ordinary lightcurves (Kaasalainen *et al.* 2001). The specific approach to the sparse data inversion was described by Kaasalainen (2004). As has been shown by Kaasalainen (2004) and Ďurech *et al.* (2006), the crucial parameters affecting results of sparse photometry inversion (uniqueness and stability of the solution) are calibration errors, the number of points in the data set, the time span of observations, and the shape of the asteroid. Observations should be carried out at the widest possible range of solar phase angle and solar elongation in order to provide good sampling of viewing/illumination geometry. We can say in general that if the noise is not higher than $\sim 5\%$, if there are more than ~ 100 points spread over at least five years, and if the shape is not too spheroidal, then a unique physical model can be derived.

We used Gaussian random shapes (Muinonen 1996) and Hapke's scattering model with uniform albedo for synthetic data generation. After adding a certain level of noise, we inverted the data using the empirical light-scattering model of Kaasalainen *et al.* (2001). As is discussed in KD, Sect. 2.2, the convex inversion guarantees that the solution is not very sensitive to the light-scattering model we use. Another advantage of the convex inversion method is the fact, that any significant asymmetric albedo variegation over the surface reveals itself as the violation of the convexity condition (see KD, Sect. 2.1, for more detailed discussion).

3.1. *Period*

The most important part of the sparse photometry inversion is the correct determination of the rotation period. Unlike in the case of ordinary lightcurves, the period is not 'visible' in the data and it must be searched over a very wide interval of expected values in every single case. Any formal time series analysis will fail to find the period because a typical sampling interval of the data is much longer than the period. However, the correct solution can be found if a realistic physical model is used.

The rotation period is searched by scanning a wide range of periods (typically 2–24 hr) with the step slightly shorter than $0.5P^2/\Delta T$, where P is the trial rotation period and ΔT is the length of the interval covered by observations. This step in period corresponds

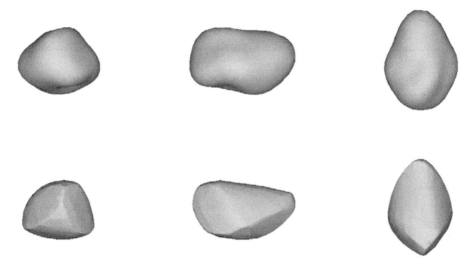

Figure 4. Original shape (top) and the model (bottom) of a MBA asteroid. There are two equatorial views 90° apart (left and middle) and one pole-one view (right) for both shapes. Pole direction of the original shape and of the model is $(348°, -19°)$ and $(347°, -15°)$, respectively. The model rotates with the same period as the original shape, $P = 17.6669$ hours.

to a typical separation of local minima in the period parameter space (Kaasalainen 2004). When the best period is found, a refined pole/shape solution is computed using a dense initial pole grid. The period search is a time consuming process because the number of trial periods is very large, typically $\sim 100\,000$ for $\Delta T = 10$ years.

The information about the period hidden in the data is so strong, that it is possible to use a very simple and inaccurate physical model of a geometrically scattering triaxial ellipsoid (rotating along its shortest axis) and still the correct solution stands out clearly. Moreover, the integrated brightness of such model can be computed analytically (see KD, eqs. (3.6)–(3.9)), which makes the period search process about hundred times faster. There is an example of a period search in Fig. 3. The rms residual of the fit is plotted against the period. Although the ellipsoidal model cannot fit the data down to the noise level (3%), the correct period (or more possible candidates) can be found and then the physical model can be derived using convex lightcurve inversion.

Our simulations show that the shape of an asteroid is crucial for the period determination. Shapes that are only slightly elongated have only small lightcurve amplitudes that are comparable to the noise and any unique period cannot be found. On the other hand, elongated asteroids have large lightcurve amplitudes and the correct period can be detected even from noisy data.

3.2. *Pole and shape*

The pole direction and the corresponding shape model can be derived once a unique period solution is found. The error in the pole direction is usually not larger than 10° of arc. There are two solutions of the pole direction for a given period in many cases. They have similar values of the ecliptic latitude β and the ecliptic longitudes λ are 180° apart. This ambiguity is inevitable for disk-integrated photometry of targets that orbit near the plane of ecliptic. See KD, Sect. 2.1, or Kaasalainen & Lamberg (2006) for more detailed explanation. An example of shape reconstruction is shown in Fig. 4. We used simulated Pan-STARRS cadence of a MBA, a nonconvex shape and Hapke's scattering

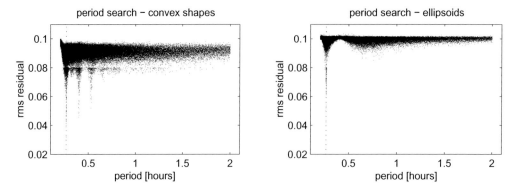

Figure 5. Period search for a fast rotator ($P = 0.264\,\mathrm{hr}$) using convex shapes (left) and triaxial geometrically scattering ellipsoids (right) . The correct period is marked with the dotted vertical line.

model to compute synthetic brightness at given epochs covering ten years of observation. Gaussian noise of 3% was added to the data. The solution of the pole direction is very close to the correct value and the convex shape model is similar to the original.

3.3. *Fast rotators*

If we process sparse data of a fast rotating asteroid, then there is no solution found within the basic interval of typical rotation periods and it is necessary to look for the correct period outside this interval. The number of trial periods increases dramatically when scanning the short period range ($\sim 1\,000\,000$ trial periods for interval 0.2–2 hr and 10 years of observation). Nevertheless, the simulations show that at least for Pan-STARRS cadences even very short rotation periods can be detected.

A representative example of the period search for a fast rotator ($P = 0.264\,\mathrm{hr}$) is shown in Fig. 5. Both inversion algorithms – using convex shapes or ellipsoids – find the correct period. An important fact is that all trial periods from the basic interval 2–24 hr give bad fit. There is only one period fitting the data well.

We assume that the rotation period is constant during the interval of observation. YORP effect that could be important at the time scale of several years for fast rotating small objects is in principle detectable from sparse data but no detailed simulations have been done so far.

3.4. *Slow rotators*

Another nonstandard case are slow rotators. Long periods can be scanned easily because the separation between subsequent trial periods increases with period. If we simulate photometric observations of an asteroid with the period P which is longer than one day, there is usually a false period detection at $P/2$ and some other harmonics (Fig. 6). However, the shape models corresponding to periods P and $P/2$ are usually completely different. The shape corresponding to the correct period almost always rotates around the principal axis of the moment of inertia (as does the original model). On the other hand, the shape corresponding to the half period is usually in a physically unacceptable rotation state – its rotation axis does not correspond with the principal axis of the moment of inertia. An example of the original shape, the model for the correct period, and the model for the half period is shown in Fig. 7.

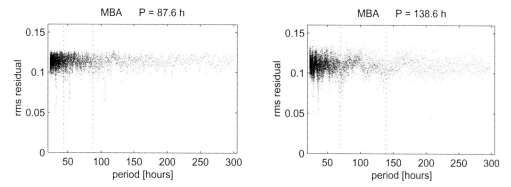

Figure 6. Period search for slowly rotating objects. Both the correct period and the half period fit the data well, they are marked with dotted lines.

3.5. *Binary asteroids*

Binary asteroids are an important part of the asteroid population. We carried out simulations with various binary systems in order to realize how the sparse photometry of such systems differ from that of single asteroids and what kind of results the formal application of lightcurve inversion to data of a binary system gives.

3.5.1. *Synchronous binary asteroids*

From a photometric point of view, a synchronous binary system behaves like a single body with one period of rotation. As has been shown by Ďurech & Kaasalainen (2003) and Kaasalainen *et al.* (2002a), results of the lightcurve inversion procedure give us some indications about the binary nature of the target – a convex shape model of a binary asteroid exhibits large planar areas and the global shape is cone-like or brick-like. The same applies to the sparse data.

3.5.2. *Asynchronous binary NEAs*

We simulated sparse photometric data of asynchronous binary NEAs and tested the possibility of detecting such objects based on sparse photometric data. Our models of binary asteroids were based on physical properties of several objects derived by Pravec *et al.* (2006). A typical binary NEA consists of a fast rotating (2–3 hr), not very elongated primary (lightcurve amplitudes 0.08–0.2 mag), and of a secondary with the size 0.2–0.5 of the primary. Secondaries are usually elongated (lightcurve amplitudes up to 0.8 mag) and their orbital periods are longer than ∼ 10 hr.

There are three effects important for the Pan-STARRS ability to detect binaries: the size of the secondary, the elongation of the primary, and the observing geometry. Mutual events are clearly visible for a big secondary and suitable geometry. If the secondary is small compared to the primary, its contribution to the brightness is small and the signal is dominated by the lightcurve of the primary.

The observing geometry is in many cases such, that there are no visible mutual events at all. The only clue that the object is 'nonstandard' is the fact that lightcurve inversion fails to find any fitting model for any single period.

Two examples of systems that could be recognized as binary from sparse data are shown in Fig. 8. Mutual events are visible in the brightness versus phase angle plot as drops in brightness and there is no single period model that would fit the data. Although sparse data are not sufficient for deriving a physical model, they can give us a strong

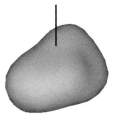  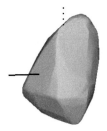

Figure 7. An example of shape reconstruction for a slow rotator. Rotation axis of the original shape (left), the shape model for the correct period (middle), and the model for the half period (right) is marked as a dotted line, while the principal axis of the moment of inertia is marked as a solid line. The axes coincide for the original shape.

indication that the object is binary. Such objects can be selected for further follow-up observations.

3.6. *Trans-Neptunian Objects (TNOs)*

TNOs create a special group of objects that differ from MBAs and NEAs. The observing geometry hardly changes due to their big distance from the Earth and from the Sun, and there are many spin/shape solutions corresponding to one period. Nevertheless, the correct period can be derived. There are often false solutions for other harmonic periods that fit the data only slightly worse than the correct period, but the shapes corresponding to those false periods are usually in a physically unacceptable rotation state (cf. Sect. 3.4 and Fig. 7) or have 'strange' triangular or hexagonal pole-on silhouettes.

3.7. *Tumbling asteroids*

We carried out several simulations with tumbling asteroids. Results of those simulations clearly show that sparse photometric data generated by a tumbler cannot be fitted with a principal axis rotator model, which means that tumblers cannot be misinterpreted as ordinary principal rotators. Although it is possible to carry out the inversion of lightcurves of a precessing asteroid and solve the inverse problem (Kaasalainen 2001), this cannot be done with sparse data. The parameter space of two periods is so huge that it is practically impossible to scan all combinations.

4. Conclusion

The era of sparse photometric data that is about to come will require automated processing of the data. With the large number of observed asteroids it will not be possible to deal with individual targets separately. The aim of our extensive simulations is to explore limits of sparse data sets and to understand results of lightcurve inversion applied to data produced by different types of asteroids. So far results show that sparse data can be sufficiently used for modelling of physical properties of asteroids. The main result is that sophisticated application of the lightcurve inversion method to sparse data does not produce false solutions. If a unique solution fitting data down to the noise level is found, it is always the correct solution. This enables us to process sparse photometric data automatically. All the cases that lead to multiple solutions that all fit the data well or such cases that cannot be fitted with any model can be removed from the processing pipeline and tagged for closer investigation and photometric follow-up.

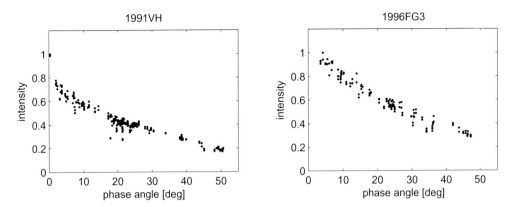

Figure 8. Intensity against the solar phase angle plot for two asynchronous binary NEAs. The decrease of intensity due to the mutual events is visible in both cases. The size ratio of the secondary to the primary is ∼ 0.4 and ∼ 0.3 for 1991VH and 1996FG3 respectively. Both primaries are nearly spheroidal objects.

Photometric follow-up observations of selected targets (suspected binaries or tumblers, extreme fast or slow rotators, etc.) will be very important. Only a few dense lightcurves can reveal the nature of such object. As is shown in KD, Sect. 4.1, inversion of combined datasets can lead to a unique physical model even if the inversion would not be possible with either dataset alone. Sparse sequences that contain few points or are too noisy to be used alone can be combined with ordinary lightcurves and a model can be derived. We can expect hundreds or thousands of asteroid models within next few years if we combine the first Pan-STARRS data with the already available database of dense lightcurves for about 2000 asteroids. Tens of thousands of models will be available within the next decade after Pan-STARRS and other surveys collect enough data.

Acknowledgements

This work was supported, in part, by CIMO and the Academy of Finland. We thank for the computer time on the computational cluster Tiger at the Astronomical Institute of the Charles University in Prague, and we thank L. Šubr, who has designed and maintained the cluster.

References

Ďurech, J. & Kaasalainen, M. 2003, *Astron. Astrophys.* 404, 709
Ďurech, J., Grav, T., Jedicke, R., Kaasalainen, M. & Denneau, L. 2006, *Earth, Moon & Planets*, in press
Kaasalainen, M. 2001, *Astron. Astrophys.* 376, 302
Kaasalainen, M. 2004, *Astron. Astrophys.* 422, L39
Kaasalainen, M. & Lamberg, L. 2006, *Inverse Problems* 22, 749
Kaasalainen, M., Pravec, P., Krugly, Y. N., Šarounová, L., Torppa, J., Virtanen, J., Kaasalainen, S., Erikson, A., Natheus, A., Ďurech, J., Wolf, M., Lagerros, J. S. V., Lindgren, M., Lagerkvist, C.-I., Koff, R., Davies, J., Mann, R., Kušnirák, P., Gaftonyuk, N. M., Shevchenko, V. G., Chirony, V. G. & Belskaya, I. N. 2004, *Icarus* 167, 178
Kaasalainen, M. & Torppa, J. 2001, *Icarus* 153, 24
Kaasalainen, M., Torppa, J. & Muinonen, K. 2001, *Icarus* 153, 37
Kaasalainen, M., Torppa, J. & Piironen, J. 2002a, *Astron. Astrophys.* 383, L19
Kaasalainen, M., Torppa, J. & Piironen, J. 2002b, *Icarus* 159, 369
Kaasalainen, S., Kaasalainen, M. & Piironen, J. 2005, *Astron. Astrophys.* 440, 1177

Marchis, F., Kaasalainen, M., Hom, E., Berthier, J., Enriquez, J., Hestroffer, D., Le Mignant, D. & de Pater, I. 2006, *Icarus* 185, 39

Muinonen, K. 1996, *Earth, Moon & Planets* 72, 339

Pravec, P., Scheirich, P., Kušnirák, P., Šarounová, L., Mottola, S., Hahn, G., Brown, P., Esquerdo, G., Kaiser, N., Krzeminski, Z., Pray, D. P., Warner, B. D., Harris, A. W., Nolan, M. C., Howell, E. S., Benner, L. A. M., Margot, J.-L., Galád, A., Holliday, W., Hicks, M. D., Krugly, Y. N., Tholen, D., Whiteley, R., Marchis, F., Degraff, D. R., Grauer, A., Larson, S., Velichko, F. P., Cooney, W. R., Stephens, R., Zhu, J., Kirsch, K., Dyvig, R., Snyder, L., Reddy, V., Moore, S., Gajdoš, Š., Világi, J., Masi, G., Higgins, D., Funkhouser, G., Knight, B., Slivan, S., Behrend, R., Grenon, M., Burki, G., Roy, R., Demeautis, C., Matter, D., Waelchli, N., Revaz, Y., Klotz, A., Rieugné, M., Thierry, P., Cotrez, V., Brunetto, L. & Kober, G. 2006, *Icarus* 181, 63

Torppa, J., Kaasalainen, M., Michalowski, T., Kwiatkowski, T., Kryszczyńska, A., Denchev, P. & Kowalski, R. 2003, *Icarus* 164, 346

Near Earth Objects, our Celestial Neighbors: Opportunity and Risk
Proceedings IAU Symposium No. 236, 2006
A. Milani, G.B. Valsecchi & D. Vokrouhlický
© 2007 International Astronomical Union
doi:10.1017/S1743921307003237

Tidal disturbances of small cohesionless bodies: limits on planetary close approach distances

Patrick Michel[1] and K. A. Holsapple[2]

[1]Côte d'Azur Observatory, UMR 6202 Cassiopée/CNRS, BP 4229, 06304 Nice Cedex 4, France
email: michel@obs-nice.fr

[2]Dept of Aeronautics & Astronautics, University of Washington, Seattle, WA 98195, USA
email: holsappl@aa.washington.edu

Abstract. The population of Near-Earth Objects contains small bodies that can make very close passages to the Earth and the other planets. Depending on the approach distance and the object's internal structure, some shape readjustment or disruption may occur as a result of tidal forces. A real example is the comet Shoemaker Levy 9 which disrupted into 21 fragments as a result of a close approach to Jupiter, before colliding with the planet during the next passage in July 1994. We have recently developed an exact analytical theory for the distortion and disruption limits of spinning ellipsoidal bodies subjected to tidal forces, using the Drucker-Prager strength model with zero cohesion. This model is the appropriate one for dry granular materials such as sands and rocks, for rubble-pile asteroids and comets, as well as for larger planetary satellites, asteroids and comets for which the cohesion can be ignored. Here, we recall the general concept of this theory for which details and major results are given in a recent publication. In particular, we focus on the definition of "material strength": while it has great implications this concept is often misunderstood in the community of researchers working on small bodies. Then, we apply our theory to a few real objects, showing that it can provide some constraints on their unknown properties such as their bulk density. In particular it can be used to estimate the maximum bulk density that a particular object, such as 99942 Apophis, must have to undergo some tidal readjustments during a predicted planetary approach. The limits of this theory are also discussed. The cases where internal cohesion cannot be ignored will then be investigated in the near future.

Keywords. Self-gravitation; asteroid, shape

1. Introduction

In our Solar System, a great number of small bodies are on trajectories that make them pass very close to a planet. This is also the case for small planetary satellites, such as the ones recently discovered by the Cassini mission around Saturn. For asteroids and comets in orbit around the sun, the close trajectory to a planet occurs in a fly-by with a short time interval. For instance, the comet Shoemaker Levy 9 made a very close approach to Jupiter which resulted in its disruption into 21 fragments, which in turn collided with Jupiter during the next passage in July 1994. A particular population of bodies whose members can undergo close Earth approaches is the population of Near Earth Objects (NEOs). which can pass as close as a few Earth radii. In this case, depending on the internal structure of the small body, tidal forces may be strong enough to cause shape readjustment or even the total disruption of the body. Tidal effects have also been considered as a potential formation mechanism of binary NEOs, although recent studies suggest that they may not be sufficient to explain the proportion of binaries in the NEO population. Since some very close approaches to the Earth can be predicted in advance for some asteroids, it is interesting to predict whether an object may undergo some shape readjustment or disruption at its closest distance; and which constraints on its internal properties, such as its bulk density, may allow or prevent such an occurrence. An important example is the well-known asteroid (99942) Apophis which is predicted to come as close as 5.6 Earth radii to the Earth. We have recently developed an exact analytical theory for the distortion or disruption limits of solid spinning ellipsoidal bodies subjected to tidal forces. The details of this theory as well as its major results are described in Holsapple & Michel (2006). Most NEOs whose shapes are known can be well approximated to at least first order by ellipsoids; that is the assumption of our model. We use the zero cohesion Drucker-Prager model of strength. That model characterizes the behavior of dry granular materials such as sands and rocks, rubble pile asteroids and comets, as well as all larger planetary satellites. We will describe in more detail the concept of strength later in this paper. We then note that our theory uses the same approach as the studies of spin limits for solid ellipsoidal bodies given in Holsapple (2001). It is a static theory that predicts conditions for break-up as well as the nature of the deformations at the limit state. However, it does not track the dynamics of the body as it comes apart. Some other studies treat this problem, and we also plan to investigate it in the future. Here, we just limit our analysis to the minimum distance at which a change can happen, but not the subsequent evolution of a body when those changes happen.

The remainder of this paper is organized as follows. Section 2 describes in some details the concept of material strength and the failure criterion used to determine whether a solid body may be affected by external forces. In section 3, we summarize our analytical theory for the tidal disruption limits of cohesionless solid spinning ellipsoids. Section 4 is devoted to the application of this theory to real cases. Conclusions and perspectives are then given in Section 5.

2. What do we mean by material strength?

The term "strength" is often used in imprecise ways, and it is important that this concept is well understood. The description provided here is largely inspired from Holsapple & Michel (2006) and Holsapple (2006). We believe that it is important to present it in different places to ensure that a large part of our community speaks the same language.

Materials such as rocks, dirt and ice, which constitute small bodies of our Solar System, are complex and characterized by several kinds of strength. Generally, the concept of "strength" is a measure of an ability to withstand stress. But stress, as a tensor, can take on many different forms. One of the simplest is a uniaxial tension, and the tensile strength is often (mis)used to characterize material strength as whole. Thus, while it is common to equate "zero tensile strength" to a fluid body, that is not correct. In fact, while "fluid" has zero tensile strength, "solids" may also. For instance, dry sand has no tensile strength. However, dry sand and granular materials in general can withstand considerable shear stress when they are under pressure: that is why we can walk on dry sand but not on water. Here comes into play a second kind of strength: the shear strength which measures the ability to withstand pure shear. The shear strength in a granular material under confining pressure comes from the fact that the interlocking particles must move apart to slide over one another, and the confining pressure resists that. Then a third kind of strength, the compressive strength, governs the ability to withstand compressive uniaxial stress. Thus, a general material has tensile strength, shear strength at zero pressure (technically the "cohesion") and compressive strength. In geological materials such as soils and rocks, the failure stresses depend strongly on the confining pressure; as a result, these three strength values can be markedly different. A cohesionless body, such as the one considered in this paper, is simply a solid body whose cohesion (shear strength at zero pressure) is null, but that does not mean it does not have any shear strength under confining pressure (provided by the self-gravity for large bodies). Hence, a body can be cohesionless but nevertheless solid.

The Drucker-Prager (DP) model is a common model for geological materials, as is the Mohr-Coulomb criterion (MC) (the difference between these two models is not relevant for our study). The DP model assumes that the allowable shear stress depends linearly on the confining pressure. The shear stress magnitude is measured by the square root of the second invariant of the deviator stress. Thus, the DP model is similar to models for linear friction and is defined by two constants: one characterizes the "cohesion" (shear strength at zero pressure); and the second characterizes the dependence on the confining pressure and is related to the so-called angle of friction. Those two constants then determine the tensile and compressive strengths. When the cohesion is zero, so is the tensile strength; but not the compressive strength. Physically, the pressure dependence is, as already explained, the consequence of the interlocking of the granular particles and not the friction of the surfaces of the particles. In fact, a closely packed mass of uniform rigid *frictionless* spherical particles has an angle of friction about $23°$, so the term angle of friction is somewhat a misnomer. Angle of *interlocking* would be more correct, but we will keep the standard name to avoid confusion.

Figure 1 gives a representation of the DP model. Using the three principal stresses σ_1, σ_2, σ_3 (positive in tension) of a general three-dimensional stress state, the pressure (positive in tension) is given as:

$$p = \frac{1}{3}(\sigma_1 + \sigma_2 + \sigma_3) \tag{2.1}$$

and the square root of the second invariant of the deviator stress is:

$$\sqrt{J_2} = \frac{1}{\sqrt{6}}\sqrt{[(\sigma_1 - \sigma_2)^2 + (\sigma_2 - \sigma_3)^2 + (\sigma_3 - \sigma_1)^2]} \tag{2.2}$$

Then, the DP failure criterion is generally given as:

$$\sqrt{J_2} \leqslant k - 3sp \tag{2.3}$$

which is illustrated as a straight line with slope $3s$ and intercept k on Fig. 1. Clearly negative pressure (compression) increases the allowable $\sqrt{J_2}$ when s is positive.

For the special case of a pure shear stress only, $\sqrt{J_2}$ is just that shear stress and the pressure p is zero. On Fig. 1, the uniaxial tension strength σ_T has $\sqrt{J_2} = 3^{-1/2}\sigma_T$ and $p = \sigma_T/3$. The uniaxial compression strength σ_c has $\sqrt{J_2} = -3^{-1/2}\sigma_C$ and $p = \sigma_C/3$.

The DP criterion can be made to match the MC one in all combinations of pressure plus uniaxial compression if the parameters s and k are related to the cohesion and the angle of friction ϕ used in the MC model. In particular, the slope s is related to the angle of friction of the MC model ϕ by:

$$s = \frac{2\sin\phi}{\sqrt{3}(3 - \sin\phi)} \tag{2.4}$$

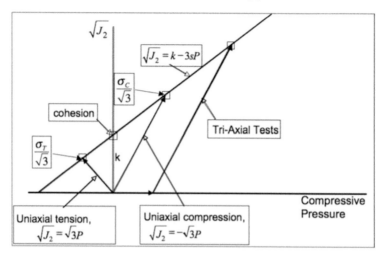

Figure 1. The Drucker-Prager failure criterion. The abscissa is positive in compression. The four small squares indicate the failure condition in, respectively from the left: tension, shear, compression and a confined compression or tri-axial test. The intercept at zero pressure at the value k is called the cohesion and the slope of the line passing through k is $3s$. From Holsapple (2006).

The intercept k of the DP model is also the shear stress τ for failure in pure shear. Technically the term "cohesion" means this intercept value of shear stress strength at zero pressure. When the cohesion is zero, so is the tensile strength, and vice versa; both cases would have the envelope starting at the origin in Fig. 1. Thus our theory applies to bodies for which those two measures are zero, while the plot shows the more general case where they are non-zero.

The MC model defines a maximum shear stress directly, which is determined by the difference of the maximum and minimum principal stresses. As a consequence, to use this criterion in algebraic manipulations involving general stress states, one must first determine the principal stresses and which is the largest and which is the smallest. The result is a difference in the algebra of the results in six different regimes, where the three principal stress components take on different orderings. An example of the six possible cases of the ordering of the stress magnitudes is given in Fig. 4 of Holsapple (2001). Moreover, there are "corners" in the curves shown where the ordering of the principal stresses changes. In contrast, the DP criterion has a single algebraic relation for all stress states. Thus, although the algebraic form of that relation is more complicated than the MC criterion, there is no need to consider the six different possibilities of the ordering of the stress magnitudes. The algebraic complexity of the DP model is of little consequence when an algebraic manipulation program such as *Mathematica* is used. So, in our theory, we use the DP failure criterion, noting that the differences between the two models are small (see, e.g. Fig. 1 in Holsapple & Michel, 2006).

3. Tidal disruption theory for cohesionless bodies

Using these concepts of strength and the DP failure criterion, we now move on to give a brief description of our theory for the limit distance for tidal readjustment or disruption of general ellipsoid solids (for details, see Holsapple & Michel, 2006).

The equilibrium problem of an ellipsoid body has been presented in Holsapple (2001). Three stress equilibrium equations must be satisfied by the stresses σ_{ij} in any body in static equilbrium with body forces b_i, which are given as (using repeated index summation convention):

$$\frac{\partial}{\partial x_j}\sigma_{ij} + \rho b_i = 0 \qquad (3.1)$$

We use an x, y, z coordinate system aligned with the ordered principal axes of the ellipsoid. In the problems here, the body forces arise from mutual gravitational forces, centrifugal forces, and/or tidal forces; they all have the simple linear forms $b_x = k_x x$, $b_y = k_y y$, $b_z = k_z z$. The full expressions of k_x, k_y and k_z are explicitly given in Holsapple & Michel (2006).

Then, for the limit states sought, the stresses must satisfy the DP failure criterion (2.3) at all points x, y and z, which, in terms of principal stresses, can be written for a cohesionless body as:

$$\frac{1}{6}[(\sigma_1 - \sigma_2)^2 + (\sigma_2 - \sigma_3)^2 + (\sigma_3 - \sigma_1)^2] = s^2[\sigma_1 + \sigma_2 + \sigma_3] \qquad (3.2)$$

Also, the surface tractions are zero on the surface points of the ellipsoidal body surface defined by: $\frac{x^2}{a} + \frac{y^2}{b} + \frac{z^2}{c} - 1 = 0$. Holsapple (2001) solves this problem, showing that the distribution of stress in that limit state just at uniform global failure has the simple form:

$$\sigma_x = -\rho k_x a^2 \left[1 - \left(\frac{x}{a}\right)^2 - \left(\frac{y}{b}\right)^2 - \left(\frac{z}{c}\right)^2\right],$$

$$\sigma_y = -\rho k_y b^2 \left[1 - \left(\frac{x}{a}\right)^2 - \left(\frac{y}{b}\right)^2 - \left(\frac{z}{c}\right)^2\right],$$

$$\sigma_z = -\rho k_z c^2 \left[1 - \left(\frac{x}{a}\right)^2 - \left(\frac{y}{b}\right)^2 - \left(\frac{z}{c}\right)^2\right], \qquad (3.3)$$

and the shear stresses in this coordinate system are all zero (note that the x, y, z stresses being principal stresses, they can be used for the 1, 2, 3 stresses in the previous equations). The body force constants k_x, k_y, k_z depend on the body forces, so those body forces must be such that the DP failure criterion is not violated. That condition determines the limit states.

Note that the fact that all the stresses have a common functional dependence on the coordinates is a necessary consequence of the fact that the DP and MC failure criteria are homogeneous functions of the stress components. Putting the expressions of these components into the DP criterion(3.2), one can see that the common functional dependence $1 - \left(\frac{x}{a}\right)^2 - \left(\frac{y}{b}\right)^2 - \left(\frac{z}{c}\right)^2$ will cancel out of the Eq. (3.2). That is because the limit stress state has simultaneous failure at all points. Thus, we can omit that functional dependence and focus on finding the combinations of the leading multipliers of the three terms of 3.3 that satisfy the failure criterion. We define the dimensionless spin and dimensionless distance by:

$$\Omega = \frac{\omega}{\sqrt{\pi \rho G}}, \qquad \delta = \left(\frac{\rho}{\rho_p}\right)^{1/3}\frac{d}{R} \qquad (3.4)$$

then, as detailed in Holsapple & Michel (2006), failure will occur when:

$$\frac{1}{6}[(c_x - c_y)^2 + (c_y - c_z)^2 + (c_z - c_x)^2] = s^2[c_x + c_y + c_z]^2 \qquad (3.5)$$

where, for arbitrary spin and when the long axis points towards the Earth:

$$c_x = \left(-A_x + \frac{1}{2}\Omega^2 + \frac{4}{3}\delta^{-3} \right),$$

$$c_y = \beta^2 \left(-A_y + \frac{1}{2}\Omega^2 - \frac{2}{3}\delta^{-3} \right),$$

$$c_z = \alpha^2 \left(-A_z - \frac{2}{3}\delta^{-3} \right) \tag{3.6}$$

and A_x, A_y and A_z are the components of the self-gravitational potential of a homogneous ellipsoidal body of uniform mass density ρ in the body coordinate system expressed as: $U = \pi\rho G(A_0 + A_x x^2 + A_y y^2 + A_z z^2)$ (e.g. Chandrasekhar 1969).

Holsapple & Michel (2006) give a similar form when the long axis points along the trajectory at its closest approach. The criterion expressed in these forms can then be used to solve for the dimensionless distances δ at the failure condition as a function of the aspect ratios α and β (which determine the A_x, A_y and A_z), the mass ratio p of the secondary to the primary, and for any value of the constant s related to the angle of friction. The solution always has the dimensionless form:

$$\delta = \frac{d}{R} \left(\frac{\rho}{\rho_p} \right)^{1/3} = F[\alpha, \beta, p, \phi, \Omega] \tag{3.7}$$

so that the mass density only occurs with this cube root. The number of independent variables is reduced by one when the scaled spin is zero or the spin-locked value, and by another one when $p = 0$, i.e. when the mass of the secondary is negligible compared to that of the primary, which is the case for an asteroid flying by a planet.

We also want to stress that the limit distance to the primary corresponds to the distance below which a secondary with a given shape cannot exist, because the failure criterion would be violated. However, it does not mean that below this distance, the secondary would disrupt, and a flow rule is required to indicate the nature of any readjustment (or disruption). Then, if those changes lead to a new configuration that is within failure at the given distance, a shape change is indicated. Otherwise, if the new shape still violates the failure criterion, a global disruption is indicated. Such analysis goes beyond the scope of this paper but is developed in Holsapple & Michel (2006).

4. Application to some real NEOs

In this section, we show how our theory can be used to estimate the tidal limit distance to Earth of some Near-Earth Objects whose required physical properties for this theory have been estimated. In Holsapple & Michel (2006), we used the asteroid Apophis as an example, and estimated its mass densities that might allow a tidal disruption or readjustment during its predicted close approach.

4.1. Limit distance to the Earth of Itokawa

The asteroid (25143) Itokawa was visited by the Japanese space mission Hayabusa during fall 2005. The goal of the Hayabusa mission was to collect some samples of this asteroid and bring them back to Earth. The return of the spacecraft is anticipated in 2010, and whether or not some samples have been collected will only be known at that time. Nevertheless, this mission has provided the first detailed images of such a small asteroid. Fujiwara et al. (2006) review the main results of the mission and indicate that the size of Itokawa is 535x294x209 meters and its bulk density is 1.9 ± 0.13 g/cm^3, which is lower than expected for coherent S-type asteroids (usually around 2.7 g/cm^3). Its rotation period of 12.1324 hours is also longer than usual ones for bodies of this size.

Thanks to this mission, we have almost all the necessary parameters to estimate the limit distance to the Earth which Itokawa can approach without undergoing any readjustment or disruption. We set $p = 0$ (Itokawa's mass is negligible with respect to Earth's mass), its aspect ratios are $\alpha = 0.39$ and $\beta = 0.55$. We also set the Earth's bulk density to $\rho_p = 5.515$ g/cm^3. Since Itokawa's density is $\rho = 1.9$ g/cm^3, the factor $(\rho/\rho_p)^{1/3}$ in the expression of δ (3.7) is equal to 0.701, and we can divide the right hand side of (3.7) to express the distance in term of d/R only. Note that the classical Roche limit (for a fluid and ignoring the fact that Itokawa does not have the required Roche fluid shape) would then equal $d/R = 2.455/0.701 = 3.50$. The only unknown is the angle of friction and we can determine the limit distance that it can approach as a function of this parameter. The results are shown in Figure 2, for the two special orientations at the closest approach: the long axis pointing towards the Earth or the long axis pointing along the trajectory. As expected, the distance decreases for increasing angle of friction. The minimum limit distance corresponds to an angle of friction of 90° and is about 1.4 Earth radii from Earth's center. But, for an angle of friction of 30°, which is typical for rocky bodies, this distance is about 2.5 Earth radii from Earth's center for either orientation, distinctly closer than the classical Roche value.

Note that both solutions approach infinity if the angle of friction is less than 15, indicating failure at any distance. That is because Itokawa could not have its observed spin even without tidal forces if the angle of friction were less than 15°: if so it would collapse into a more spherical body; see Holsapple (2001). The case with the long axis pointed down actually has near-Earth solutions down to about 7°. That is because at those distances the tidal forces can add additional tensile stresses along the long axis, partially offsetting the excessive compression from self-gravity. However, there is no way that portion of the curve can be accessed, since at further distances where the tidal forces are not yet effective, the asteroid has to have the higher friction angle of 15°.

Also we note that a close approach could change the spin of the asteroid. However, as shown in Holsapple & Michel (2006), the spin has little effect on the closest approach distance as long as it is not close to the limit spin.

Thus, since all currently possible predicted Earth approaches of Itokawa until 2100 are always situated well above those distances (see the NEODys web site), tidal readjustments or disruptions due to Earth approaches are not expected for this object in the near future.

4.2. *Limit distance to the Earth of Geographos*

The asteroid (1620) Geographos was the target of radar observations in 1994 when the asteroid was 7.2 million kilometers from Earth. A planetary radar instrument at the Deep Space Network's facility in Goldstone was used and the details have been published by Ostro *et al.* (1996). The images show a highly elongated object whose representation by an homogeneous ellipsoid leads to a size approximately $4.7 \times 1.9 \times 1.9$ km, as estimated by Hudson & Ostro (1999). Its rotation period is 5.22 hours and it is an S-type Near-Earth asteroid, so that we assume a typical bulk density for non-porous bodies of this taxonomic type, about 2.7 g/cm^3. Of course, if porous, the density could be less.

We have thus again all the parameters required to repeat the same exercise as for Itokawa.

First, we can determine the required properties at its present state removed from external gravitational forces. With the aspect ratio of 0.35 and the period of 5.22 hours, the required angle of friction to withstand gravitational collapse along the long axis is

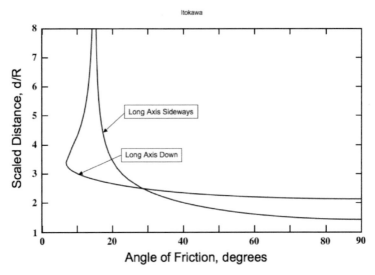

Figure 2. Limit distance (in Earth radii) of Itokawa to the Earth as a function of its angle of friction for two different orientations at closest approach.

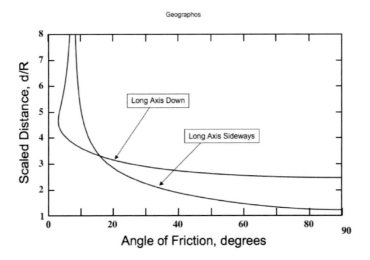

Figure 3. Limit distance (in Earth radii) of Geographos to the Earth as a function of its angle of friction.

7.7°. Note that this is a very different answer than the points plotted for Geographos in Fig. 6 of Holsapple (2001). That is a consequence of using much better dimensions for Geographos: in the region in spin-shape space occupied by this very elongated asteroid, a small difference in shape makes a large difference in required friction angle.

Then we consider a close approach to Earth. The results are shown in Figure 3.

The minimum limit distance corresponds always to an angle of friction of 90° and for the long axis pointing down, is about 2.3 Earth radii from Earth's center. But for an angle of friction of 30°, this distance is about 2.9 Earth radii from Earth's center. For the case of a sideways closest approach, the distance is a mere 1.2 radii, or almost impacting. At the friction angle of 7.7° both curves approach infinity; again that is the free-space value

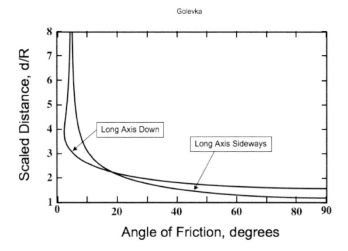

Figure 4. Limit distance (in Earth radii) of Golevka to the Earth as a function of its angle of friction.

required for its spin. Again these values for Geographos rule out any tidal disruption in the near future.

4.3. *Limit distance to the Earth of Golevka*

Radar ranging from Arecibo (Puerto Rico) has been performed by Chesley *et al.* (2003) for the near-Earth asteroid (6489) Golevka. For the first time, the magnitude of the thermal Yarkovsky effect was measured for a small asteroid. Since this effect is a function of the asteroid's mass and surface thermal characteristics, its direct detection helped constrain the physical properties of Golevka, such as its bulk density, and refine its orbital path. Based on the strength of the detected perturbation, the authors estimated the bulk density of Golevka to be about 2.7 g/cm^3. Its size is estimated to be about $0.35 \times 0.25 \times 0.25$ km and its rotation period is 6.0264 hours.

So, for Golevka, we can again apply the same exercise as previously. However, it should be noted that this asteroid has a very strange angular shape, and the use of an average ellipsoid certainly introduces some error. In any case, Figure 4 shows the results. For an angle of friction of 30° the asteroid can approach as close as 1.8–2.0 Earth radii from Earth's center, depending on the orientation, suggesting that this asteroid is also safe with respect to tidal disruption by the Earth in the near future.

5. Conclusion and Perspectives

We have developed a theory for cohesionless ellipsoids that gives explicit and exact results for any ellipsoidal shape and combination of gravity, spin, tidal forces, giving closed-form solutions for the limit distances to a planet for such bodies. We find that a body with the expected physical properties of a geological solid for which the cohesion can be ignored can exist in arbitrary ellipsoidal shapes and much closer to a primary than a fluid body. Applying our theory to a few NEOs whose physical properties have been determined, we find that none of these examples should undergo a tidal disruption or readjustment in the near future, as their predicted future Earth approaches are well above the limit distance for such an event.

We want to stress that our theory is static and is therefore limited to the determination of the conditions for the onset of disruption. Thus, the nature of the resulting dynamics of a body if it disintegrates is not addressed by this theory. In particular, the resulting motions are affected by how the body breaks, and the resultant history of the broken particles. For instance, although a body may disrupt once the limit distance is reached, it may reassemble as a result of gravitational reaccumulation into a single piece, so that the final outcome is not the disruption that was predicted by our theory. Some indications of such behaviors are given by Walsh & Richardson (2006) and we also plan to address this problem in future work. Our next step in the theory will finally be to consider small bodies for which the cohesion cannot be neglected, as has be done recently by Holsapple (2006) for the case of pure spin limits.

References

Chandrasekhar, S. 1969, *Ellipsoidal Figures of Equilibrium*, Dover, New York

Chesley, S.R., *et al.* 2003, *Science* 302, 1739

Fujiwara, A., *et al.* 2006, *Science* 312, 1330

Holsapple, K.A. 2001, *Icarus* 154, 432

Holsapple, K.A. 2004, *Icarus* 172, 272

Holsapple, K.A. 2006, *Icarus*, accepted

Holsapple, K.A. & Michel, P. 2006, *Icarus* 183, 331

Hudson, & Ostro, S.J. 1999, *Icarus* 140, 369

Ostro, S.J., *et al.* 1996, *Icarus* 121, 44

Wlash, K.J. & Richardson, D.C. 2006, *Icarus* 180, 201

Near Earth Objects, our Celestial Neighbors: Opportunity and Risk
Proceedings IAU Symposium No. 236, 2006
A. Milani, G.B. Valsecchi & D. Vokrouhlický, eds.
© 2007 International Astronomical Union
doi:10.1017/S1743921307003249

Earth's 2006 encounter with comet 73P/Schwassmann-Wachmann: Products of nucleus fragmentation seen in closeup

Zdeněk Sekanina

Jet Propulsion Laboratory, California Institute of Technology, Pasadena, CA 91109, USA
email: zs@sek.jpl.nasa.gov

Abstract. The large numbers of nucleus fragments observed are a spectacular illustration of the process of cascading fragmentation in progress, a concept introduced to interpret the properties of the Kreutz system of sungrazers and comet D/1993 F2. The objective is to describe the fragmentation sequence and hierarchy of comet 73P, the nature of the fragmentation process and observed events, and the expected future evolution of this comet. The orbital arc populated by the fragments refers to an interval of 3.74 days in the perihelion time. This result suggests that they all could be products (but not necessarily first-generation fragments) of two 1995 events, in early September (involving an enormous outburst) and at the beginning of November. The interval of perihelion times is equivalent to a range of about 2.5 m/s in separation velocity or 0.00012 the Sun's attraction in nongravitational deceleration. Their combined effect suggests minor orbital momentum changes acquired during fragmentation and decelerations compatible with survival over two revolutions about the Sun. Fragment B is a likely first-generation product of one of the 1995 events. From the behavior of the primary fragment C, 73P is not a dying comet, even though fragment B and others were episodically breaking up into many pieces. Each episode began with the sudden appearance of a starlike nucleus condensation and a rapidly expanding outburst, followed by a development of jets, and a gradual tailward extension of the fading condensation, until the discrete masses embedded in it could be resolved. In April-May, this debris traveled first to the southwest, but models show their eventual motion toward the projected orbit. Fainter fragments were imaged over limited time, apparently because of their erratic activity (interspersed with periods of dormancy) rather than improptu disintegration. A dust trail joining the fragments and reminiscent of comet 141P/Machholz suggests that cascading fragmentation exerts itself profoundly over an extremely broad mass range of particulate debris.

Keywords. Comet, fragmentation; data analysis

1. Introduction

Only since recently has it been recognized that fragmentation is an omnipresent process among comets that proceeds at all heliocentric distances. Fragmentation is also increasingly perceived as the dominant process of cometary demise, likely to account in most (though not necessarily all) cases for the end state.

Perhaps the most fascinating research opportunity that cometary fragmentation offers to a scientist is the benefit to examine, at no extra cost, the interior of the nucleus as subsurface areas suddenly become exposed to direct solar radiation and other outer-space effects. The fragmentation process itself is also of much interest, especially for comets that break up *nontidally* (Sekanina 1997), having experienced no close approach to the Sun or Jupiter in the past. Fragmentation of a comet's nucleus is facilitated by its extremely low mechanical strength (e.g., Whipple 1950, 1963, A'Hearn *et al.* 2005) and is probably also aided significantly by major variations in the mechanical-strength distribution throughout the nucleus interior. It appears that *nontidal* fragmentation is

211

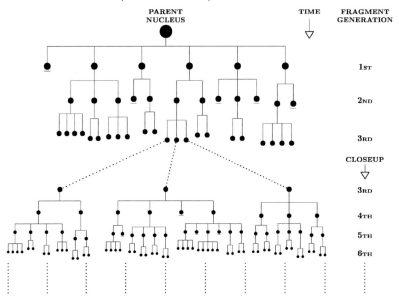

Figure 1. Schematic representation of the process of cascading fragmentation, proposed for cometary nuclei. The underscored symbols identify the surviving fragments whose (incomplete) disintegration has terminated. The first three generations of fragments are displayed in the upper part; a detail of the fourth, fifth, and sixth generations is in the lower part.

triggered by one or more of three possible mechanisms: rotational tension, thermal stress, and pressure of outflowing gases from discrete sources, especially when the volatiles are trapped beneath the surface.

Effects on split comets can be discriminated into two categories: nondestructive and cataclysmic. An event of the first category is survived by at least one fragment nearly unaffected, so that the comet's life goes on. By contrast, an event of the second category destroys the comet completely on a very short time scale. In this paper I offer some early results on nondestructive fragmentation of comet 73P/Schwassmann-Wachmann 3.

2. Cascading Fragmentation

In 1996, my interest in cometary fragmentation was aroused by the astonishingly high discovery rate of dwarf comets of the Kreutz sungrazer system, all of which, before reaching perihelion, fade and vanish while imaged with two coronagraphs onboard the NASA/ESA *Solar and Heliospheric Observatory* (SOHO). Following an earlier investigation of comet D/1993 F2 (Sekanina *et al.* 1998) that first split and eventually collided with Jupiter, I proposed a concept of *cascading fragmentation* (Sekanina 2002) to explain the observed sequence of events (Figure 1). In this scenario, the original parent comet continues to break up over and over again, with an ever larger number of fragments of ever smaller size being generated episodically.

In the case of the Kreutz system, the inevitability of this scenario has amply been documented by two facts: (i) all minor sungrazers, discovered coronagraphically (mostly with the SOHO instruments) fail to survive their perihelion passage, implying that their existence as separate objects cannot predate the previous perihelion passage and their parent bodies, in order to survive, must have been substantially (orders of magnitude)

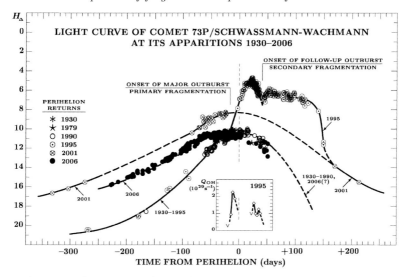

Figure 2. Light curve of comet 73P during the six apparitions 1930–2006. The plotted magnitude H_Δ was corrected for effects of geocentric distance and normalized to the visual system. The onset times of the two outbursts in 1995 and their apparent coincidence with the times of the primary and secondary events of nucleus fragmentation are depicted. During the 2001 and 2006 apparitions the magnitudes plotted refer to the primary fragment C. The inset shows the parallel 1995 variations in the hydroxyl production rate, measured by Crovisier *et al.* (1996).

more massive; and (ii) the minor sungrazers have a strong tendency to arrive at the Sun in pairs or clusters, moving along trajectories that are similar but by no means identical. The differences indicate orbital perturbations caused by separation velocities of a few m/s acquired during fragmentation events far from the Sun, where the orbital velocity does not exceed a few tens m/s.

In the work on comet D/1993 F2 (Sekanina *et al.* 1998), we were able to determine the family tree of the fragmentation products. By the time the comet collided with Jupiter two years after the initial, tidally-triggered breakup, fragments of the first, second, and third generations were identified. With many dozens of fragments now observed in the orbit of comet 73P, a new opportunity is presented to test the concept of cascading fragmentation in a case where the initial breakup was nontidal in nature.

3. Brief History of Comet 73P/Schwassmann-Wachmann

This comet is a member of the Jupiter family of short-period comets, with an orbital period of 5.4 years and perihelion near 1 AU. The 2006 apparition is the comet's sixth observed return to the Sun. Its light-curve evolution, displayed in Figure 2, had been unexciting until early September 1995, when an enormous outburst began about two weeks before perihelion. The event was first detected by Crovisier *et al.* (1996) as an OH production increase at 18 cm. The optical confirmation came several days later. Figure 2 shows that the first outburst was followed by a second, smaller one nearly two months later and that the comet's brightness remained elevated not only for the rest of the 1995 apparition, but also in 2001 and 2006.

The multiplicity of the nucleus was first detected by Boehnhardt & Käufl (1995) at the European Southern Observatory (ESO) on December 12, more than three months after the onset of the first outburst. Three optically detected fragments were aligned in a nearly rectilinear chain about 4″ long. According to the notation by Marsden (1996a), the westernmost fragment became known as A, the easternmost as C, and the middle, initially the faintest one, as B. A fourth fragment, called D, was reported independently by J. V. Scotti and A. Galád only on 1995 December 27–29 (Marsden 1996a).

Subsequent close inspection of the ESO images taken by K. Reinsch on November 28 and by J. Storm on December 2 revealed that the comet was already double (Boehnhardt *et al.* 1996). Nucleus C was clearly the primary (and presumably the most massive) fragment (Marsden 1996a). Fragments A, B, and C were observed until mid-February 1996. After conjunction with the Sun, in late August 1996, only C and one companion were detected and observed for more than three months (Marsden 1996b, 1997).

When the comet was recovered in November 2000, the primary and two companions were detected; one of them, officially designated E, appeared to be a new fragment (Green 2000). After perihelion, from July to December 2001, the primary and a single companion were under observation at ESO (Boehnhardt *et al.* 2002).

A recent effort to sort out the identity of the fragments and to establish their fragmentation sequence and hierarchy (Sekanina 2005) produced surprising results. All examined companions were found to have separated from the parent comet they shared with the main fragment C rather than from one another. Fragment A was short lived, seen only in the late 1995 and early 1996, separating in late October 1995 and moving rapidly away from C. Fragment E, introduced as a new fragment in 2000, turned out to be identical with the companion from the late 1996; it was found to have separated from the parent at the onset time of the major 1995 outburst. Fragment B could not satisfactorily be linked with the 2001 companion, nor with the other 2000 companion. These two could, however, be linked together and identified as a new fragment F, which separated from what was left of the parent nucleus at the onset time of the follow-up outburst (see Figure 2). It was concluded that fragment B was observed only in the late 1995 and early 1996 and that it separated probably during the major outburst. Thus, a strong correlation has been established between the fragmentation events and the outbursts. The 2006 ephemerides for companions E, F, and B were calculated, with the proviso that the ephemeris for B was very uncertain because of severe extrapolation.

4. Current Return of Comet 73P: A String of Nucleus Fragments

The main comet (fragment C) was recovered by C. Hergenrother with 1.2-meter reflector at Mount Hopkins on 2005 October 22 (Green 2005), 227 days before perihelion. The light curve available at the time of this writing (the beginning of August 2006) shows that along much of the preperihelion arc of the orbit this surviving fragment was still brighter than the parent comet in 1930–1995 but that near and several weeks after perihelion the light curve of C was running at or slightly below that of the parent (Figure 2).

The first companion to 73P during the current return was discovered by J. A. Farrell with his 0.41-meter reflector on 2006 January 6, or 151 days before perihelion of fragment C; I tentatively identified it with fragment B from 1995–1996 (Green 2006a). Next came R. A. Tucker's and E. J. Christensen's independent discoveries of fragment G on 2006 February 20–24 (Green 2006b). Figure 3 indicates that large numbers of additional fragments were discovered starting from March 4, most of them with the 1.5-meter reflector at Mount Lemmon (Green 2006c). An official count — 65 including fragment C — is incomplete, because none of the several dozen minifragments seen in the images of companions B and G taken with the Hubble Space Telescope† (HST) on April 18–20 has been accounted for. Similarly, the high-resolution images taken with the Very Large

† http://hubblesite.org/newscenter/newsdesk/archive/releases/2006/18

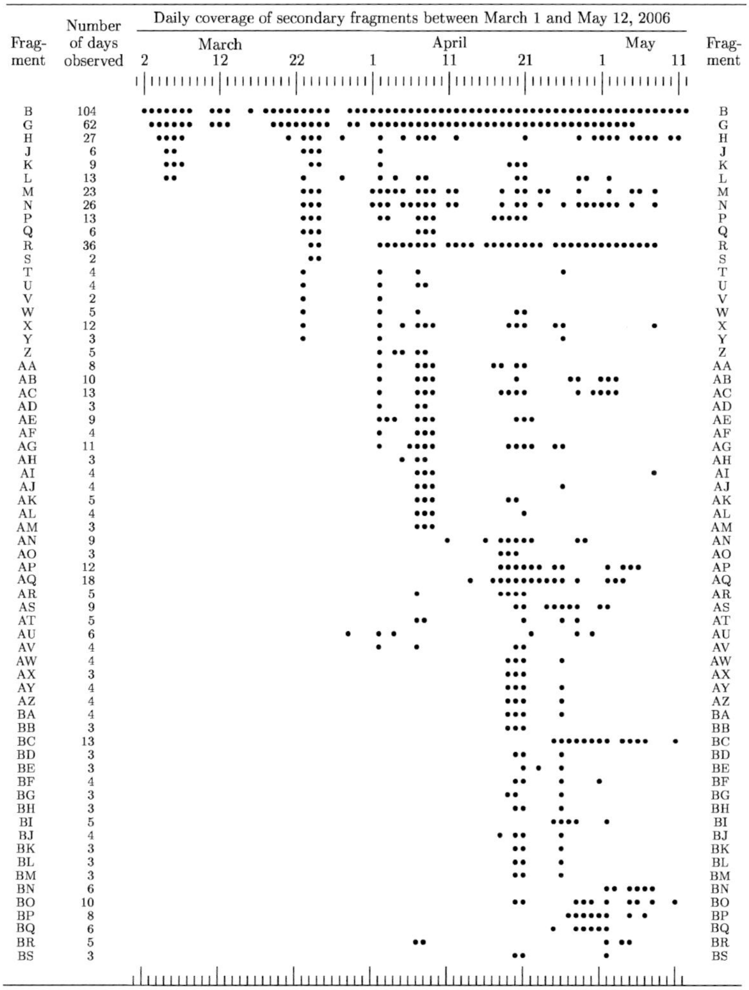

Figure 3 (chart):

Daily coverage of secondary fragments between March 1 and May 12, 2006

Fragment	Number of days observed
B	104
G	62
H	27
J	6
K	9
L	13
M	23
N	26
P	13
Q	6
R	36
S	2
T	4
U	4
V	2
W	5
X	12
Y	3
Z	5
AA	8
AB	10
AC	13
AD	3
AE	9
AF	4
AG	11
AH	3
AI	4
AJ	4
AK	5
AL	4
AM	3
AN	9
AO	3
AP	12
AQ	18
AR	5
AS	9
AT	5
AU	6
AV	4
AW	4
AX	3
AY	4
AZ	4
BA	4
BB	3
BC	13
BD	3
BE	3
BF	4
BG	3
BH	3
BI	5
BJ	4
BK	3
BL	3
BM	3
BN	6
BO	10
BP	8
BQ	6
BR	5
BS	3

Figure 3. Astrometric observations of fragments of 73P between March 1 and May 12, 2006.

Telescope (Kueyen) at ESO's Cerro Paranal Station on April 23/24 depicted seven faint fragments closely trailing companion B‡, while C. Hergenrother¶ and others noticed short-lived fragments near B and G on numerous occasions in April and May. Additional minor fragments briefly accompanying or involved with the condensations H, M, N, R, etc. were also reported by observers from time to time. All the minifragments imaged near B have in the official count been "represented" by a single generic fragment AQ, whereas those near G and other companions have mostly been ignored.

Figure 4 compares the light curves of three fragments. It is noticed that the light curve of the main fragment C is very smooth, while that of B, the second brightest fragment, has a distinctly steeper slope with three outbursts before perihelion. The entire light curve of N, one of the fainter companions, consists entirely of rapid fluctuations. It seems that the fainter (and, presumably, the smaller and less massive) the fragment is, the more erratic its activity, which apparently implies the object's lesser textural homogeneity and mechanical stability, and therefore its shorter lifetime.

‡ http://www.eso.org/outreach/press-rel/pr-2006/pr-15-06.html
¶ http://www.lpl.arizona.edu/~chergen/73P.html

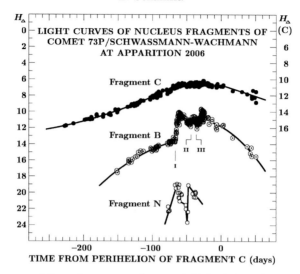

Figure 4. Light curves of three fragments of comet 73P in 2006. The plotted magnitude H_Δ was normalized to 1 AU and to the visual system. The smoothness of the curve for fragment C contrasts with the three outbursts of fragment B, marked as I (with the onset on April 1), II (April 24), and III (May 2), and with the ragged curve of fragment N. The light curve for the primary fragment C was moved 4 magnitudes up (right scale) to avoid a congestion of the curves for B and C. At the peak of outbursts I and III, fragment B was slightly brighter than C.

A fragmentation episode, like those experienced by companion B, begins with the sudden appearance of a starlike nucleus condensation and an outburst, followed by a development of jets and by a gradual tailward extension of the fading condensation, until the discrete, boulder-sized masses (minifragments) embedded in it are resolved. The described phenomena are products of a rapidly expanding cloud of microscopic dust that makes up the lower end of the size spectrum of the released debris. A procession of minifragments was observed to follow each of the outbursts of companion B as well as the flare-ups of numerous other companions.

Figure 5 presents simple fragmentation models (Sec. 5), which show a range of possible scenarios for the April–May time frame. It is noted that in this period of time, freshly released fragments traveled, relative to their parent, first to the southwest and only later to the north, toward the projected orbit. On the other hand, very old fragments traveled essentially along the orbit, explaining the observed string of lined-up fragments.

The motions of these fragments are crudely described by the orbit of C with shifted perihelion times. Figure 6 shows a peculiar effect: the expected perihelion times of the two companions from 2000/2001 coincide with the gaps in the histogram. Is this telling us that E and F have disintegrated into the observed populations of fragments? Their majority should indeed pass through perihelion at slightly later times. The range of perihelion times of the examined fragments, 3.74 days, is equivalent to an orbital-velocity increment of about 2.5 m/s or to a differential deceleration of 0.00012 the Sun's gravitational acceleration for a single event having occurred in September–November 1995. These values suggest that as few as 1–2 episodes per fragment would suffice to explain the entire span of perihelion times.

The complexities of the spatial distribution of fragments are illustrated by their four subsets on four dates in Figures 7–10. The apparent resiliency of a number of fragments and the fact that fragment C continues to be in good health suggest that 73P is not yet a dying comet, contrary to recently expressed opinions in some magazines.

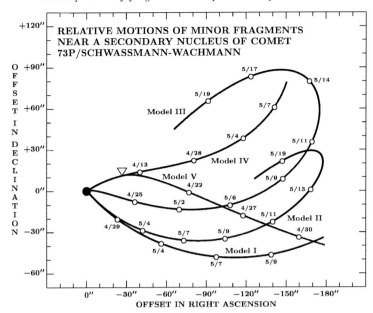

Figure 5. Models for relative motions of higher-generation fragments in the proximity of a first-generation (reference) fragment of 73P during April and May 2006. Times of separation from the reference fragment and decelerations (in units of 10^{-5} the Sun's gravitational acceleration): 2006 Apr 1 and 40 units for Model I; 2006 Apr 1 and 30 units for Model II; 2006 Jan 1 and 6 units for Model III; 2001 Jan 1 and 0.02 units for Model IV; in Model V, the fragment separated from Model IV fragment (rather than from the reference fragment) on 2006 Apr 1 at a location marked with a triangle, with a deceleration of 60 units.

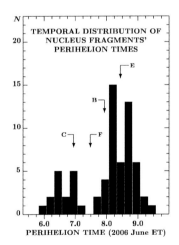

Figure 6. Histogram of the temporal distribution of perihelion times for 65 nucleus fragments of 73P with official designation. The times span an interval of 3.74 days. The times of the main fragment C and the brightest secondary fragment B and the predicted locations of fragments E and F (observed during the previous apparitions but not in 2006) are marked with the arrows. Note that the times for E and F match the gaps in the temporal distribution.

One of the products of the process of cascading fragmentation is the formation of a dust trail, the phenomenon investigated for a number of periodic comets (e.g., Sykes & Walker 1992). The trail of coarse-grain and pebble-sized debris of 73P detected in the

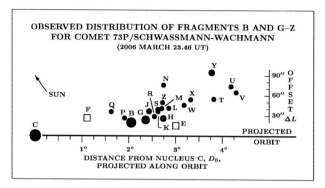

Figure 7. The locations of fragments B and G through Z relative to the main fragment C in projection onto the plane of the sky on 2006 March 23.46 UT. Offset ΔL is a displacement from the direction of the projected orbit behind fragment C. The magnitude of this displacement is magnified by a factor of 50 relative to the scale along the orbit. The circle size corresponds approximately to the peak brightness reported. The squares are the predicted locations of fragments E and F. The direction to the Sun is also shown.

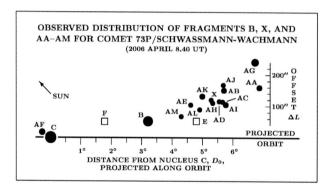

Figure 8. The locations of fragments AA–AM relative to fragment C in projection onto the plane of the sky on 2006 April 8.40 UT. Fragments B and X are plotted to allow comparison with Figure 7. Offset ΔL is magnified by a factor of 36 relative to the scale along the orbit. For more information, see the caption to Figure 7.

infrared by the Spitzer Space Telescope† is apparently of the same nature as features observed in other split comets in the optical wavelengths, namely, a sheath of material in the sungrazer C/1882 R1 (e.g., Kreutz 1888) and a dust filament or trail in D/1993 F2 (e.g., Weaver *et al.* 1994) and 141P/Machholz (cf. Sekanina 1999).

5. Analysis, Early Results, and Conclusions

Only preliminary results are available at present, based on nearly 8000 astrometric positions between January 6 and the end of July 2006, extracted from the *Minor Planet Electronic Circulars* (MPECs) 2006-B20 through 2006-P18 published by the IAU Minor Planet Center.‡ The same data are subsequently published, once a month, in the *Minor Planet Circulars*. There are currently over 1600 positions for fragment B, over 800 positions for G, over 400 for R, etc. The 2006 light-curve data are mostly from the web site of the *International Comet Quarterly*,¶ but some are from the MPECs.

† http://www.spitzer.caltech.edu/Media/releases/ssc2006-13
‡ http://www.cfa.harvard.edu/mpec/RecentMPECs.html
¶ http://cfa-www.harvard.edu/icq/CometMags.html

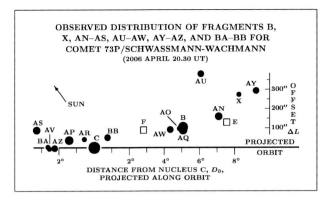

Figure 9. The locations of fragments AN–AS, AU–AW, AY–AZ, and BA–BB relative to fragment C in projection on the plane of the sky on 2006 April 20.30 UT. Fragments B and X are plotted to allow comparison with Figures 7 and 8. Offset ΔL is magnified by a factor of 38.4 relative to the scale along the orbit. For more information, see the caption to Figure 7.

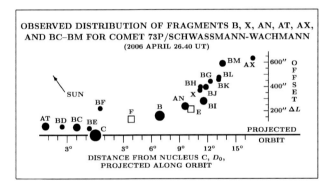

Figure 10. The locations of fragments AT, AX, and BC–MN relative to fragment C in projection onto the plane of the sky on 2006 April 26.40 UT. Fragments B, X, and AN are plotted to allow comparison with Figures 7–9. Offset ΔL is magnified by a factor of 45. For more information, see the caption to Figure 7.

The extensively tested fragmentation model (Sekanina 1978, 1982) is used to fit the relative motion of fragments in each examined pair by employing their offsets in right ascension and declination. The model allows the user to solve for up to five parameters: the time of fragmentation; the radial, transverse, and normal components of the separation velocity (referred to the orbit plane of the shared parent and aligned with the Sun-comet direction); and the differential nongravitational deceleration. The procedure involves an iterative, least-squares, differential-correction algorithm that searches for an optimum fit. One can solve for any combination of fewer than the five parameters. Because of the long periods of time involved in the case of 73P (1995–2006), the differential planetary perturbations are accounted for in a code's version that I developed in a joint effort with P. W. Chodas. With 65 fragments there are more than 2000 pair combinations possible, although many can be ruled out as implausible. A poor distribution of residuals from a solution means that the fragments in the chosen pair do not share a common parent.

The separation velocity, which is particularly important for understanding the existence of fragments on the leading side of C (such as AT in Figure 10), is interpreted as an extra momentum acquired during breakup by the smaller fragment in a pair relative to the reference mass. In reality, of course, either fragment acquires a net orbital momentum change, albeit the one for the reference mass is much smaller. By the same token, if B had been a first-generation fragment before the three outbursts (Figure 4), it became a second-generation fragment after outburst I, a third-generation fragment after outburst II, and a fourth-generation fragment after outburst III, even though it was then still called B and its net orbital momentum changed hardly at all due to the three events.

The early results show that, as expected, companion B separated from C in 1995, most probably in early November, and that the cluster of minifragments near B observed with the HST on April 18–20 (Sec. 4) was released on about April 1 (outburst I in Figure 4), thus illustrating (with many other similar events) the process of cascading fragmentation in progress and confirming a correlation between outbursts and fragmentation. Finally, it has by now become clear that the process begun with the fragmentation events in 1995 has not been cataclysmic and that 73P shows at present no signs of being a dying comet.

Acknowledgements

This research was carried out at the Jet Propulsion Laboratory, California Institute of Technology, under a contract with the National Aeronautics and Space Administration.

References

A'Hearn, M. F. and the Depp Impact Team 2005, *Science* 310, 258
Boehnhardt, H., & Käufl, H. U. 1995, *IAU Circ.* 6274
Boehnhardt, H., Holdstock, S., Hainaut, O., Tozzi, G. P., Benetti, S., & Licandro, J. 2002, *Earth Moon Plan.* 90, 131
Boehnhardt, H., Käufl, H. U., Goudfrooij, P., Storm, J., Manfroid, J., & Reinsch, K. 1996, *ESO Messenger* No. 84, p. 26
Crovisier, J., Bockelée-Morvan, D., Gérard, E., Rauer, H., Biver, N., Colom, P., & Jorda, L. 1996, *AA* 310, L17
Green, D. W. E., ed. 2000, *IAU Circ.* 7534
Green, D. W. E., ed. 2005, *IAU Circ.* 8623
Green, D. W. E., ed. 2006a, *IAU Circ.* 8659, 8660
Green, D. W. E., ed. 2006b, *IAU Circ.* 8679
Green, D. W. E., ed. 2006c, *IAU Circ.* 8685, 8692, 8693, 8703, 8704, 8709, 8715
Kreutz, H. 1888, *Publ. Kiel Sternw.* No. 3, p. 1
Marsden, B. G., ed. 1996a, *IAU Circ.* 6301
Marsden, B. G., ed. 1996b, *Minor Planet Circ.* 28339
Marsden, B. G., ed. 1997, *Minor Planet Circ.* 28917
Sekanina, Z. 1978, *Icarus* 33, 173
Sekanina, Z. 1982, in: L. L. Wilkening (ed.), *Comets* (Tucson: Univ. Arizona), p. 251
Sekanina, Z. 1997, *AA* 318, L5
Sekanina, Z. 1999, *AA* 342, 285
Sekanina, Z. 2002, *ApJ* 566, 577
Sekanina, Z. 2005, *Internat. Comet Quart.* 27, 225
Sekanina, Z., Chodas, P. W., & Yeomans, D. K. 1998, *Planet. Space Sci.* 46, 21
Sykes, M. V. & Walker, R. G. 1992, *Icarus* 95, 180
Weaver, H. A., Feldman, P. D., A'Hearn, M. F., Arpigny, C., Brown, R. A., Helin, E. F., Levy, D. H., Marsden, B. G., Meech, K. J., Larson, S. M., Noll, K. S., Scotti, J. V., Sekanina, Z., Shoemaker, C. S., Shoemaker, E. M., Smith, T. E., Storrs, A. D., Yeomans, D. K., & Zellner, B. 1994, *Science* 263, 787
Whipple, F. L. 1950, *ApJ* 111, 375
Whipple, F. L. 1963 in: B. M. Middlehurst & G. P. Kuiper (eds.), *The Moon, Meteorites, and Comets* (Chicago: Univ. Chicago), p. 639

Part 4

Surfaces and Composition

Near Earth Objects, our Celestial Neighbors: Opportunity and Risk
Proceedings IAU Symposium No. 236, 2006
A. Milani, G.B. Valsecchi & D. Vokrouhlický, eds.
© 2007 International Astronomical Union
doi:10.1017/S1743921307003262

Collision and impact simulations including porosity

Willy Benz[1] and Martin Jutzi[1]

[1]Physikalisches Institut, University of Bern, Siedlerstrasse 5, 3012 Bern, Switzerland
email: willy.benz@space.unibe.ch

Abstract. The Smooth Particle Hydrodynamics (SPH) impact code (Benz & Asphaug 1994) has been developed for the simulation of impacts and collisions involving brittle solids in the strength- and gravity-dominated regime. In the latter regime, the gravitational overburden is used to increase the fracture threshold. In this paper, we extend our numerical approach to include the effect of porosity at a sub-resolution scale by adapting the so-called P - α model (Herrman 1969). Using our extended 3D SPH impact code, we investigated collisions between porous bodies to examine the sensitivity of collisional outcomes to the degree of porosity. Two applications that illustrate the capabilities of our approach are shown: 1) the modeling of a Deep Impact-like impact and 2) the computation of the amount of momentum transferred to an asteroid following the impact of a high velocity projectile.

Keywords. Hydrodynamics; shock wave; equation of state

1. Introduction

The evidence of the ubiquity of porosity in the structure of small bodies in the present day solar system is rapidly mounting. Spacecraft missions and ground based observations are providing increasing evidence that many or even most asteroids must be, to some degree or another, porous bodies (e.g. Housen & Holsapple 2003; Britt *et al.* 2006). Other small bodies in the solar systems like comets are also thought to have highly porous structures. In parallel, the dissipative properties of porous media are invoked more and more as playing an important role in the formation of early planetesimals (e.g. Wurm *et al.* 2006). Hence, porosity emerges slowly as playing a major role from the time of the formation of the planets to the collisional evolution of the present day solar system.

Despite the growing focus on porosity, our ability to model its effect on the outcome of impacts and collisions remains limited. Wuennemann *et al.* (2006) proposed recently a strain based model suitable for the use in hydrocodes. In this paper we propose an alternative approach based on the so-called P - α model (Herrmann 1969) and show how to incorporate it into our 3D Smooth Particle Hydrodynamics (SPH) code. We begin by recalling the equations we use in our modeling of brittle solids (section 2) before deriving the new equations and relevant modifications in order to extend our material description towards porous media (section 3). Finally two applications are presented pertaining to impact cratering and momentum transfer in collisions (section 4).

2. Equations

Our numerical tool is based on the on the Smooth Particle Hydrodynamic (SPH) method. Since the basic method has already been described many times (see for examples reviews by Benz 1990; Monaghan 1992) we refer the interested reader to these earlier papers.

The standard gas dynamics SPH approach was extended (see for example Libersky & Petschek, 1990) to include an elastic-perfectly plastic material description and a fracture model in order to model brittle solids (Benz & Asphaug, 1994). As our porosity model interfaces with this material description, we begin with a short review of our approach.

2.1. Elastic perfectly plastic strength model

The equations to be solved are the well-known conservation equations of elasto-dynamics; they can be found in most standard textbooks. The mass conservation can be written as (using the implicit repeated index summation rule):

$$\frac{d\rho^\kappa}{dt} + \rho\frac{\partial v^{\kappa\lambda}}{\partial x^\lambda} = 0 \tag{2.1}$$

where d/dt is the Lagrangian time derivative, ρ the mass density, v the velocity and x the position. The conservation of momentum has the following form:

$$\frac{dv^\kappa}{dt} = \frac{1}{\rho}\frac{\partial \sigma^{\kappa\lambda}}{\partial x^\lambda} \tag{2.2}$$

where $\sigma^{\kappa\lambda}$ is the stress tensor given by

$$\sigma^{\kappa\lambda} = S^{\kappa\lambda} - P\delta^{\kappa\lambda} \tag{2.3}$$

and P is the pressure, $\delta^{\kappa\lambda}$ the Kroneker symbol and $S^{\kappa\lambda}$ the (traceless) deviatoric stress tensor. Finally, the conservation of energy is given by the equation

$$\frac{dE}{dt} = -\frac{P}{\rho}\frac{\partial}{\partial x^\kappa}v^\kappa + \frac{1}{\rho}S^{\kappa\lambda}\dot{\epsilon}^{\kappa\lambda} \tag{2.4}$$

where $\dot{\epsilon}$ is the strain rate tensor given by

$$\dot{\epsilon}^{\kappa\lambda} = \frac{1}{2}\left(\frac{\partial v^\kappa}{\partial x^\lambda} + \frac{\partial v^\lambda}{\partial x^\kappa}\right). \tag{2.5}$$

In order to specify the time evolution of the deviatoric stress tensor $S^{\kappa\lambda}$ we adopt Hooke's law and define the time evolution of the deviatoric stress tensor as:

$$\frac{dS^{\kappa\lambda}}{dt} = 2\mu\left(\dot{\epsilon}^{\kappa\lambda} - \frac{1}{3}\delta^{\kappa\lambda}\dot{\epsilon}^{\nu\nu}\right) + S^{\kappa\lambda}\Omega^{\lambda\nu} + S^{\lambda\nu}\Omega^{\kappa\nu} \tag{2.6}$$

with μ is the shear modulus, and Ω is the rotation rate tensor:

$$\Omega^{\kappa\lambda} = \frac{1}{2}\left(\frac{\partial v^\kappa}{\partial x^\lambda} - \frac{\partial v^\lambda}{\partial x^\kappa}\right). \tag{2.7}$$

Finally, plasticity is treated using the von Mises yielding criterion.

In order to close the system of equations, an equation of state (EOS) has to be specified which relates density, specific energy and pressure:

$$P = P(\rho, E) \tag{2.8}$$

For the simulations presented in this paper we use the so-called Tillotson equation of state (e.g. Melosh 1989).

2.2. Fracture

Brittle materials cannot be modeled using elasticity and plasticity alone because these materials break under tension or shear stress. To take this behavior into account, we use the fracture model based on explicit incipient flaws (Benz & Asphaug 1995). In this model it is assumed that the number density of active flaws at strain ϵ is given by a Weibull distribution

$$n(\epsilon) = k\epsilon^m \tag{2.9}$$

where k and m are the material dependent Weibull parameters. When the local tensile strain has reached the flaw's activation threshold, a crack is allowed to grow at a constant velocity c_g which is some fraction of the local sound speed. The half length of a growing crack is therefore

$$a = c_g(t - t') \tag{2.10}$$

where t' is the crack activation time. Crack growth leads to a release of local stresses. To model this behavior, we follow Benz & Asphaug (1995) and introduce a state variable D (damage) which expresses the reduction in strength under tensile loading:

$$\sigma_D = \sigma(1 - D) \tag{2.11}$$

where σ is the elastic stress in the absence of damage and σ_D is the damage-relieved stress. The state variable D is defined locally as the fractional volume that is relieved of stress by local growing cracks

$$D = \frac{\frac{4}{3}\pi a^3}{V} \tag{2.12}$$

where $V = 4/3\pi R_s^3$ is the volume in which a crack of half length R_s is growing. Using equation (2.10) and (2.12) we get the following differential equation for the damage growth that we integrate together with the other variables

$$\frac{dD^{1/3}}{dt} = \frac{c_g}{R_s} \tag{2.13}$$

3. Extension of our numerical method: Including a porosity model

While porosity at large scales can be modeled explicitly by introducing macroscopic voids, porosity on a scale much smaller than the numerical resolution has to be modeled through a different approach. Our porosity model is based on the so called P-α model initially proposed by Herrmann (1969) and later modified by Carroll & Holt (1972). The model provides a description of microscopic porosity with pore-sizes beneath the spatial resolution, which is homogeneous and isotropic on the scales we resolve.

3.1. *P-α model*

The basic idea of the P-α model is to separate the volume change in porous material into two parts: the pore collapse on the one hand and compression of the matrix material on the other hand. This separation can be done by introducing the so called distention parameter α which is defined as

$$\alpha \equiv \frac{\rho_s}{\rho} \tag{3.1}$$

where ρ is the density of the porous material and ρ_s the density of the corresponding solid (matrix) material. Distention is related to porosity as $1 - 1/\alpha$. Using the distention parameter α, the equation of state can be written in the general form:

$$P = P(\rho, E, \alpha) \tag{3.2}$$

According to Carroll & Holt (1972), the EOS of a porous material is explicitly given by:

$$P = \frac{1}{\alpha}P_s(\rho_s, E_s) = \frac{1}{\alpha}P_s(\alpha\rho, E) \tag{3.3}$$

where $P_s(\rho_s, E_s)$ represents the EOS of the solid phase of the material (the matrix). A central assumption in this model is that the pressure depends on density of the matrix material. The specific internal energy E is assumed to be the same in the porous and the solid material ($E = E_s$). In the P-α - model, the distention is solely a function of the pressure P:

$$\alpha = \alpha(P) \tag{3.4}$$

where $P = P(\rho, E, \alpha)$. The relation between distention and pressure is often assumed to have the following quadratic form:

$$\alpha = 1 + (\alpha_0 - 1)\frac{(P_s - P)^2}{(P_s - P_e)^2}. \tag{3.5}$$

where P_e and P_s are constants. This is obviously a very simple model, however, it is appropriate enough for many applications. A more realistic relation can be obtained experimentally by a one dimensional static compression of the porous material during which the actual distention α_m is measured as a function of the applied pressure P_m. The resulting so-called crush-curve $\alpha_m(P_m)$ then provides the required relation between distention and pressure and can be used directly in the code.

In a quasi-static compression, the energy contribution to the pressure is very small and in the porous regime (where $\alpha > 1$), even mostly negligible. Therefore, the pressure can be approximated by $P = P(\rho, \alpha)$ and consequently, we can transform the function $\alpha(P[\rho, \alpha])$ into $\alpha = \alpha(\rho)$. This transformation can be done in both cases where either analytical relations $(\alpha(P))$ or crush-curves measured experimentally $\alpha_m(P_m)$ have been used. As we will see below, the use of $\alpha(\rho)$ instead of $\alpha(P)$ has some significant advantages. Therefore, in practice we are actually using $\alpha = \alpha(\rho)$ Using $\alpha(\rho)$, the time evolution of the distention parameter is simply given by

$$\dot{\alpha} = \frac{d\alpha}{d\rho}\dot{\rho} \tag{3.6}$$

which is pressure independent. The corresponding relation for the case where the energy contribution cannot be neglected, $\alpha = \alpha(P)$, is given by

$$\dot{\alpha}(t) = \frac{\dot{E}\left(\frac{\partial P_s}{\partial E_s}\right) + \alpha\dot{\rho}\left(\frac{\partial P_s}{\partial \rho_s}\right)}{\alpha + \frac{d\alpha}{dP}\left[P - \rho\left(\frac{\partial P_s}{\partial \rho_s}\right)\right]} \cdot \frac{d\alpha}{dP} \tag{3.7}$$

which depends upon the pressure. This pressure-dependence of $\dot{\alpha}$ can cause numerical instability because the pressure itself depends on α (equation 3.3). For this reason, other alternative models such as the strain based porosity model (Wuennemann et al. 2006) have been explored. However, we did not find these instabilities when using $\alpha = \alpha(P)$. The main reason that we use $\alpha = \alpha(\rho)$ instead of $\alpha = \alpha(P)$ is its applicability to relate the distention α to strength.

3.2. Porosity and strength

As we have discussed in the previous section, the pressure is calculated using the matrix density ρ_s instead of ρ. Therefore, the deviatoric stress tensor has to be modified as well. In order to compute the time evolution of $S^{\kappa\lambda}$ as a function of the matrix variables, we introduce the following correction factor:

$$f \equiv \frac{\dot{\rho}_s}{\alpha\dot{\rho}}. \tag{3.8}$$

In fact, this factor relates the velocity divergence of the matrix to that of the porous material:

$$[\vec{\nabla}\vec{v}]_s = f\vec{\nabla}\vec{v} \tag{3.9}$$

where $[\vec{\nabla}\vec{v}]_s$ is defined to satisfy the continuity equation of the matrix:

$$\dot{\rho}_s = -\rho_s[\vec{\nabla}\vec{v}]_s \tag{3.10}$$

Using $\alpha = \alpha(\rho)$ and $\dot{\rho}_s = \alpha\dot{\rho} + \dot{\alpha}\rho$, f has the simple form:

$$f = \frac{\rho}{\alpha}\frac{d\alpha}{d\rho}. \tag{3.11}$$

Using $\alpha = \alpha(P)$, we obtain $f = f(P, \rho, \dot{\rho}, E, \dot{E})$ which (in the case of a small $\dot{\rho}$ and a large $\dot{E}$) can lead to numerical instabilities. This is the main reason that we are actually using $\alpha = \alpha(\rho)$. The factor f is used to compute the corrected time evolution of $S^{\kappa\lambda}$:

$$\frac{dS^{\kappa\lambda}}{dt} \rightarrow f\frac{dS^{\kappa\lambda}}{dt} \tag{3.12}$$

In addition, the deviatoric stress tensor $S^{\kappa\lambda}$ is multiplied by α^{-1} as it is done with the pressure P for the same reason (see Carroll & Holt 1972. We finally write the time evolution of $S^{\kappa\lambda}$ in the following form:

$$\frac{d}{dt}\left[\frac{1}{\alpha}S^{\kappa\lambda}\right] = \frac{1}{\alpha}\frac{dS^{\kappa\lambda}}{dt} - \frac{1}{\alpha^2}S^{\kappa\lambda}\frac{d\alpha}{dt} \tag{3.13}$$

where $dS^{\kappa\lambda}/dt$ is modified according to (3.12).

3.3. *Porosity and damage*

Porosity does not only affect the stress behavior it also has to be taken into account to compute the state variable damage.

Compression of a porous material beyond the elastic limit is accompanied by breaking cell walls. We model this crushing behavior by relating the distention to the state variable damage (D). Since both, damage D and distention α, are defined as a volume ratio (equations 2.12 and 3.1, respectively) we assume for simplicity (other forms will be investigated in the future) a linear relation between D and α. The conditions: $D = 0$ at $\alpha = \alpha_0$, and $D = 1$ when all pores have been crushed ($\alpha = 1$), lead to the following expression:

$$D = 1 - \frac{(\alpha - 1)}{(\alpha_0 - 1)}. \tag{3.14}$$

The time evolution of $D^{1/3}(\alpha)$ is given by

$$\frac{dD^{1/3}}{dt} = \frac{dD^{1/3}}{d\alpha}\frac{d\alpha}{dt} \tag{3.15}$$

and using equation (3.14) we obtain

$$\frac{dD^{1/3}}{dt} = \frac{1}{3} - \left[\frac{\alpha - 1}{\alpha_0 - 1} + 1\right]^{-\frac{2}{3}}\frac{1}{\alpha_0 - 1}\frac{d\alpha}{dt}. \tag{3.16}$$

We now have two equations describing damage growth: the first treats damage under tension (2.13) while the second (3.16) is related to the compression of the (porous) material. In order to get the total grow of damage, we build the sum of these two differential equations:

$$\left[\frac{dD^{1/3}}{dt}\right]_{total} = \left[\frac{dD^{1/3}}{dt}\right]_{tension} + \left[\frac{dD^{1/3}}{dt}\right]_{compression} \tag{3.17}$$

3.4. Material parameters

All parameters used by our porosity model are material parameters which can in principle be measured; some of them quite easily (e.g. the crush-curve) even though this is rarely done in practice. Others, such as Weibull parameters, shear strength etc. are more difficult to measure.

The lack of an experimentally determined reliable database of relevant material parameters is actually one of the most limiting factor in our model. In particular, the thorough testing of the model by comparison with experiments is rendered particularly difficult if all material properties have not been measured properly. Freely choosing the missing values so as to match the experiment is not a satisfactory approach for an *ab initio* method such as ours. Unfortunately, this is often the only alternative we have.

4. Applications of our model

Although our method still needs to be rigorously tested, the first calculations show very encouraging results. For instance, we simulated the laboratory impacts in porous material carried out by Housen & Holsapple (2003). We were able to reproduce the shape of the crater produced in these experiments (Jutzi 2004).

As an illustration of the capabilities of our code, we show in this paper the results of two sets of simulations. In the first one, we computed the impact of a Deep Impact-like projectile onto a icy target of different porosity. In this calculation, we did not try to obtain an exact match by fiddling with all the parameters but rather used standard inputs to see how porosity was affecting the results. The second example illustrates our efforts to compute the amount of momentum transferred to an asteroid by the impact of a high speed projectile as a function of the target porosity.

4.1. Modeling a Deep Impact-like experiment

We carried out simulations of a Deep Impact-like impact on comet Temple-1 using three different initial porosities (0%, 33% and 67%) for the target comet. Only a small part of Temple-1 was modeled (half sphere with a radius of 22 m). As target material we used pre-damaged (strenghtless) ice with a corresponding initial distention of $1.0, 1.5$ and 3.0. The impactor was modeled as a 370 kg aluminum sphere impacting at an angle of 30 degree (from horizontal) with a velocity of 10 km/s. For all these simulations we used $700'000$ particles for the target and 20 particles for the projectile. This results in a spatial resolution of 36 cm (target) and 22 cm (projectile), and particles masses of 9.5 to 29 kg (target) and 19 kg (projectile).

Figure 1 shows the outcome of the simulation after 50 ms. In these 2-D slices of the 3-D target, the color shows the vertical velocity of the particles (cm/s). Note that the escape velocity of Temple 1 is about 200 cm/s. Obviously, there is much more material ejected in the non-porous than in the porous cases. The difference of ejected material in the simulations with porosity is only small. However, since for $\alpha_0 = 1.5$, the particle mass is twice the mass of the particles with $\alpha_0 = 2.0$, slightly more mass is ejected in the simulation with the lower porosity. For illustration, we also show the actual value of the distention α in the case $\alpha_0 = 3.0$ (figure 2).

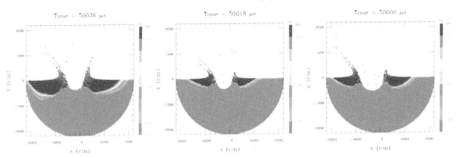

Figure 1. Simulation of a Deep Impact-like impact using different initial distention: $\alpha_0 = 1.0$ (left), $\alpha_0 = 1.5$ (middle), $\alpha_0 = 3.0$ (right). The colors label the z-component of the velocity, red indicates a velocity higher than the escape velocity.

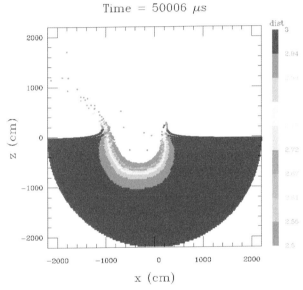

Figure 2. A measure of the compaction is provided by the actual value of the distention α. The initial distention of $\alpha_0=3$ is decreased to $\alpha=1$ in a small zone around the crater. With increasing distance from the crater, distention increases until $\alpha = \alpha_0$

A quantity of interest is the amount of mass ejected with a velocity higher than a certain velocity. Figure 3 shows the corresponding relation obtained in our simulations. If we assume the escape velocity to be $v_{esc} = 2$ m/s, we get the total mass ejected: 3.20×10^6 kg, 0.52×10^6 kg, and 0.49×10^6 kg for $\alpha_0 = 1.0, 1.5$ and 3.0 respectively. Again there is a big difference between the nonporous and porous simulations and a smaller difference between the two porous simulations. The reason could be that we only changed one parameter in our porosity model (α_0). Further investigations will be done to examine the sensitivity of the simulation outcome on the model parameters (crush-curve).

4.2. Momentum transfer

The change in orbit of an asteroid resulting from the impact of a projectile depends upon the momentum transferred during the collision. To compute this amount is not straight forward as it is largely determined by the amount of material ejected from the impact crater. This amount as well as its velocity distribution are functions of the internal structure and material characteristics of the target (at least in the impact area). Hence,

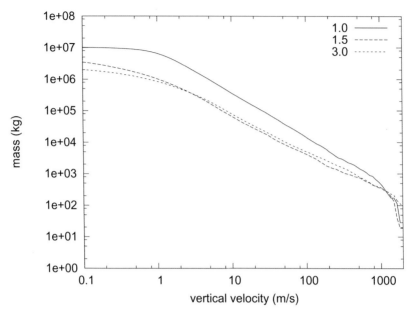

Figure 3. Deep Impact-like simulation: cumulated mass as a function of the ejection velocity for different initial distentions α_0

the degree of porosity of the asteroid will play a key role in determining the size of impact required to achieve a given orbital change.

To examine how porosity might affect the momentum transfer achieved in a given collision, we simulate impacts of a space-craft with a velocity of 10 km/s into an asteroid. The space-craft is modeled as a spherical solid body with a mass of 400 kg and a density of 5.5 g/cm³. Since the expected crater is small compared to the overall size of the asteroid, we only model a small fraction of the target (half sphere of 40 m radius). The material type is taken to be basalt and targets with 0%, 9.1%, 33% and 67% porosity (which corresponds to a initial distention of $1.0, 1.1, 1.5, 3.0$) are investigated. We use 1'400'000 particles for the target and 15 for the projectile, resulting in a spatial resolution of 51 cm (target) and 19 cm (projectile) and a particle mass of $85 - 256$ kg (target) and 26 kg (projectile). The high mass of target particles is a result of the large target size which is required to make sure that the reflection of the shock-wave at the border of the half sphere does not affect the results.

To compute the momentum transferred to the target, we determined the mass and velocity of the material reaching velocities equal or in excess of the escape velocity for each simulation. Using momentum conservation, the transferred momentum is then given by $P_{trans} = P_{proj} + P_{ejecta}$, or in units of of P_{proj}: $P_{trans} = 1 + P_{ejecta}$.

Figure 4 shows P_{trans} obtained in our simulations as a function of time. There are two main differences between the non-porous and porous simulations. First, there is a peak at early times in the non-porous case which does not occur in porous targets. This peak is caused by the acceleration of particles behind the shock-front due to the rarefaction wave. After a certain time (indicated by the curve's peak) the velocity caused by the rarefaction wave drops below the ejection limit (escape speed), hence the rapid decrease. Using porous material, the shock wave is strongly damped and it also travels much slower. Therefore, the described effect does not occur.

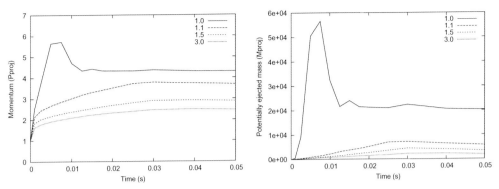

Figure 4. *Left:* The amount of momentum transferred in an impact to the target decreases with increasing porosity (i.e., increasing distention). *Right:* The small amount of momentum transferred to porous targets can be explained by the small amount of ejected mass.

The most important difference between the non-porous and porous simulations is the final amount of transferred momentum which (as expected) decreases with increasing porosity. The lower amount of transferred momentum using porous material can be explained by the smaller amount of mass ejected from targets with increasing porosity. This is shown in figure 4 where the potentially ejected mass ($v_z > v_{esc}$) is plotted as a function of time.

In order to show that the different behavior of non-porous and porous material is not merely due to the different initial density, we compare a simulation where $\alpha_0 = 1.5$ with a simulation where we do not model porosity but use the same the initial density: $\rho_0 \to \rho_0/1.5$. As it can be seen in figure 5, the effect porosity is not to simply change the bulk density but porosity really affects the dynamics in the sense that crushing the pores represents an important sink of energy.

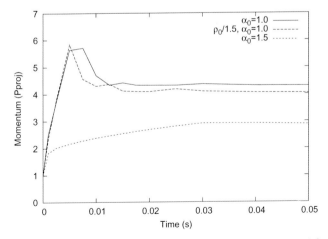

Figure 5. Comparison of a simulation with $\alpha_0 = 1.5$ and a simulation with the same initial density $\rho_0 \to \rho_0/1.5$ but without explicit modelling of porosity.

5. Conclusions

In this paper, we have presented a new approach to model small scale porosity in brittle solids that can be coupled to a 3D SPH hydrocode in order to simulate impacts and collisions involving porous bodies. In practice, the implementation of our model does not consume excessive CPU time and is easily implemented in a parallel code (porosity is a local property) so that simulations involving multi-million particles can readily be performed. In fact, extensive testing has shown that high resolution is really needed to obtain converged solutions in the case of simulations involving fracturing and/or porosity.

Testing of the code including porosity is made difficult by the lack of well described experiments in which all relevant material parameters have been measured and published. More often than none, the missing parameters can be chosen freely. The fact that, with a reasonable choice for these, a good agreement can be found lends some confidence to the model but is not completely satisfactory.

Despite these difficulties, the relative effect of porosity on the outcome of an impact or collision can be studied by keeping, for example, all other parameters fixed. By means of two examples, we illustrate how strongly porosity can affect the dynamics leading to important changes in the event's outcome. In the first example, we show how the amount of ejecta following a Deep Impact-like event is decreasing with increasing porosity. Our results are indeed consistent (given the simplicity of the model) with values inferred from the experiment (see A'Hearn this volume). The second example deals with the amount of momentum transferred to a target asteroid by the impact of a high-speed projectile. We showed that this amount is considerably reduced (more than a factor 2) by increasing distension from 1 to 3. The cause of this reduction is to be found in the corresponding decrease of ejected matter with increasing distension. We also show that taking into account the dissipation provided by porosity is important. Simply reducing the bulk density is not a proper approach to model porous media.

Acknowledgements

The authors gratefully acknowledge partial support from the Swiss National Science Foundation.

References

Benz, W. 1992, in: J.R. Büchler (ed.), *The numerical modelling of nonlinear stellar pulsations: Problems and prospects* (Springer: Berlin), p. 269.
Benz, W. & Asphaugh, E. 1994, *Icarus* 107, 98
Benz, W. & Asphaugh, E. 1995, *Comp. Phys. Comm.* 87, 253
Britt, D. T., Consolmagno, G. J. & Merline, W. J. 2006, *37th Annual Lunar and Planetary Science Conference*, 2214
Carroll, M. M. & Holt, A. C. 1972, *J. Appl. Phys.* 43, 759
Herrmann, W. 1969, *J. Appl. Phys.* 40, 2490
Housen, K. R. & Holsapple K. A. 2003, *Icarus* 163, 102
Jutzi, M. 2004, Diploma thesis, University of Bern
Libersky, L. D. & Petschek, A. G. 1990, in *Lecture Notes in Physics* 395, 248
Melosh, H. J. 1989, *Impact cratering: A geologic process* (Oxford: Oxford University Press)
Monaghan, J. J. 1992, *Annu. Rev. Astron. Astrophys.* 30, 543
Wuennemann, K., Collins, G. S. & Melosh, H. J. 2006, *Icarus* 180, 514
Wurm, G., Paraskov, G. & Krauss, O. 2005, *Icarus* 178, 253

Near Earth Objects, our Celestial Neighbors: Opportunity and Risk
Proceedings IAU Symposium No. 236, 2006
A. Milani, G.B. Valsecchi & D. Vokrouhlický, eds.
© 2007 International Astronomical Union
doi:10.1017/S1743921307003274

Space weathering and tidal effects among near-Earth objects

S. Marchi[1], P. Paolicchi[2], D. Nesvorný[3], S. Magrin[1] and M. Lazzarin[1]

[1]Dipartimento di Astronomia, Università di Padova, Vicolo dell'Osservatorio 2, I-35122 Padova, Italy
email: simone.marchi@unipd.it; sara.magrin.1@unipd.it; monica.lazzarin@unipd.it

[2]Dipartimento di Fisica, Università di Pisa, largo Pontecorvo 3, I-56127 Pisa, Italy
email: paolicchi@df.unipi.it

[3]Department of Space Studies, Southwest Research Institute, 1050 Walnut St., Boulder, CO 80302, USA
email: davidn@boulder.swri.edu

Abstract. The effect of the space weathering on the spectral properties of the S–complex asteroids (both Main Belt bodies and near–Earth asteroids) has been widely discussed in recent times. It has also shown that the evolution of spectral properties of planet–crossing bodies, and in particular of near–Earth asteroids (NEAs), is also affected by other physical processes, such as tidal resurfacing due to close encounters with planetary bodies. In this paper we show how to combine previous analyses with the purpose of obtaining a global model for NEAs space weathering.

Keywords. asteroid, surfaces; asteroid, spectra; space weathering

---

1. Introduction

The action of the space environment on optical properties of asteroid surfaces has been put into evidence in several recent papers (Hapke 2001; Hiroi & Sasaki 2001; Clark *et al.* 2002; Chapman 2004). In particular, the analysis of the so–called *space weathering* has been thorough and systematic for what concerns the S–complex asteroids: older asteroids are expected to be redder and darker. After the work of Binzel *et al.* (2004) devoted to NEAs, a colour–age relation has been suggested within an analysis devoted to Main Belt family asteroids (Jedicke *et al.* 2004; Nesvorný *et al.* 2005). More recently, a general relation has been shown to hold for all S–complex asteroids (Marchi *et al.* 2006a, hereinafter Paper I). It has been also shown that the most significant parameter to be correlated with the colour change (in technical terms, the observed visible spectral slope) is not simply the diameter (as shown in Binzel *et al.* 2004) nor the age (as in Jedicke *et al.* 2004) but rather a combination of age and orbital parameters, namely the *exposure* defined as

$$Exposure = \int \frac{dt}{r(t)^2} \simeq \frac{age}{a^2\sqrt{1-e^2}}$$

i.e. the age times the inverse squared mean distance from the Sun (function of the semi–major axis, a, and of the eccentricity, e). The relevance of the exposure parameter entails the dominance, at least for distances smaller than about 3–4 AU, of space weathering effects connected to the Sun (such as, for instance, the ion bombardment; see also Lazzarin *et al.* 2006 and Marchi *et al.* 2005).

In a following paper (Marchi *et al.* 2006b, hereinafter Paper II) a relation has been evidenced to hold for NEAs and Mars crossers (MCs), involving the perihelion distance and the spectral slope. The underlying idea is that a smaller perihelion distance increases the probability of having undergone a recent close encounter with one of the inner planets (Nesvorný *et al.* 2005). The tidal effects of a deep encounter might severely affect the surface, introducing, among the others, a sort of *de–weathering*, i.e. a rejuvenation of the asteroidal surface. In reality the weathering and the de–weathering effects act simultaneously on the asteroids, but the range of application of the tidal–triggered de–weathering is essentially limited to the inner bodies.

In this preliminary paper we try to combine the analyses introduced in Papers I and II. We will use an updated database and updated age estimates. The details of these improvements are presented in Paolicchi *et al.* (2006).

2. Summary of updates

We closely follow the methodology presented in Paper I. In particular the basic method to estimate the ages is the same, taking into account the collisional lifetime only for what concerns Main Belt asteroids (and MCs) and adding a correction due to Yarkovsky effect for the age computation of NEOs. However we use here a larger database and introduce several relevant corrections:

(*a*) The formation of a family is assumed to come from a violent collisional process (Zappalà *et al.* 2002), usually catastrophically breaking both projectile and target involved in the impact, sometimes (for instance, in the case of Vesta family) only creating a large crater on the –massive– target body. In all cases the process converts internal parts of the parent body into surface regions of the resulting fragments. Also the –space weathered– surface of the parent body is presumably deeply shaken, and thus it may appear as rejuvenated when observed as surface of the largest remnant. Thus the age of the family is an upper limit to the age of its members; this value has thus to be used also to constrain the time a family asteroid has been exposed to space–weathering. We have used the new estimates of family ages summarized in Nesvorný *et al.* (2006).

(*b*) We used the proper orbital elements, when available, instead of the osculating orbital elements: they are more stable over time and hence more meaningful for estimating the exposure.

(*c*) The ages of MCs, computed with the same method as the other Main Belt asteroids (MBAs), may have been overestimated: they are in a region subject to fast dynamical processes. Thus their ages should be, in mean, smaller than those of MBAs of similar size. Moreover, the relevance of the –relatively– young (Nesvorný *et al.* 2006) Flora cluster in this region may affect their age. In agreement with these theoretical considerations, we find that MCs slopes are, on average, lesser than MBAs slopes. A first order estimate of the age correction might be introduced assuming that the smaller mean spectral slope of MCs compared to slopes of Main Belt asteroids is fully due to their typical smaller age. The mean slope of MCs is $\simeq 0.447 \ \mu\mathrm{m}^{-1}$, compared to the MBAs value of $\simeq 0.503$ $\mu\mathrm{m}^{-1}$. Therefore we estimate that the age of MCs has to be reduced by a factor of $\simeq 0.3$. This point is discussed in detail in Paolicchi *et al.* (2006).

(*d*) Here, we use an updated database, including also the new spectroscopic data obtained by $\mathrm{S}^3\mathrm{OS}^2$ survey (about 190 objects were not present in previous works; Lazzaro *et al.* 2004) for a grand total of 1026 S–complex asteroids, most of which are from SMAS-SII.

3. The new slope–exposure relation

We are now ready to include all S–complex asteroids (NEOs, MCs and MBAs) in a unique plot (Fig. 1). We can also try to plot a best fit curve. We overplot a linear fit, as we did in the previous analysis (Paper I). The linear fit, represented in the figure, is highly significant. However, taking into account some general physical considerations (see later) and the fit presented by Nesvorný *et al.* (2005) for what concerns family asteroids, we decided to look also for a logarithmic slope–exposure one. The best fit curve is represented in the figure. It corresponds to a correlation coefficient equal to 0.375. In terms of the RMS deviation of the data from the fit we pass from a value $RMS = 0.221$ (linear fit) to a value $RMS = 0.205$ (dashed line logarithmic fit; note that the number of parameters is the same in the two cases). Thus, combining the physical suggestions and this –moderate– statistical improvement, we decided to go on with the logarithmic fit. In the figure we also show the finally corrected logarithmic fit (see next section for details on the correction).

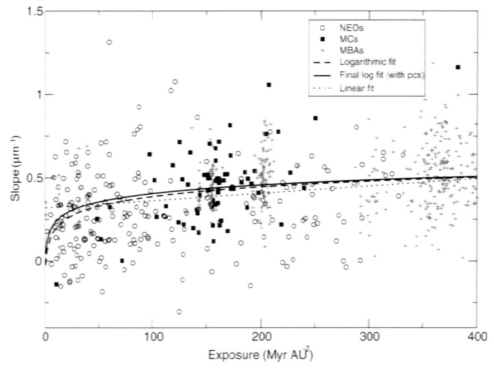

Figure 1. The slope–exposure plot including part of MBAs, MCs and NEOs. The figure has been limited to an exposure value of 400 My AU^{-2} to give a better representation of the NEO region. A linear fit (obtained for all bodies) is plotted (dotted line). A logarithmic plot, more meaningful (see the text for discussion), is also represented (dashed line). The solid–line curve represent the final corrected logarithmic plot (see text).

4. The slope vs. perihelion correction

According to Paper II, there is a significant evidence for the existence of a slope–perihelion correlation, in the range of NEOs and MCs. It should be taken into account, not to mix different effects. In order to do so, we represent the exposure corrected slope (ECS) vs. perihelion for all asteroids in our sample. The ECS has been computed (see

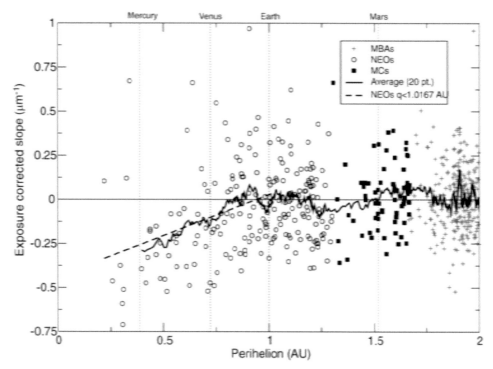

Figure 2. Using the logarithmic fit presented in the Fig. 1, we have corrected the observed slope. We plot the residual slope value (i.e. the difference between the real and the fit values) as function of the perihelion (limited to 2 AU). The existence of a significant relation between the residual slope and the perihelion is apparent at least for a semi–major axis smaller than about 1 AU.

above) subtracting from the actual spectral slope the mean slope obtained from the best fit curve of Fig. 1 and corresponding to the same exposure. The plot is represented in Fig. 2.

As we can see, there is a systematic and significant residual deviation for perihelia smaller than –about– 1 AU. The figure reproduces the slope–perihelion *real* relation, and should be seen as an improved version of Fig. 1, Paper II. No significant deviation is present for perihelia larger than Mars semi–major axis. The intermediate region is somehow less clear, but the presence of a correlation is not evident. Thus we conclude that the slope–perihelion relation is robust within distances smaller than 1.016 AU (namely Earth's aphelion). The relation is such as

$$slope = -0.44 + 0.47 \cdot q$$

and can be used to introduce a perihelion correction to the slope exposure relation. The simplest way to do it is simply to compute the perihelion corrected slope (*PCS*), by subtracting the perihelion correction (if applicable) from the observed slope. The result is represented in Fig. 1, and corresponds to a relation such as:

$$PCS = 0.07 + 0.17 \cdot \log(exposure).$$

The variance is slightly smaller than that obtained above ($RMS = 0.200$). The exposure range is from about 0.7 to 800 My AU^{-2}. At 1 AU this corresponds to an excursion from 0.7 My to 800 My; or 2.8 My to 3.2 Gy at 2 AU.

In this range of exposure, *PCS* varies from 0.044 μm^{-1} to 0.559 μm^{-1}, namely it increases by more than one order of magnitude. Notice that the reddening is very steep at the beginning, and that, for instance, the 80% of the excursion is reached at about 200 My at 1 AU; or 800 My at 2 AU.

5. Discussion

The fit obtained in the previous Section is not the only possible one, but is fully representative of the two major features of the data: the steep increase in the slopes for very short exposures, and the nearly–saturated trend for large exposures. Both features are qualitatively observed also in laboratory experiments. On the other hand, the log–fit cannot be significant (due to its divergent behavior at the $t = 0$ limit) to reproduce the very beginning of the slope–exposure relation, in particular for exposure ~ 0. Apart the mathematical difficulties, maybe a physical problem remains open. The astronomical space weathering (for what comes out from the observations) seems a bit slower than one might expect on the basis of laboratory experiments: in particular one should expect a faster saturation.

The most immediate suggestion may be to invoke the regolith mixing. In analogy with what observed on the Moon, regolith evolution is a complex process, which involves gardening (stirring of grains by micrometeorites), erosion (from impacts and solar wind sputtering), maturation (exposure on the bare lunar surface to solar winds ions and micrometeorite impacts) and comminution of coarse grains into finer grains, blanket deposition of coarse-grained layers, and other processes. As a result, the degree of maturation of the regolith varies with depth, and reaches depths by far exceeding the tens of nanometers attained in laboratory experiments. Therefore the observed discrepancy would be the result of the reddening and the mixing of the upper layer due to meteoritic impacts (see Paolicchi *et al.* 2006 for a thorough discussion). Unfortunately, the verification or falsification of a such a possibility is presently out of reach, due to the lacking of detailed knowledge of regolith properties on asteroids.

6. Conclusions

The combination of the slope–exposure and slope–perihelion relations presented in Paper I and II has been performed, also with the use of new data and of an improved analysis of the age of MBAs and MCs. The presented final slope–exposure relation qualitatively confirms the results, and the underlying physical ideas, discussed in the previous papers. However, the slope–exposure relation is less steep than previously discussed, and may be properly fitted with a logarithmic relation. Moreover, we find again a perihelion effect which is safely identified below the Earth's aphelion. As a side result, we have found that the MCs are presumably –in mean– younger by about a factor 3 than their MB siblings of the same size.

However, the space weathering timescale that we obtain from our observational sample are systematically larger than one might expect, on the basis of laboratory experiments, even if a progressive saturation has to be taken into account. A complex asteroidal regolith evolution, as found for the Moon, might be invoked to solve the discrepancy.

References

Binzel, R.P., Rivkin, A.S., Stuart, J.S., Harris, A.W., Bus, S.J. & Burbine, T.H. 2004, *Icarus* 170, 259

Clark B.E., Hapke B., Pieters C. & Britt D. 2002, in: W.F. Bottke, A. Cellino, P. Paolicchi & R.P. Binzel (eds.), *Asteroids III* (Tucson: University of Arizona Press), p. 585

Chapman, C.R. 2004, *Annu. Rev. Earth Planet. Sci.* 32, 539

Hapke, B. 2001, *J. Geophys. Res.* 106(E5), 10039

Hiroi, T. & Sasaki, S. 2001, *Meteor. Planet. Sci.* 36, 1587

Jedicke, R., Nesvorný, D., Whiteley, R., Ivezić, Z. & Jurić, M. 2004, *Nature* 429, 275

Lazzarin, M., Marchi, S., Moroz, L.V. , Brunetto, R., Magrin, S., Paolicchi, P. & Strazzulla, G. 2006, *Astrophys. J.* 647, L179

Lazzaro, D., Angeli, C. A., Carvano, J. M., Mothé-Diniz, T., Duffard, R., & Florczak, M. 2004, *Icarus* 172, 179

Marchi, S., Paolicchi, P., Lazzarin, M. & Magrin, S. 2006a, *Astron. J.* 131, 1138

Marchi, S., Magrin, S., Nesvorný, D., Paolicchi, P. & Lazzarin, M. 2006b, *MNRAS* 368, L39

Marchi, S., Brunetto, R., Magrin, S., Lazzarin, M., & Gandolfi, D. 2005, *Astron. Astrophys.* 443, 769

Nesvorný, D., Jedicke, R., Whiteley, R.J. & Ivezić, Z. 2005, *Icarus* 173, 132

Nesvorný, D., Bottke, W. F., Vokrouhlický, D. , Morbidelli, A. & Jedicke, R. 2006, in: D. Lazzaro, S. Ferraz-Mello & J.A. Fernandez (eds.), *Asteroids, Comets and Meteors* (Cambridge: Cambridge University Press), p. 289

Paolicchi, P., Marchi, S., Nesvorný, D., Magrin, S. & Lazzarin, M. 2006, *Astron. Astrophys.*, submitted

Zappalà, V., Cellino, A., Dell'Oro, A. & Paolicchi, P. 2002, in: W.F. Bottke, A. Cellino, P. Paolicchi & R.P. Binzel (eds.), *Asteroids III* (Tucson: University of Arizona Press), p. 619

Near Earth Objects, our Celestial Neighbors: Opportunity and Risk
Proceedings IAU Symposium No. 236, 2006
A. Milani, G.B. Valsecchi & D. Vokrouhlický, eds.

© 2007 International Astronomical Union
doi:10.1017/S1743921307003286

Space weathering of asteroids: similarities and discrepancies between Main Belt asteroids and NEOs

S. Marchi[1], P. Paolicchi [2], M. Lazzarin [1] and S. Magrin[1]

[1]Dipartimento di Astronomia, Università di Padova, Vicolo dell'Osservatorio 2, I-35122 Padova, Italy
email: simone.marchi@unipd.it; sara.magrin.1@unipd.it; monica.lazzarin@unipd.it

[2]Dipartimento di Fisica, Università di Pisa, largo Pontecorvo 3, I-56127 Pisa, Italy
email: paolicchi@df.unipi.it

Abstract. A sample of 35 C–complex objects is present among near–Earth objects. In spite of the poor statistics, some striking differences compared to Main Belt asteroids can be established: for instance the percentage of near–Earth objects (NEOs) showing hydration features is very small. Moreover the spectral slope of C–complex NEOs seems to be anti–correlated with the exposure to the ion flux coming from the Sun, in contrast with the general behavior of C–complex Main Belt asteroids (and of most asteroids, in general). We discuss some possible implications and suggest some preliminary partial explanations.

Keywords. asteroids, surface

1. Introduction

In two recent papers (Marchi *et al.* 2006a and Lazzarin *et al.* 2006) it has been shown that the mean visible spectral slope of most Main Belt asteroids (S–, C– and X–complexes) is positively correlated with the exposure to the weathering effects due to the Sun. These findings have relevant consequences for what concerns the chemistry, the evolutionary history and even the taxonomical classification of Main Belt asteroids (MBAs). Notice that the dominance of reddening effects within the C–complex is effective for what concerns the bulk of MBAs, but does not hold for some subgroups, such as some families or bodies in the outer Main Belt.

Among the C–complex MBAs some exhibit hydration features, some others do not. This occurrence indicates the presence of some intrinsic compositional variation or the presence of some selective process –not yet well understood– in the history of C–complex asteroids; however we remark that the presence of hydration features seems to be not relevant for what concerns the properties of the space weathering: both samples redden with the exposure in a similar way.

2. Anomalies among C–complex NEOs

According to the currently accepted scenario of near–Earth objects (NEOs) origin, two main source regions have to be taken into account. The Main Belt (MB), from which about 80% of NEOs have been originated, according to the current estimates, is the principal one; the outer Solar System (SS) is by far less important ($\sim 20\%$). Dynamical models (Morbidelli *et al.* 2002) also indicate which are the injection main "channels", i.e. the regions from which various dynamical effects can cause a strong and relatively fast evolution of the orbital parameters, causing the injection of some –or most– bodies into the near–Earth region or into a dynamical trail with the same final destination. Seven main channels are listed: for most NEOs it is possible to estimate the probability of having been originated from each of them. The scenario presented in the above quoted paper and in the following improvements is capable to explain many observed features, and is considered as rather reliable.

The analysis of primitive NEOs (i.e. belonging to C–complex; up to now 35 objects belong to this complex) is somehow problematic. Due to a lack of distinctive spectral features, the real composition of these objects is not well understood, and there is a sort of "degeneracy" among C–complex MB–related NEOs and outer SS–related ones: in other words, on a spectral basis, C–complex NEOs may originate from the MB or the outer SS. Notice that this does not happen for the –by far more numerous– S–complex NEOs. Indeed, with an average probability for the JFC channel of about 4%, the majority of C–complex NEOs are thought to originate from the MB (see Fig. 1). Notice that, according to the discussion presented by Lazzarin *et al.* (2006) the taxonomic separation among objects with relatively featureless spectra might be not completely meaningful, being mainly due to a different reddening history.

The C–complex NEOs are different –at least for a pair of statistical properties– from the analogue MBAs:

(*a*) Aqueous alteration signatures have been observed on several MBAs (about 40% of C–complex MBAs are aqueously altered). In the visible range the hydration signatures consist of several absorption bands, the 0.7 μm band being the most prominent. It must be noted that this feature is not easily detected in the case of a low S/N value of the spectrum, such as in the case of some small MBAs or NEOs. The hydration percentage, among MBAs, does not depend on the size (see, however, the discussion in the following). We cannot find a systematic effect connected to the family membership. Some families (i.e. (24) Themis) are characterized by a very small percentage of hydrated bodies, some others (i.e., (668) Dora) behave in the opposite way, without any apparent rule. Note also that some Themis family members show an anomalous quasi–cometary activity (Hsieh & Jewitt 2006). The combination of this feature with the overall scarcity of hydrated bodies in the same family might be strongly constraining for a theoretical explanation. The properties of the C–complex MBAs and the distribution of origin channels (Fig. 1) should entail that a considerable fraction (a conservative estimate yields $> 22\%$, i.e. something as $> 8 \pm 1$ out of 35 objects) of NEOs should be hydrated. In reality only 2 NEOs show clear hydration features. Notice that the above estimate takes into account the actual abundance of hydrated bodies in the proximity of the channels.

(*b*) Another difficulty comes from the analysis of space weathering properties. While the C-complex MBAs redden with increasing exposure to Sun ion flux, with a behavior which is rather similar to what happens to silicate–rich MBAs (and NEOs), the C–complex NEOs behave in the opposite way (Fig. 2). This result stands even if we exclude in the sample all the NEOs which may have been originated from the outer SS.

3. Discussion

A number of tentative explanations to the former problems can be suggested, but none (until now) completely satisfactory. The full range of possibilities will be discussed in detail in a forthcoming paper.

(*a*) Hydrated NEOs actually do not exists: only non–hydrated bodies enter the channels and become NEOs. This would be the case if hydrated bodies would not exist among very small asteroids. We remind that all C–complex NEOs have diameter (D) less than

10 km. Only few MB objects have been observed in this size range; among them 10 non–hydrated and 2 hydrated MBAs, with D<10 km, have been observed so far. The present values are consistent both with the independence of hydration properties on size and with a deficit of hydrated small asteroids. In principle, the dehydration might be induced by high temperature shocks which originate the small fragments, and, in particular, the NEOs. On the other hand the mineralogical product of water alteration has been found on many primitive meteorites (CCs). CCs of type 1 contain limited, but significant, evidence of aqueous alteration; while types 2 and 3 ubiquitously present products of aqueous alteration. If these products are so common on CC meteorites, why could hydration be absent among C–complex NEOs? Both the observational evidence and the theoretical explanation are severely challenged.

(*b*) Another possibility is that the percentage of NEOs coming from the channels at the extremities of the alteration region (Hungaria (H in Fig. 1), ν_6, JFC; where hydrated MBAs are not present) is larger than expected. However, H and ν_6 can be excluded because of the paucity of C–complex asteroids in their neighborhood. A larger than expected fraction of primitive NEO should come from the outer SS. However, we have no dynamical reason to cast doubts on a model which seems to work properly.

(*c*) Another possibility is that hydrated asteroids are indeed injected into the usual delivery channels and evolve to NEOs, but during this evolution they have lost the visible spectral signatures of aqueous alteration. What could be the cause(s) for this behavior? Indeed, Marchi *et al.* (2006b) have shown that planetary close encounters may affect the surface properties of S–complex NEOs. May a similar process be responsible of the lack of aqueous alteration among NEOs? Note that, in this case, the mixing of surface material with underlying layers will not cancel the alteration features, unless the body is assumed as not homogeneous, with its aqueous alteration depending on the depth. Considering that NEOs are small, and that the alteration is generally believed to be a large scale effect, it is difficult to support this suggestion. A different mechanism should be at work. Some experimental works (Hiroi *et al.* 1996) suggested that the heating of hydrated material, at a temperature in the range $400 - 600$ K, would cancel out the visible spectral signatures of aqueous alteration. A possible source of heating could be the Sun radiation. For very dark objects (such as C–complex bodies are), the sub–solar temperature (namely the maximum temperature achievable) scales as $5777\,(R_\odot/a)^{1/2}$ K, which entails a temperature of about 400 K at 1 AU from the Sun. Therefore, Sun heating may represent an efficient mechanism for the surface dehydration of NEOs with perihelion distance of the order or less than about 1 AU. The Sun–heating mechanism is, for the moment, the most promising one among those we discussed, and should be analysed in more detail, even with a systematic scrutiny of the dynamical properties of all the objects in the sample.

However, at the moment, the problem of explaining the strong deficit of hydrated NEOs remains open.

Moreover, the inverse behavior of the slope vs. exposure should be understood. In the case of family bodies, according to Lazzarin *et al.* (2006) it might be due to a sample selection effect. In this case the reason might be similar, since the sample is rather small. Or may some of the physical processes discussed above affect also the space weathering properties? In both cases we do not have at disposal any explanation, for the moment. In principle, we think that the capability to explain this anomaly together with the lack of hydrated NEOs should be the marker of a –future– inspired and successful theoretical model.

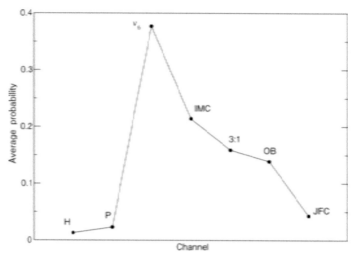

Figure 1. Average probability of origin, for all C–complex NEOs, as a function of the main different channels of origin (data by A. Morbidelli). H means Hungaria, P Phocaea, IMC, OB and JFC are respectively Mars Crossers, Outer Belt bodies and Jupiter Family Comets.

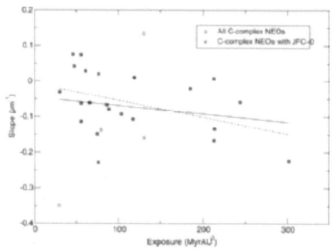

Figure 2. The slope–exposure plot for C–complex NEOs. The dashed line represents the best linear fit including only the bodies (labelled as JFC=0) which cannot, according to the channel–model (see Fig. 1), have been originated in the Jupiter–family cometary region, while the solid line corresponds to all bodies. Note that in the plot we represent only those NEOs (27 of 35) for which the slope can be estimated from the spectrum.

References

Hiroi, T., Zolensky, M.E., Pieters, C.M. & Lipschutz, M.E. 1996, *M&PS* 31, 321

Hsieh, H. & Jewitt, D. 2006, *Science* 312, 561

Lazzarin, M., Marchi, S., Moroz, L.V., Brunetto, R., Magrin, S., Paolicchi, P. & Strazzulla, G. 2006, *ApJ* 647, L179

Marchi, S., Paolicchi, P., Lazzarin, M. & Magrin, S. 2006a, *AJ* 131, 1138

Marchi, S., Magrin, S., Nesvorný, D., Paolicchi, P. & Lazzarin, M. 2006b, *MNRAS* 368, L39

Morbidelli, A., Bottke W.F., Froeschlé, Ch. & Michel, P. 2002, in: W.F. Bottke, A. Cellino, P. Paolicchi & R.P. Binzel (eds.), *Asteroids III* (Tucson: University of Arizona Press), p. 619

Near Earth Objects, our Celestial Neighbors: Opportunity and Risk
Proceedings IAU Symposium No. 236, 2006
A. Milani, G.B. Valsecchi & D. Vokrouhlický, eds.

© 2007 International Astronomical Union
doi:10.1017/S1743921307003298

X-ray fluorescence modelling for Solar system regoliths: Effects of viewing geometry, particle size, and surface roughness

**Jyri Näränen[1]†, Hannu Parviainen[1]
and Karri Muinonen[1]**

[1] Observatory, Tähtitorninmäki (PO Box 14), FI-00014 University of Helsinki, Finland

Abstract. Soft X-ray fluorescent emission from the surfaces of asteroids and other atmosphere-less solar-system objects is studied using ray-tracing techniques. X-ray observations allow the assessment of the elemental composition and structure of the surface. The model regolith is assumed to consist of close-packed uniformly distributed spherical particles of equal size. The surface is also assumed to be rough according to a fractional-Brownian-motion model. The fluorescent X-ray emission from regolith surfaces is simulated in order to better understand the contribution of viewing-geometry -related phenomena on the signal obtained from, e.g., orbiting platforms. The first results are presented and the applicability of the methods to the interpretation of future asteroid and Mercury mission X-ray data (e.g., BepiColombo) is discussed.

Keywords. spectroscopy, Moon: asteroids, surface; X-rays, scattering

1. Introduction

Several atmosphereless solar-system bodies, including the Moon (Foing *et al.* 2006), and asteroids (433) Eros (Nittler *et al.* 2001) and (25143) Itokawa (Okada *et al.* 2006), have been studied in soft X-ray wavelengths, i.e., at energies $\sim$0.1-10 keV. The main objective since the first planetary X-ray observations with the Apollo program (Adler *et al.* 1972) has been to obtain elemental ratios (e.g., Mg/Si, Al/Si) and absolute abundances from fluorescent emission induced by solar irradiation. These measurements allow the determination of the elemental composition of the surface layer of the object which is crucial for establishing the mineralogy of the surface.

The past studies have been somewhat limited in spatial and spectral resolution because of the detector technology available. As a consequence, viewing-geometry-related phenomena have not been studied nearly as much as in, e.g., the visible wavelengths. With the advent of new CCD-based X-ray spectrometers (Grande *et al.* 2002) and microchannel optics suitable for space missions, the situation is changing. The future missions (especially to objects close to the Sun, e.g., NEOs and Mercury) will deliver much more detailed data than has been available thus far and, as a result, more detailed analytical tools are needed. The ESA mission to Mercury, BepiColombo, for example, will carry an X-ray spectrometer capable of sub-kilometer resolution at best.

In the visual wavelengths, the use of viewing-geometry-related phenomena for derivation of physical parameters, e.g., particle size and surface roughness, has been extensively studied. Some research has also been carried out to understand how viewing geometry affects the soft X-ray signal, but so far the data has not been good enough to give

† email: naranen@astro.helsinki.fi

conclusive results. The study by Okada & Kuwada (1997) shows that the roughness and size distribution of the surface regolith affect the signal at large phase angles. Okada (2004) has also shown that the NEAR-Shoemaker X-ray analysis on the mineralogy of asteroid 433 Eros can be improved if the phase angle of the observations is taken into account. The characteristic elemental fluorescent lines are affected at different rates, making the relative abundance studies more complex than previously considered.

Much of the background in planetary X-ray data is caused by different forms of viewing-geometry-dependent scattering (both elastic and inelastic). In this paper, we will not concentrate on scattering, but it will be an important addition to our studies in the near future.

In the present paper, we describe first results from our studies of particle size, surface roughness, and viewing-geometry effects on soft X-ray fluorescence from a realistic medium. We employ a ray-tracing method to simulate the fluorescence from a semi-infinite plane-parallel medium consisting of spheres of equal size and fixed volume fraction. Surface roughness is approximated by weighting the ray-tracing results with a shadowing function produced with a fractional-Brownian-motion surface model.

First, we outline the methods employed in our studies. We then proceed to presenting the first results. We conclude by discussing the importance of this study to future space missions and by outlining the plans for future work.

2. Theoretical and numerical models

We employed two different approaches in this study to replicate the fluorescent signal expected from a regolith surface of an atmosphereless object. First, the signal itself was produced by using ray-tracing simulations on a surface model consisting of spheres of equal size with known volume fraction (ratio of the volumes of the particles and to the full medium). Second, we took into account the more macroscopic effect of surface roughness. This we achieved by producing shadowing functions from a fractional Brownian-motion-surface model.

Here we concentrate on fluorescent emission and omit, e.g., Bragg scattering. Also secondary fluorescence, i.e., fluorescence induced by fluorescent radiation of higher energy, is not included.

2.1. Ray-tracing fluorescence simulations

Fluorescent emission was simulated by tracing rays through a medium made of spherical particles of equal size. The method is similar to that by Muinonen *et al.* (2001) with the largest differences in the interaction of rays with the particles. In the study at hand, the free path of the ray in the medium was Monte Carlo sampled and whenever the ray encountered a particle the occurrence of absorption was studied. In the case of absorption, a fluorescent yield factor was applied. If fluorescence happened the ray was emitted isotropically and traced until it was absorbed again or it escaped from the medium, in which case the emitted direction and flux density was registered (Muinonen *et al.* 2004).

The volume fraction of the medium was kept at 0.3 in all of the simulations. As a light source we used simulated solar spectrum during X-flare (cut-off at 30 keV) (Juhani Huovelin, private communication). The light source was assumed to be in nadir. This allowed the averaging of the results over all azimuth angles.

The number of particles per surface elements was adjusted according to the size of individual particles to stop rays from going through the medium. The average number of particles was 2000-10000 per surface element. To prevent artificial features from structures in the simulated medium, one hundred random media with fixed volume fraction were created for each simulation, i.e. each of these media were targets for 1% of the total number of rays used. To get good statistics, several million rays are required. This requirement was fulfilled for all the other particle sizes except 1 μm for which we had to contend with 750 000 rays due to computational limitations.

As for the fluorescent material, we concentrated on two elements that have been found in planetary soft X-ray spectra, iron (Fe) and calcium (Ca). However, other materials can also be easily implemented.

We validated the simulation algorithm by comparing it with radiative transfer results for isotropic scatterers (a modification of the work by Muinonen 2004).

2.2. Rough surface modelling

Surface roughness can have a significant effect on the observed radiation interacting with the surface. The surface features can shadow the surface from incident radiation, and mask the escaping radiation from the irradiated parts. For a homogeneous isotropic surface with certain roughness statistics, the geometric self-shadowing and self-masking can be combined into a shadowing function $S(\theta_i, \phi_i, \theta_e, \phi_e)$. The shadowing function gives the probability for a point visible to the observer to be illuminated as a function of the incident angles (θ_i, ϕ_i) and the emergent angles (θ_e, ϕ_e) (Shepard & Campbell 1998).

Exact formulation of the combined shadowing and masking function is currently possible only for a few simplified surface-roughness models, but true natural rough surfaces show complex roughness features in all scales of examination. This sort of self-affine fractal behaviour cannot currently be handled with analytical methods, and models based on numerical simulations are required.

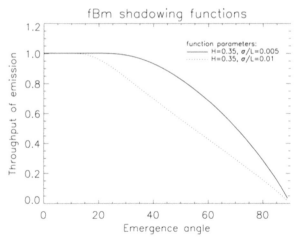

Figure 1. Shadowing functions used in our study. The emergence angle is the angle measured from the surface normal in which the ray emerges from the medium.

For creating shadowing functions for this study we employed a fractional-Brownian-motion surface model by Parviainen & Muinonen (2006); see also Peitgen & Saupe (1988). The parameters used in the model represent the standard deviation of the surface heights normalized to element width (σ/L) and Hurst exponent (H) which determines the self-affine fractal roughness features of all the length scales.

Figure 2. An illustration of the rough-surface models. The Hurst exponent H and height standard diviation σ/L are as follows: $H = 0.35$, $\sigma/L = 0.01$ (left); $H = 0.35$, $\sigma/L = 0.005$ (right).

3. Results

3.1. *Effect of grain size*

For the study of the effect of grain size on the fluorescent emission, we report results produced with iron (Fe) particles, but similar effects were also seen for calcium (Ca). We noted overall increase in the fluorescent radiation with decreasing particle size. The increase is most notable in particle sizes of 10 μm and less. For example, the fluorescent emission at $60°$ emergence angle is ~2.6 times greater for 1 μm particles than for 250 μm particles.

A nonlinear brightening near the opposition geometry was also observed. We will address this opposition effect later on. There is also a notable nonlinear brightening in larger emergence angles. This increase is also observed in the case of isotropic scattering in the visible wavelengths. It will, however, cancel out once rough surfaces are introduced into the simulation as can be seen below.

The apparent scatter in the emission curves in Figure 3 is due to computational noise and not a real effect.

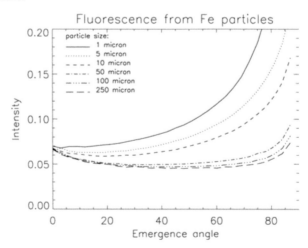

Figure 3. Fluorescent emission from iron-particle media of different particle sizes.

3.2. *Variation of apparent elemental ratios with changing viewing geometry*

The NEAR-Shoemaker X-ray spectrometer (XRS) obtained data of asteroid 433 Eros in a large variety of different viewing conditions (Trombka *et al.* 2000). The initial analysis did not, however, produce conclusive results for mineralogical composition. This can be due to various reasons, but Okada (2004) has shown that the analysis can be improved if the viewing-geometry-dependent effects of particle size and mutual shadowing of particles (micro-scale surface roughness) are taken into account.

In our study, we assumed two media of the same dimensional properties but different elemental composition. The first consisted of iron and the second of calcium. We then produced the Fe/Ca emission ratio as a function of viewing geometry for the case of equally sized particles and the case where the iron particles were 10 times larger. No shadowing functions were included in this part of the study, since macro-scale shadowing is independent of particle sizes and elemental compositions.

In both, cases a change in the elemental ratio as a function of viewing geometry was seen. This can be understood by the different mean free paths of radiation for the different elements in the media, i.e., iron absorbs photons more efficiently thus making it easier for fluorescent photons to escape in small emergence angles in comparison to larger angles. In the case of unequal particle sizes, an enhancement of the effect was observed due to the particle-size effect on fluorescence. The two cases presented here illustrate well the complexity of the analysis of elemental ratios from X-ray spectrometer data.

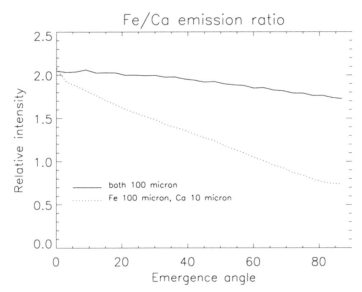

Figure 4. The Fe/Ca ratio of emitted fluorescent radiation as a function of viewing geometry. The two cases are: 1) both elements have equal particle size; 2) Ca particles are smaller by an order of magnitude.

3.3. *Opposition effect*

When the viewing geometry is close to the opposition geometry, i.e., the observer and the illumination source are in the same direction, a nonlinear brightening was observed for all grain sizes. This is explained by the fact that, in opposition, the particles in the surface offer no obstruction for the escaping photons. The situation is analogous to the shadow-hiding opposition effect in visual wavelengths.

3.4. *Rough-surface shadowing*

The application of two shadowing functions obtained from the fractional-Brownian-motion surface model clearly smoothens out the brightening in the larger viewing angles. The effect can be almost an order of magnitude in the larger emergence angles (ratio of the intensities at $80°$, when shadowing is and is not accounted for is ~ 7).

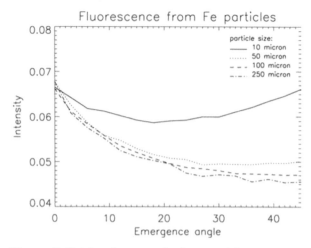

Figure 5. Brightening towards the opposition geometry.

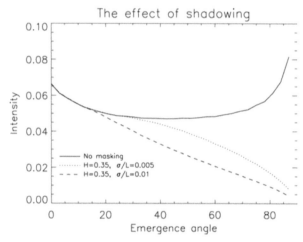

Figure 6. The effect of shadowing on the fluorescent emission of 100 μm Fe particles. The parameters for the two shadowing functions are given in the figure.

4. Conclusions and future work

We have presented the first results from our study of viewing-geometry, particle-size and surface-roughness effects in soft X-ray wavelengths. The results show that the physics of soft X-ray fluorescence are complicated. Change in particle size can affect the amount of fluorescent emission in a given viewing geometry by as much as 100% or more. The ratio of fluorescent emission of two identical media made of different elements is a function of viewing geometry. It also depends on the sizes of the elemental particles.

This shows that the analysis of soft X-ray data obtained with spectrometers in the future with good spatial and spectral resolution requires more elaborate models than used until now. This is especially important for missions to asteroids, where the viewing geometry can change rapidly and cover a large number of viewing angles.

In addition to creating new problematics for detailed soft X-ray spectrometry, there is also potential for using viewing-angle-dependent phenomena as data for inversion of physical parameters, e.g., surface roughness and particle sizes, of the surface.

One of the most important improvements to our model in the future will be the implementation of secondary fluorescence, i.e., in allowing the possibility of fluorescent photons to be absorbed and re-emitted at lower energy levels. This also allows us to consider particles consisting of several elements thus providing a more realistic fluorescing medium. Developing a medium of particles with different size distributions for the simulations is planned. The effect of volume fraction needs to be studied in more detail. The effect of different types of Solar fluxes will also be addressed. It will be also interesting to study the NEAR Shoemaker XRS data and verify that the results reported by Okada (2004) can be replicated with our model.

Laboratory measurements with a soft X-ray spectrometer capable of changing the viewing geometry would provide important ground reference to our studies and is under consideration.

Scattering, both elastic and inelastic, forms a major contribution in the measured background. Thus, it will be important in the future to also include scattering in our modeling. Several forms of scattering need to be investigated including, e.g., Bragg scattering and classical electromagnetic scattering (in the softest part of the spectrum).

Acknowledgements

This work has been funded by the Academy of Finland grant 210607. We thank George Fraser for insightful discussions on the subject. Juhani Huovelin helped with the solar simulation. We would also like to thank Hsiang-Kuang Chang for a helpful review.

References

Adler, I., Trombka, J., Gerard, J., Lowman, P., Schmadebeck, R., Blodget, H., Eller, E., Yin, L., & Lamothe, R. 1972, *Science* 175, 436

Bielefeld, M. 1977, in: 8th. Lunar Science Conference, Proc. Volume 1. (A78-41551 18-91) (New York: Pergamon Press, Inc.), p. 1131

Foing, B.H., Racca, G.D., Marini, A., Evrard, E., Stagnaro, L., Almeida, M., Koschny, D., Frew, D., Zender, J., Heather, J., Grande, M., Huovelin, J., Keller, H.U., Nathues, A., Josset, J.L., Mälkki, A., Schmidt, W., Noci, G., Birkl, R., Iess, L., Sodnik, Z., & McManamon, P. 2006. *AdSpR* 37, 6

Grande, M., Dunkin, S., Heather, D., Kellett, B., Perry, C. H., Browning, R., Waltham, N., Parker, D., Kent, B., Swinyard, B., Fereday, J., Howe, C., Huovelin, J., Muhli, P., Hakala, P.J., Vilhu, O., Thomas, N., Hughes, D., Allayne, H., Grady, M., Russell, S., Lundin, R., Barabash, S., Baker, D., Clark, P.E., Murray, C.D., Christou. A., Guest, J., Casanova, I., d'Uston, L.C., Maurice, S., Foing, B., & Kato, M. 2002, *AdSpR* 30, 1901

Muinonen, K., Nousiainen, T., Fast, P., Lumme, K., & Peltoniemi, J. 1996, *JQSRT* 55, 577

Muinonen, K., Stankevich, D., Shkuratov, Yu.G., Kaasalainen, S., & Piironen, J. 2001, *JQSRT* 70, 787

Muinonen, K. 2004, *Waves in Random Media* 14, 365

Muinonen, K., Huovelin, J., Alha, L., Lagerkvist, C.-L., Warell, J., Koskinen, H., Hämäläinen, K., Laukkanen, J., Nenonen, S., Andersson, H., Ritala, J., Stenberg, J., Korpela, S., & Heilimo, J. 2004, in: Proceedings of the XXXVIII Annual Conference of the Finnish Physical Society (Department of Physical Sciences, University of Oulu, Finland, March 18-20, 2004), 4.18., abstract

Nittler, L., Starr, R., Lim, L., McCoy, T., Burbine, T., Reedy, R., Trombka, J., Gorenstein, P., Squyres, S., Boynton, W., McClanahan, T., Bhangoo, J., Clark, P., Murphy, M.E., & Killen R. 2001, *MaPS* 36, 1673

Okada, T. & Kuwada, Y. 1997, in: 28th. L&PSC, abstract no. 1708

Okada, T. 2004, in: 35th. L&PSC, abstract no. 1927

Okada, T., Shirai, K., Yamamoto, Y., Arai, T., Ogawa, K., Hosono, K., & Kato, M. 2006, *Science* 312, 1338

Parviainen, H. & Muinonen, K. 2006, *JQSRT*, submitted

Peitgen, H.O. & Saupe, S. 1988, *The Science of Fractal Images*, Springer-Verlag

Shepard, M. & Campbell, B. 1998, *Icarus* 134, 279

Trombka, J., Squyres, S., Brückner, J., Boynton, W., Reedy, R., McCoy, T., Gorenstein, P., Evans, L., Arnold, J., Starr, R., Nittler, L., Murphy, M., Mikheeva, I., McNutt Jr, R., McClanahan, T., McCartney, E., Goldsten, J., Gold, R., Floyd, S., Clark, P., Burbine, T., Bhangoo, J., Bailey, S., & Petaev, M. 2000, *Science* 289, 2101

Near Earth Objects, our Celestial Neighbors: Opportunity and Risk
Proceedings IAU Symposium No. 236, 2006   © 2007 International Astronomical Union
A. Milani, G.B. Valsecchi & D. Vokrouhlický eds.   doi:10.1017/S1743921307003304

Near-Earth objects as principal impactors of the Earth: Physical properties and sources of origin

Dimitrij F. Lupishko[1], Mario Di Martino[2] and Richard P. Binzel[3]

[1]Institute of Astronomy of Karazin Kharkiv National University, Kharkov, Ukraine
email: lupishko@astron.kharkov.ua

[2]INAF-Osservatorio Astronomico di Torino, Pino Torinese, Italy
email: dimartino@oato.inaf.it

[3]Massachusetts Institute of Technology, Boston, USA
email: rpb@mit.edu

Abstract. Near-Earth objects (NEOs) are objects of a special interest from the point of view not only of cosmogonic problems of the Solar system, but of the applied problems as well (the problem of asteroid hazard, NEOs as the potential sources of raw materials, etc.). They are much smaller in sizes than main-belt asteroids (MBAs), very irregular in shape and covered with a great number of craters of different sizes. Most of NEOs are covered by regolith of low thermal inertia and different thickness. Objects with complex non-principal axis rotation (tumbling bodies) and with super-fast rotational periods have been detected among them. The new data, based on photometric and radar observations, evidence that about $15-20$ % of NEOs could be binary systems. Most of the classified NEOs fragments of differentiated assemblages of S- and Q-types. Analysis of physical properties of NEOs clearly indicates that the asteroid main-belt is the principal source of their origin and only about 10 % of NEOs have a cometary origin.

Keywords. transport of meteorites; asteroid, orbit; asteroid, shape; asteroid, spectra

1. Introduction

Starting from Shoemaker *et al.* (1979), the objects belonging to the NEA population, which can approach and cross the Earth's orbit, are traditionally divided into three groups (the relative abundances have been estimated by Bottke *et al.* (2002) and refer to a modeled debiased population):

$$\text{Amor} \qquad a \geqslant 1.0 \text{ AU} \qquad 1.017 < q < 1.3 \text{ AU} \qquad (32 \pm 1\%)$$

$$\text{Apollo} \qquad a \geqslant 1.0 \text{ AU} \qquad q < 1.017 \text{ AU} \qquad (62 \pm 1\%)$$

$$\text{Aten} \qquad a < 1.0 \text{ AU} \qquad Q > 0.983 \text{ AU} \qquad (6 \pm 1\%)$$

There is also an additional group of rather dangerous asteroids whose orbits reside entirely inside that of the Earth ($Q < 0.983$ AU). These objects can become Earth-crossers without having been previously spotted, because they are usually difficult for observations. According to Bottke *et al.* (2002), objects of this inner-Earth asteroid (IEA) group can constitute about 2% of the total NEO population.

Milani *et al.* (1989) proposed a new classification of the NEOs. On the basis of orbital evolution analysis, they named 6 dynamical classes after the most representative object in each class, i.e. Geographos class, Toro class, Alinda class, Kozai class, Oljato class and Eros class.

Table 1. Bulk density estimates currently available for NEOs. *) radar data.

Asteroid	Density, g/cm^{-3}	Type	Reference
433 Eros	2.67 ± 0.03	S	Yeomans et al. (2000)
6489 Golevka*	$2.7(+0.4, -0.6)$	Q	Chesley et al. (2003)
25143 Itokawa	1.95 ± 0.14	S,Q	Abe et al. (2006)
1999 KW4	1.97 ± 0.24 (primary)	S	Ostro et al. (2006)
	$2.81(+0.82, -0.63)$ (secondary)		Ostro et al. (2006)
2100 Ra-Shalom*	$1.1 - 3.3$	C	Shepard et al. (2000)
1996 FG3	1.4 ± 0.3	C	Mottola & Lahulla (2000)
2000 DP107	$1.6(+1.2, -0.9)$	–	Hilton (2002)
2000 UG11	$1.5(+0.6, -1.3)$	–	Hilton (2002)

More than 4,000 NEOs have been discovered so far. They are the objects of special interest from the point of view not only of basic science but applied science as well (the problem of asteroid hazard, NEOs as the potential sources of raw materials in the nearest to the Earth, etc.). NEOs are the principal cosmic bodies which strike our planet occasionally and therefore they are a real threat to the humankind.

2. Sizes and Densities

The principal distinctions between NEOs and MBAs are in their orbits and sizes. NEOs are much smaller in size in comparison with MBAs. Amor asteroid (1036) Ganymed, with a diameter of 38.5 km, is the largest among NEOs. Two objects, 433 Eros and (3552) Don Quixote, are about 20 km in diameter and all others are smaller than 10 km. Fortunately, the three largest NEOs belong to the Amor group, that is, they are not dangerous bodies (at least now) because they can only approach the Earth but not cross its orbit. Among Earth-crossers, (1866) Sisyphus is the largest object with a diameter of about 9 km. The smallest known NEOs are about 10 m across (2003 SQ$_{222}$).

The size distribution of NEO population can be approximated by a power law

$$N(> D) = kD^{-b}$$

with an exponent $b = 1.65 - 2.0$ (Morbidelli & Vokrouhlický (2003); Stuart & Binzel (2004)) and is similar, and only slightly steeper, than that of the MBAs.

Unfortunately, the data on densities are available only for a few NEOs and the most accurate bulk density (see Table 1) was obtained for (433) Eros from the successful NEAR-Shoemaker mission. Similar values were obtained for other S- or Q-type objects (silicate types). Comparing their bulk densities with a density of S-asteroid meteorite analogues (ordinary chondrites) we have to suppose about 30% for the NEO porosity. Approximately the same situation is with low-albedo C-type objects. It means that some NEOs are not monolithic bodies but "rubble-pile" structures, which have no coherent tensile strength and are weakly held together by their own mutual gravity. One example of a "rubble-pile" body is the Apollo-object (25143) Itokawa (Fujiwara et al. (2006)).

For the largest Apollo-object 1866 Sisyphus (S-type, $D = 9$ km), considering a density of 2.67 g/cm^{-3}, the mass is 10^{12} tons. The collision of it with the Earth at a velocity of 20 km/s would release an energy of about 5×10^7 MT, that is, about 106 times the Tunguska explosion. However, the frequency of such event is about 10^{-7} - 10^{-8} year^{-1}.

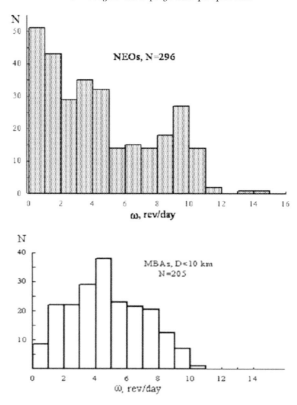

Figure 1. Distribution of the rotation rates of NEOs and small ($D \leqslant 10$ km) MBAs.

3. Shapes and Axis Rotation

Table 2 contains the average measured amplitudes (corrected to zero phase angle) of asteroid lightcurves, which characterize the shape elongation of the body. It is important to know that corrected amplitudes of the NEOs and MBAs are the same. Thus, NEOs on the average are elongated to the same extent as those of MBAs of corresponding sizes. But observations showed a striking diversity of NEOs shapes from nearly spherical (1943 Anteros, 2102 Tantalus) to very elongated (1620 Geographos, 1865 Cerberus) and to bifurcated (4179 Toutatis) and contact-binary ones (4769 Castalia, 2005 CR$_{37}$). The most elongated asteroid among observed NEOs is 1865 Cerberus ($D = 1.2$ km), the axis ratio $a : b$ of its figure is estimated to be equal to 3.2. The opinion that NEOs have more exotic shapes than MBAs may belong only to the large MBAs, because apart from lightcurves we know practically nothing about the real shapes of km-sized main-belt objects.

Figure 1 shows the histograms of distribution of the rotation rates (ω) of NEOs in comparison with that for small MBAs in the same intervals of ω. As one can see, they are quite different. It is an observational result which needs to be explained. But it is rather complex task, because there exist several reasons for that difference, among them the influence of the rotational parameters of binaries, the difference in asteroid diameter distributions within these two asteroid populations, possible influence the radiation pressure torques (YORP-effect), imperfect data statistics and maybe some selection effects. However, the whole interval of NEO rotation periods ranges over four orders of magnitudes from 500-600 hrs ((96590) 1998 XB and 1997 AE$_{12}$) to 1.3 min (2000 DO$_8$).

Table 2. Mean values of asteroid lightcurve amplitudes.

Population	Amplitude measured mag	Amplitude corrected mag	$a:b$	N
NEOs	0.53 ± 0.03	0.27	1.3	292
MBAs, $D < 10$ km	0.32 ± 0.02	0.26	1.3	205
MBAs, $D > 130$ km	0.22 ± 0.01	0.19	1.2	100

Among the km-sized NEOs the fastest rotators have rotation periods about $1-2$ hrs, for example,

$$(1566) \text{ Icarus - 2.273 hrs}$$

$$2000 \text{ EB}_{14} \text{ - 1.79 hrs}$$

$$(23714) \text{ 1998 EC}_3 \text{ - 1.2 hrs}$$

but the slowest ones rotate with the periods equal to $1-2$ hundred hrs and even more (up to $500-600$ hrs):

$$(4179) \text{ Toutatis - 129.8 hrs}$$

$$(3102) \text{ Krok - 147.8 hrs}$$

$$(1998) \text{ QR}_{52} \text{ - 235 hrs}$$

There exist two peculiarities of NEO rotation, which are not discovered elsewhere among MBAs. The first: recently among the small NEOs the objects with superfast rotation were discovered, among them

$$2000 \text{ PH5 - 0.203 hrs } (D \sim 100 \text{ m})$$

$$2000 \text{ AG6 - 0.077 hrs } (D \sim 30 \text{ m})$$

$$2000 \text{ DO8 - 0.022 hrs } (D \sim 40 \text{ m})$$

which have rotation periods within $2-20$ min Harris & Pravec (2006). It is clear that such fast-spinning bodies are beyond the rotational breakup limit for aggregates like "rubble piles" and they are monolithic fragments.

The second peculiarity is that among this population there are objects with very complex and non-principal axis rotation (so-called "tumbling" asteroids). They are usually slowly spinning objects. An example is the Apollo object (4179) Toutatis ($D \sim 3$ km), which rotates around the longest axis with a period 129.8 hrs, and has a long axis precession with a period of 176.4 hrs (Spencer *et al.* (1995); Hudson & Ostro (1995)). Other examples include 3288 Seleucus, (4486) Mithra, 2002 TD$_{60}$, and other NEOs that show two or more harmonic frequencies in their lightcurves. Among them there is only one MBA, (253) Mathilde, which is suspected to be tumbling (Pravec *et al.* (2005)). Tumbling objects may have experienced a recent collision and their internal stresses try to re-align the rotation axis with the principal axis to restore a non-tumbling state of the body.

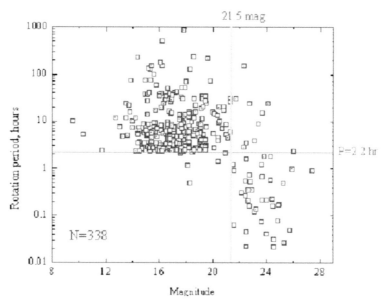

Figure 2. Rotation periods versus absolute magnitude for NEOs (from Krugly (2003)).

Figure 2 shows the rotation periods of NEOs versus their absolute magnitudes similar to the Figure 3 in Harris & Pravec (2006) for MBAs, NEOs and comets, but instead D (km) H (mag) is used. This diagram indicates the critical limit magnitude $H = 21.5$ which (at the bias-corrected mean albedo for the NEOs $pv = 0.14 \pm 0.02$, Stuart & Binzel (2004)) corresponds to the boundary diameter equal to 180 m. Objects below 180 m appear to be capable of fast rotation, indicating they must have an internal tensile strength (monolithic bodies). There is also the rotation "speed limit", corresponding 2.2 hours (as pointed out by Harris & Pravec (2006) as the "rubble pile spin barrier"). Thus, the region in rotation-diameter space where "rubble piles" can exist is limited by rotation period $P \geqslant 2.2$ hrs and $D \geqslant 180$ m. One can note that knowledge of the rotation period and size sometimes can provide important initial information on the nature and internal strength of a NEO.

4. Taxonomy and Mineralogy

Practically all Tholen's taxonomic classes of MBAs are represented among classified NEOs (Figure 3), including the low-albedo P- and D-types most commonly found in the outer asteroid belt among the Hilda and Trojan groups. Binzel *et al.* (2004) in their spectroscopic survey of 252 NEOs and Mars-crossers noted that 25 of 26 Bus' taxonomic classes of main belt are represented among the NEO population. Once more the principal question of the NEO taxonomy is the relative abundance of the two most numerous super-classes: C (carbon) and other low-albedo classes and S-Q (silicate) classes. About 70 % of the classified NEOs belong to S- and Q-classes, and the observed number of these objects exceeds the number of low-albedo ones (C and others) by as much as a factor $4-5$. Stuart & Binzel (2004) modeled the bias-corrected distribution of taxonomic classes and obtained that C and other low-albedo classes consist of 27% while S+Q classes consist of 36% of the NEO population. At the same time, Bus & Binzel (2002) obtained the bias-corrected distribution of asteroids ($D \geqslant 20$ km) of taxonomic complexes S an C in main belt ($2.25 - 3.25$ AU). Their data show that the ratio of asteroid number of C-complex to that of S-complex is about 1.8 (see Figure 19 in their paper). It means that the relative number of low-albedo objects among near-Earth population is 2.4 times less than in the main asteroid belt. It is very important result of NEO taxonomy and the most immediate explanation could be that NEOs are coming preferentially from the inner regions of main belt (Bottke *et al.* (2002)), where the relative abundance of low-albedo asteroids is small.

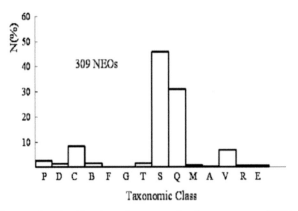

Figure 3. Distribution of taxonomic classes among NEOs.

Both taxonomy and mineralogic interpretation of spectra show evidence of genetic relationship between NEOs and MBAs. Many of the NEOs represent differentiated matter. Among them, there are objects with monomineral silicate composition and purely metallic ones. For example, small asteroid (1915) Quetzalcoatl does not contain olivine, thus diogenitic meteorites (Mg-pyroxenes) are the best analogs for it. A contrary example is (3199) Nefertity, which has no pyroxene and its composition corresponds to that of stony-iron meteorites - pallasites, that is, olivine and iron. There is a representation of unusual types such as R-types, which contain an olivine-pyroxene mixture. Three NEOs belong to the M-class, one of them, 6178 1986 DA, has a radar albedo (0.58) clearly indicating metallic composition for this asteroid. (3103) Eger with very high albedo (0.53) corresponds to assemblages of iron-free silicate minerals, such as enstatite. 22 NEOs classified as V-class, have spectra identical to those of main-belt asteroid (4) Vesta, which is known to be a differentiated body and is covered by basaltic (pyroxene-rich) material. About 30 % of classified NEOs belong to the Q-class which are the ordinary chondrite-like objects. So, the problem of finding parent bodies for the most common class of meteorites, the ordinary chondrites, now does not exist. Observing smaller and smaller S-objects Binzel *et al.* (1996, 2001) showed a continuous range of NEO spectra from those of S-types to ordinary chondrites (Figure 4). That is, there is a continuous transition (some continuum) from spectra of S-types to those of Q-types. At the same time Q-objects are smaller in size and brighter than S-objects, that is, their surfaces are "younger and fresher". Therefore, this continuum is interpreted as a result of a space weathering process, that is, a process of alterating the young surface of Q-asteroid to look more and more like an S-type surface. Figure 5 gives the spectral slope versus diameter for Q and S-asteroids. A running box mean shows that the spectral slope of Q-asteroids increases with diameter and at $D = 5$ km it becomes equal to the slope of S-types. That is, 5 km may represent a critical size in the evolution from an ordinary chondrite-like surface (Q-type) to an S-type surface. This means that objects of this size and larger have sufficient age for complete space weathering of their surface. Independently, Cheng (2004) found that $D = 5$ km may mark the boundary between primordial survivors and multi-generation fragments among the asteroids.

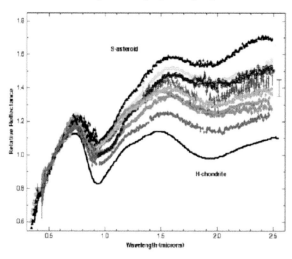

Figure 4. Continuum of spectral properties between S-type asteroids and ordinary chondrite meteorites.

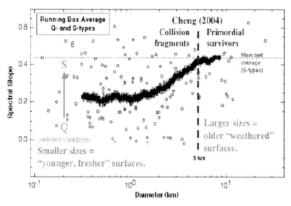

Figure 5. Spectral slope versus diameter for Q- and S-asteroids. A running box mean shows that spectral slope of Q-objects increases with diameter and at $D = 5$ km it become equal to slope of S-types.

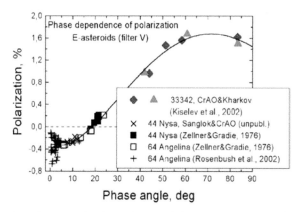

Figure 6. Polarization-phase angle dependence of Aten object 33342 (1998 WT$_{24}$) at large phase angles let us to obtain the complete phase dependence of polarization for E-type asteroids (Kiselev *et al.* (2002)).

Table 3. Mean optical parameters of S-type asteroids in V-band (Binzel *et al.* (2002)).

Parameter	NEAs	N	MBAs $(D > 100km)$	N
Albedo polarim.	$0,183 \pm 0,011$	9	$0,177 \pm 0,004$	28
Albedo radiom.	$0,190 \pm 0,014$	23	$0,166 \pm 0,006$	27
U-B (mag)	$0,445 \pm 0,013$	30	$0,453 \pm 0,008$	28
B-V (mag)	$0,856 \pm 0,013$	31	$0,859 \pm 0,006$	28
β (mag/deg)	$0,029 \pm 0,002$	9	$0,030 \pm 0,006$	18
P_{min} (%)	$0,77 \pm 0,04$	3	$0,75 \pm 0,02$	28
h (% /deg)	$0,098 \pm 0,006$	9	$0,105 \pm 0,003$	23
α_{inv} (deg)	$20,7 \pm 0,2$	6	$20,3 \pm 0,2$	18

5. Optical Properties and Surface Structure

Analysis of photometric, polarimetric, radiometric and other observational data clearly demonstrates that the surfaces of NEOs display in general the same optical properties as the surfaces of MBAs (see Table 3). The whole range of NEO albedos (from 0.05 to about 0.50) is basically the same as that of MBAs and it corresponds to the same in general mineralogy within these two populations. But the strict similarity of the other photometric and polarimetric parameters (such as phase coefficient, polarization slope and others, which are related to surface structure) gives evidence of the similar surface structures at a submicron scale.

Having similar optical properties, NEOs help us to some extent to study the MBAs, since, approaching the Earth, they let us to observe objects in a wide range of geometries of illumination and observation. For example, polarimetry of the high-albedo Aten object 33342 (1998 WT$_{24}$) at large phase angles (see Figure 6) allows us to obtain the complete phase dependence of polarization for E-type asteroids (Kiselev *et al.* (2002)). It was quite unexpected to obtain the maximum positive polarization $P_{max} = 1.7$ % at a phase angle $a_{max} \approx 76$ deg, whereas for S-asteroids $P_{max} = 8.5 - 10$ % and $a_{max} = 100 - 110$ deg.

The data of radiometry, polarimetry and direct imaging of 433 Eros and 25143 Itokawa, obtained by the NEAR-Shoemaker and Hayabusa missions, show that most observed NEOs are covered with regolith. But the conditions of formation, accumulation and evolution of regolith on NEOs are different from those on MBAs because of the much smaller gravity NEOs, the higher flux of impactors in the main belt than in the near-Earth region ($1-3$ orders of magnitude), and the difference in intensity of solar wind. As a result, the regolith of NEOs tends to be more coarse-grained than that of MBAs and still more coarse-grained than the lunar one. The recent studies of NEO thermal IR emission showed that the thermal inertia of the observed NEOs is 550 ± 100 Jm^{-2}s$^{-0.5}$K^{-1}, that is about 11 times that of the Moon (Delbò (2004)). It means that regolith of the NEOs is really coarser than the lunar one and, it is very likely, than that of MBAs.

Finally, radar data reveal that NEO surfaces are rougher than surfaces of large MBAs at the scale length of decimeters and meters and the porosity of NEO surface matter is about $30 - 50$ %, corresponding to a porosity of the top 5 to 10 cm of lunar regolith. The images of one of the largest NEOs, (433) Eros, and the rather small NEO (25143) Itokawa showed surfaces with variety of characteristics ranging from craters and boulders to perfectly smooth "ponds" of fine-grained dust. Ground-based radar observations also showed that even the relatively small NEO 4179 Toutatis and 1999 JM$_8$ ($D \sim 3$ km both) are cratered at about the same extent as MBAs (951) Gaspra and (243) Ida.

6. On the Origin of NEOs

The short typical lifetime of NEOs $\sim 10^6 - 10^7$ years implies that the currently present population must be continually supplied. The study of physical and mineralogical properties of NEOs clearly indicate that the main asteroid belt is the principal source of their origin. It means that most of the NEOs are the fragments of main-belt asteroids ejected into their current orbits by processes of collisions and chaotic dynamics (Farinella *et al.* (1993)). On the other hand, the identification of a few objects with extinct or dormant comets (118401, 133P/Elst-Pizarro, P/2005 U1) does not exclude the cometary origin of some of them. Hence, the problem is the determination of the relative contributions of both sources.

The candidates for comet origin should be low-albedo objects of D, P and C-types, with lower rotation and more elongated shapes as compared with asteroids of corresponding sizes; they should also have atypical for asteroids orbits (comet-like, with $Q \geqslant 4.5$ AU) and association with meteor streams.

The most recent estimates of the contributions of the cometary origin of NEO are:

Lupishko & Lupishko (2001): $\leqslant 10\%$ of NEOs
Fernandez *et al.* (2001): "at least 9 % of NEOs are cometary nuclei"
Whiteley (2001): "~ 5 % rather than 50 % of the cometary origin".
Bottke *et al.* (2002): ~ 6 % of the NEO comes from the Jupiter-family comet region
Binzel *et al.* (2004)): ~ 15 % of the NEO may be extinct comets.

Apparent difference between the two last quantities (6% and 15%) is due to the estimate in Bottke *et al.* (2002) is given for a H-limited sample, while in Binzel *et al.* (2004) for a size-limited sample. When the correction for albedo is done, the two fractions are actually the same.

Thus, the recent estimates give the contribution of cometary origin of the NEO population to be on average about 10 %. This conclusion does not contradict the results of dynamical considerations (Menichella *et al.* (1996); Bottke *et al.* (2002)), according to which the main asteroid belt can supply a few hundred km-sized NEAs per 1 Myr, a rate sufficient to sustain the current NEO population.

7. Summary

The discovery rate of NEOs has increasd greatly over the last few years due to new observational programs. It is a very positive result but a new problem arises: the rate of physical studies of the NEOs remains behind of the rate of their discovery. Therefore our knowledge with respect to the discovered NEO population are becoming more and more scanty. That is why the study of the nature and physical properties of NEOs remains one of the priority directions of Solar system investigations which is necessary for addressing both fundamental scientific problems and applied problems of the humanity survival.

References

Abe, S., Mukai, T., Hirata, N., Barnouin-Jha, O.S., Cheng, A.F., Demura, H., Gaskell, R.W., Hashimoto, T., Hiraoka, K., Honda, T., Kubota, T., Matsuoka, M., Mizuno, T., Nakamura, R., Scheeres, D.J., & Yoshikawa, M. 2006, *Science* 312, 1344
Binzel, R.P., Bus, S.J., Burbine, T.H., & Sunshine, J.M. 1996, *Science* 273, 946
Binzel, R.P., Harris, A.W., Bus, S.J., & Burbine, T.H. 2001, *Icarus* 151, 139

Binzel, R.P., Lupishko, D.F., Di Martino, M., Whiteley, R.J., & Hahn, G.J. 2002, in: W. Bottke, A. Cellino, P. Paolicchi, & R. Binzel, (eds.), *Asteroids III* (Univ. Arizona Press, Tucson), p. 255

Binzel, R.P., Rivkin, A.S., Scott, J.S., Harris, A.W., Bus, S.J., & Burbine, T.H. 2004, *Icarus* 170, 259

Bottke, W.F.Jr., Morbidelli, A., Jedicke, R., Petit, J.-M., Levison, H.F., Michel, P., & Metcalfe, T.S. 2002, *Icarus* 156, 399

Bus, S.J. & Binzel, R.P. 2002, *Icarus* 158, 146

Cheng, A. F. 2004, *Icarus* 169, 357

Chesley, S.R., Ostro, S.J., Vokrouhlický, D., Čapek, D., Giorgini, J.D., Nolan, M.C., Margot, J.-L., Hine, A.A., Benner, L.A.M., & Chamberlin, A.B. 2003, *Science* 302, 1739

Delbò, M. 2004, *PhD. Thesis*, Freie Universitat Berlin (Germany).

Farinella, P., Gonczi, R., Froeschlé, Ch., & Froeschlé, C. 1993, *Icarus* 101, 174

Fernandez, Y.R., Jewitt, D.C., & Sheppard S.S. 2001, *Astrophys. J.* 553, L197

Fujiwara, A., Kawaguchi, J., Yeomans, D.K., Abe, M., Mukai, T., Okada, T., Saito, J., Yano, H., Yoshikawa, M., Scheeres, D.J., Barnouinjha, O., Cheng, A.F., Demura, H., Gaskell, R.W., Hirata, N., Ikeda, H., Kominato, M.T., Miyamoto, H., Nakamura, A.M., Nakamura, R., Sasaki, S., & Uesugi, K. 2006, *Science* 312, 1330

Harris, A.W. & Pravec, P. 2006, in: D. Lazzaro, S.Ferraz Mello, & J.A. Fernandez (eds.) *Asteroids, Comets, Meteors* (Cambridge Univ. Press), p. 439

Hilton, J.L. 2002, in: W. Bottke, A. Cellino, P. Paolicchi, & R. Binzel, (eds.), *Asteroids III* (Univ. Arizona Press, Tucson), p. 103

Hudson, R.S. & Ostro, S.J. 1995, *Science* 270, 84

Kiselev, N.N., Rosenbush, V.K., Jockers, K., Velichko, F.P., Shakhovskoj, N.M., Efimov, Yu.S., Lupishko, D.F., & Rumyantsev, V.V. 2002, *Asteroids, Comets, Meteors 2002* (Techn. Univ., Berlin), p. 887

Krugly, Yu.N. 2003, *PhD. Thesis*, Kharkiv University (Ukraine).

Lupishko, D.F. & Lupishko, T.A. 2001, *Solar System Research* 35, 227

Menichella, M., Paolicchi, P. & Farinella, P. 1996, *Earth, Moon, and Planets* 72, 133

Milani, A., Caprino, M., Hahn, G., & Nobili, A. 1989, *Icarus* 78, 212

Morbidelli, A. & Vokrouhlický, D. 2003, *Icarus* 163, 120

Mottola S. & Lahulla F. 2000, *Icarus* 146, 556

Ostro, S.J., Benner, L.A.M., Nolan, M.C. Magri, C., Giorgini, J.D., Scheeres, D.J., Broschart, S.B., Kaasalainen, M., Vokrouhlický, D., Chesley, S.R., Margot, J.L., Jurgens, R.F., Rose, R., Yeomans, D.K., Suzuki, S., & De Jong E.M. 2004, *Meteoritics and Planetary Science* 39, 407

Ostro, S.J., Margot, J.L., Benner, L.A.M., Giorgini, J.D., Scheeres, D.J., Fahnestock, E.G., Broschart, S.B., Bellerose, J., Nolan, M.C. Magri, C., Pravec, P., Scheirich, P., Rose, R., Jurgens, R.F., De Jong, E.M., & Suzuki, S. 2006, *Science*, 12 Oct. 2006 (DOI: 0.1126/science.1133622)

Pravec, P., Harris, A.W., Scheirich, P., Kusnirak, P., Sarounova, L., Mottola, S., Hicks, M., Masi, G., Krugly, Yu., Shevchenko, V., Nolan, M., Howell, E., Galad, A., Brown, P., Hergenrother, C.W., DeGra, D.R., Lambert, J.V., & Foglia, S. 2005, *Icarus* 173, 108

Shepard, M.K., Benner, L.A.M., Ostro, S.J., Harris, A.W., Rosema, K.D., Shapiro, I.I., Chandler, J.F., & Campbell, D.B. 2000, *Icarus* 147, 520

Shoemaker, E.M., Williams, J.G., Helin, E.F., & Wolf, R.F. 1979, in: T. Gehrels (ed.), *Asteroids* (Univ. of Arizona Press, Tucson), p. 253

Spencer, J.R., Akimov, L.A., & Angeli, C. and 45 colleagues. 1995, *Icarus* 117, 71

Stuart, J.S. & Binzel, R.P. 2004, *Icarus* 170, 295

Yeomans, D.K., Antreasian, P.G., Barriot, J.-P., Chesley, S. R., Dunham, D.W., Farquhar, R.W., Giorgini, J.D., Helfrich, C.E., Konopliv, A.C., McAdams, J.V., Miller, J.K.,Owen, W.M.Jr., Scheeres, D.J., Thomas, P.C., Veverka, J., & Williams, B.G. 2000, *Science* 289, 2085

Whiteley, R.J. 2001, *PhD Thesis*, University of Hawaii (USA).

Near Earth Objects, our Celestial Neighbors: Opportunity and Risk
Proceedings IAU Symposium No. 236, 2006   © 2007 International Astronomical Union
A. Milani, G.B. Valsecchi & D. Vokrouhlický, eds.   doi:10.1017/S1743921307003316

Itokawa: The power of ground-based mid-infrared observations

Thomas G. Müller[1], T. Sekiguchi[2], M. Kaasalainen[3], M. Abe[4] and S. Hasegawa[4]

[1]Max-Planck-Institut für extraterrestrische Physik,
Giessenbachstraße, 85748 Garching, Germany;
tmueller@mpe.mpg.de

[2]National Astronomical Observatory of Japan, 2-21-1 Osawa, Mitaka, Tokyo 181-8588, Japan;
t.sekiguchi@nao.ac.jp

[3]Department of mathematics and statistics, Gustaf Hallstromin katu 2b, P.O. Box 68,
FIN-00014 University of Helsinki, Finland;
mjk@rni.helsinki.fi

[4]Institute of Space and Astronautical Science, Japan Aerospace Exploration Agency, 3-1-1
Yoshinodai, Sagamihara, Kanagawa 229-8510, Japan;
abe@planeta.sci.isas.jaxa.jp; hasehase@isas.jaxa.jp

Abstract. Pre-encounter ground-based thermal observations of NEA 25143 Itokawa at $10\,\mu$m led to a size prediction of $520(\pm 50) \times 270(\pm 30) \times 230(\pm 20)$ m, corresponding to an effective diameter of $D_{eff}^{TPM} = 318$ m (Müller *et al.* 2005). This is in almost perfect agreement with the final in-situ results $535 \times 294 \times 209$ m ($D_{eff}^{Hayabusa} = 320$ m; Demura *et al.* 2006). The corresponding radar value, based on the same shape model (Kaasalainen *et al.* 2005), were about 20% too high: $594 \times 320 \times 288$ m ($D_{eff}^{Radar} = 379$ m; Ostro *et al.* 2005). The very simple N-band observations revealed a surface which is dominated by bare rocks rather than a thick regolith layer. This prediction was nicely confirmed by the Hayabusa mission (e.g., Fujiwara *et al.* 2006; Saito *et al.* 2006). The ground-based measurements covered three different phase angles which enabled us to determine the thermal properties with unprecedented accuracy and in excellent agreement with the results from the touch-down measurements (Okada *et al.* 2006; Yano *et al.* 2006). These thermal values are also key ingredients for high precision Yarkovsky and YORP calculations (mainly the rotation slowing) for Itokawa (e.g., Vokrouhlický *et al.* 2004; Vokrouhlický *et al.* 2005). In addition to the above mentioned properties, our data allowed us to derive the surface albedo and to estimate the total mass. We believe that with our well-tested and calibrated radiometric techniques (Lagerros 1996, 1997, 1998; Müller & Lagerros 1998, 2002; Müller 2002) we have tools at hand to distinguish between monolithic, regolith-covered and rubble pile near-Earth objects by only using remote thermal observations. This project also emphasizes the high and so far not yet fully exploited potential of thermophysical modeling techniques for the NEA/NEO exploration.

Keywords. asteroid, surface; infrared, thermal; photometry

1. Introduction

The Japanese Hayabusa spacecraft made a successful rendezvous with the target asteroid (25143) Itokawa in late 2005. It investigated the properties of this Apollo-type near-Earth asteroid during several months and performed two touch-downs to collect surface samples. The return of the samples to Earth is currently foreseen for 2010.

In preparation of this interplanetary mission, we performed ground-based thermal observations at around $10\,\mu$m during the 2001 and 2004 oppositions to study the thermophysical behavior and to determine size, albedo and surface properties of this small asteroid. The goal of our work was to support the Hayabusa mission with information on physical and thermal properties of Itokawa well before the encounter and touch-down phases. As a side effect, we wanted to validate and, if necessary, improve our modeling techniques to maximize the outcome of future ground-based mid-IR observing programs to study other potentially hazardous asteroids.

Müller *et al.* (2005) listed all ground-based mid-IR observations of 25143 Itokawa, including data points by Delbó (2004) and by Sekiguchi *et al.* (2003), and describe the data reduction and calibration in detail. Here, we briefly present our thermophysical model together with the Itokawa input parameters (Sec. 2) which we used to interpret our observations. The results of our modeling efforts are then discussed in the light of the Hayabusa findings (Sec. 3).

2. Thermophysical Model (TPM) and input parameters

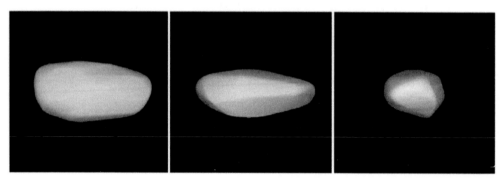

Figure 1. Shape model of 25143 Itokawa through light-curve inversion technique (Kaasalainen *et al.* 2003).

We applied the TPM by Lagerros (1996, 1997, 1998) to investigate the physical and thermal properties of the asteroid 25143 Itokawa. On the large scale, the TPM considers the asteroid size, the global shape and spin vector and the actual observing and illumination geometry at the time of an observation. On the small micrometer scale, the TPM takes into account the reflected, absorbed and emitted energy, and also the heat conduction into the surface regolith. The albedo and emissivity control the energy balance and thereby the surface temperature. The thermal inertia in combination with the rotation period and the orientation of the spin vector influence the diurnal temperature variations. As a result, the thermal inertia is strongly connected to the interpretation of mid-IR observations, namely when comparing before and after opposition observations at large phase angles with very different temperatures of the terminator. Moreover, the thermal inertia determines the amplitude of the thermal light-curve for a given aspect angle. The beaming model, described by ρ, the r.m.s. of the surface slopes and f, the fraction of the surface covered by craters, accounts for the non-isotropic heat radiation, noticeable at phase angles close to opposition. The parameters are dimensionless and they could describe a surface with large craters as well as a rough surface due to bolders of different sizes in combination with micro-impact structures. The beaming model mainly influences the shape of the spectral energy distribution in the mid-IR. But our data are all taken in N-band and slope effects in the spectral energy distribution are not relevant. Our standard set of beaming parameters was therefore the most obvious solution. Various TPM applications for NEAs and main-belt asteroids are described in e.g., Müller & Lagerros (1998, 2002), Müller & Blommaert (2004), Müller *et al.* (2004) or Müller *et al.* (2005).

The following input parameters have been used for the modeling: (i) The H (19.9 mag) and G (0.21) values describe the absolute visual brightness and the brightness change with phase angle; (ii) We assume a constant emissivity of 0.9 at all mid-IR wavelengths; (iii) We applied the pre-encounter shape-model by Kaasalainen *et al.* (2005), which is based on large datasets of photometric observations (Fig. 1); the pole solution from light-curve inversion techniques was $\beta_{pole} = -89°$; $\lambda_{pole} = 331°$; $P_{rot} = 12.13237$ hours; $T_0 = 2451933.95456$ and $\phi_0 = 0.0$ (zero-points in time and phase); (iv) We assumed "default" surface roughness parameters (Müller 2002), with $\rho = 0.7$ (r.m.s. of the surface slopes) and f = 0.6 (fraction of the surface covered by craters).

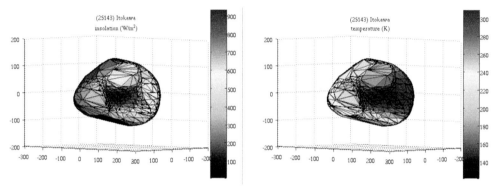

Figure 2. Left: Solar insolation (as seen from the Sun) in [W/m^2]; Right: Surface temperature (as seen from the Sun) in [K]. The temperature distribution reflects one specific TPM solution for a given set of thermal parameters in combination with the shape model by Kaasalainen *et al.* (2005): thermal inertia $\Gamma = 750$ J m^{-2} s$^{-0.5}$ K^{-1}, "standard" beaming model, emissivity $\epsilon = 0.9$. The x-, y- and z-axes are given in metres, the rotation axis is in z-direction (vertical).

Based on these input parameters and the above mentioned thermal observations, we calculated the effective diameters and albedo values for a range of thermal inertias. The effective diameter is the diameter corresponding to the equal volume sphere and represents the absolute scale factor for the Kaasalainen-shape model. If the shape and thermal models are correct, the effective diameter and albedo values should be independent of the observed wavelength, rotational phase, phase angle or aspect angle (Müller 2002). Müller *et al.* (2005) demonstrated for 25143 Itokawa the variations of the diameter and albedo solutions as a function of these parameters. We also investigated how the uncertainties of the various input parameters would influence our results. Fig. 2 shows for a thermal inertia $\Gamma = 750$ J m^{-2} s$^{-0.5}$ K^{-1} the temperature distribution on the surface. In the asymmetric temperature distribution one can easily recognize the combined rotation and thermal inertia effect. Observing such distributions under large phase angles constrain the interpretation with respect to size and albedo dramatically: The standard deviation of the 20 calculated diameter (or albedo) values changes more than a factor of two for different thermal inertias. Figure 3 illustrates this effect on basis of the σ_{albedo}/albedo values.

3. Results and Discussion

The best fit to all our thermal data resulted in an effective diameter of $D_{eff} = 0.32 \pm 0.03$ km and an albedo of $p_V = 0.19^{+0.11}_{-0.03}$. Taking only the highest quality photometric

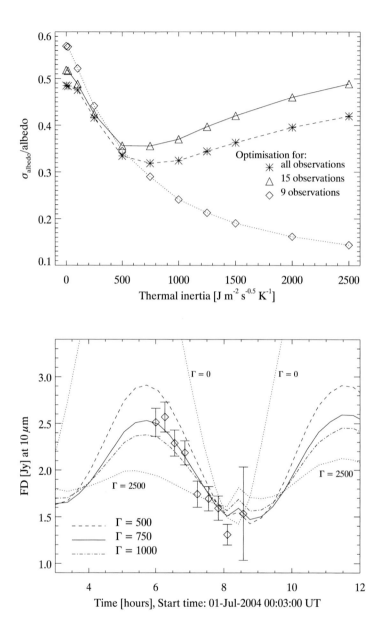

Figure 3. Top: Thermal inertia optimization for the individual TPM albedos and their standard deviation, using 20/15/9 individual observations (dashed/solid/dotted lines). Using only observations from a very small phase angle range (dotted line) does not provide a solution in this optimisation process. Bottom: Predicted thermal light-curve at 10.0 μm. The original measurements were transported to the 10.0 μm wavelength via the TPM. Predictions and measurements are shown with their absolute values. Note that a metallic surface (very high Γ-values) would produce a very small lightcurve amplitude, while a thick regolith on the surface would increase the amplitude significantly.

data would reduce the size uncertainty to only ± 0.01 km, i.e., 10 m! The transfer to the slightly asymmetrical and flattened ellipsoid input shape-model gives an absolute size of $520(\pm 50) \times 270(\pm 30) \times 230(\pm 20)$ m. Our size prediction was within 2% of the final

Hayabusa results ($535 \times 294 \times 209$ m; $D_{eff}^{Hayabusa} = 320$ m; Demura *et al.* 2006). Such an enormous precision was only possible due to the fact that our observations covered a very wide range of phase angles, including measurements at phase angles of $110°$ where only a fraction of the surface was illuminated by the sun. Such observing geometries are crucial for thermophysical techniques to see how much a cold or warm terminator contribute to the total measured flux density. It is also interesting to note that the derived size from good quality radar images (Ostro *et al.* 2005) was overestimated by more than 15% for unknown reasons. A comparison between our albedo value and the true surface albedo was so far not possible due to the lack of suitable information from the Hayabusa mission.

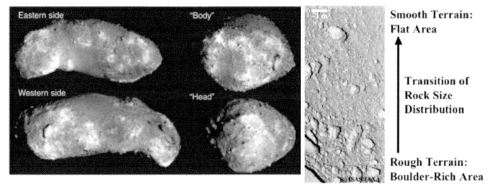

Figure 4. Left: Hayabusa composite color image of Itokawa (Saito *et al.* 2006). Right: "rocky surface"; image taken from `http://www.isas.ac.jp/e/news/2006/0602.shtml`

Our final model solution is closely connected to a relatively high thermal inertia of 750 J m^{-2} s$^{-0.5}$ K^{-1}. Only in this way it was possible to find a robust diameter/albedo solution which would fit all observed fluxes (see Fig. 3). A repetition of our optimization calculations with different sub-sets of data confirmed this solution (see solid, dashed and dotted lines in Fig. 3). Robust solutions are only found if the observational data are of similar quality and taken at different phase angles (solid and dashed lines). However, the high thermal inertia was a clear indication for a bare rock dominated surface. A thick dust regolith (typical inertia values of below 50 J m^{-2} s$^{-0.5}$ K^{-1}) could be excluded as well as a metallic surface which would have Γ-values above $10\,000$ J m^{-2} s$^{-0.5}$ K^{-1} and would produce a very small thermal lightcurve amplitude (see bottom of Fig. 3). This prediction was nicely confirmed by the Hayabusa mission (e.g., Fujiwara *et al.* 2006; Saito *et al.* 2006, see also Fig. 4). The ground-based measurements covered three very different phase angles which enabled us to determine the thermal properties with unprecedented accuracy and in excellent agreement with the results from the touch-down measurements (Okada *et al.* 2006; Yano *et al.* 2006). These thermal values are also key ingredients for high precision Yarkovsky and YORP calculations (mainly the rotation slowing) for Itokawa (e.g., Vokrouhlický *et al.* 2004; Vokrouhlický *et al.* 2005). Based on our modeling and the results from Hayabusa, we also believe that such thermal properties might be an indications for rubble pile structure of a body: monolithic rocks will have very high thermal inertias, while regolith-covered bodies have very low values. We interpret the intermediate range as indicator for an unstable internal structure. The surface of a rubble-pile asteroid will be rearranged by close encounters with large bodies, impacts and vibrational effects. Therefore, the surface conditions are renewed frequently and an Itokawa-like thermal behaviour is the result.

In Fig. 1 the shape model is shown from different orientations; let us compare with two Hayabusa images (Fig. 4). The overall model shape, just based on light-curve inversion methods, agrees very well. The same is true for the spin vector. Kaasalainen *et al.* (2005) derived values of $\beta_{pole} = -89°$, $\lambda_{pole} = 331°$ and $P_{rot} = 12.13237$ hours, in very good agreement with the Hayabusa results of $\beta_{pole} = -89.66°$, $\lambda_{pole} = 128.5°$ (3.9° error margin; Demura *et al.* 2006) and $P_{rot} = 12.1324 \pm 0.0001$ hours (Fujiwara *et al.* 2006).

4. Conclusions

Simple observations at thermal mid-IR wavelengths are the key element for the determination of high quality albedo, size and surface properties. Such measurements are easily possible from ground. And, with state-of-the-art instrumentation on large telescopes, even small near-Earth objects can be observed. Shape and spin vector are usually available from visual light-curve inversion techniques. Thermophysical model investigations contribute then the size, albedo and regolith properties. We believe that with our well-tested and calibrated radiometric techniques we have tools at hand to distinguish between monolithic, regolith-covered and rubble pile near-Earth objects by only using remote thermal observations. This project also emphasizes the high and so far not yet fully exploited potential of thermophysical modeling techniques for the NEA/NEO exploration.

Acknowledgements

We would like to thank F. Hormuth for his support in the data analysis of the TIMMI2 observations and J. Lagerros for his modifications in the TPM code to allow a proper use of the Kaasalainen shape models. M. Delbó supported our re-evaluation of the TIMMI2 data of his thesis work.

References

Delbó, M. 2004, PhD thesis, FU Berlin, `http://www.diss.fu-berlin.de/2004/289/index.html`
Demura, H., Kobayashi, S., Nemoto, E., *et al.* 2006, *Science* 312, 1347
Fujiwara, A., Kawaguchi, J., Uesugi, K., *et al.* 2006, *Science* 312, 1330
Kaasalainen, M., Abe, M., Byron, J., *et al.* 2005, Proceedings of the 1st Hayabusa Symposium, *ASP Conf. Series*, submitted
Lagerros, J. S. V. 1996, *A&A* 310, 1011
Lagerros, J. S. V. 1997, *A&A* 325, 1226
Lagerros, J. S. V. 1998, *A&A* 332, 1123
Müller, T. G. & Lagerros, J. S. V. 1998, *A&A* 338, 340
Müller, T. G. & Lagerros, J. S. V. 2002, *A&A* 381, 324
Müller, T. G. 2002, *Meteor. & Planet. Sci.* 37, 1919
Müller, T. G. & Blommaert, J. A. D. L. 2004, *A&A* 418, 347
Müller, T. G., Sterzik, M. F., Schütz, O., Pravec, P., & Siebenmorgen, R. 2004, *A&A* 424, 1075
Müller, T. G., *et al.* 2005, *A&A* 443, 347
Okada, T., Yamamoto, Y., Inoue, T., *et al.* 2006, *LPS* XXXVII, 1965
Ostro, S. J., Benner, L. M., Magri, C., *et al.* 2005, *Meteor. & Planet. Sci.*, 40, 1563
Saito, J., Miyamoto, H., Nakamura, R., *et al.* 2006, *Science* 312, 1341
Sekiguchi, T., Abe, M., Böhnhardt, H., *et al.* 2003, *A&A* 397, 325
Vokrouhlický, D., Čapek, D., Kaasalainen, M. & Ostro, S. J. 2004, *A&A* 414, L21
Vokrouhlický, D., Čapek, D., Chesley, S. R. & Ostro, S. J. 2005, *Icarus* 173, 166
Yano, H., Kubota, T., Miyamoto, H., *et al.* 2006, *Science* 312, 1350

Part 6

Surveys: Observatories and their Performances

Near Earth Objects, our Celestial Neighbors: Opportunity and Risk
Proceedings IAU Symposium No. 236, 2006
A. Milani, G.B. Valsecchi & D. Vokrouhlický, eds.
© 2007 International Astronomical Union
doi:10.1017/S174392130700333X

Single and multiple solution algorithms to scan asteroid databases for identifications

Maria E. Sansaturio and O. Arratia

E.T.S. de Ingenieros Industriales, University of Valladolid, Paseo del Cauce s/n 47011
Valladolid, Spain
email: genny@pisces.eis.uva.es

Abstract. The process of cataloguing the minor planet population of the Solar System has experienced a great advance in the last decades with the start-up of several surveys. The large volume of data generated by them has increased with time and given rise to huge databases of asteroids with uneven qualities.

In fact, a significant fraction of these objects have not been enough observed, thus leading to the computation of very poor quality orbits as to carry out useful predictions of the positions of such asteroids. As a result, some objects can get lost, which is particularly embarrassing for those with Earth crossing orbits.

When this situation persists for a long time, the aforementioned databases end up contaminated in the sense that they contain more than one discovery for the same physical object and some kind of action must be taken. The algorithms for asteroid identifications are thought precisely to mitigate this problem and their design will depend upon the quality of the available data for the objects to be identified.

In this paper we will distinguish two cases: when both objects have a nominal orbit and when one of them lacks it. In addition, when the available data poorly constrain the solution, other orbits in the neighbourhood of the nominal one are also compatible with the observations. Using these alternative orbits allows us to find many identifications that otherwise would be missed. Finally, we will show the efficiency of all these algorithms when applied to the datasets distributed by the Minor Planet Center.

Keywords. Asteroids, catalogs; orbit determination; multiple solutions; identification

---

1. Introduction

For the last decade, the current surveys have generated a large volume of data, which is increasing with time, and given rise to huge databases of asteroids. However, a large fraction of these asteroids have only been observed over an arc of a few days to a few weeks. When the orbit determination procedure is applied to these data, only low quality orbits can be got. As a consequence, as time goes by and this situation goes on and on for a long time, such asteroids are to be considered lost, that is, they cannot be recovered by pointing the telescope at the predicted position. Therefore, the asteroid databases get polluted with data that belongs to the same physical object, without having an a priori notice of this fact.

This situation requires some action to be taken and the algorithms for asteroid identifications can play an important role. In order to design such algorithms, one needs to take into account the quality, as well as the amount and distribution in time of the available data for the objects to be identified. When there is sufficient observational data to separately solve for two orbits, one for each arc, we will call it orbit identification. In this case the input data are two sets of orbital elements and the identifications are proposed taking into account the similarity of the orbits according to a suitable metric, which is not just the size of the difference in the orbital elements (Milani *et al.* 2000). If the available observational data only allows for the computation of the orbit for one of the two arcs, while the other lacks it, we have what we call the observation attribution problem. That is, we try to attribute the observations of the arc for which it has not been possible to compute a reliable orbit to the other arc which indeed has an orbit. This is done by performing a comparison between the predicted observations, based on the known orbit, and the observed data of the other arc (Milani *et al.* 2001).

On the other hand, when an asteroid has just been discovered, its orbit is poorly constrained by observations spanning only a short arc. Thus, even if the least squares procedure provides a nominal solution, other orbits are also possible solutions, in the sense that they are compatible with the observations and the corresponding rms of the residuals are not too far from the minimum. All these orbits belong to the so-called confidence region and what we would like is to find an efficient way of sampling it. To this end we use a one-dimensional segment of a curved line called Line Of Variations (LOV). It is possible to define the LOV in different ways, but the general idea is that it is a kind of spine of the confidence region. Here, we use the set of points of convergence of the constrained differential corrections to define it. The idea of "multiple solutions" sampling the LOV was introduced by Milani (1999) and has been used in different contexts. In this case, we are showing its application in asteroid identifications.

In this paper we will try to provide an idea about how we have contributed both to the theoretical understanding and to the solution of the asteroid identification problem. To this end, it is organized as follows. Section 2 is devoted to review the single solution identification methods, which include two main cases: the orbit identification and the observation attribution algorithms, as well as the interaction between the two of them: the recursive procedure, and the results obtained with their application on the Minor Planet Circular batches of data distributed by the Minor Planet Center (MPC). In Section 3 we describe the multiple solutions method and its application to the asteroid identification problem. Finally, in Section 4 we draw some conclusions about the main characteristics of the algorithms presented.

2. Single solution algorithms

In general, the asteroid identification problem deals with two separate sets of astrometric observations and can take different forms depending upon the amount and distribution in time of the available observations and upon the time interval between these two sets of data.

2.1. *Orbit identification algorithm*

A first case appears when each of the two sets contains enough information to solve for all the orbital elements by a least squares fit to the observations. Then, we have what we call the orbit identification problem. In such a case, the question is to find the minimum of the target function for a single orbit fitting all the observations.

That is, let us assume that we have two solutions X_1 and X_2, with their corresponding normal matrices C_1 and C_2, of two separate orbit determination processes. The cost functions can be expanded as

$$Q_i(X) = Q_i(X_i) + \frac{1}{m_i}(X - X_i)^T C_i (X - X_i) + \cdots = Q_i^* + \Delta Q_i \qquad i = 1, 2.$$

For them to represent the same object, we have to find the minimum of the joint target function $Q(X)$, which is the sum of the two separate target functions

$$Q = \frac{1}{m}(m_1 Q_1 + m_2 Q_2) = Q^* + \frac{\chi^2}{m},$$

where Q^* is the value corresponding to the sum (with suitable weighting) of the two separate minima.

Then, if the linear approximation applies, we can approximate each of them by a quadratic form

$$\chi^2 \simeq (X - X_1)^T C_1 (X - X_1) + (X - X_2)^T C_2 (X - X_2),$$

which results in the joint target function being also approximated by a non-homogeneous quadratic form

$$\chi^2 \simeq (X - X_0)^T C_0 (X - X_0) + K,$$

The most important feature of this procedure is that we are not only computing the most likey solution for the joint problem through this expression

$$X_0 = C_0^{-1}(C_1 X_1 + C_2 X_2),$$

where $C_0 = C_1 + C_2$, but that we are also computing the minimum identification penalty which allows us to assess the uncertainty of the identified solution by means of the following formula

$$K = \Delta X^T C \Delta X,$$

where

$$C = C_1 - C_1 C_0^{-1} C_1.$$

A more detailed description of the how to obtain these formulae can be found in Milani *et al.* (2001), Section 2.2.

It is clear that we are dealing with a very difficult problem. First, because the number of asteroids is very large and so we cannot effectively apply differential corrections to all the possible pairs, which can be of the order of billions. A second fact generating difficulties is that even a complete orbit, which is a solution of a least squares fit to all the observations of a given arc, is not the only possible orbit and hence the distance, defined by a suitable metric, between the two orbits must be computed using the uncertainty of both orbits.

Thus, with all this in mind, the orbit identification algorithm proposes identifications using a cascade of four filters, the last three being based upon identification metrics that take into account the difference in the orbits weighted with the uncertainty. Figure 1 shows the scheme of such a procedure.

One thing that needs to be kept in mind when trying to identify asteroids observed over short arcs and/or very separated in time is that the problem can become highly non-linear. In order to lessen this non-linearity several strategies are followed. As a general rule, we use equinoctial elements, instead of Keplerian elements, because in this way we avoid the non-linearity arising from the singularities for null inclination and eccentricity. We also replace the semimajor axis by the mean motion, since the latter is less sensible to non-linear effects. Finally, once one catalog of orbits has been computed with the full Minor Planet Center batches of data, we propagate it to five different epochs. When we start the comparison in the first filter, a simple distance between the two nominal orbits without taking into account their uncertainty, we select those in a catalog whose epoch is the closest to the arithmetic mean of the central times of the observations of each orbit.

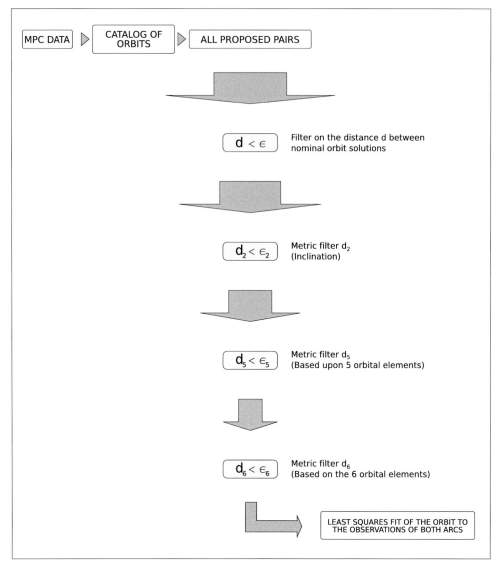

Figure 1. Orbit identification problem – The procedure.

All the pairs passing this first test are sent to the subsequent filter, which essentially looks for similarity in inclination. Later on, a metric based upon five orbital elements is applied and, finally, the last filter that takes into account all the six orbital elements establish the pairs which are sent to the least squares fit of the observations of both arcs.

2.2. *Observation attribution algorithm*

Another case in the asteroid identification problem appears when the amount of data, insufficient to compute an orbit for one arc, is compared to the orbit already computed for the other arc. This is what we call observation attribution problem.

The mathematical formulae for this algorithm are pretty similar to those presented in the previous case, the main difference being that in the orbit identification method we are working in the six-dimensional space of elements, while now we have switched to the four-dimensional space of observations, and so we will not include them again (for a complete description of the formulation of this problem, the reader should see Milani *et al.* 2001).

The underlying idea of the algorithm is easily explained as follows. Each observation has a footprint on the sky (we are assuming it to be a 2 arcsec radius circle). Likewise, any observation prediction starting from an already computed orbit has an uncertainty ellipse: if these two regions overlap, both in the space of positions on the celestial sphere and in that of the apparent motions, we can rate it as a likely identification and submit it to the differential corrections procedure.

The main point here lies in finding a suitable "observation" which can represent all the observations of the second arc. To this end, we define what we call an attributable, which contains information not only on the position of the asteroid on the sky, but also on its apparent motion. More specifically, the attributable contains the time t_m that will be referred to as *the time of the attributable* and is defined as the arihmetic mean of the times of all the observations, the position and proper motion at epoch t_m and the position of the asteroid on the sky at a rounded epoch t_r, which is defined as $t_r = k\Delta t$, k being an integer and Δt a fixed quantity (10 days as default value).

As in the previous algorithm, we also use a cascade of filters, as described in Figure 2, so that the first only compares the position on the sky at the rounded epoch t_r, while the second performs the comparison in both the position space and the apparent motion space at the exact epoch t_m of the attributable.

2.3. Recursive algorithm

Once some identifications have been found with either of the previous two algorithms and the new multi-opposition (or at least improved) orbits are available, we are in a better position to get further identifications for them by applying the observation attribution method again. Figure 3 depicts the logic of this algorithm.

As an example of such a procedure, we are presenting here the best result we have got so far: the identification of seven observation sets

$$1998\ SP_{61} = 2000\ CK_{108} = 1972\ LH = 1984\ HF_2 = 1986\ VQ_6 = 1988\ GL_1 = 1990\ UQ_1$$

corresponding to the main-belt asteroid (15427) Shabas.

2.4. Results

Table 1 shows a summary of the number of identifications obtained by means of the observation attribution method (first column), the orbit identification algorithm (third column) and the recursive procedure (second and fourth columns). The last column gives the total number of identifications which have been submitted to and published by the MPC.

Table 1. Overall Results

	Short arc obs. attributions	Attributions to the attributions	Orbit Identifications	Attributions to the orbit ids.	TOTAL
Submitted	4,976	433	5,070	638	11,117
Published	4,509	400	4,587	627	10,123

Table 2 provides a detail for the attribution results considering whether the attributed observations where One Night Stands (ONS) or observations obtained during multiple nights.

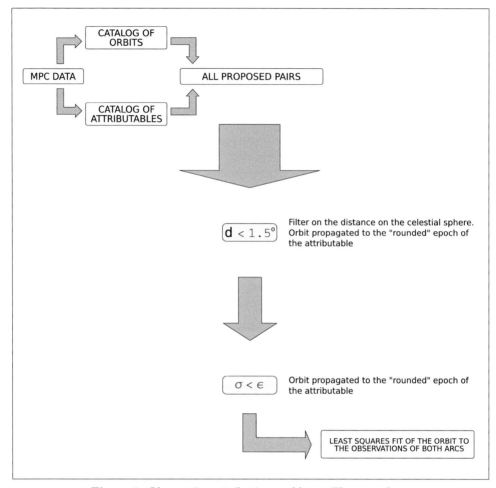

Figure 2. Observation attribution problem – The procedure.

Table 2. Attribution Results

	ONS attributions	Multiple-Nights attributions	TOTAL
Submitted	2,344	3,065	5,409
Published	1,963	2,949	4,912
Pending	95	26	121

All these identifications have been obtained by using as input the Minor Planet Circular batches of data released by the MPC. In order to apply them to the set of One Night Stands, also distributed by the MPC, we first need to assign a uniform designation to the ONS since, as distributed by the MPC, they keep the original name chosen by the observer at the moment of submission.

Once the ONS have been provided with a designation suitable for the OrbFit software, the observation attribution algorithm can be used to look for ONS attributions. When applied to known NEAs we need to take into account their bigger apparent motion so that, in the first filter of the attribution method, we use a smaller Δt value (0.001 days).

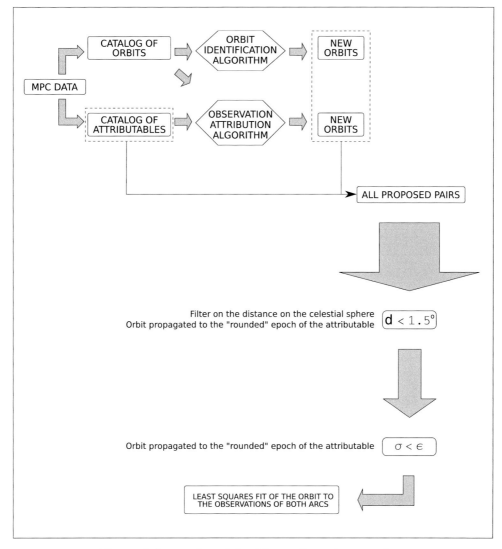

Figure 3. Interaction of algorithms – Recursive procedure.

Proceeding in this way we have obtained a total of about $2,000$ ONS attributed both to numbered and unnumbered asteroids.

3. Multiple solution identification methods

This Section describes the multiple solutions method, a one dimensional sampling of the orbital confidence region, and its applications in the identification problem. The underlying idea is to record multiple solutions along a one-dimensional subset of the confidence region, known as Line Of Variations (or LOV).

To explain mathematically what the LOV is, let us start by fixing the notation in relation with the well known computation of the nominal solution. The starting point is a fixed data set formed by m measurements of an asteroid. Then, for each orbit we can compute the residuals vector $\Xi = (\xi_1, \ldots, \xi_m)$ that measures the deviation between measured and computed values. The cost function

$$Q = \frac{1}{m} \Xi^T W \Xi,$$

built using the residuals vector, is a sum of cuadratic terms in which a weighting can be included through the symmetric, positive-definite matrix W.

The least squares fit consists, basically, in finding the stationary points of the cost function. The solution of the corresponding non-linear equations

$$\frac{\partial Q}{\partial X} = 0$$

are usually found using some iterative procedure. The standard Newton's method involves the computation of the second derivatives of the residuals with respect to the orbital elements, which is computationally expensive. Hence, a variant known as differential correction is generally used. In this method the correction $X \rightarrow X + \Delta X$ is computed as

$$\Delta X = C^{-1} D,$$

where the normal matrix C and the vector D are defined using the design matrix $B = \frac{\partial \Xi}{\partial X}$ as

$$C = B^T W B \quad \text{and} \quad D = -B^T W \Xi. \tag{3.1}$$

Whichever procedure is used, if the iterative method converges, the limit X^* is a stationary point of the cost function. If, in addition, X^* is a local minimum of $Q(X)$, it is called a best-fitting or nominal solution.

3.1. *The confidence region*

Now, let us suppose that for some data set the method of differential corrections has been succesful in providing a nominal orbit. A confidence region $Z(\chi)$ is defined by setting an upper limit to the penalty ΔQ, the increment in the cost function with respect to its value on the nominal solution:

$$Z(\chi) = \{X | \Delta Q(X) = Q(X) - Q(X^*) \leqslant \chi^2/m\}.$$

When the confidence region and the residuals are small, the cost function can be well approximated using a second order development and under these assumptions the confidence ellipsoid

$$Z_L(\chi) = \{X | \Delta X^T C(X^*) \Delta X \leqslant \chi^2\}$$

provides a good estimation of the confidence region. The longest semiaxis of this ellipsoid has the direction of the unit vector V_1, which is an eigenvector of the normal matrix $C(X^*)$ computed at the nominal solution that corresponds to its smallest eigenvalue λ_1. We say that V_1 defines the *weak direction* because in that direction the cost function exhibits the smallest variation and hence represents a weak restriction for the potential solutions of the least squares fit.

3.2. *The Line Of Variations*

The dicussion above could be equally performed when we choose any other point in the confidence region different from the nominal X^* and this allows us to define the weak direction vector field as $F = kV_1$, where $k = 1/\sqrt{\lambda_1}$. Note that the eigenspace associated to λ_1 does not define the unit vector V_1 uniquely. However, on a simply connected set it is always possible to select a continuous section of a one dimensional distribution. This allows for the practical construction of F by selecting arbitrarily one out of the two possible values of V_1 at a given point X and extending the vector field by continuity.

In this context, the LOV is defined as the integral curve of the weak direction vector field with initial condition equal to the nominal solution. Intuitively, the LOV can be thought of as the dorsal spine of the confidence region. If the linear approximation applies, it is the major axis of the confidence ellipsoid. When the linear approximation does not apply, the LOV is indeed curved and can only be computed by numerical integration of the differential equation defining it. The main disadvantage of this definition of the LOV is that it relies heavily on the existence of a nominal solution, which is not guaranteed for all cases and, in addition, there exist a lot of instabilities in its numerical computation which cannot be easily solved by brute force.

So, the question is, how to address the problems associated to the non-existence of a nominal solution. A way out is to consider the algorithm of constrained differential corrections [Milani *et al.* 2005], which consists in performing differential corrections on the orthogonal hyperplane to the weak direction

$$H(X) = \{Y|(Y - X) \cdot V_1(X) = 0\}.$$

In this case the correction is computed by means of

$$\Delta X = C^{-1}\pi_{H(X)}D(X),$$

where $\pi_{H(X)}$ is denoting the orthogonal projection on the hyperplane $H(X)$. The convergence of this method is equivalent to the D vector, defined in the algorithm of ordinary differential corrections (3.1), having the same direction as the vector V_1 giving the weak direction. This fact suggests to define the LOV as the set of points in which D is parallel to the weak direction:

$$\text{LOV} = \{X|D(X)||V_1(X)\}. \tag{3.2}$$

This new definition matches the previous one when there exists a nominal solution but has the advantage that the non-existence of the latter is no longer an intrinsical obstruction for the construction of the LOV.

So, how do we sample and parameterize the LOV? The condition (3.2), which defines the LOV, effectively imposes five restrictions on the six orbital elements allowing for a one-dimensional solution, that is, a differentiable curve. Unfortunately, there is no direct method to obtain analytically the solution of those equations. In practice, what we do is schematically depicted in Figure 4: we start at a point $X(\sigma)$ (it could be the nominal) and perform an Euler step to get a point X', which is not on the LOV (unless it is a straight line). Then we take X' as initial condition for the constrained differential corrections that at convergence will provide another point X'' on the LOV and iterate the process.

Another related problem is the selection of a natural parameterization of the LOV. If σ is the value of the parameter corresponding to X then we can use $\sigma + \delta\sigma$ to approximate the parameter value for X''. Obviously, the situation is easier to handle when there is a nominal solution available. In such a case it is possible to use the quantity χ, associated to the penalty in the cost function, to construct a canonical parameterization for the LOV. In this situation, the probabilistic interpretation of the χ parameter also provides a simple criteria to decide the moment in which the sampling of the LOV must be terminated. In the linear approximation, both parameters, σ and χ, are essentially the same.

3.3. *Results of the tests of identification*

Let us see how the two procedures introduced above, the constrained differential corrections and the multiple solutions method, can be applied in the asteroid identification problem. It is obvious that the more orbits we have computed, the better results can be obtained with the orbit identification method. Thus, what we have done is to adapt the

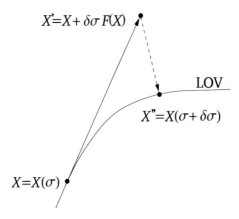

Figure 4. Parameterizing the LOV

orbit identification algorithm described in Subsection 2.1 to also be used with multiple solutions.

In order to quantify the improvement provided by the new techniques we have carried out three full computations using a given data set of unnumbered asteroids. The results of these tests are summarized in Figure 5.

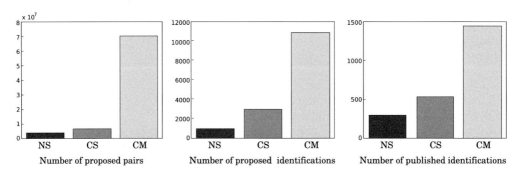

Figure 5. Summary of the tests' results

In each plot, the first bar (labeled NS for Nominal–Single solution), refers to the test that uses the orbit identification method on a catalog of nominal orbits computed with the classical differential correction method. The second bar (labeled CS for Constrained–Single solution) represents the results of the test in which the orbit identification algorithm is applied to a catalog of orbits built with the constrained differential corrections method. Finally, the third bar (labeled CM for Constrained–Multiple solutions) gives the numbers for the test in which the multiple solution orbit identification algorithm has been used on the same extended catalog of orbits. For each test, the height of the bars depicts the number of proposed pairs (pairs passing the three filters), number of proposed identifications (i.e., passing differential corrections) and number of published identifications.

All three plots in Figure 5 clearly state the superiority of the method using multiple solutions. Leaving apart its good performance in the number of proposed identifications, the achievements obtained in the published identifications plot are really conclusive: CM multiplies by a factor of three the results of CS and by five those of NS. Therefore, the tests are showing the positive effect of the combination of the two new techniques for the identification task.

Table 3. Analysis of the identifications proposed by the CS and CM methods.

PROP. IDS.	Only in CS	Only in CM	In CS & CM	TOTAL
	7.0%	73.9%	19.1%	100%
Not in NS	6.5%	71.9%	13.5%	91.9%
1 C	4.5%	31.7%	7.9%	44.1%
2 C	0.8%	4.4%	1.0%	6.2%

Table 3 highlights the role played by the constrained differential corrections. The third and fourth rows show the percentage of identifications in which one or both of the involved asteroids have an orbit computed by means of that procedure. A significant fraction of the identifications obtained with the methods based on constrained solutions involve constrained orbits. As a matter of fact, in more than half of the proposed identifications not detected by the NS method one or both of the orbits were constrained. When we restrict our attention to the published identifications, as shown in Table 4, the corresponding fraction reaches 40%, which is still a substantial contribution.

Table 4. As in Table 3, but for published identifications.

PUB. IDS.	Only in CS	Only in CM	In CS & CM	TOTAL
	2.7%	63.2%	34.1%	100 %
Not in NS	2.3%	60.9%	16.6%	79.8 %
1 C	1.5%	20.6%	7.7%	29.8 %
2 C	0.07%	2.0%	0.3%	2.4 %
Credited	1.3%	35.6%	13.6%	50.5 %

A comparative analysis of the behavior of the filters exhibited by the different algorithms shows a sensible variation in the values associated to identifications simultaneously found by more than one method. Generically speaking, there is an overall decreasing in the filter values obtained by the CM procedure with respect to the CS one. This feature becomes more remarkable as the quality of the filters increases, that is, towards the end of the filter procedure. In practical terms, it implies a significant reduction in the optimal cutting values associated to the more selective filters of the CM method. This means that the multiple solutions method notably reduces the identification penalty between orbits, which also implies a reduction in the non-linearity of the problem and finally helps to understand the success of the CM test.

4. Conclusions

We have presented several algorithms to deal with the asteroid identification problem discussing their range of applicability and their practical implementation. We have also pointed out the most relevant results obtained using these techniques and we have compared its performance.

The following is a summary of the main conclusions that can be drawn from the results:

• The orbit identification method is more powerful than the observation attribution algorithm, mainly due to the fact that it makes use of more information, since we are working in the six-dimensional space of orbital elements, whereas in the observation attribution method we are in the four-dimensional space of the observations.

- The constrained differential correction method greatly improves the orbital determination process, which also means better results when looking for new asteroid identifications.

- The multiple solutions algorithm significantly decreases the non-linearity of the asteroid identification problem and this implies a greater number of successful identifications. It is worth mentioning that the results got with this algorithm are better both from a quantitative and a qualitative point of view, since it not only finds more identifications, but also allows us to obtain some difficult ones which otherwise would be missed.

Despite the advances introduced by the new identification methods, the problem is far from being completely solved. Time performance of the algorithms and reliability of the proposed identification are some of the areas that need to be improved to face the challenge the new large asteroid surveys, promising mammoth quantities of nightly generated data, will pose in the near future.

Acknowledgements

This research has been funded by the *Observatorio de Mallorca* (OAM).

References

Milani, A. 1999, *Icarus* 137, 269

Milani, A., La Spina, A., Sansaturio, M. E. & Chesley, S. R. 2000, *Icarus* 144, 39

Milani, A., Sansaturio, M. E. & Chesley, S. R. 2001, *Icarus* 151, 150

Milani, A., Sansaturio, M. E., Tommei, G., Arratia, O. & Chesley, S. R. 2005, *Astron. Astrophys.* 431, 729

Near Earth Objects, our Celestial Neighbors: Opportunity and Risk
Proceedings IAU Symposium No. 236, 2006
A. Milani, G.B. Valsecchi & D. Vokrouhlický, eds.
© 2007 International Astronomical Union
doi:10.1017/S1743921307003341

Near-Earth-Object identification
over apparitions using n-body ranging

Mikael Granvik and Karri Muinonen

Observatory, P.O. Box 14, 00014 University of Helsinki, Finland
email: mikael.granvik@helsinki.fi

Abstract. Under ideal conditions, Earth-based telescopes can observe near-Earth objects (NEOs) continuously from a few days to months during each apparition. Due to the usually complicated dynamics of the Sun-Earth-NEO triplet, the time interval between consecutive apparitions typically ranges from months to several years. On these time scales, exiguous single-apparition sets (SASs) of observations having short observational time-intervals lead to substantial orbital uncertainties. Linking of SASs over apparitions thus becomes a nontrivial task. For example, of a total of roughly 4,100 NEO observation sets, or orbits, currently known, some 2,300 are SASs, for which the observational time interval is less than 180 days. Either these SASs have not been observed at an apparition following the discovery apparition or the linkage of SASs has failed, an option which should preferably be eliminated. As a continuation to our work on the short-arc linking problem at the discovery moment (Granvik & Muinonen, 2005, *Icarus* 179, 109), we have investigated the possibility of using a similar method for linking exiguous SASs over apparitions. Assuming that the observational time-interval for SASs of NEOs is typically at least one day (minimum requirement set by the Minor Planet Center), the orbital-element probability-density function is constrained as compared to the typical short-arc case with an observational time interval of only a few tens of minutes. Because of the smaller orbital-element uncertainty, we can use the short-arc method (comparison in ephemeris space) for longer time spans, or even do the comparison directly in the orbital-element space (Cartesian, Keplerian, equinoctial, etc.), thus allowing us to assess the problem of linking SASs of NEOs. Due to possible close approaches with the Earth and other planets, and substantial propagation intervals, we have developed new n-body techniques for the orbit computation.

Keywords. Identification; statistical methods; data analysis

1. Introduction

An observation set of an asteroid is typically classified as belonging to a certain asteroid group based on a single point estimate of the orbital-element probability-density function (p.d.f.), which is usually obtained via either the least-squares solution assuming Gaussian statistics, or a single Väisälä solution based on human judgement. For short observation time intervals leading to wide, clearly non-Gaussian, and strongly nonlinear orbital-element p.d.f.'s, the use of point estimates for classification is prone to errors. When searching for near-Earth objects (NEOs), an observation set may, for instance, seem a probable main-belt object (MBO; which in this particular case is uninteresting), because the point estimate places it within the main asteroid belt at the observation date. However, in reality, the observation set may belong to a near-Earth object that only spends part of its time in the main asteroid belt. Currently, some 50,000 provisionally designated observation sets in the Minor Planet Center's (MPC) observation database span less than 48 hours. Most of these single-apparition sets (SASs) of observations have been observed for at least two nights, as the MPC guidelines require to get a provisional designation. The overwhelming majority of these so-called two-nighters are currently classified as MBOs. For the NEO suspects in the two-nighter data, the linkage to other observation sets may have failed with the current methods. This could, for instance, happen for fast-moving objects on high-eccentricity orbits as these objects also spend a substantial fraction of their time in the main asteroid belt around aphelion, that is, masquerading themselves as MBOs.

For the current long-term linking problem, we will use the same scheme that has successfully been applied by Granvik & Muinonen (2005) in the short-term case. However, in the long-term NEO case, some of the assumptions are no longer valid. The main obstacle is the two-body approximation, which can be acceptable even for NEOs in the short-term case assuming current orbital uncertainties, but not in the long-term case. Even though the differences between the two-body and the n-body p.d.f.'s are negligible around the inversion epoch (assuming, in turn, that the inversion epoch is close to the observation dates), the nonlinear propagation of the p.d.f.'s to a comparison epoch, often several years from the inversion epoch, will generally lead to notable differences between the two cases. The methods that are used to solve the two-point boundary-value problem (the problem of solving an orbit from two heliocentric Cartesian positions) in Ranging Virtanen, Muinonen & Bowell (2001) have so far been either the p-iteration method or the continued-fraction method for details, see, e.g., Danby (1992), both of which are based on the two-body approximation. Until now, the use of two-body solutions in Ranging has been acceptable because, for example, the observational data sets have spanned time intervals short enough for perturbations to be negligible Virtanen & Muinonen (2006) e.g.,. However, n-body perturbations have been accounted for in the propagations for the computation of collision probabilities.

Another assumption used in the short-term case, but proven problematic in the long-term case, is that finding a common orbital solution for two observation sets separated in time is straightforward and can be done efficiently with, for instance, Ranging. For the long-term, it is clear that the least-squares method Danby (1992) see, e.g., is both a valid and an optimal tool for the verification of trial linkages. But finding an initial orbit for the method is cumbersome; one could think of using sample orbits computed with, e.g., Ranging for the separate observation sets, but it turns out that these orbits are often too far off at either end to allow the least-squares method to converge, even if the linkage would be correct. Another possibility would be to use a semi-analytical method like the one by Kristensen (1995) to find an initial orbit by using observations from both sets simultaneously. However, the semi-analytical methods are problematic because they, again, use the two-body approximation which is not valid over long time intervals.

New n-body methods are thus needed both for the solution of the two-point boundary-value problem, and for the generation of initial orbits to be used by the least-squares method. Our aim is to (i) present new methods for these two specific tasks, (ii) present the overall structure of a new sampling method which can link exiguous observation sets over several apparitions, and (iii) apply the linking method to simulated data to prove its feasibility. The paper is organized as follows. In Sect. 2, we present the methods and techniques, while Sect. 3 explains the generation of the simulated data. We present and discuss the results in Sect. 4, and finally give our conclusions in Sect. 5.

2. Methods

The current long-term linking method uses two filters; the first one is used to find a substantially reduced set of trial linkages (as compared to all possible trial linkages) worth to be analyzed in detail. The second filter tries to find an orbital solution which ties two separate observation sets together assuming realistic observation uncertainties.

If successful, the second step thus implies a linkage between the two sets. Before getting into details of the linking method, we need to go back to the basics of orbit computation, and develop a robust method that solves the two-point boundary-value problem using an n-body dynamical model.

2.1. *Robust n-body solution to the two-point boundary-value problem*

Starting with two Cartesian heliocentric positions $\mathbf{r}_0(t_0)$ and $\mathbf{r}_1(t_1)$ (where t_0 and t_1 indicate the corresponding epochs), the solution to the two-point boundary-value problem is a velocity $\mathbf{v}_0(t_0)$ such that once an orbit $\mathbf{P}(t_0) = (\mathbf{r}_0(t_0), \mathbf{v}_0(t_0))$ at epoch t_0 is propagated to the epoch t_1, the distance

$$d_{\mathrm{Car}} = |\mathbf{r}_1(t_1) - \mathrm{pos}[\mathbf{P}(t_1)]| < \epsilon_{\mathrm{acc}}, \tag{2.1}$$

where ϵ_{acc} is an adjustable parameter that defines the accuracy of the solution and the operator $\mathrm{pos}[\mathbf{P}]$ gives the Cartesian position vector of orbit $\mathbf{P}$. A suitable fixed value for ϵ_{acc} could be, for instance, 10 m or 100 m. Optimally, ϵ_{acc} should be adjusted according to both the observation accuracy and the topocentric distance of the observed object. Several solutions to the inverse problem can be found in the literature (see, e.g., Danby (1992)) but we are not aware of any that would not require the validity of the two-body approximation. Note, however, that a problem resembling the current two-point boundary-value problem has been solved for collision orbits by Muinonen (1999) and Muinonen, Virtanen & Bowell (2001). In essence, the two-point boundary-value problem is an optimization problem where the three components of the velocity vector $\mathbf{v}_0(t_0)$ are free parameters, and the distance d_{Car} has to be minimized. By assuming that the difference between the two-body solution and the n-body solution is fairly small, we initialize a simplex optimization routine (for a description of a simplex method, see, e.g., Press, Teukolsky, Vetterling, *et al.* (1992), Press, Teukolsky, Vetterling, *et al.* (1999)) with the two-body solution. The simplex routine then uses the full n-body approach to make a correction to the initial two-body solution. On one hand, the n-body solution will be found substantially slower for close-approaching orbits as compared to the two-body solution. On the other hand, only one n-body propagation is needed if the object is far from perturbers, or over very short time intervals, because the n-body correction turns out to be negligible.

2.2. *Phase-space address comparison*

In practice, the first filter requires that the p.d.f.'s of the orbital elements (or any quantities derived from them) computed from the two separate observation sets have to overlap at one or more epochs. see Granvik & Muinonen (2005). For exiguous observation sets, a rigorous sampling of the orbital-element p.d.f. is critical to be able to link observation sets over long time intervals. For the inversion of the SASs, we use either stepwise Ranging Granvik & Muinonen (2005), or the Volume-of-Variation method (VoV; Muinonen, Virtanen, Granvik *et al.* (2006)), depending on the number of observations and the length of the observational time interval. Cartesian orbital elements, and a complete n-body approach, including the n-body solution of the two-point boundary-value problem, is used during the inversions. The inversion epoch of each observation set is the midnight (TT) closest to the mid-date of the observations.

All orbital-element p.d.f.'s are then propagated to the comparison epoch by using a full n-body dynamical model. We tried several different comparison variables (Cartesian orbital elements, Keplerian orbital elements, equinoctial orbital elements, Poincaré variables, the angular momentum vector, heliocentric spherical coordinates, etc.) and a few of their combinations. The number of comparison epochs was also altered when spherical

Table 1. Current phase-space discretization parameters for the address comparison algorithm. The parameters and their bin sizes will be further optimized in the future.

	a [AU]	e	i [°]	Ω [°]	ω [°]
Lower limit	0.0	0.0	0.0	0.0	0.0
Upper limit	500.0	1.0	180.0	360.0	360.0
Bin size	0.1	0.05	0.2	5.0	5.0

coordinates were tested. Since most of these choices performed more or less equally well, we decided to use Keplerian orbital elements at one comparison epoch and without the mean anomaly for the time being. We chose to use a comparison epoch after the last observations, which, in the future, will allow us to efficiently compare "new" discoveries with old SASs. However, the comparison epoch can easily be changed were there more optimal choices found later.

In practice, we do not compare smooth volumes in the phase space, but use a large amount, say 5,000, discrete orbits that sample the true orbital-element p.d.f.. For the comparison phase, the five-dimensional comparison vector is squeezed into a single scalar, which permits fast comparison of different orbits (for details, see Granvik & Muinonen (2005), Muinonen, Virtanen, Granvik *et al.* (2005)). Values that we currently use for the discretization of the phase space are shown in Table 1. Once identical scalars (integers) are found for orbital-element p.d.f.'s originating from two separate observation sets, we conclude that the observation sets give rise to similar orbits thus implying a potential linkage. The maximum deviation of the two orbits in comparison space is explicitly given by the bin sizes. The next step is to try to find a single orbital solution tying both of the corresponding observation sets together assuming realistic observational uncertainties.

2.3. Initial orbits through optimization

As stated earlier, the optimal way to scrutinize linkages between SASs in the second filter is to use the least-squares method. The reason why the least-squares method is the best choice is also its weak point. While the inversion of exiguous SASs of observations typically produces wide orbital-element p.d.f.'s (Ranging or VoV applicable), the inversion of a combination of two SASs separated by several years produces very constrained p.d.f.'s (least squares applicable). Typically, the difference in the orbital uncertainties are several orders of magnitude. It is thus clear that using a sample orbit from either one of the SASs as an initial orbit for the least-squares method will usually fail to converge, simply because the initial orbit is too far from the final solution and the linearity assumption does not apply.

The orbits computed for the SASs naturally provide a good first approximation and should hence be used. The apparent thing to do is to find the two orbits corresponding to the two SASs that are closest to one another in the orbital-element space at some common epoch, here, naturally, the comparison epoch. After tests on different orbital elements, we decided to use the Keplerian ones. We then tested several metrics, all of which are based on the work by Southworth & Hawkins (1963), and found the (a,e,i)-version used by Nesvorny, Bottke, Dones, *et al.* (2002) to be adequate for our use:

$$d_{\mathrm{Kep}} = na\sqrt{C_a(\delta a/a)^2 + C_e(\delta e)^2 + C_i(\delta \sin i)^2}\,, \qquad (2.2)$$

where na is the heliocentric velocity of an asteroid on a circular orbit with semimajor axis a, $\delta a = |a_1 - a_2|$, $\delta e = |e_1 - e_2|$, $\delta \sin i = |\sin i_1 - \sin i_2|$. For the constants C_a, C_e, and C_i we use 1.25, 2, and 2, respectively. Note that Nesvorny, Bottke, Dones, *et al.* (2002) used proper elements while trying to find fragments from a single break-up event, whereas we are comparing osculating elements (at a common epoch) that have been derived from two different observation sets and (might) refer to the same object.

Even though the resulting orbits are superficially very similar, the least-squares method does not usually yet converge. Before improving the orbits with least squares, they thus need to be slightly optimized with a robust nonlinear method for which partial derivatives are not needed. To solve the optimization problem, we, again, use a simplex method. The six Cartesian orbital elements are now the free parameters, and the resulting χ^2 with respect to the observations is minimized. To initialize the simplex we use the three closest pairs of orbits between the sets, while the seventh orbital-element set is the arithmetic mean of the elements of the closest pair of orbits. The simplex naturally uses the full n-body model. When one of the seven orbits reaches some predefined χ^2-value, it is used as the initial orbit for the least-squares method, which now converges (at least for correct linkages). The final acceptance for the linkage is given if the residuals of the least-squares fit conform to the assumed observational uncertainties. The verification of the correctness of an accepted linkage can only be made by additional, archive or new, observations.

3. Simulated observations

Simulated observations of NEOs were generated using the ASurv software (see, e.g., Granvik & Muinonen (2005)), which randomly draws uncorrelated orbital elements and absolute magnitudes from specified p.d.f.'s. Here we used NEO p.d.f.'s based on the work by Jedicke, Larsen & Spahr (2002) and an upper limit of $H = 18$ mag for the absolute magnitude. Positions and apparent magnitudes for these random objects are then computed for specified observation dates. If the position falls inside the observation window and the apparent magnitude is lower than a given threshold (here, $V_{\text{lim}} = 19$ mag), the observation is accepted. Here we allowed the target to be anywhere on the sky with the only limit being a minimum solar elongation of $45°$. The dynamical model used in propagations between observation dates took into account perturbations induced by all planets as well as the dwarf planet Pluto. Finally, random Gaussian noise with standard deviation of $\sigma = 0.5''$ was added to the observations.

We used a cadence of two observations separated by one hour on two consecutive nights, which was repeated after a time interval of 2800 days, or roughly seven years and eight months. For a single object, we thus got a maximum of $2 \times 2 \times 2 = 8$ observations, which was split up into two two-nighters.

The previous parameters resulted in 211 two-night observation sets and 188 different objects. The maximum number of trial linkages between the observation sets is thus $211 \times (211 - 1) = 44310$, while the number of correct linkages is 23. The number of observation sets that can be linked assuming realistic observation uncertainties, is somewhere between these two extremes, usually closer to the lower number.

4. Results and Discussion

To show how the method performs for a correct linkage, we randomly chose the following two simulated geocentric observation sets which we know belong to the same object (designation, date, R.A., Dec., magnitude):

00107	2004 01 20.24925	16 27 20.254	-08 59 15.78	18.88V
00107	2004 01 20.28925	16 27 43.694	-09 00 01.92	18.89V
00107	2004 01 21.24925	16 36 58.287	-09 18 15.96	18.97V
00107	2004 01 21.28925	16 37 21.076	-09 19 00.34	18.97V
00111	2011 09 20.24924	03 45 55.689	-59 03 24.40	16.32V
00111	2011 09 20.28924	03 45 08.261	-59 02 59.67	16.32V
00111	2011 09 21.24924	03 26 44.342	-58 47 45.73	16.31V
00111	2011 09 21.28924	03 25 59.625	-58 46 55.75	16.31V

Orbital-element p.d.f.'s were then obtained by solving the inverse problem for both observation sets at local inversion epochs, that is, 2004/1/21.0 TT for 00107 and 2011/9/21.0 TT for 00111. The resulting marginal distributions are shown in Figs. 1 and 2. After

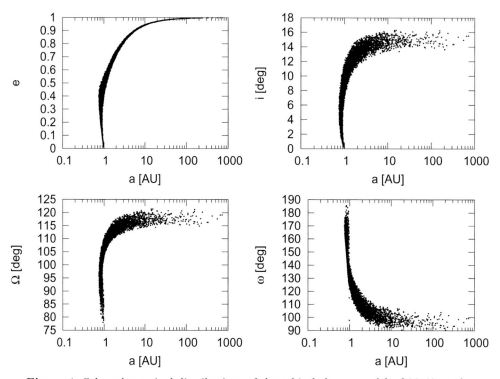

Figure 1. Selected marginal distributions of the orbital-element p.d.f. of 00107 at the inversion epoch 2004/1/21.0 TT.

the inversion, the p.d.f.'s were propagated to the comparison epoch 2012/1/1.0 TT (see Fig. 3). Recalling that the uncertainty in Cartesian orbital elements at the inversion epoch lies mainly along the line of sight (thus producing a long, narrow cone in the configuration space), the nonlinearity induced by the propagation to the comparison epoch is clearly visible in Fig. 4.

The comparison algorithm finds similar orbits between the sets, which correctly implies a possible linkage. A more elaborate inspection reveals astonishing similarity of the orbits considering the exiguous data and the substantial length of the propagation (see Table 2). In the current example, the least-squares method would actually converge, if the fourth orbit in Table 2 is fed into the least-squares routine. However, to make sure that the

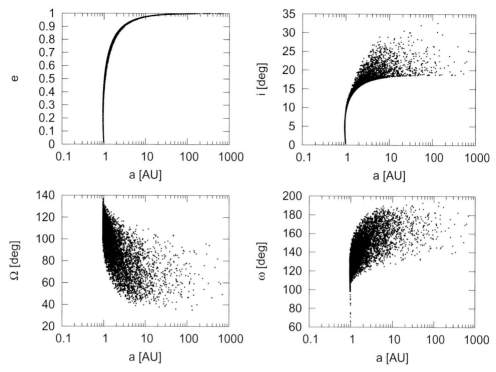

Figure 2. Selected marginal distributions of the orbital-element p.d.f. of 00111 at the inversion epoch 2011/9/21.0 TT.

Table 2. The three closest pairs of orbits at the comparison epoch. Note that there are only two different orbits shown for 00107.

SAS	a [AU]	e	i [°]	Ω [°]	ω [°]	M [°]
00107	1.17397	0.597785	12.5345	111.997	130.005	60.178
00111	1.17365	0.597630	12.5302	120.544	117.272	128.100
00107	1.03496	0.495752	9.5710	106.713	128.354	155.926
00111	1.03520	0.495986	9.5595	106.815	128.113	161.302
00107	1.03496	0.495752	9.5710	106.713	128.354	155.926
00111	1.03521	0.495296	9.5622	107.184	127.803	161.326

method converges for all correct linkages, the six orbital elements can also be further optimized with the simplex method described in Sect. 2.3. To induce as much stability as possible into the system, the inversion epoch is, again, the midnight (TT) closest to the mid-date of the observations, which in this particular case was 2007/11/21.0 TT. When the optimization reaches a certain level, the orbits are fed into the least-squares routine. The convergence is fast due to the optimized initial orbit, and the results (with a comparison to the ground truth) are given in Table 3. The orbital uncertainties for the combined SASs are several orders of magnitude lower than for the SASs separately (for example, for the semimajor axis roughly by a factor of 10^{-6}!). The small uncertainty of the resulting orbit makes verification by additional observations fairly easy.

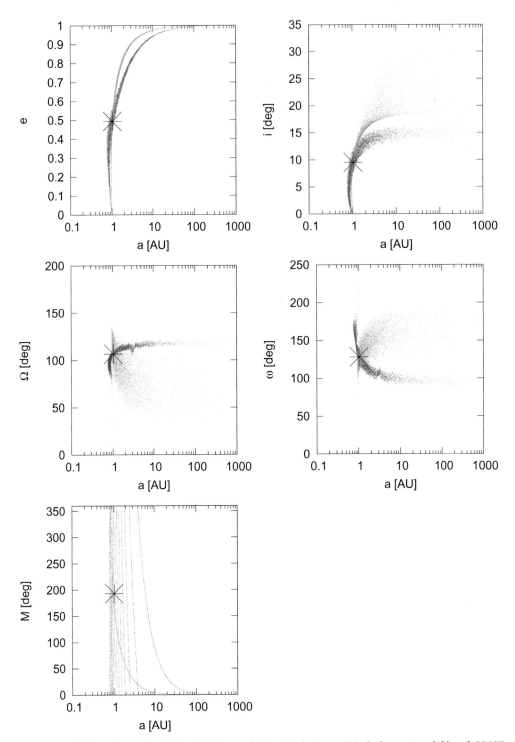

Figure 3. Selected marginal distributions of the Keplerian orbital-element p.d.f.'s of 00107 and 00111 at the comparison epoch 2012/1/1.0 TT. The star shows the ground-truth orbital elements based on which the observation sets were generated.

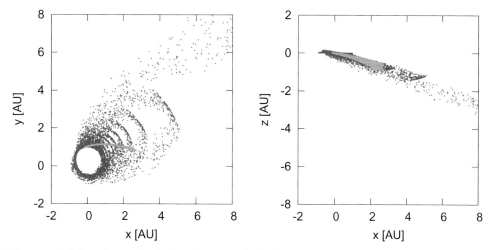

Figure 4. Selected marginal distributions of the Cartesian orbital-element p.d.f.'s of 00107 and 00111 at the comparison epoch 2012/1/1.0 TT. Note that the p.d.f.'s have been cut off at $x = y = 8$.

Table 3. The resulting orbital elements for the combined SASs accompanied with the uncertainties (the standard deviation) at the inversion epoch 2007/11/21.0 TT. The third line shows the ground-truth orbit from which the simulated observations were originally generated. The $\Delta\chi^2$ between the the ground-truth orbit and the least-squares solution (that is, the nominal orbit and the corresponding covariance matrix) is approximately 12.1, which, for six degrees of freedom, means that the solution is within the boundaries corresponding to 95.4% of the total probability mass (for a Gaussian distribution this equals to the 2-σ boundaries).

	a [AU]	e	i [°]	Ω [°]	ω [°]	M [°]
00107+00111 / elem.	1.03360694	0.4937476	9.51334	106.6827	128.2433	192.90980
00107+00111 / unc.	0.00000021	0.0000040	0.00047	0.0027	0.0030	0.00058
ground truth	1.03360674	0.4937444	9.51414	106.6857	128.2401	192.91025

5. Conclusions

The new methods presented in previous chapters have been succesfully applied to case studies, one of which has been presented here. In the very near future we will apply the method simultaneously to the entire set of simulated observations to estimate the accuracy and efficiency of the linking method. In connection with the application of the linking method to larger populations, we will also investigate possibilities of further optimizations or even approximations to reduce the need for CPU time. Our goal, however, is to have a method which will find virtually all correct linkages having time intervals of at least, say, 10–15 years. In the future, we will also apply the method to all known SASs having observational time intervals less than 48 hours.

With the development of the new n-body techniques, we now have the means to completely eliminate the two-body approximation from the Ranging results. Besides allowing long-term linking, this will also refine future collision probability computations.

Acknowledgements

We thank the referee, Giovanni F. Gronchi, for the constructive criticism. Research supported, in part, by the Academy of Finland. MG is grateful for funding from the foundations of Alfred Kordelin and Magnus Ehrnrooth.

References

Danby, J. M. A. 1992, *Fundamentals of Celestial Mechanics*, 2. edn, Willman-Bell, Inc., Richmond, Virginia, U.S.A.

Granvik, M. & Muinonen, K. 2005, *Icarus* 179, 109

Granvik, M., Muinonen, K., Virtanen, J., Delbó, M., Saba, L., De Sanctis, G., Morbidelli, R., Cellino, A., & Tedesco, E. 2005, in: Z. Knežević & A. Milani (eds.), *IAU Colloquium 197: Dynamics of Populations of Planetary Systems*, (Cambridge: Cambridge University Press), p. 231

Jedicke, R., Larsen, J. & Spahr, T. 2002, in: W.F. Bottke, A. Cellino, P. Paolicchi & R. P. Binzel (eds.), *Asteroids III*, University of Arizona Press, p. 71

Kristensen, L. K. 1995, *Astron. Nachr.* 316(4), 261

Muinonen, K. 1999, in: B. A. Steves & A. E. Roy (eds.), *The Dynamics of Small Bodies in the Solar System*, Kluwer Academic, Dordrecht Norwell, MA, p. 127

Muinonen, K., Virtanen, J. & Bowell, E. 2001, *CMDA* 81, 93

Muinonen, K., Virtanen, J., Granvik, M., & Laakso, T. 2005, in: M. Perryman (ed.), *Three-Dimensional Universe with Gaia*, ESA Special Publications SP-576, Noordwijk, p. 223

Muinonen, K., Virtanen, J., Granvik, M., & Laakso, T. 2006, *MNRAS* 368(2), 809

Nesvorny, D., Bottke Jr., W. F., Dones, L., & Levison, H. F. 2002, *Nature* 417, 720

Press, W. H., Teukolsky, S. A., Vetterling, W. T., & Flannery, B. P. 1992, *Numerical Recipes in Fortran The Art of Scientic Computing*, 2. edn, Cambridge University Press

Press, W. H., Teukolsky, S. A., Vetterling, W. T., & Flannery, B. P. 1999, *Numerical Recipes in Fortran 90 The Art of Parallel Scientic Computing*, 2. edn, Cambridge University Press

Southworth, R. B. & Hawkins, G. S. 1963, *Smiths. Contr. Astrophys.* 7, 261

Virtanen, J., Muinonen, K. & Bowell, E. 2001, *Icarus* 154(2), 412

Virtanen, J. & Muinonen, K. 2006, *Icarus* 184(2), 289

Near Earth Objects, our Celestial Neighbors: Opportunity And Risk
Proceedings IAU Symposium No. 236, 2006
A. Milani, G.B. Valsecchi & D. Vokrouhlický, eds.
© 2007 International Astronomical Union
doi:10.1017/S1743921307003353

Low solar elongation searches for NEO: a deep sky test and its implications for survey strategies

Andrea Boattini[1], A. Milani[2], G. F. Gronchi[2], T. Spahr[3] and G. B. Valsecchi[4]

[1]Department of Physics, University of Tor Vergata, and OAR-INAF, Via Frascati 33, 00040 Monteporzio Catone (Roma), Italy, email: boattini@iasf-roma.inaf.it
[2]Department of Mathematics, University of Pisa, Piazza Pontecorvo 5, 56127 Pisa, Italy email: milani@dm.unipi.it
[3]Minor Planet Center, Smithsonian Astrophysical Observatory, 60 Garden Street, Cambridge, MA 02138, USA
[4]IASF-INAF, Via Fosso del Cavaliere 100, 00133 Roma, Italy

Abstract. A survey for NEO aiming at 90% completeness for a given size range cannot ignore that a significant fraction of the population is observable essentially only at low solar elongation, in the so called "sweet spots". There are several penalties for such low elongation: poorer observing conditions imply a lower limiting magnitude, shorter available time in each night and a more difficult orbit determination. Our aim is to show that these difficulties can be overcome. We have tested the observation procedures and the mathematical methods of orbit determination on two sweet spot test runs. One was performed at ESO La Silla in Jan–Feb 2005, the other at Mauna Kea in Sept–Dec 2005. The results of the tests are presented in this paper; the observed area was not large enough (especially at Mauna Kea) to discover a significant number of new NEO, the purpose was rather to identify the problems. These tests have allowed us to identify all the key elements to be accounted for in the strategy for a successful sweet spot NEO survey. When very short arc observations from different nights have to be identified, a specific difficulty occurs at the sweet spots: the same set of observations from three nights can be fitted to two incompatible orbits, in most cases including one NEO and one MBA. This can lead to two different failures (false positive, false negative) in deciding whether a NEO has been discovered. The classical theory of preliminary orbits shows that three observations at an elongation less than 116.5° can be compatible with two different orbits. From this theory we have derived an algorithm to find the alternate solution, if it exists, when only one is available. In this way we have generated a set of examples of possible discoveries with two well determined but incompatible solutions. Most of the MBA-NEO alternatives have been solved by finding a known MBA which could be identified; in two cases the MBA solution has been confirmed by a later observation.

Keywords. surveys, orbit determination, identification

---

1. Purpose and Method

A survey for NEOs aiming at 90% completeness for a given size range cannot ignore that a significant fraction of the population passes in the neighborhood of opposition either never or very rarely or only in very poor observing conditions (Stokes *et al.* 2003). This implies that there is a bias against the discovery of Inner Earth Objects (IEO), Atens and other NEOs with larger eccentricity and/or inclination.

To compensate for this bias a fraction of the available telescope time needs to be used at low solar elongation (around or below 90°), in the so called *sweet spots*. Many potential Earth impactors can be discovered only by scanning the sweet spots

(Chesley & Spahr 2002). Indeed, observing this part of the sky – that is along the Earth's orbit – with powerful telescopes and appropriate orbital discrimination and calculation tools can yield a large number of Potentially Hazardous Asteroids (PHA). A good example of this is the discovery at a sweet spot of (99942) Apophis, which so far is the best and more challenging example for impact calculation studies.

We have studied the observing strategy and data processing methods which could be used to implement such a *sweet spot survey*, either as a separate program or as a subset of a next generation survey. The specific difficulties of such an approach are:

1) Lower limiting magnitude at the sweet spots (for the same telescope and camera) than near opposition, due to larger airmass, poorer seeing and higher sky background.

2) Rapid variability of the observing conditions and shorter total time available for observations in each night.

3) Larger average dispersion in proper motion, making discrimination between NEOs and Main Belt Asteroids (MBAs) more difficult.

4) Orbit Determination subject to problems from double and multiple solutions, from false identifications and computational overload.

Difficulty 1) can be overcome with adequate equipment at a high quality observing site. For 2) we propose using a short observing time each night, near sunrise and/or sunset, for many nights, exploiting the lower competition for telescope time under conditions inappropriate for other programs. We also use a short Transient Time Interval (TTI) between frames covering the same area (e.g., 15 minutes) to get more similar conditions.

Difficulty 3) can be overcome if the survey covers a large enough area to follow up automatically most objects, including most NEOs, as will be the case for the next generation all sky surveys. For small field surveys, targeted follow up is practically impossible if discrimination of NEOs is not achieved from the very first detections. As we will see, this is the main difficulty of a small scale test such as the one presented below. On the other hand, our purpose was not to obtain a number of discoveries comparable to large field of view surveys, which is anyway impossible, but to obtain enough discoveries and recoveries to validate the procedures.

Difficulty 4) was not enough considered in previous discussions of the sweet spot strategies; we will address this in detail in Section 2.

The difficulties seen above are somewhat compensated by significant advantages.

i) comparatively small NEOs can be detected at the achievable limiting magnitudes; the same objects would be either not observable or much dimmer near opposition.

ii) the MBAs are much dimmer than they are at opposition: this implies that most MBAs imaged at the sweet spots are already known, a prior information which is very effective in discriminating from NEOs.

iii) distant bodies, like Trans Neptunians, move so slowly that confusion is unlikely.

iv) even if the total number of discoveries in the sweet spots is lower than in a comparable telescope time at opposition, the number of *interesting* objects (Atens, IEOs, PHAs, even Virtual Impactors like Apophis) is very significant.

This paper discusses the specific orbit determination problems and their solution, presents our test sweet spot survey and discusses the result, allowing us to draw conclusions on the strategy to be used for future operational surveys.

2. Identification and Orbit Determination Problems

The problems of orbit determination occurring in a sweet spot (sub)survey are three: multiple, or anyway weakly determined, solutions; double solutions; false identifications.

2.1. *Weakly determined solutions*

Even after accumulating observations of the same object over several nights, the orbit may be weakly determined, because the observed arc is still too short. In such cases there is typically a *weak direction* (in orbital elements space) along which the RMS of the residuals increases very little even by moving very far from the nominal, minimum RMS solution. As described in (Milani *et al.* 2005a), it is possible to define a continuous line of orbital elements, the *Line Of Variations* (LOV), containing *constrained least squares solutions*, in intuitive terms solutions with just 5 free parameters and a sixth parameter fixed at an arbitrary value (the one along the weakly constrained direction).

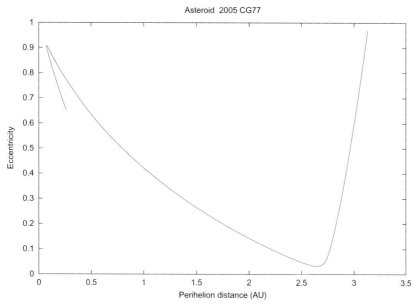

Figure 1. For 2005 CG$_{77}$ eccentricity and peculiar distance of the solutions along the LOV. Note that all the solutions in this plot correspond to residuals with a RMS less than 0.25 arcsec, as shown in Figure 4. Given that low eccentricity MBA solutions are possible, most likely they are the real ones, but there is no way to draw such a conclusion based on the available observations only.

The changes of the elements along the LOV can be extreme, as in the example of 2005 CG$_{77}$, one of the asteroids discovered in our test survey, with observation in three consecutive nights. As shown by the plot of 600 LOV solutions in Figure 1, there are solutions fitting in a satisfactory way the observations corresponding to MBA, NEO and even comet-like orbits (not shown, along the continuation of the LOV on the right of the figure). Targeted recovery in the next opposition would have required an inordinate use of telescope resources; as a matter of fact, it has not been recovered. The set of 10 observations in 3 nights forms an observed arc of type 2, according to the definition by (Milani *et al.* 2006), that is, the information contained in the data is equivalent to two arcs each without significant curvature (aligned, within the astrometric error, along a great circle). Roughly speaking, this 3-night orbit is not significantly better than in the case of two well separated nights of observation.

Indeed, all the cases in which an object has been observed only in two nights are of the same type, with largely undetermined orbit. In our opinion this type of designations should not be considered *discoveries*. The conclusion relevant for planning surveys is that the use of consecutive nights on the same area has to be avoided.

2.2. *Double solutions*

The classical theory of preliminary orbits (methods of Laplace and Gauss) shows that three observations at an elongation less than $116°.5$ can be compatible with two different orbits, depending upon the values of elongation, distance and curvature of the observed arc. The classical analysis of the problem, based on the solutions of a degree 8 polynomial equation (Plummer 1918; Danby 1989), allows us to describe the behavior of the preliminary orbit algorithms in a plane with polar geocentric coordinates (ρ, ϵ) (with $\rho =$ distance from Earth, $\epsilon =$ solar elongation). Note that this theory uses the approximation neglecting the topocentric correction.

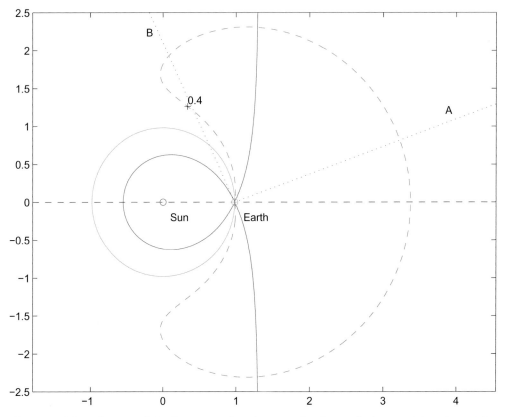

Figure 2. The solutions of the first approximation equation for preliminary orbits, given either three observations (method of Gauss) or the observations with their first and second derivatives (method of Laplace). When observing near opposition (line A: elongation $160°$) there can be only one solution, while in the sweet spots (line B: elongation $65°$) there can be two solutions, as well as none if the direction and curvature are incompatible.

Figure 2 shows as continuous lines the boundary between the regions with 1 and with either 0 or 2 solutions. The actual solutions for a given elongation and curvature appear in the figure as intersections of a dotted line representing the direction of observation with a dashed level line of a quantity related to curvature. Double solutions can occur at $|\epsilon| \leqslant 116°.5$, on the left of the loop-shaped *limiting curve* and outside of the zero curvature circle with radius equal to the Sun-Earth distance. Inside the circle and the loop there

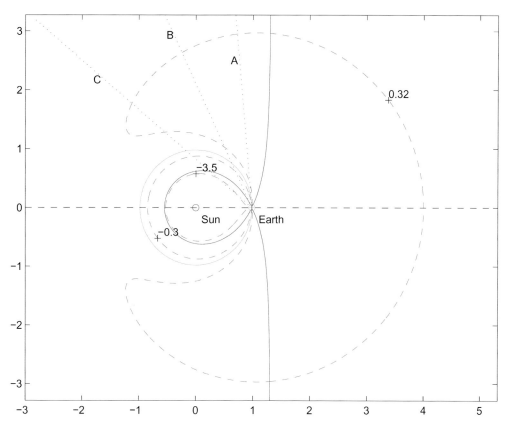

Figure 3. As in the previous Figure, showing a double solution near quadrature (line A: elongation 85°). The possibility of both single and double solutions depending upon the value of curvature is shown for line B (elongation 65°) and line C (elongation 40°).

is another region with double solution, which can be reached only for $|\epsilon| \leqslant 63°.5$: these very low elongations are practically reachable only from very good sites like Mauna Kea. For $|\epsilon| \leqslant 90°$ both one and two solutions are possible, depending upon the curvature, that is upon the actual heliocentric distance (see Figure 3).

The double solutions for the preliminary orbit, when used as first guess for a standard differential correction procedure, become double local minima of the target function (weighed sum of squares of the residuals) (See Milani *et al.* 2005a, Figure 11). If there are only three nights of observations, with a short time span for each night (as it is unavoidable for low elongation), the two minima are low and both correspond to acceptable solutions; if there are data over more than three nights and/or the data in each night have a sufficient time span and good accuracy, one of the two minima corresponds to a much higher RMS and it may be possible to discard it.

As an example, the least squares fit for 2005 CG$_{77}$ has two minima corresponding to an Apollo orbit with RMS 0.069 arcsec and an Aten orbit with RMS 0.126 arcsec. Not only both are acceptable fits, but also the solutions along the LOV in between the two and well beyond are acceptable, as is shown by Figure 4. For better determined discoveries, e.g., with arcs of type 3, only two small segments along the LOV near the two local minima have acceptable RMS.

The most embarrassing case for surveys is when one of the two least squares solutions is a NEO orbit and the other is of MBA type, which happens very often. Then it is not

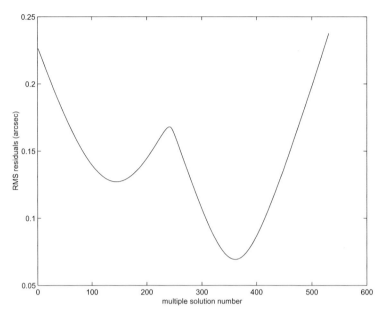

Figure 4. For the object 2005 CG$_{77}$, 600 alternate orbit solutions along the Line Of Variations have been computed. The RMS of the residuals (arcsec) is plotted against the solution number, showing two local minima.

possible to conclude that the object is a NEO; actually, because of known population statistics in most cases it is an MBA. On the other hand, it is not possible to exclude that it is a NEO, and by not following it up the risk is taken of missing an important discovery, defeating the very purpose of a sweet spot survey. If the object actually is a NEO, it may not pass at the opposition (under possible observing conditions) in the same year and for a long time. The radical solution is to obtain 4 well spaced nights of observations.

2.2.1. *False identifications*

To avoid weakly constrained orbits, the interval between nights observing the same area should be increased. On the other hand, this increases the likelihood of false identifications between short arcs of different nights. In fact, two night identifications with well separated nights are so unreliable to be useful almost only as intermediary step for finding 3 nighters (Milani *et al.* 2005b). Surprisingly, even 3 night identifications can be false and be fit with very small residuals and a well determined, spurious orbit.

The most damaging property of these spurious orbits is that they are very often spurious NEOs. Actually, in our test all the 3-night identifications we found and which we could prove false had only NEO type orbits, and were formed with data from two different MBAs. We do not have a formal proof that this must always be the case. Without a method to dismiss these false identifications, we would be forced to waste resources in follow up activities bound to fail.

The conclusion is that it is necessary to select an interval between the nights of observations of the same area which is of intermediate duration. The optimal value would depend upon the achievable limiting magnitude (for deeper surveys the number of false identifications grows with the square of the detections number density) and upon the astrometric accuracy (the better the astrometry, the longer the useful interval). The upper limit is set by the need to have at least 4 nights of observation in the same lunation.

3. The experimental survey

We have tested the observation procedures and the mathematical methods of identification/orbit determination on two sweet spot test runs. One was performed at ESO La Silla in Jan–Feb 2005, using the Wide field Imager on the 2.2 m telescope (field of view 34×33 arcmin, pixel scale in 3×3 binning mode 0.71 arcsec); for the use of this instrument, see (Boattini *et al.* 2004). This test covered 181 square degrees at low elongation over the first two and last two hours of the observing night, in two runs of 10 nights each in January and February. Observations began 20 minutes before the end of the astronomical twilight, and ended symmetrically after the start of twilight. The observation were unfiltered with 60 s exposure, and the limiting magnitude practically achieved was equivalent to 21.5 in the R band. The astrometric accuracy was about 0.2 arcsec in all nights.

With the goal of probing even deeper limiting magnitudes, we made a second test at Mauna Kea in 7 nights of Sept-Dec 2005, using the MegaCam on the CFHT 3.6 m telescope with an R filter and 60 s exposure. The MegaCam is a mosaic of 40 2048×4612 pixel CCDs, with a field of view of 1×1 degree. We even used a portion of one night of SupremeCam at the Subaru 8.3 m (field of view 34×27 arcmin). For both telescopes the pixel scale in 2×2 binning mode was 0.37 arcsec and the time used was only the last 20 minutes before the start of the astronomical twilight. The total coverage at low elongation was 35 square degrees over 8 nights, with limiting magnitude equivalent to 22.5 in the R band. The astrometric accuracy was particularly good, with a random component below 0.1 arcsec. The astrometric catalog used was USNO-B1.0 (Monet *et al.* 2003); systematic effects at this level of accuracy are difficult to assess.

The data were processed in near real time only to look for the fast moving objects. The complete astrometric reduction was performed offline. The data was then passed through the identification and orbit determination computer code which has been developed, within the OrbFit system, for the simulations of the Pan-STARRS survey; for the algorithms see (Milani *et al.* 2004; Milani *et al.* 2005b). However, additional code had to be written to handle the problem of double solutions of Section 2.2; the same code has been later retrofitted to the Pan-STARRS simulation code.

The next processing step has required a significant interaction with the Minor Planet Center (MPC). Of course, a large survey would only need to compare with its own observations near opposition, to recognize the known MBAs among the observations near quadrature. For a small survey it is essential to consult the largest possible database of known asteroids. We have used for this purpose the MPChecker online service†. In some difficult cases, the interaction with the MPC databases was more complicated and will be described in the next Section.

4. Results

4.1. *The La Silla test*

The observational data from La Silla amounted to a total of more than 5,000 detections of moving objects. From the point of view of the observation strategy, the results from La Silla were affected by two problems:

• In the January 2005 run, most of the objects detected and measured were observed only once, with the purpose of covering "fresh sky" as much as possible. During the run in February 2005 we revisited the same fields a number of times, looking for identifications.

• Even when the observed arcs included more than one night (mostly in the February

† http://scully.harvard.edu/cgi/CheckMP

run) the total time span was short, usually 2–4 days; this was due to the short total duration of the run. The good side of short time spans is that the identifications from different nights turned out to be all correct (that is, they belong to the same object) apart from just one case.

Thus the results of January are based on the traditional method of following up the fast moving objects, and indeed a new NEO was found in this way. The February run was more interesting for the development of a new sweet spot surveying strategy. Table 1 shows the results from the test at La Silla, essentially only the February run. The cases listed are only those in which a NEO solution was possible; some objects with only Main Belt solutions were also detected.

Table 1. Summary of the objects detected at La Silla exhibiting multiple orbit solutions. Long-LOV indicates cases in which the multiple solutions along the Line-Of-Variations (LOV) span different orbital type (NEA, MBA, Trojan, etc.). In addition to those shown, there were 6 cases of single orbit solution, 4 of which correspond to known MBAs, 1 to a new Trojan, and the last was found to be a false identification.

Observations	Multiple solutions	Identification
4 nights	2 (MBA-NEA)	2 known MBAs
3 nights	13 (MBA-Aten)	12 known MBAs + 1 new MBA
	14 long LOV	10 known MBAs + 1 new MBA

Although the results are negative, in that no new NEO moving as main belters was discovered, this test pointed out the importance of the double and multiple solutions problem (Table 1). There were 13 cases of 3-nighters with double solutions, including one NEO; the same even in 2 cases of 4-nighters. There were 14 long LOV solutions, too weak to exclude a NEO solution.

In 24 out of these 29 cases the problem was solved by recognizing the detected object as a known MBA (with MPChecker). Two other cases (one double solution and one long LOV) have been solved by recovery and identification. In three cases a designation was assigned by the MPC: 2005 CL_{61}, 2005 DE_3 and 2005 CG_{77} (see Figure 1), but the objects have been neither recovered nor identified yet, thus strictly speaking we do not yet know if they are NEOs, although of course they are more likely to be MBAs.

4.2. *The Mauna Kea test*

At Mauna Kea we were able to fully implement a new observing strategy and distribute the observations over an arc of about 30–40 days; in this way, we could obtain many identifications and good orbits from the $\simeq 3,700$ detections of moving objects. Weak solutions, with long LOV not allowing us to discriminate NEOs from MBAs, did not occur in any 3-night orbit, because of the longer time spans. Nevertheless, the number of nights of observations was not enough to solve all the problems of double solutions and false identifications.

In this observation run we intentionally included a known Aten (1998 XE_{12}), with an apparent motion compatible with an MBA, especially in the first two nights (proper motion $0°.49/d$ and $0°.52/d$; the third night had $0°.60/d$, which is indicative of a nearer object. The identification and orbit determination software had no problem in finding two solutions, both of NEA type (one Aten and one Apollo).

Table 2 shows the results from the run at Mauna Kea. There were 16 cases with double solutions, all with very good fits (maximum RMS of the residuals was 0.10 arcsec) and

Table 2. Summary of the objects detected at Mauna Kea exhibiting multiple orbit solutions. In addition to those shown, there were 5 cases of single orbit solution for detection over 4 nights, 3 of which correspond to known MBAs, and 2 to new MBAs, and 13 cases of single orbit solution for detection over 3 nights; all of these come from false linking of uncorrelated detections, and in most cases correspond to a NEA orbit solution.

Observations	Multiple solutions	Identification
3 nights	16 (MBA-NEA)	15 known MBAs + 1 new MBA
	1 (NEA-NEA)	known Aten inserted on purpose

with one MBA and one NEA solution. The special case of 1998 XE$_{12}$ had RMS of 0.23 arcsec (before removing the outlier) for the best NEA solution.

Using MPChecker we were able to solve 15 out of these cases, identifying them with known MBAs. There were no false identifications among these.

In one case, no identification with known MBAs was found, and the designation 2005 SW$_{277}$ was assigned by the MPC. Even if MBAs are on average brighter by two magnitudes at opposition, a sweet spot survey with limiting magnitude 22.5 may occasionally find an MBA not yet discovered by the present surveys. However, it was later possible to attribute some LONEOS observations from a single night (taken four months later) to 2005 SW$_{277}$; the 4-night orbit was uniquely determined and was of type MBA.

Besides the results listed in Table 2, the OrbFit software proposed 13 additional 3-night identifications with a single orbital solution and RMS of the residuals between 0.14 and 0.24 arcsec. The solutions with RMS > 0.24 arcsec have been automatically discarded. In all these 13 cases it was possible to identify at least one of the 3 nights with a known MBA, in some cases 2-3 nights with different MBAs, thus confirming that all these proposed identifications were false.

The result which may appear disturbing is that in 13 out of 13 cases the spurious solutions were NEO type orbits. Moreover, the reason why there was no double solution was, in 10 cases out of 13, that the preliminary orbit was also unique (in 2 cases the preliminary orbit algorithm gave 0 solutions, in 1 case there were 2 solutions). The sample is too small to draw statistically significant conclusions, but it is enough to suggest some correlation between the false identification and the orbital class.

A qualitative explanation can be as follows. If three single night arcs are wrongly identified, it is very unlikely that they are well aligned along a single great circle, thus the curvature of the whole observed arc will be comparatively large. The curvature goes to $-\infty$ as the heliocentric distance goes to zero. Thus, if the sign happens to be right, a large spurious curvature can be interpreted by the preliminary orbit algorithm as if the object is closer to the Sun than the Earth, and this implies a NEO orbit. Moreover, if the elongation is more than 63°.5, the objects belongs to the region where there is a single solution (see Section 2.2 and Figure 3).

Thus the problem of false identifications is especially critical at the sweet spots, because it can result in false positive NEO discoveries. The other problem, the double solutions, can result in false negatives unless both solutions are reliably computed.

5. Conclusions

We have performed a small, but complete test of a *sweet spot survey.* We have identified the specific difficulties, including the ones occurring in identification and orbit determination which previously had not been investigated. In particular, the weakly determined (long LOV) solutions can be avoided by properly spacing the observing nights. We have not determined an optimal spacing, for now we can only conclude that it should be between 2 and 6 days.

A difficulty not found near opposition is the occurrence of double solutions. A possible remedy is to exploit the prior knowledge of the MBAs observable in the sweet spots; this is possible if the sweet spot survey is a component of a larger survey covering also the opposition region up to the same limiting magnitude. Another solution is scheduling the sweet spot survey to guarantee at least 4 nights of observations; this is possible because of the reduced competition for telescope time at times near the astronomical twilight.

The third difficulty, which occurs also near opposition, is due to false identifications. In the sweet spots, a false identification often results in spurious NEO orbits, and such false positives could be the source of a waste of astronomical resources. If the false identifications combine observations of different MBAs, again with enough prior information collected at opposition these spurious cases can be discarded, as it happened in our test. However, we cannot exclude that in a larger experiment this problem could lead to dubious cases.

The problems with the observation strategy and scheduling are very severe for a survey covering a small area in the sky, while they are easy to solve for the wide field surveys and when dedicated follow up resources are available. What needs to be investigated is a possible strategy for a narrow field survey going much deeper than the next generation all sky surveys.

Acknowledgements

This research has been funded by: the Italian *Ministero dell'Università e della Ricerca Scientifica e Tecnologica*, PRIN 2004 project "The Near Earth Objects as an opportunity to understand physical and dynamical properties of all the solar system small bodies". The observations from Mauna Kea were taken within the UHAS survey, in collaboration with D. Tholen and F. Bernardi; the ones from La Silla were made possible by allocation of telescope time by the standard ESO procedure. We thank D. Tholen, F. Bernardi, N. Kaiser, R. Jedicke, L. Denneau and O. Hainaut for useful discussions and suggestions.

References

Boattini, A., *et al.* 2004, *Astron. Astrophys.*, 418, 743

Chesley, S. & Spahr T., 2002, in *NASA Workshop on Scientific Requirements for Mitigation of Hazardous Comets and Asteroids*, Arlington, VA, September 3-6, 2002

Danby, J. M. E. 1989, *Fundamentals of Celestial Mechanics*, Willmann-Bell, Richmond (VA)

Milani, A., Gronchi, G. F., de' Michieli Vitturi, M. & Knežević, Z. 2004, *Celest. Mech. Dyn. Astron.* 90, 59

Milani, A., Sansaturio, M. E., Tommei, G., Arratia, O. & Chesley, S. R. 2005a, *Astron. Astrophys.* 431, 729

Milani, A., Gronchi, G. F., Knežević, Z., Sansaturio, M. E. & Arratia, O. 2005b, *Icarus*, 179, 350

Milani, A., Gronchi, G. F. & Knežević, Z. 2006, *Earth, Moon and Planets*, in press

Monet, D. *et al.* 2003, *Astron. J.*, 125, 984

Plummer, H. C. 1918, *An introductory treatise on dynamical astronomy*, Dover Publications, New York.

Stokes, G. H., *et al.* 2003, Report of the Near-Earth Object Science Definition Team. August 22, 2003 (available at `http://neo.jpl.nasa.gov/neo/report.html`).

Near Earth Objects, our Celestial Neighbors: Opportunity and Risk
Proceedings IAU Symposium No. 236, 2006
A. Milani, G.B. Valsecchi & D. Vokrouhlický, eds.
© 2007 International Astronomical Union
doi:10.1017/S1743921307003365

Initial linking methods and their classification

Leif Kahl Kristensen[1]

[1]Department of Physics and Astronomy, University of Aarhus, DK-8000 Aarhus C, Denmark
email: LKK@PHYS.AU.DK

Abstract. The problem of initial linking of asteroids is of increasing interest for the next generation surveys. During the first week after discovery elliptical elements are very uncertain and other methods are used. A summary is given of 7 initial linking methods. There are two different types: In one, a search area is computed on a second night from the known and undoubtedly linked positions, typically on the first night. The other type assumes candidates which are then checked by the computation of O – C residuals of an orbit. Computations may be classified as belonging to the 3-dimensional space or the 2-dimensional sky-plane. A new basis, with a simpler computational algorithm, is given for the widely used Väisälä method. For a new N-Observation Orbit method a simple, efficient PC-programme is given.

Keywords. Orbit determination; identification; ephemerides

1. Introduction

Different methods can be used for linking a great number of asteroids during their first critical lunation. When 3–4 nights of observations are secured during 2–3 weeks it makes sense to compute elliptical elements but different methods are needed at start. Here we try to list the 7 known methods but will not give exhaustive references. For familiar methods we only give remarks not available elsewhere.

Focus is on computational efficiency, so statistical (Monte Carlo) methods based on many orbits are excluded in advance.

There are essentially two problems:

1) Some observations are given for a definite object but not enough for the determination of an orbit. A typical example is 1–5 observations on a **single** night. The problem is to predict a search area another night in which the candidates may be found.

2) **Two** groups of observations are given which may possibly belong to the same object. Typically two nights with daily motions. The problem is to check if an orbit exists which gives O - C residuals consistent with observational errors.

Computing methods are of two types:

A) Computing takes place in 3-dimensional space. Topocentric coordinates (X,Y,Z) of the observer is for instance used.

B) All computations occur in the 2-dimensional sky-plane. These methods are based on the smoothness of the apparent motion and use interpolation formulae.

In this scheme the widely used (generalized) Väisälä method is classified 1A and some aspects are discussed in Sec. 7.

Type 1B methods are: extrapolation, (T,I)-coordinates and distant objects, to be discussed in Sec. 2, 3 and 4.

Type 2B is L - R method and daily motion method in Sec. 5 and 6.

Type 2A is the N-Observation method discussed in Sec. 8. Its use for impact monitoring is given in Sec. 9. A (Pascal) PC programme is given in the Appendix.

2. Extrapolation (Type 1B)

To link the positions each individual night we "predict" second positions at $t_2 = t_1 + h$ from the first at t_1 by the local mean daily motion and its scatter. For main belt objects near opposition we have the daily motions in ecliptical longitude and latitude:

$$\lambda' = 0.225 \pm 0.0315 \,°/d, \quad \beta' = 0 \pm 0.045 \,°/d \tag{1}$$

With $h = 1/24\,d$ the area to be searched is 1.1×10^{-5} sq. deg. Assume a density of objects $\mu = 900$ per square degree which corresponds to a mean distance $1'(= \frac{1}{2}/\sqrt{\mu})$ to the nearest neighbour. The probability to find **two** objects, the correct one plus a random straggler, is only about 1%. The moving objects are assumed distinguished from the much larger density of stars by trailed images. There seems to be little difficulty at this first step.

The following nights objects are obtained by linear extrapolation based on Lagrange formula with remainder term (Abramowitz & Stegun 1970, 25.2.1):

$$f(t) = f(t_1)\frac{t_2 - t}{t_2 - t_1} + f(t_2)\frac{t - t_1}{t_2 - t_1} + (t - t_1)(t - t_2)\frac{f''(\xi)}{2} \tag{2}$$

Here $t_1 \leqslant \xi \leqslant t$ if $t_1 \leqslant t_2 \leqslant t$. The remainder term can be estimated from the second differences Δ^2 in the 10-days ephemerides, conveniently in EMP1988 (the last volume stating differences). A sample gave typical values of $\Delta^2\alpha$ to be of order -2^m before and $+2^m$ after opposition. Similarly $\Delta^2\delta \approx \pm 15'$. After 5 days the area to be seached is thus $1/8°$ by $1/16°$. The observational error $\pm\sigma$ (here assumed $0.15''$) gives

$$\pm\sigma\frac{\sqrt{(t - t_1)^2 + (t - t_2)^2}}{t_2 - t_1} \tag{3}$$

With $\frac{1}{2}$ hour between exposures this gives after 5 days $\pm 0.8'$.

All interpolation methods are affected by the short-period parallax which gives an error in the daily motion of order

$$\cos\delta\alpha' = -2\pi\frac{8''80}{\Delta} \tag{4}$$

where Δ is the geocentric distance. For main belt objects $1/\Delta = 0.75 \pm 0.36$. After 5 days there is a systematic error $-3.5' \pm 1.7'$ in α. The corresponding effect in δ is negligible. Adding the above errors ($\pm 3.8'$ from (2), $\pm 0.8'$ from (3) and $\pm 1.7'$ from (4)) gives the uncertainty in α: $\pm 4.2'$. Similarly in δ adding $\pm 1.9'$ from (2) and $\pm 0.8'$ from (3) gives $\pm 2.1'$. The 1σ limit area is thus 0.0098 square degree, corresponding to an average of 8.8 erroneous objects. Including the object in question there are around 10 candidates for which trial orbits must be found. The example illustrates the relative importance of the approximations which scales like T^4, T^2 and T respectively.

3. (T,I) coordinates (Type:1B)

If we assume circular orbits with $a = 2.64$ AU and longitude in orbit $90°$ we can compute the inclination (I) and the opposition date (T) for any **single** position. This corresponds to a coordinate transformation from (α, δ) to (T,I) (Kristensen 1990). For the "mean" main belt asteroid T and I are constant and for the neighboring real asteroids T and I are slowly varying and the mean of their daily motion relative to the center of the belt is zero. There is thus no mean motion depending on the position in the sky as the constants in the ordinary coordinates in (1). The main irregularities in the apparent motion is caused by the reflex motion of the Earth which is much reduced and the retrograde loop entirely removed. The parallax is partly corrected for to the mean distance to the main belt. The method has the same accuracy as Väisälä's perihelion orbits with $q = a$ and gives an alternative justification of this method. Väisälä assumes essentially, as did Thiele (1883), a smooth motion on a sphere with radius q. In (T,I) the computations are easier and based on individual rather than pairs of observations. Each observation (α, δ) is transformed to the slowly varying (T,I). This makes interpolation methods more efficient. A trial with 4 synthetic plates with a density $\mu = 900$ objects/deg.sqr. (Kristensen 2005, unpublished) has shown that **single** main belt positions each night can be effectively linked. Only algorithms well developed in computer science are needed, such as finding all objects in given search areas, combining positions and checking differences.

4. Distant objects (Type:1B)

In classical stellar astrometry the elements are position at the epoch, proper motion, radial velocity and the (yearly) parallax. Compared with orbit determination for main belt asteroids this is much simplified! Surprisingly, the transition region between these two methods is rather near, say at 20 AU. For a few years the simple method ignoring acceleration and using only plate coordinates may be used for TNO's (Kristensen 2004).

5. L-R methods (2B)

Assume 4 positions at times: $-\frac{1}{2}h, +\frac{1}{2}h, T - \frac{1}{2}h$ and $T + \frac{1}{2}h$. Use the formulae 25.1.5 and 25.1.10 in Abramowitz & Stegun (1970) and re-arrange, this gives:

$$\frac{f(t_2)(T + h) - f(t_1)(T - h)}{2h} + \frac{f(t_4)(T - h) - f(t_3)(T + h)}{2h} = \frac{T(T^2 - h^2)}{12} f^{(3)}(\xi) \quad (5)$$

The first term is linear extrapolation from the first night to $T/2$ and the second term extrapolates the second night to the same instant. From the first and second nights we construct a fictitious plate at $T/2$ on which adjoining objects can be identified.

This method is a special case of the L-R method (Kristensen 1992). It is denoted the "midpoint" method by Kubica (2005) who tested it. The remainder term can be estimated by the third differences of the 10-days ephemerides. For $T = 7$ the error is of order $0.5'$. In α there is still the systematic parallax displacement. The search area scales as T^6.

6. Daily motion integration (Type 2B)

To check if observations a second night (at $t = T$) belong to those the first night (at $t = 0$) the daily motions must agree. The numerical integration of, say $d\delta(t)/dt$, gives (Abramowitz & Stegun 1970, 25.4.1)

$$\int_0^T \delta'(t)dt = \delta(T) - \delta(0) = \frac{T}{2}(\delta'(T) + \delta'(0)) - \frac{T^3}{12}\delta^{(3)}(\xi) \quad (6)$$

from which

$$\delta'(T) = \frac{2(\delta(T) - \delta(0))}{T} - \delta'(0) + \frac{T^2}{6}\delta^{(3)}(\xi) \quad (7)$$

As an example (the main belt object (2156) Kate in 2002, Appendix B, Table 1).

Table 1. Daily motion for object (2156) Kate

t	δ
- 0.05	11.040279°
0.05	11.044407°
6.95	11.276349°
7.05	11.278964°

This gives the daily motion $\delta'(0) = +0.041280°/d$ and from the 7-days interval $+0.033616$. By (7) is obtained $\delta'(7.0) = +0.025952°/d$ to be compared with the true value $+0.026150$, the error is $O - C = +0.000198$. The difference between the daily motions of two objects taken at random $\pm 0.070°/d$ due to the scatter (1) from the mean motion. It is thus unlikely that the two nights refer to different objects. Formula (7) is equivalent to (5) in the limit $h \rightarrow 0$.

7. The generalized Väisälä method (Type 1A)

This well known method gives search areas from two observations under different assumptions, partial knowledge or a priori information. Marsden (ACM 1991, 395) assumes limits on eccentricity e, semimajor axis a and peri- and aphelion distances (q and Q). Kristensen (2002) assumes intervals for the geocentric distance (Δ) and $r' = dr/dt$. Milani *et al.* (2005) assume r and r' limited by energy. We shall not comment on these methods but only show, that the demarcation line of the search area can be delineated very simply for main belt objects at opposition by only **one single** Väisälä perihelion orbit.

Formula (5) in Kristensen (2002) shows that there are two non-linear terms in the apparent motion. One is due to the attraction of the Sun and, because $r^3 >> R^3$, is proportional to $1/\Delta$. The other is a perspectivistic effect along the daily motion. In Fig.1 we denote by V the position by assuming $\Delta = 1.0$ and $r' = 0$. The position L is obtained by linear extrapolation along the great circle between position 1 and 2 and corresponds to $\Delta \rightarrow \infty$. The perihelion orbits are approximately on the line connecting L and V and subdivided as $1/\Delta$. For constant values of Δ the curves of varying r' are situated on great circles through position 2 and along the direction of motion. For $\Delta = 1$ and $r' = ke/r \approx 0.0024$ (assuming $e = 0.20$) formula (5) in Kristensen (2002), the "foreshortening" term, is a product of the motion from the Nov. 30.94 position (2) to position 3 at Nov. 57.76 and the factor $-\Delta' t/\Delta \approx -tr'/\sqrt{1 + \Delta}$:

$$(-5.4°, +1.3°) \times (-16.82) \times 0.0024 = (+0.22°, -0.05°) \tag{8}$$

The three lines of constant r' in the figure are great circles through position 2.

A sphere with radius q osculates perihelion orbits and to a good approximation the motion is uniform. Thiele (1883) assumed uniform great circle motion on this sphere. There is thus little difference between these methods for small arcs.

8. N-Observation Orbits (Type 2A)

The importance of using **all** available observations for the initial orbit is stressed by Branham (2005). The first and last positions are urgent to increase the arc but any number in between improves the critical geocentric distance Δ – especially by parallax

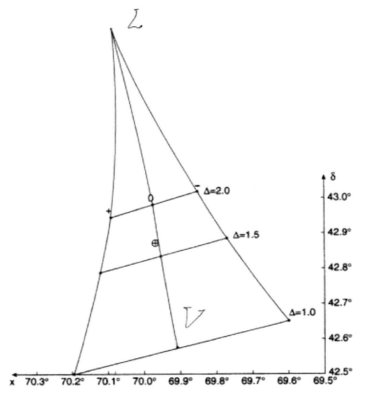

From the two discovery positions of 1984 WH Nov. 25.95 and 30.94 (MPC 9263, 9326) the recovery position on Dec. 17.76 is predicted by assuming $\Delta = 1.0$, 1.5, and 2.0. Trial values for r' were 0 and $\pm 0.20/\sqrt{1+\Delta}$. The r' variation lines are segments of great circles all passing through the first position as a "radiant." This is a perspectivistic effect of the velocity being parallel to n_1 by Eq. (38). The search area is limited by curves corresponding to receding ($r' > 0$) and approaching ($r' < 0$) heliocentric motion. Midpoints give the Väisälä (1939) variation curve of perihelion orbits ($r' = 0$). The area is only 0.13 square degrees. It is indicated by the six points which can easily be computed by a programmable pocket calculator. The object happened to be near perihelion and was actually found at the indicated point near the Väisälä line of variation.

Figure 1. Search area from two positions. The vertex L is obtained by linear extrapolation and corresponds to $\Delta = \infty$ and V assumes $\Delta = 1.0$ and $r' = 0$. The three lines of variation of r' are great circles through the first position (see text). The Väisälä locus connects L and V and is subdivided as $1/\Delta$. The search area is essentially determined by one single perihelion orbit

from different sites. The present method (Kristensen 2005) gives positions as a strictly linear function of the coordinates x and velocities x' at the epoch To in terms of the f and g coefficients. Good approximations for f and g are obtained by a trial value for the heliocentric distance r. Adopting $r = 2.30$ gives 3-4 correct figures for main belt objects during the first month, – the shorter the arc the better the approximation! Given f and g each observation gives stricly linear equations of condition which are adjusted rigorously by a least squares solution for the elements x and x' by the classical method (Watson 1868) of Gauss elimination. We note especially that the square sum (χ^2) is obtained **before** the solution which may be skipped if a statistical χ^2-test indicates inconsistent observations. Realistic observational error estimates may be obtained from Bykov *et al.* (2002) and Carpino *et al.* (2003). Regardless of any errors in f and g the O-C residuals by 3 observations are exactly zero. For observations on 3 nights the daily motion has the 3-4 figure accuracy. This makes the method very suitable for the initial linking problem. Due to the linear equations there is always a single solution and no need for differential corrections. It may, however, be necessary to iterate $r = |\mathbf{x}|$, especially if four nights are observed.

My first step towards this method was the projection matrix (18) in Kristensen (2002) but I could not reduce the right hand sides of the normal equation before I noted the ingenious trick of T.N. Thiele's "fabricated observations" (Thiele 1889). After submission of my paper (Kristensen 2005) I found a similar method by Neutsch (1981). Some advantages of these methods are: all available observations are used from the start, two simultaneous positions from different sites gives Δ by parallax, Radar surveillance observations can be included, the intermediate step of differential corrections of elliptical elements is not used so the possible lack of convergence in this step is avoided, computations are strictly linear in degenerate cases and convergence problems are avoided, within 2-3 weeks for main belt asteroids iterations are not necessary, which is especially useful for initial linking and, finally, impact monitoring is facilitated.

Rather than explaining how the coefficients of the normal equations are derived is here given a simplified PC-programme in Pascal in the Appendix.

1) The main belt object (2156) Kate observed 2 nights at times $JD = 2452200.5 + t$. Maximum O – C is −0.005" so the two short arcs are undoubtedly linked. The advantage of an initial orbit based on 4 - rather than 3 - observations is obvious.

2) The Centaur (60558) 2000 EC$_{98}$ discussed by Milani *et al.* (2005). There are 3 nights: Feb. 5, Mar. 3 and 4 with daily motions. The mean positions on 3 nights are always fitted exactly but it is unlikely that the motions also agree if not referring to the same object. Rounding to 4 decimals in degrees corresponds to a mean observational error ±0.10". The recovery position April 4 is predicted with errors +22.970" and −13.076" and a motion accurate enough for identification. There is no need to repeat the computation based on $r = 2.30$ with the improved value $r = 13.446$, obtained from the solution: $x = (−13.07, +2.93, +1.14)$.

3) The last example is the NEO (85640) 1998 OX$_4$ at 6 ephemeris positions 2006 Aug. 24.0 + t. The first run with $r = 2.30$ indicates an NEO with $r = 1.20$. The orbit is well determined and the f and g coefficients should be improve by rigorous expressions, but this facility is removed from the present abbreviated programme.

9. Impact monitoring

If the elements are coordinates and velocities at the epoch, x and x', a (Radar) geocentric distance Δ or a (Doppler) radial velocity Δ' at time t gives a linear equation of condition which contributes directly to the normalequations. However, these facilities have been removed from the simplified programme in the Appendix. Two positions give 4 conditions and adopting trial values for Δ and Δ' the orbit is determined and also the coordinates $x(\Delta, \Delta'; t)$ et.c. Earth coordinates are $X(t)$ and a collision at time T gives 3 equations:

$$x(\Delta, \Delta'; T) = X(T) \quad etc. \tag{9}$$

for the unknowns Δ and Δ' and T.

Mathematical solutions of (9) are very frequent if the daily motion and Δ are small (Kristensen 2002b). Nothing about a collision risk may be said before enough observations have been secured to make sense for Δ. The 2004 AS$_1$ false alarm was due to the indeterminateness of Δ and the adoption of a too small value.

In the case of (85640) with the 3-days arc in Appendix B there is a close passage at $t = 16$ (EMP 2006, p.680). To check if it actually is a hit we assume a Radar observation with $\Delta = 0$ at $t = 16$ and re-adjust. The assumption is not consistent with the optical positions. The O - C residuals are max. -0.099" in Appendix B but is forced to 1472".

By series of trial times T it can be checked if a collision is imminent. This gives a 1-dimensional set rather than a 2-dimensional set of trial orbits.

Acknowledgements

The author is much indebted to Mikael Julian Jørgensen for the LATEX manuscript and to the referee, Giovanni F. Gronchi, for improvements of the text.

References

Abramowitz, M. & Stegun, I. 1970, *Handbook of Math. Functions*, Nat. Bureau of Standards, Washington.

Branham, R. L. 2005, *Celest. Mech. Dyn. Astron.* 93, 53

Bykov, O. P., L'vov, V. N., Izmailov, L. S., & Sumzina, N. K. 2002, in: Proceedings of ACM 2002, ESA-SP-500

Carpino, M., Milani, A. & Chesley, S. 2003, *Icarus* 166, 248

Ephemerides of minor Planets for 1988, Institute for Theoretical Astronomy, Leningrad 1987

Kristensen, L. K. 1990, *Astron. Nachr.* 311, 133

Kristensen, L. K. 1992, *Astron. Astrophys.* 262, 606

Kristensen, L. K. 2002, *Icarus* 159, 339

Kristensen, L. K. 2002, in: Proceedings of ACM 2002, ESA-SP-500

Kristensen, L. K. 2004, *Astron. J.* 127, 2424

Kristensen, L. K. 2005, N-Observations and Radar Orbits (Submitted)

Kubica, J. 2002, Efficient discovery of spatial associations and structure, Robotic Institute, Carnegie Mellon univ., Pittsbúrgh, Pennsylvania

Marsden, B. G. 1992, in: Proceedings Asteroids, Comets and Meteors, Houston, Texas, p. 395

Milani, A. & Knežević Z. 2005, *Celest. Mech. Dyn. Astron.* 92, 1

Neutsch, W. 1981, *Astron. Astrophys.* 102, 59

Thiele, T. N. 1889, *Almindelig Iagttagelseslre.* (Reitzel, Copenhagen) (In Danish, English translation in E. L. Lauritzen: Thiele – Pioneer in Statistics. Oxford Univ. Press, 2002)

Thiele, T. N. 1883, *Astron. Nachr.* 107, 291, 357

Thiele, T. N. 1903, *Theory of Observations*, Charles & Edwin Layton, London

Watson, J. C. 1868, *Theoretical Astronomy*, J. B. Lippincott and Co., reprinted by Dover Publ. New York 1964

Appendix A. N-Observation Orbits PC programme

The elements are x and x' at epoch T0. Positions are expressed in terms of the $f \approx 1$ and $g \approx t$ coefficients. Four figure accuracy is obtained the first month by assuming $r = 2.30$. Given f and g the least squares adjustment is stricly linear in the elements x and x'. The coefficients of the normalequations are stored in a 7×7 matrix. The programme is a simplified version with Radar observations and computation of f and g from the elements by exact formulae removed. Perturbations and light-time are ignored.

The first line in input has 4 numbers: N observations, epoch T0, adopted r and observational mean error (m.e.). The following N lines gives times of observation, coordinates (X, Y, Z) of the observatory, observed (true) RA and Decl in units of degrees, and finally a weight w. The program stops when a first line is $-9 - 9 - 9 - 9$.

The output repeats the first line in input. The next line gives the square sum of the residuals (AO[7,7] or χ^2) before and after a Gauss elimination and the number of degrees of freedom. The N lines give: t, w, observed and computed RA and Decl (in degs) and $O - C$" (in seconds of arc).

The $1\frac{1}{2}$ page source code in Turbo-Pascal, the executable PC programme and the sample Input/Output may be requested from the author by e-mail.

Near Earth Objects, Our Celestial Neighbors: Opportunity And Risk
Proceedings IAU Symposium No. 236, 2006
A. Milani, G.B. Valsecchi & D. Vokrouhlický, eds.
© 2007 International Astronomical Union
doi:10.1017/S1743921307003377

Spins, shapes, and orbits for near-Earth objects by Nordic NEON

Karri Muinonen[1]†, Johanna Torppa[1], Jenni Virtanen[1], Jyri Näränen[1], Jarkko Niemelä[1], Mikael Granvik[1], Teemu Laakso[1], Hannu Parviainen[1], Kaare Aksnes[2], Zhang Dai[2], Claes-Ingvar Lagerkvist[3], Hans Rickman[3], Ola Karlsson[3], Gerhard Hahn[4], René Michelsen[5], Tommy Grav[6], Petr Pravec[7], and Uffe Græe Jørgensen[8]

[1]Observatory, Kopernikuksentie 1, P.O. Box 14, FI-00014 University of Helsinki, Finland
email: Karri.Muinonen@helsinki.fi

[2]University of Oslo, Institute of Theoretical Astrophysics, PB 1029 Blindern, 0315 Oslo, Norway

[3]Uppsala Astronomical Observatory, Box 515, S-75120 Uppsala, Sweden

[4]German Aerospace Center (DLR), Institute of Planetary Research, Rutherfordstrasse 2, D-12489 Berlin, Germany

[5]Ørsted DTU, Technical University of Denmark, Elektrovej, bldg. 327, DK-2800 Kgs. Lyngby, Denmark

[6]University of Hawaii, Institute for Astronomy, 2680 Woodlawn Drive, Honolulu, Hawaii 96822-1897, USA

[7]Astronomical Institute AV CR, Fricova 298, 25165 Ondrejov, Czech Republic

[8]Astronomical Observatory, Niels Bohr Institute, Juliane Maries Vej 30, DK-2100 Copenhagen, Denmark

Abstract. The observing program of the Nordic Near-Earth-Object Network (NEON) accrues knowledge about the physical and dynamical properties of near-Earth objects (NEOs) using state-of-the-art inverse methods. Photometric and astrometric observations are being carried out at the Nordic Optical Telescope. Here, the NEON observations from June 2004–September 2006 are reviewed. Statistical orbital inversion is illustrated by the so-called Volume-of-Variation method. Statistical inversion for spins and shapes is carried using a simple triaxial shape model yielding analytical disk-integrated brightnesses for both Lommel-Seeliger and Lambert scattering laws. The novel approach allows spin-shape error analyses with the help of large numbers of sample solutions. Currently, such spin-shape solutions have been derived for 2002 FF_{12}, 2003 MS_2, 2003 RX_7, and 2004 HW. For (1862) Apollo, an unambiguous spin-shape solution has been obtained using the conventional, convex inversion method and, for (1685) Toro and (1981) Midas, the conventional method has been applied repeatedly to map the regime of possible solutions.

Keywords. Identification; statistical methods; data analysis; astrometry; photometry; asteroid, rotation; asteroid, lightcurves

---

1. Introduction

The Near-Earth-Object Network (NEON) carries out coordinated observations of NEOs in order to contribute to the study of their physical and dynamical properties;

† Present address: Observatory, Kopernikuksentie 1, P.O. Box 14, FI-00014 University of Helsinki, Finland.

at the moment we concentrate on photometric and astrometric observations, yielding spin states, shapes, and orbits for NEOs. NEON was initiated by the Nordic Group for Small Planetary Bodies (NGSPB) which represents collaboration between asteroid and comet researchers from mainly Nordic countries. The vast majority of NEON observations have been carried out at the Nordic Optical Telescope (NOT) which is a 2.56-m telescope located at Roque de los Muchachos on La Palma (Canary Islands, Spain).

In the photometric part of the NEON program, the first and second priorities are given to potentially hazardous NEOs with and without earlier photometric data, respectively, whereas the third priority is given to NEOs with earlier photometric data. Note that NOT allows photometric observations and spin-shape analyses of faint fast-moving objects, thus extending the NEO physical studies towards smaller objects.

The primary objective of the astrometric part of the NEON program is the recovery and follow-up of faint potentially hazardous NEOs, that is, securing orbits for critical objects with short observational arcs. Astrometric observations are carried out both for potentially hazardous NEOs not observed lately, which are thus difficult to recover, and for newly discovered objects at risk of becoming lost. Accurate orbit computation is needed both for evaluating the potential collision risk as well as for planning the photometric observations.

During April 2004-September 2006, the total amount of visitor-mode observing time has consisted of 37 nights, of which 27 have been photometric. The visitor-mode time has been mainly used for photometric lightcurve observations. In addition, roughly once a month, two-hour-long service mode/target-of-opportunity observation time slots have been used for astrometric observations of faint NEOs. The service-mode time of $31 \times 2 = 62$ hours has yielded over 40 useful hours for astrometry. For more information about the NEON observing program, we refer the reader to Torppa, Virtanen & Muinonen *et al.* (2006).

The statistical inversion of asteroid astrometric observations for orbits has been reviewed by Bowell, Virtanen, Muinonen, *et al.* (2002) and Virtanen, Tancredi, Bernstein *et al.* (2006). They provide methods for objects observed over short, moderate as well as long time intervals. In particular, for moderate time intervals, Muinonen, Virtanen, Granvik *et al.* (2006) have recently offered a solution using what they called volumes of variation in the six-dimensional orbital-element phase space.

The inversion of asteroid photometric observations for spins, shapes, and overall scattering properties of the surfaces have been reviewed by Kaasalainen, Mottola & Fulchignoni (2002). Whereas the convex inversion method – hereafter refered to as the conventional method – is currently well matured and shown to yield realistic spin and shape solutions, the error analysis of the spin-shape solutions, in particular, is still in its infancy: steps towards a statistical treatment of the inverse problem are taken in the present article.

In Sect. 2, we present the NEON observing program with a summary of the photometric and astrometric observations made so far. Section 3 includes the orbital inverse methods and highlights of their application to NEON astrometry, in particular, in connection to NEON recoveries. Section 4 presents the inverse methods for spins and shapes, culminating, on one hand, in the statistical inversion of single-lightcurve data and, on the other hand, on the conventional inversion of the spin and shape for (1862) Apollo. We close the article by conclusions and future prospects in Sect. 5.

2. Nordic-Optical-Telescope Observations

2.1. *Photometry*

We have carried out the observations using the Andalucia Faint Object Spectrograph and Camera (ALFOSC), which is a 2048×2048 CCD camera with an effective field-of-view of $\sim 6' \times 6'$. We have used the Bessell R-filter for the observations. It provides the best signal since asteroids are mainly at their brightest in the R-band. In addition, we have used 2×2 binning of the pixels to increase the signal-to-noise ratio (S/N) and to decrease the readout times. Care has been taken that no photometric pixel undersampling occurs even with extraordinary seeing conditions.

In the present paper, we include those photometric lightcurve observations in June 2004-September 2006 that have provided more accurate or new results for the spin states of NEOs. Three of our objects – (1685) Toro, (1862) Apollo, and (1981) Midas – have observations published by Lagerkvist, Piironen & Erikson (2001) and by Torppa & Muinonen (2005)), and four – 2002 FF$_{12}$, 2003 MS$_2$, 2003 RX$_7$, and 2004 HW – have no earlier lightcurve observations. For example, Figure 4 depicts the lightcurve observed for 2003 MS$_2$.

When selecting objects for each observing run, there are three particular issues to be considered: (1) A lightcurve provides new information about the object if it significantly increases the total time range of observations, since the longer the total time range is, the more accurate period determination we get; (2) in addition, for shape and spin axes determination we need observations carried out at different observing geometries, i.e., we need to observe the asteroid from both its northern and southern hemispheres – this is ensured by observing the object at different Earth-centered ecliptic longitude and latitude; (3) shape information is also increased when we have observations from a wide range of phase angles, since shadowing effects become more dominant at large phase angles.

The observations included here are tabulated in Table I. We have mainly used relative photometry (i.e., used field stars as a reference for brightness change in the object) due to its simplicity and the weather conditions not being acceptable for absolute photometry throughout the night. However, we have also observed appropriate Landolt photometric standard stars whenever reasonable.

For spin state and shape determination, relative photometry is sufficient, provided that the weather conditions remain stable for at least half the rotation of the object and that the object remains visible. In the case where the period of the object is long, and half the period cannot be covered during one night, absolute photometry is required to be able to obtain a full lightcurve.

2.2. *Astrometry*

For the follow-up and recovery observations of NEOs, we have been running a monitoring program (corresponding to observatory code J50) that has mostly been operated in service mode but including also observations carried out during the visitor runs. The service mode observations have been obtained in two-hour slots on a monthly basis. Using ALFOSC with 2×2 binning has allowed efficient use of the two-hour runs by providing fast pointing and read-out times combined with acceptable astrometric accuracies (on the 0.1-arcsec level).

Our observing strategy has been to concentrate on the faint end of objects that are not likely to be observed by most amateur telescopes or automated surveys with relatively bright limiting magnitude (around $V = 20$ mag), but which provide the bulk of NEO observations. Highest priority has been given to potentially hazardous objects. Thus,

Table 1. Lightcurves included in the present study (Torppa, Virtanen & Muinonen *et al.* (2006)). All but three curves have been observed at the NOT, one at Dk1.54 and two at Ondrejov. The columns contain the name of the object, date of observation, observing site, phase angle, the filters used and the duration of observations during each night.

Asteroid	Date	Obs site	phase angle	Filters	Duration (h)
(1685) Toro	Jun 18 2004	La Palma, NOT	45	R(rel)	2.9
	Jun 19	La Palma, NOT	45	R(rel)	3.4
	Jun 27	La Silla, Dk1.54	47	R(rel)	3.4
(1862) Apollo	Mar 20 2005	La Palma, NOT	9	R(rel)	3.5
	Mar 31 1998	Ondrejov	5	R	3.5
	Apr 20 1998	Ondrejov	22	R(rel)	3.9
(1981) Midas	Sep 14 2004	La Palma, NOT	26	R(rel)	2.6
	Sep 15	La Palma, NOT	26	R(rel)	4.0
2002 FF12	Sep 16 2004	La Palma, NOT	6	R(rel)	5.8
2003 MS2	Jan 14 2005	La Palma, NOT	17	R(rel)	5.9
2003 RX7	Jun 19 2004	La Palma, NOT	19	R(rel)	6.0
2004 HW	Aug 15 2004	La Palma, NOT	44	R(rel)	3.5
	Aug 17	La Palma, NOT	43	R(rel)	2.3
	Sep 14	La Palma, NOT	22	R(rel)	4.7
	Sep 15	La Palma, NOT	23	R(rel)	4.2
	Sep 16	La Palma, NOT	23	R(rel)	2.7

reasonable object selection is a key part of the observation planning process. To create priority lists of observable NEOs, we have been making use of the orbit computation tools presented below, as well as several web-based asteroid observing services, such as the Lowell Observatory Asteroid Data Services, Minor Planet Center ephemerides, and the ESA Spaceguard Central Node.

For short-arc objects with large sky-plane uncertainties, the above on-line services currently do not give reliable error estimates for the position. Thus, for objects whose positional uncertainties are on the order of, or larger than, the instrument field of view, we rely on our nonlinear methods for uncertainty propagation.

In connection to the astrometric observations, we have obtained altogether 313 Bessell R-filter magnitudes of total 75 objects (data stored in Standard Asteroid Photometric Catalog SAPC http://www.astro.helsinki.fi/SAPC/)). The magnitudes are instrumental magnitudes that have been corrected for the photometric zero point according to the NOT zero-point monitoring program, but they have not been corrected for extinction, since we do not have standard star observations from the service-mode observations. Also, S/N required for astrometry is not necessarily good enough for reliable photometry. This gives 1-σ error estimates of ±0.5 mag for the majority of the objects and ±1.0 mag for the faintest objects (R>22 mag).

2.3. *Reductions*

The data reductions have been accomplished using IRAF (APPHOT/PHOT and custom-made astrometric routines) and Astrometrica (Raab (2004)). Bias and flat-field corrections were always carried out by using calibration images taken on the same night. For photometry, six to ten reference stars in the field were used to derive the relative magnitudes of the objects. The reference stars were monitored for variability. In the cases of faint astrometric objects, we utilized the stacking technique, where several images are added to obtain one, more accurate position (higher S/N).

3. Orbits

3.1. *Inverse methods*

The statistical orbital inversion methods consist of the nonlinear least-squares method with linearized covariances (LSL; Bowell, Virtanen, Muinonen, *et al.* (2002), Muinonen & Bowell (1993)), the Volume-of-Variation method (VoV; Muinonen, Virtanen, Granvik *et al.* (2006)), and orbital ranging (Ranging;Virtanen, Muinonen & Bowell (2001) and Muinonen, Virtanen & Bowell (2001)). They have been developed in the framework of statistical inversion theory which aims at characterizing the full probability-density functions for the parameters of the inverse problem.

Typically, LSL is applicable to objects with long observational time intervals and large numbers of observations. The solution of the statistical inverse problem is specified by the least-squares orbital elements and their covariance matrix computed at the least-squares point in the orbital-element phase space. It is partly the point-estimate characteristics of LSL that limit its applicability.

VoV offers a cure to the limitations of LSL by mapping the local least-squares solutions in the phase space as a function of one or more of the parameters. The local least-squares solutions of lesser dimensions than the original one allow local sampling of orbital elements. Trial orbital elements qualify for sample elements if they produce statistically acceptable fits to the observational data. VoV is limited by the requirement that partial derivatives need to be computed at the observation dates with subsequent matrix inversion for local covariance matrices. Should the inverse problem be poorly enough defined, VoV runs into difficulties with the matrix inversion.

Ranging offers a rigorous solution to the statistical inverse problem by exploring the full plausible orbital-element phase space for sample orbital elements. By exploring the topocentric ranges and angular elements (Right Ascension and Declination) at two observation dates, Ranging manages to map the full permissible region of the orbital elements without relying on partial derivatives, thus without facing numerical instabilities. Again, trial orbital elements qualify for sample elements if they produce acceptable fits to the data.

3.2. *Results*

During the course of the program (as of November 2006), four NEOs have been recovered, while improved orbits have been obtained for more than 76 objects. One of the follow-up objects was the PHO $2004\,AS_1$ whose drastic discovery-night prediction implied a possibility for a short-term (within 48 hours) Earth impact (Virtanen, Muinonen, Granvik *et al.* (2005), Virtanen & Muinonen (2006)) but which was quickly ruled out with new observations. The newly developed VoV technique was tested and proved useful in the follow-up work.

The MBO 2004 QR constitutes a serendipitous discovery by the NEON program. It was discovered on Aug. 15, 2004, and thereafter followed up on Aug. 17 (observational interval of 2 days; 4 observations), Aug. 22 (7 days; 10 observations), and Sept. 16 (31 days; 13 observations). The full observational interval comprises 19 observations over 39 days. Follow-up observations were carried out by making ephemeris predictions from the discovery night onwards using the statistical techniques: Ranging for the first three data sets and VoV for the last.

Figure 1 shows the time evolution of ephemeris predictions with increasing numbers of observations for the MBA 2004 QR discovered and followed up during the NEON

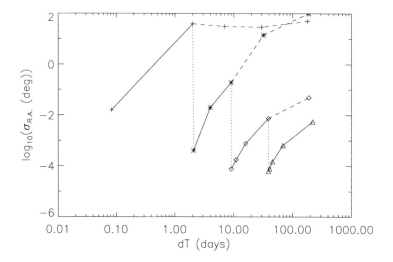

Figure 1. Time evolution of ephemeris uncertainty for 2004 QR. Standard deviation of the R.A. p.d.f.'s as a function of time elapsed from discovery for different lengths of the observational time interval, top to bottom: < 1 day (crosses), 2.0 (stars), 7.2 (diamonds), and 31 days (triangles). Solid-dashed curves show the evolution of ephemeris prediction, dashed part corresponds to the hypothetical evolution without the new observations (timing of which indicated with vertical dotted lines).

program. The collapses (dotted lines) mark the times of the new observations added to the data set. The uncertainty region becomes well-constrained after a week of observations, enabling follow-up observations to be planned for several months ahead. Note the decreasing slope in the increasing ephemeris uncertainty after each time new observations are added.

4. Spins and shapes

4.1. *Inverse methods*

A simple convex shape model is here constructed from the four quadrants of a sphere, the four halves of two cylinders, and two plane elements (SCyPe; Fig. 2). In the principal-axes reference frame of the shape model, the two plane elements are located on the southern and northern polar regions, being parallel to the equatorial plane defined by the two longest principal axes. The cylinder halves are located on four sides so that their axes are aligned with either one of the two longest axes of the shape. Finally, the four quadrants of the sphere join together the four cylinder halves. There are two parameters in the resulting shape model: the two aspect ratios b/a and c/a among the three principal axes $a \geqslant b \geqslant c$. Due to the convexity of the resulting shape, the disk-integrated brightness is simply the sum of the brightnesses from the components, that is, it is an analytical formula consisting of the disk-integrated brightness of the single sphere, the two cylinders, and the two plane elements.

For a semi-infinite plane-parallel medium of scatterers, the reflection coefficient R relates the incident flux density πF_0 and the emergent intensity I as

$$I(\mu, \mu_0, \phi) = \mu_0 R(\mu, \mu_0, \phi) F_0, \qquad (4.1)$$

where $\mu_0 = \cos \iota$ and $\mu = \cos \epsilon$, ι and ϵ being the angles of incidence and emergence as measured from the outward normal vector of the surface element, and where ϕ is the azimuthal angle of emergence (the azimuthal angle of incidence $\phi_0 = \pi$). The Lambert (subscript "L") and Lommel-Seeliger ("LS") reflection coefficients are

$$R_{\mathrm{L}}(\mu, \mu_0, \phi) = 1,$$

$$R_{\mathrm{LS}}(\mu, \mu_0, \phi) = \frac{1}{4}\tilde{\omega}P_{11}(\alpha)\frac{1}{\mu + \mu_0}, \qquad (4.2)$$

where $\tilde{\omega}$ is the single-scattering albedo, P_{11} is the scattering phase function, and α is the phase angle. The Lambert reflection coefficient is applicable to bright scattering media, even though it cannot be derived mathematically from, e.g., the radiative transfer theory. The Lommel-Seeliger reflection coefficient – as the first-order multiple-scattering approximation from the radiative transfer theory – is applicable to dark scattering media: the intensity terms $[\tilde{\omega}^k]$, $k \geqslant 2$ are assumed negligible. Note that a user-friendly realistic rough-surface scattering model is offered for future work by Parviainen & Muinonen (2006).

The disk-integrated brightness L of an asteroid equals the surface integral (e.g., Muinonen (1998))

$$L(\alpha) = \int_{\iota, \epsilon > 0} dA\, \mu I(\mu, \mu_0, \alpha)$$

$$= \int_{\iota, \epsilon > 0} dA\, \mu\mu_0 R(\mu, \mu_0, \alpha) F_0. \qquad (4.3)$$

For an irregularly shaped asteroid, L depends on the orientation of the asteroid with respect to the scattering plane, where L is measured.

The plane-element disk-integrated brightnesses for the Lambert and Lommel-Seeliger scattering laws follow, in a straightforward way, from Eqs. 4.2 and 4.3. For a spherical object (diameter D) with Lambert and Lommel-Seeliger reflection coefficients, the disk-integrated brightnesses are

$$L_{\mathrm{Ls}}(\alpha) = \frac{1}{6\pi}\pi F_0 D^2[\sin\alpha + (\pi - \alpha)\cos\alpha],$$

$$L_{\mathrm{LSs}}(\alpha) = \frac{1}{32}\pi F_0 D^2\tilde{\omega}P_{11}(\alpha)\left[1 - \sin\frac{1}{2}\alpha\,\tan\frac{1}{2}\alpha\,\ln\left(\cot\frac{1}{4}\alpha\right)\right]. \qquad (4.4)$$

For a cylindrical envelope (diameter D and length h, cylinder ends excluded) in its own natural reference frame ($x'y'z'$; with its axis along the z'-axis) and with Lambert and Lommel-Seeliger reflection coefficients, we obtain the disk-integrated brightnesses

$$L_{\mathrm{Lc}}(\alpha) = F_0\frac{1}{2}Dh\left[(\epsilon'_x\iota'_x + \epsilon'_y\iota'_y)\frac{1}{2}(\phi_2 - \phi_1) + (\epsilon'_x\iota'_x - \epsilon'_y\iota'_y)\frac{1}{4}(\sin 2\phi_2 - \sin 2\phi_1)\right.$$

$$\left. - (\epsilon'_x\iota'_y + \epsilon'_y\iota'_x)\frac{1}{4}(\cos 2\phi_2 - \cos 2\phi_1)\right],$$

$$L_{\mathrm{LSc}}(\alpha) = \frac{1}{4}\tilde{\omega}P_{11}(\alpha)F_0\frac{Dh}{2A}$$

$$\cdot\left[C_1(\sin\lambda_2 - \sin\lambda_1) - C_2(\cos\lambda_2 - \cos\lambda_1) + C_3\ln\frac{\cot(\frac{\pi}{4} - \frac{1}{2}\lambda_2)}{\cot(\frac{\pi}{4} - \frac{1}{2}\lambda_1)}\right],$$

$$A = \sqrt{(\epsilon'_x + \iota'_x)^2 + (\epsilon'_y + \iota'_y)^2}, \quad \cos\tilde{\phi} = \frac{\epsilon'_x + \iota'_x}{A}, \quad \sin\tilde{\phi} = \frac{\epsilon'_y + \iota'_y}{A},$$

$$\lambda_{1,2} = \phi_{1,2} - \tilde{\phi}, \qquad (4.5)$$

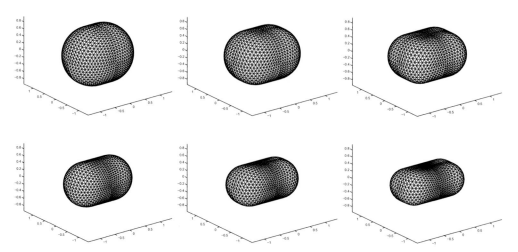

Figure 2. Sphere-Cylinder-Plane-element (SCyPe) example shapes: upper row, aspect ratios $b/a = c/a = 0.8$ (left), $b/a = 0.8$ and $c/a = 0.7$ (middle), $b/a = 0.8$ and $c/a = 0.6$ (right); lower row, $b/a = c/a = 0.6$ (left), $b/a = 0.60$ and $c/a = 0.525$ (middle), $b/a = 0.60$ and $c/a = 0.45$ (right).

where ϕ_1 and ϕ_2 denote the azimuths of the terminators on the cylindrical envelope and

$$
\begin{aligned}
C_1 &= (\epsilon'_x \iota'_x - \epsilon'_y \iota'_y)\cos 2\tilde{\phi} + (\epsilon'_x \iota'_y + \epsilon'_y \iota'_x)\sin 2\tilde{\phi}, \\
C_2 &= -(\epsilon'_x \iota'_x - \epsilon'_y \iota'_y)\sin 2\tilde{\phi} + (\epsilon'_x \iota'_y + \epsilon'_y \iota'_x)\cos 2\tilde{\phi}, \\
C_3 &= \epsilon'_x \iota'_x \sin^2 \tilde{\phi} + \epsilon'_y \iota'_y \cos^2 \tilde{\phi} - (\epsilon'_x \iota'_y + \epsilon'_y \iota'_x)\sin \tilde{\phi} \cos \tilde{\phi}.
\end{aligned}
\tag{4.6}
$$

Finally, in order to compute the total disk-integrated brightness adhering to the SCyPe model, the various rotations need to be accounted for. The present numerical implementations have been carefully checked both internally and against other independent implementations to compute disk-integrated brightnesses.

The statistical inversion of observed photometric brightnesses for spins and shapes is carried out by sampling the spin parameters and by fitting the shape parameters for each trial spin solution. When the trial spin-shape solution provides a statistically acceptable fit to the observations, it qualifies for a sample solution. Obtaining a large number of sample solutions allows one to characterize the solution space within the present SCyPe model. The complete probabilistic treatment requires, additionally, the derivation of weights for the sample solutions.

It is clear that SCyPe will not generally lead to rms fits as good as those from conventional inversion (Kaasalainen, Torppa & Muinonen (2001), Kaasalainen, Mottola & Fulchignoni (2002), Torppa (1999)). However, the application of SCyPe models is some two orders of magnitude faster than the application of the conventional methods.

4.2. Results

In the case of the previously observed Apollo asteroids – (1685) Toro, (1981) Midas, and (1862) Apollo – the amount of data is large enough for using the conventional inversion technique. All three objects represent moderate albedos (Toro and Midas as S-type and Apollo as Q-type). The previous photometric observations were extracted from Standard Asteroid Photometric Catalog SAPC http://www.astro.helsinki.fi/SAPC/). We have compared our results to the previous ones available at the European Asteroid Research

Node's (EARN) Near-Earth Asteroid Database (http://earn.dlr.de/nea/), which is an update and extension of that by Binzel, Lupishko, Di Martino, *et al.* (2002). For the four new objects (2002 FF$_{12}$, 2003 MS$_2$, 2003 RX$_7$, 2004 HW), the spin and shape were studied using SCyPe.

In what follows, we summarize the results for the NEOs included in this paper, with pole directions in ecliptic longitude λ and latitude β:

(1685) Toro

In addition to the two new curves from this observing program, we used one unpublished curve observed with the La Silla Danish 1.54-m telescope and ten previously observed lightcurves by Dunlap, Gehrels & Howes (1973). There exist also five lightcurves from 1988 by Hoffmann & Geyer (1990), but it was impossible to get a good fit when they were combined with either the Dunlap data or our NOT and La Silla observations. Also, since the new lightcurves are from the same view point and observing geometry as Hoffmann and Geyer's data, we left the latter unused. There were thus 13 lightcurves available from a time interval of 31 years. The Earth-centered ecliptic longitude of Toro ranges 15° for NEON observations and 80° for Dunlap's observations. No unambiguous pole solution could be obtained, and the distribution of the possible spin states was obtained using the conventional inversion method. Results with rms error greater than 0.03 mag can be discarded due to the high accuracy of the data. The pole latitude was constrained to negative values, i.e., retrograde rotation, whereas the pole longitude is restricted to $\lambda \in [250°, 120°]$. Except for the period, the previous pole solution by Dunlap, Gehrels & Howes (1973) ($\beta = 55°$, $\lambda = 200°$ and $P = 10.196$ h) disagrees with our findings.

(1862) Apollo

For the analysis of Apollo, we used 23 previously observed lightcurves by Harris, Young, Goguen, *et al.* (1987) and Hahn (1983) (UAPC), two unpublished curves from the Ondrejov NEO program (Pravec, Wolf & Šarounová (1998)), and one new lightcurve from this program. The Earth-centered ecliptic longitude of the previously observed data sets range 65° and 85°, and the new curve from the NOT observations as well as the Ondrejov curves provide information from two more view points. The total time range of the observations increased to 25 years along with the new lightcurve, thus increasing also the accuracy of the period solution. We used the conventional inversion method for the data analysis, and one spin solution produced clearly the best fit with an rms of 0.036 mag being of the same order as the noise of the data. Spin values for this solution were $\beta = -50°$, $\lambda = 20°$, and $P = 3.0662$ h, which is not far from the previous solution $\beta = -26°$, $\lambda = 56°$ and $P = 3.065$ h by Harris, Young, Goguen, *et al.* (1987). Although the two rms minima in the period plot are equal, the difference becomes larger when fitting the final shape model. The nominal shape solution is shown in Fig. 3 – the detailed error analysis is left for the future.

(1981) Midas

For the analysis of Midas, we used six previously observed lightcurves: one by Wisniewski, Michalowski, Harris *et al.* (1997) and five by Mottola, de Angelis, di Martino *et al.* (1995). The two new lightcurves from this program increased the total time range of the observations to 18 years, and the number of observing geometries to three. A distribution for the possible spin solutions was obtained using the conventional inversion method. The former period solution is 5.22 hours (Wisniewski, Michalowski, Harris *et al.* (1997) and Mottola, de Angelis, di Martino *et al.* (1995)), which is in the error bars of the one (5.215 ± 0.035 h) obtained by us. In terms of rms of the fits, there exist no clear

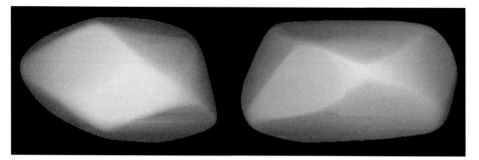

Figure 3. Convex shape model for (1862) Apollo (Torppa, Virtanen & Muinonen *et al.* (2006)).

minima. Due to the low accuracy of the data, we have to accept all the solutions below the rms-value of 0.04 mag.

2002 FF$_{12}$

We had one 5.8-h lightcurve of 2002 FF$_{12}$, which by chance covers at least part of one maximum. From these data we can see only by inspecting the lightcurve, without any specific model, that the period is most probably more than 8.5 hours. The SCyPe method was applied including also period determination with sampling from 8 to 20 h. The forbidden regions in the pole space turn out to be small. The shape is not constrained either and values up to 0.95 are allowed for the aspect ratios due to the small amplitude of the lightcurve (0.3 mag). However, the dependence of the shape on the pole direction is noted. The smallest rms values are obtained with periods from nine to ten hours, but the distribution is flat and values larger than ten hours have to be taken into account as well.

2003 MS$_2$

We observed one 5.9-h lightcurve of 2003 MS$_2$ (Fig. 4), and the period according to this is approximately 7 hours. The distribution of possible spin-axis solutions obtained with the SCyPe model are also depicted in Fig. 4. There are two large forbidden regions in β vs. λ: the possible solutions are constrained into two narrow rings around the forbidden regions. Note that the density of the points does not correspond to the probability of the solution. In addition, the shape is quite well constrained due to the large amplitude of the lightcurve (0.7 mag); maximum value for b/a is 0.67. However, since only one curve, assumingly over one rotation, was available for the analysis, one must be cautious when using these spin and shape estimates for further studies. The results are described in more detail in Torppa, Virtanen & Muinonen *et al.* (2006); with the present pilot study, we want to show that it is indeed possible to derive information about an object from a small amount of data.

2003 RX$_7$

We observed one 6-hour lightcurve of 2003 RX$_7$, which shows a period of about 2.6 hours. The SCyPe method was applied to obtain distributions of the spin and shape parameters. The results are quite similar to those of 2002 FF$_{12}$, but the forbidden regions in β vs. λ are more clear. The amplitude of the lightcurve is 0.2 mag, and thus the aspect ratios are allowed to obtain values up to 0.92. Best-fit solutions are distributed evenly all over the region of possible solutions.

2004 HW

Five lightcurves of 2004 HW were observed showing the period of 2.52 hours. In the curve observed on Sept. 15, we see an increasing trend in the magnitude. This may be

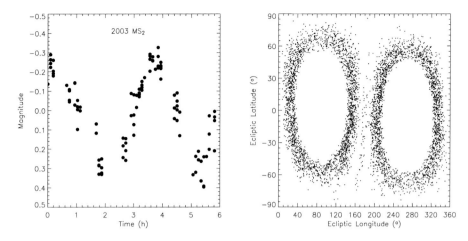

Figure 4. Relative R-filter photometry (left) and spin-axis solutions for 2003 MS$_2$ (right).

due to a satellite (cf. Pravec, Harris & Warner (2006); or cometary activity), since no change in magnitude should be present due to scattering behaviour; the change in phase angle at epoch of observations was 0.5° per day. The lightcurves were observed within a short time (one month apart), and a distribution of possible pole and shape solutions was obtained by applying the SCyPe method. Best-fit solutions settle mostly around pole longitudes $\lambda = 200°$ and $\lambda = 20°$. Shape is strongly dependent on the pole solution.

5. Conclusions

We have reviewed the progress of the NEON program for characterizing physical and dynamical properties of potentially hazardous near-Earth objects. Observations at the Nordic Optical Telescope have resulted in improved understanding of the spins, shapes, and orbits of a number of near-Earth objects. In particular, R-filter relative magnitudes have been obtained for a considerable number of objects, keeping the door open for an absolute calibration in the future.

A novel statistical method based on SCyPe shape models (Sphere-Cylinder-Plane-element) has been outlined for the inversion of spins and shapes from photometric observations, allowing for detailed error analyses. Furthermore, SCyPe modeling can allow the optimization of future photometric observations. The forbidden zones from single lightcurves suggest that the subsequent observations could be made at ecliptic longitudes and latitudes covered by the permissible zones. If such observational circumstances are available, one may obtain additional constraints on the pole orientation.

Acknowledgements

We thank the Nordic Optical Telescope (NOT) and Dk1.54 for granting us observing time. Most of the data utilized have been obtained using ALFOSC, which is owned by the Instituto de Astrofisica de Andalucia (IAA) and operated at the Nordic Optical Telescope under agreement between IAA and the NBIfAFG of the Astronomical Observatory of Copenhagen. The study has been funded by Vilho, Yrjö and Kalle Väisälä foundation, the Academy of Finland, and the Finnish Graduate School in Astronomy and Space Physics. Finally, we would like to thank the anonymous reviewer for constructive criticism.

References

Alvarez-Candal, A., Duffard, R., Angeli, C. A., Lazzaro, D., & Fernandez, S. 2004, *Icarus* 172, 388

Binzel, R. P., Lupishko, D. F., Di Martino, M., Whiteley, R. J., & Hahn, G. J. 2002, in: W. F. Bottke, A. Cellino, P. Paolicchi, R. P. Binzel (eds.), *Asteroids III* (University of Arizona Press), p. 255

Bowell, E., Virtanen, J., Muinonen, K., & Boattini, A. 2002, in: W. F. Bottke, A. Cellino, P. Paolicchi, R. P. Binzel (eds.), *Asteroids III* (University of Arizona Press), p. 27

Dunlap, J. L., Gehrels, T. & Howes, M. L. 1973, *Astron. J.* 78, 491

Granvik, M. & Muinonen K. 2005, *Icarus* 179, 109

Hahn, G. 1983, in: C.-I. Lagerkvist, H. Rickman (eds.), *Asteroids, Comets, Meteors* (Uppsala University), p. 35

Harris, A. W., Young, J. W., Goguen, J., Hammel, H. B., Hahn, G., Tedesco, E. F., Tholen, D. J., *et al.* 1987, *Icarus* 70, 246

Hoffmann, M. & Geyer, E. H. 1990, *Acta Astronomica* 40, 389

Kaasalainen, M. 2004, *A&A* 426, 1103

Kaasalainen, M. & Torppa, J. 2001, *Icarus* 153, 24

Kaasalainen, M., Torppa, J. & Muinonen, K. 2001, *Icarus* 153, 37

Kaasalainen, M., Mottola, S. & Fulchignoni, M. 2002, in: W.F. Bottke, A. Cellino, P. Paolicchi, R.P. Binzel (eds.), *Asteroids III* (University of Arizona Press), p. 139

Lagerkvist, C.-I., Piironen, J. & Erikson A 2001, *Uppsala Asteroid Photometric Catalogue*, Fifth update (Uppsala Astronomical Observatory)

Mottola, S., de Angelis, G., di Martino, M., Erikson, A., Hahn, G. & Neukum, G. 1995, *Icarus* 117, 62

Muinonen, K. 1998, *A & A* 332, 1087

Muinonen, K. & Bowell, E. 1993, *Icarus* 104, 255

Muinonen, K., Virtanen, J. & Bowell, E. 2001, *Cel. Mech. Dyn. Astron.* 81(1), 93

Muinonen, K., Virtanen, J., Granvik, M., & Laakso, T. 2005, in: *Proceedings of the symposium Three Dimensional Universe with Gaia* (ESA SP-576), p. 223

Muinonen, K, Virtanen, J., Granvik, M., & Laakso, T. 2006, *MNRAS* 368, 809

Parviainen, H. & Muinonen, K. 2006, *J. Quantit. Spectrosc. Radiat. Transf.*, in press

Pravec, P., Wolf, M., & Šarounová, L. 1998, *Icarus* 136, 124

Pravec, P., Harris, A. W. & Warner, B. 2006, in: A. Milani, G. B. Valsecchi, D. Vokrouhlický (eds.), *Proceedings of the IAU Symposium 236: Near-Earth Objects, our Celestial Neighbors: Opportunity and Risk*, in press

Raab H. 2004, Astrometrica, http://www.astrometrica.at/

Tedesco, E., & Zappalà, V. 1980, *Icarus* 43, 33

Torppa J., 1999, Asteroid lightcurve inversion: methods for obtaining a unique and stable shape solution, MSc thesis, Univ. Helsinki

Torppa, J. & Muinonen, K. 2005, in: *Proceedings of the symposium Three Dimensional Universe with Gaia* (ESA SP-576), p. 321

Torppa, J., Kaasalainen, M., Michalowski, T., Kwiatkowski, T., Kryszczyńska, A., Denchev, P., & Kowalski, R. 2003, *Icarus* 164, 346

Torppa, J., Virtanen, J., Muinonen, K., Laakso, T., Niemelä, J., Näränen, J., Aksnes, K., Dai, Z., Lagerkvist, C.-I., Rickman, H., Hahn, G., Michelsen, R., Grav, T., Pravec, P., & Jörgensen, O. G. (2006), *Icarus*, submitted

Virtanen, J. 2005, *Asteroid orbital inversion using statistical methods*, PhD thesis (University of Helsinki)

Virtanen, J. & Muinonen, K. 2006, *Icarus* 184, 289

Virtanen, J., Muinonen, K. & Bowell, E. 2001, *Icarus* 154, 412

Virtanen, J., Tancredi, G., Muinonen K., & Bowell, E. 2003, *Icarus* 161, 419

Virtanen, J., Muinonen, K., Granvik, M., Laakso, T. 2005, in: Z. Kneževič & A. Milani (eds.), *Proceedings of IAU Colloquium 197: Dynamics of Populations of Planetary Systems*, p. 239

Virtanen, J., Tancredi, G., Bernstein, G. M., Spahr, T., & Muinonen, K. 2006, in: A. Barucci, H. Boehnhardt, D. Cruikshank, A. Morbidelli (eds.), *Kuiper Belt*, submitted

Wisniewski, W. Z., Michalowski, T. M., Harris, A. W., & McMillan, R. S. 1997, *Icarus* 126, 395

Part 5

Surveys: Orbit Determination and Data Processing

Near Earth Objects, our Celestial Neighbors: Opportunity and Risk
Proceedings IAU Symposium No. 236, 2006
A. Milani, G.B. Valsecchi & D. Vokrouhlický, eds.
© 2007 International Astronomical Union
doi:10.1017/S1743921307003390

Current NEO surveys

Stephen Larson

Lunar and Planetary Laboratory, University of Arizona, Tucson, AZ 85721, USA
email: slarson@lpl.arizona.edu

Abstract. The state and discovery rate of current NEO surveys reflect incremental improvements in a number of areas, such as detector size and sensitivity, computing capacity, detection software efficiency and availability of larger telescope apertures. The result has been an increase in the NEO discovery rate. There are currently eight telescopes ranging in size from 0.5-1.5 meters carrying out full- or part-time systematic surveying in both hemispheres. The sky is covered 1-2 times per lunation to V$\tilde{1}$9, with a band near the ecliptic to V$\tilde{2}$0.5. We review the current survey programs and their contributions towards the Spaceguard goal of discovering at least 90% of the NEOs larger than 1 km.

Keywords. Surveys; telescopes; discoveries

1. Establishment and Evolution of NEO Surveys

The identification of the Chicxulub impact crater as the source of the iridium anomaly at the Cretaceous-Tertiary boundary (e.g., Hildebrand *et al.* 1991) provided strong evidence of the ongoing process of impacts as an important agent of evolution of life on Earth. This, coupled with the emergence of appropriate technology to detect moving asteroids, triggered a directive from NASA to quantify the impact threat of objects large enough to create global consequences. The result was a congressional mandate to discover and catalog to the 90% confidence level near-Earth objects one kilometer in diameter and larger by the end of 2008-the Spaceguard Survey.

The original Spaceguard Survey study recommended the construction of six 2.5-m telescopes in a coordinated international network in both northern and southern hemispheres (Morrison 1992) at an estimated cost of US $50M to build and US $10M per year to operate. A second NASA workshop report (Shoemaker 1995) studied in greater detail existing telescopes and more NASA/USAF cooperation, and effectively defined the budget level of the NASA Near Earth Objects Observation Program (US $4 Myr). The available funds dictated the use of existing, little-used telescopes outfitted with modern CCDs and modern computer control, which has resulted in various evolutionary paths and timescales among the surveys.

The characteristics of the current NASA-supported NEO surveys are a result of a range of entrepreneurial approaches and technology development to carry out the mandate. With attention turned towards the next phase of deep, extensive surveys for smaller NEOs that can cause regional damage, changes to the current surveys have slowed, and the characteristics of the current Spaceguard Survey capabilities may represent a stable end state for the existing surveys.

2. Current Survey Characteristics

We briefly summarize the current characteristics of the surveys in order in which they originally came on line with details listed in Table 1. They can be compared with

Tables 1 and 2 in the chapter in Asteroids III by Stokes *et al.* (2002). The URL for the various surveys are included for more information.

2.1. *Spacewatch*

The first CCD small-bodies survey was begun by the University of Arizona's Lunar and Planetary Laboratory Spacewatch group using the Steward Observatory 0.9-m reflector on Kitt Peak. It was originally used in 1984 in a drift-scan mode using a single CCD, but was upgraded in 2002 with a mosaic camera and new telescope optics to provide a larger field. A 1.8-m telescope was completed in 2001, which is used mostly for follow-up of fainter NEOs (Spacewatch.lpl.arizona.edu).

2.2. *NEAT*

The Near Earth Asteroid Tracking project of the Jet Propulsion Laboratory utilizes the Oschin 1.2-m Schmidt telescope with its wide-field Quest camera on Palomar Mountain for the first half of each lunation. The NEAT survey originally used 1.0-m GEODSS and 1.2-m MOTIF telescopes on the Air Force Maui Optical Station. It developed autonomous data acquisition, reduction, and detection software that allowed remote operation and vetting of NEO candidates from JPL. The survey also developed the Sky-Morph online archive, which facilitates searches for pre-discovery images of new NEOs (neat.jpl.nasa.gov; skys.gsfc.nasa.gov/; skymorph/skymorph.html).

2.3. *LONEOS*

The Lowell Observatory Near Earth Object Survey uses a 0.6-m wide-field Schmidt telescope at the Anderson Mesa site for dedicated, full-time NEO searching. The mosaic camera gives a large 8.3 square degree field (asteroid.lowell.edu/asteroid/loneos/loneos1.html).

2.4. *LINEAR*

The Lincoln Lab's Near Earth Asteroid Research Program uses two identical 1.0-m GEODSS telescopes at the Experimental Test Site at the north end of the White Sands Missile Range near Socorro, New Mexico. It utilizes very fast, frame-transfer readout CCD arrays to cover large swaths of sky each night. LINEAR became the dominant NEO survey in 1998, and is responsible for the vast majority of NEO discoveries. It utilizes five visits per field, and is the only survey that regularly searches in the galactic plane and high north ecliptic latitudes (www.ll.mit.edu/LINEAR/).

2.5. *Catalina Sky survey*

The University of Arizona's Lunar and Planetary Laboratory Catalina Sky Survey (CSS) uses a wide-field 0.7-m Schmidt and 1.5-m reflector in the Santa Catalina Mountains north of Tucson, and the 0.5-m Uppsala Schmidt in Siding Spring Observatory in New South Wales, Australia. These three components provide complementary characteristics in terms of field, depth, and sky coverage, while sharing the same control and detection software. Since being upgraded with thinned, sensitive CCDs in late 2004, the CSS has led in the discovery of NEOs. The CSS relies heavily on the observer to make real-time decisions on where to survey, and to validate the reality of NEO candidates flagged by the software. Software tools help the observer make same-night follow-up of likely NEOs to check validity of the objects and extend the observed arc for subsequent follow-up (www.lpl.arizona.edu/css).

Program	Spacewatch	NEAT	LONEOS	LINEAR	Catalina	Siding Spr.	Mt.Lemmon
Observatory	Kitty Peak	Palomar	Lowell	Socorro	Catalina	Siding Spr.	Mt.Lemmon
Aperture	0.93 m	1.2 m	0.6 m	1.0 m	0.68 m	0.5 m	1.5 m
f ratio	3.0	2.5	1.9	2.2	1.8	3.4	2.0
FOV	2.9	7.0	8.3	2.0	8.2	4.2	1.2
No. CCD	4	112	2	1	1	1	1
CCD size K	2×4	2.4×0.6	2×4	2×2.6	4×4	4×4	4×4
V limit	21.7	22.0	18.9	19.0	19.5	19.0	21.5
No. visits	3	3	4	5	4	4	4
Exposure, s	120	60	45	8	30	30	20
Coverage rate	15	70	110	120×2	120	60	18
Recent results 1/2005–6/2006							
No. all NEOs	128	41	55	191	230	83	210
No. > 1 km	12	9	5	25	27	8	9
No. PHAs	16	9	10	29	35	25	13

Table 1. Characteristics and recent 18 month results of the current NEO search program telescopes.

3. Need for Follow-up

Because it typically requires 24-48 hours of observation to reasonably define an NEO orbit, rapid follow-up is an integral part of NEO discovery. The efforts of many amateur observers worldwide provide the bulk of follow-up positions, and are important in preventing NEOs from becoming lost. With sensitive, commercial-science-grade CCD cameras and sophisticated, computer-controlled, and sometimes robotic telescopes, amateur astronomers can do what professionals could not do 10 years ago. There are some amateurs who can regularly reach V=21 with modest apertures using stack-and-add techniques.

The MPC NEO Confirmation Page (cfa-www.harvard.edu/iau/NEO/ToConfirm.html) and the Minor Planet Mailing List are powerful communications tools for both amateur and professional observers. It is fair to say that without the drive and dedication of these many volunteer observers, the Spaceguard goal would be out of reach.

For the fainter objects, the JPL Table Mountain 0.6-m, Mt. John 0.64-m, Klet Observatory Klenot 1.1-m, Spacewatch 1.8-m, and Mt. Lemmon 1.5-m are used regularly for follow-up.

Extended follow-up on timescales of weeks and months is usually required for subsequent return recoveries, and may become critical in PHAs not becoming lost.

4. Results

4.1. Discovery Rate

As Figure 1 shows, the discovery rate varies with time with each survey according to its technical status. Taken as an ensemble, the plot of all NEOs shows an increasing trend in discovery rate throughout the Spaceguard period, while the $H < 18$ NEOs show the expected decrease as an increasing proportion of the population becomes known.

4.2. Coverage

The number and efficiency of the surveys means that the sky is being covered almost twice per clear lunation, with the ecliptic covered more often (Fig. 2). There is currently little coordination between the surveys with LINEAR systematically covering the observable sky in a pre-planned sequence (unless affected by weather), while Spacewatch

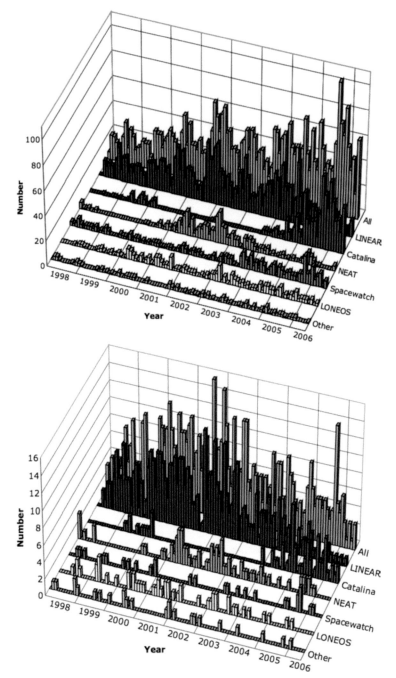

Figure 1. Monthly discoveries of all NEOs (upper) and $H < 18$ NEOs (lower) for the surveys.

concentrates in the opposition regions, and the Catalina Sky Survey make nightly decisions based on covering areas not recently observed. Movies of the nightly build-up of coverage for some example lunations can be found on the CSS web site.

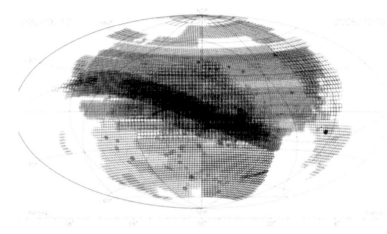

Figure 2. Sky coverage for all surveys during the 2006 September 10 to October 7 lunation.The ecliptic plane near opposition is covered multiple times, while the galactic plane is covered only by LINEAR. This plot is courtesy of the Minor Planet Center.

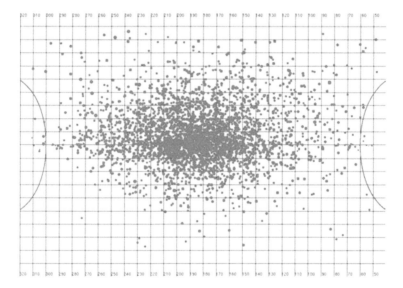

Figure 3. The discovery locations for all NEOs in ecliptic coordinates with respect to the Sun. Arcs at each edge are 60° from the Sun.

4.3. *Discovery Circumstances*

The discovery circumstances are dictated by a complicated convolution of coverage, limiting magnitude, and the intrinsic distribution of NEOs and their phase effects. Figure 3 shows the distribution of the discovery location of all NEOs in ecliptic longitude and latitude with respect to the Sun. The N-S asymmetry results from the greater coverage in the north, while the higher density near opposition is a combination of opposition effect and increased coverage. The expected increase near sun "sweet spots" is not apparent for the magnitude range or coverage represented here.

4.4. *Survey Efficiency*

Significant advances in computing power, detector sensitivity and effective array sizes have made the current survey telescopes as efficient as those recommended in the

Spaceguard Report. Although there may be incremental improvements in detection software in the remainder of the Spaceguard Survey, it is not likely that more aperture or larger detector arrays will come on line as most of the improvement in NEO surveying is being directed towards the next goal of finding and cataloging NEOs down to 140-m sizes.

Although it is generally accepted that the ideal survey telescope is fast (to minimize trailing losses during exposure), large field, and minimal cycle time, the current discovery results shown in Table 1 fail to show a clear correlation among the survey systems.

4.5. *Prospects for Attaining the Spaceguard Goal*

Studies are underway to estimate the population of $H < 18$ NEOs based upon current statistics and of the discovery/re-observation ratio. As of this writing, there are 845 such NEOs out of an estimated $1050+/-60$ (Boattini *et al.* 2006). Given the relatively high efficiency the surveys have attained, it may still be possible to satisfy the Spaceguard goal in the two remaining years of the survey.

Acknowledgements

The author thanks the various surveys for their updates and input for this review. Alan Chamberlin (JPL) provided the monthly statistics in fig.1, and Gareth Williams (MPC) provided a new tool in the MPC NEO Coverage tool that allowed the generation of Fig. 2. Ed Beshore, Eric Christensen and Rik Hill provided help with the graphics. This work is carried out through the support of the NASA NEO Program.

References

Boattini, A, Abramo, G. D., Harris, A. W. & Valsecchi, G. B. 2006, *BAAS* 38, 581

Hildebrand, A. R., Penfield, G. T., Kring, D. A., Pilkington, M., Zanoguera, A. C., Jacobsen, S. B., & Boynton, W. V. 1991, *Geology* 19, 867–871

Morisson, D. 1992, *The Spaceguard Survey. Report of the International Near-Earth-Object Detection Workshop*, NASA Office of Space Science, Solar System Exploration Office.

Shoemaker, E. M. 2002, *Report of the Near-Earth Objects Survey Working Group. NASA Office of Space Science, Solar System Exploration Office.*

Stokes, G. H., Evans, J. B. & Larson, S. M. 2002, in: W. F. Bottke, A. Cellino, P. Paolicchi & R. P. Binzel (eds.), *Asteroids III* (University of Arizona Press, Tucson), p. 45

Near Earth Objects, our Celestial Neighbors: Opportunity and Risk
Proceedings IAU Symposium No. 236, 2006
A. Milani, G.B. Valsecchi & D. Vokrouhlický, eds.
© 2007 International Astronomical Union
doi:10.1017/S1743921307003407

Spacewatch preparations for the era of deep all-sky surveys

Robert S. McMillan[1] and The Spacewatch Team[1]

[1]Lunar and Planetary Laboratory and Steward Observatory, University of Arizona, Tucson, AZ 85721, USA
email: bob@lpl.arizona.edu

Abstract. The Spacewatch Project at the University of Arizona uses a 0.9-meter and a 1.8-meter telescope to search for new Near-Earth Objects (NEOs) and make astrometric followup measurements of known ones. Among the presently operational asteroid astrometry programs, Spacewatch is uniquely suited to support discoveries by the planned deep all-sky surveys. The Spacewatch 1.8-meter telescope is the largest in the world that is used exclusively for observations of asteroids and comets. Since 2003 January 1, Spacewatch has made ∼2400 separate-night detections (discoveries plus followup) of NEOs with absolute magnitude H⩽22, including 117 fresh discoveries of NEOs with H⩽22 and ∼900 separate-night detections of Potentially Hazardous Asteroids (PHAs). Objects have been recovered at V=23 and at elongations less than 60 degrees from the Sun. Spacewatch followup observations have contributed to the removal of 137 objects from JPL's impact risk website. Examples of notable recoveries by Spacewatch include the extension of orbital arcs from one month to multi-opposition orbits, and a successful targeted search for a large PHA (1990 SM) with 80 degrees of uncertainty. Spacewatch has been making as many observations of PHAs with H⩽22 and V>21 as all other followup stations combined. Followup of NEOs while they are not near Earth provides better leverage on orbital elements and will be well suited to follow up some of the discoveries by the larger-scale, deeper sky surveys: both ground- and space-based. Spacewatch is collaborating with the Panoramic Survey Telescope and Rapid Response System (PS) of the University of Hawaii's Institute for Astronomy. Each lunation, Spacewatch sends its listings of point sources detected in survey images for PS's moving object detection team to test their software. Spacewatch is also prepared to follow up objects of special interest, fast motion, or less than three nights of observations by PS itself. Spacewatch's current equipment is only a few years old, but there is still room to improve limiting magnitude & time efficiency.

Keywords. Telescopes; catalogs; survey; astrometry; discoveries

---

1. GOALS and CONTRIBUTIONS of SPACEWATCH

The goals of the Spacewatch Project are to discover and observe small bodies in the solar system and to analyze the distributions of their orbits and absolute magnitudes. Astrometric imaging observations are scheduled an average of 24 nights per lunation with the 0.9-meter and 1.8-meter Spacewatch Telescopes at Steward Observatory on Kitt Peak mountain in the Tohono O'odham Nation, Arizona. Spacewatch discoveries and detections provide information about the dynamical history of the solar system. Analyses of the Centaur (Jedicke & Herron 1997), Main-Belt (Jedicke & Metcalfe 1998), Trans-Neptunian (Larsen *et al.* 2001; Roe *et al.* 2005; Larsen *et al.* 2006a); and Near-Earth Object (NEO) populations (Rabinowitz 1991; Rabinowitz 1993; Jedicke 1996; Bottke *et al.* 2002; Morbidelli *et al.* 2002; McMillan, Block & Descour 2005; Larsen *et al.* 2005; Larsen *et al.* 2006b) have used Spacewatch observations. Spacewatch also finds and follows up potential targets for interplanetary spacecraft missions (McMillan 1999a; McMillan 1999b) and radar observations (Ostro *et al.* 2003), and finds and follows objects that might present a hazard to the Earth. Spacewatch provides hundreds of thousands of astrometric and photometric observations of asteroids annually, and recovers and does astrometry of high-priority comets and asteroids that are too faint for most other asteroid observing stations. This report is focused on the recent and future contributions of Spacewatch to campaigns to characterize the threat of impacts of asteroids on the Earth.

2. FACILITIES and METHODS

Site: Moving objects are discovered by imaging the sky with CCD detector arrays on a 0.9-meter telescope and a 1.8 meter telescope on Kitt Peak. This location has several advantages. Except in the summer when the ecliptic is unfavorably placed anyway, the climate rarely causes cloudy conditions for more than 2 or 3 days in a row. This helps the study of time dependent phenomena. The proximity to Tucson causes ∼0.5 mag loss of sensitivity on moonless nights, but it also makes excellent infrastructure possible. The relatively short commuting time from Tucson is favored by daytime technical support personnel.

Usage: The two Spacewatch telescopes complement each other, with the 0.9-m and its wider field of view operating in a systematic search pattern near opposition and the narrower-field 1.8-m concentrating on followup of specific targets. The wider field of the mosaic of CCDs on the 0.9-meter telescope has also recovered lost PHAs. The image scale on both telescopes is 1.0 arcsec per pixel, which samples the typical seeing at this site well. Three images, or "passes", are made at short intervals to reveal moving objects.

Optics: R. A. Buchroeder designed the coma corrector/field flattener for the 1.8-meter f/2.7 paraboloidal primary mirror, based on a modified Klee prescription that prevents light reflected off the CCD from forming a ghost image of the telescope entrance pupil onto the CCD. Because the CCDs manufactured by Tektronix in the 1990s were not flat, but slightly convex, the optical prescription for the corrector at the 1.8-meter telescope accommodates the measured spherical component of the shape of the CCD surface. *This fact must be remembered if the detector at the 1.8-meter telescope is ever changed!* The lenses were fabricated by Tucson Optical Research Corporation. Buchroeder also designed the new (2002) f/3 optics for the 0.9-meter telescope, which includes a spin-cast primary mirror from Wangsness Optics of Tucson, AZ and a multi-element field lens fabricated by Cumberland Optics of Marlow Heights, MD. Unlike the 1.8-m mirror, which has a conventional coating of evaporated aluminum, the 0.9-m primary is silvered and protected by a red-optimized overcoating by Denton Vacuum of Moorestown, NJ. Rayleigh Optics of Baltimore, MD figured the 0.9-m mirror to a hyperboloidal surface.

Filtering: Both telescopes are filtered with Schott OG-515 filters that transmit from 515 nm to the long-wavelength cutoff of the CCDs. The effective wavelength on typical asteroids is ∼700 nm. However, Spacewatch photometry is still calibrated to the V bandpass for consistency with the absolute magnitude system used by the asteroid community. Reasons for the yellow-orange filter are primarily simplicity of optical prescription and ease of fabrication of the field correction lenses, allowing the use of flint glass for achromatism and all spherical surfaces in the lens designs. The filtering also provides cleaner images at high airmass without the need for atmospheric dispersion compensation, suppresses the mostly-blue twilight and scattered moonlight, and suppresses color equation between stars and asteroids in astrometric field modeling at high airmass. The filtering does not cost much light from asteroids shortward of 515 nm, especially when atmospheric dispersion of the shorter wavelength light is taken into account. Finally, glass colloidal filters are more stable and less expensive than interference filters, and are better suited to fast f/numbers.

Detectors: The back-illuminated, antireflection-coated 2048x2048 CCDs we used until 2002 with great success at the 0.9-m telescope and which we still use at the 1.8-m telescope were made by Tektronix (later Scientific Imaging Technologies, SITe®) of Beaverton, OR. They have high quantum efficiency, noise well below the sky background, and have never malfunctioned for us. For the 0.9-m telescope we now have six grade-one back-illuminated, antireflection-coated 4608x2048 CCDs from Marconi Applied Technologies (later EEV or E2V) of Chelmsford, Essex, UK, from which we selected the best four for our mosaic system.

0.9-meter Telescope:

In 1982 the Director of the Steward Observatory allocated the Observatory's 0.9-m telescope exclusively to the Spacewatch Project on a long-term basis, on the condition that technical support and maintenance of the telescope and dome be funded by grants obtained by and for the Spacewatch Project. Spacewatch personnel have rebuilt and upgraded many components and subsystems of this telescope over the years, making it a world-class tool for solar system research. In late 2002 the Spacewatch mosaic camera (McMillan *et al.* 2000) replaced our earlier drift-scan system (McMillan & Stoll 1982; McMillan *et al.* 1986; Gehrels *et al.* 1986; Gehrels 1991; Scotti 1994) on the same telescope and boosted our rate of detection of NEOs with that telescope by a factor of six. The mosaic of four CCDs on the 0.9-meter telescope covers a solid angle of 2.9 deg^2 (Figures 1 and 2). The history of hardware and software used on the 0.9-meter telescope is tabulated at http://spacewatch.lpl.arizona.edu/history.html .

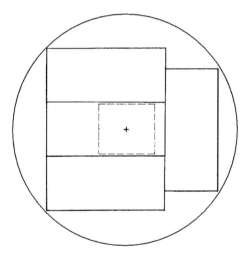

Figure 1. Comparison of the sizes and layouts of the CCDs on the 0.9-m and 1.8-m telescopes. The mosaic of four CCDs on the 0.9-m covers 2.9 square degrees, nine times larger than the area of the 2Kx2K CCD (small square) previously used on the 0.9-m and currently used on the 1.8-m.

Observations are made in the tracked "staring" mode because imagers this large are incompatible with drift scanning. The cycle goes as follows. We expose for two minutes on each position. Each exposure is followed by a 1.5-2 min interval to read the CCDs and slew and settle the telescope and dome to the next center. While the exposure time of 2 minutes may seem long for NEO work, it is the best compromise between sky coverage and open-shutter on-sky efficiency. Reading and slewing cannot be simultaneous owing

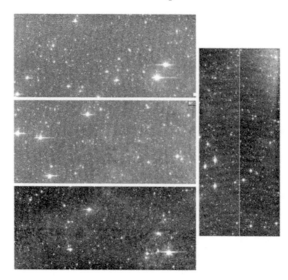

Figure 2. A star field imaged by the mosaic. The limiting V magnitude is 21.7.

to electronic noise from the telescope driving motors. With 2 minute exposures, the duty cycle of open-shutter on-sky time is 50%, as low as we choose to work. The period of the telescope's right ascension worm gear is also 2 minutes, so worm errors affect each 2-minute exposure the same way. Including overhead such as focusing, longer slews between regions, and other operations, we average a 4 min cycle per exposure. It takes ∼0.4 hours to cycle once through seven such pointings. We return to each of the seven pointing centers in a region three times over ∼1.3 hours. Thereby we search with a time baseline of about 0.9 hours for detecting motion, reaching V=21-22 mag on objects moving slower than ∼1 arcsec per minute. The 0.9-m telescope and camera were completely automated in 2005, and early in 2006 we began operating both telescopes with one observer.

Software: Two programs written by J. A. Larsen run on a cluster of computers at the 0.9-m telescope site: MOSAF (MOSaic Astrometry Finder) and MOSSUR (MOSaic SURvey). The catalog-based search is fairly efficient in terms of execution time per candidate found. MOSAF performs all flat field, dark, bias, and fringe corrections, creates an object catalog of detections in a manner similar to Sextractor (Bertin & Arnouts 1996), and produces both raw and processed, astrometrically calibrated MEF FITS images using the CFITSIO libraries of Pence (1999), the WCSLIB libraries of Mink (2002) and the USNO-A2.0 astrometry catalog (Monet *et al.* 1998). (Conversion to USNO-B1.0 on both telescopes is pending in the winter of 2006-2007.) The catalogs of detected objects contain many image parameters such as the shape, position, flux, moments and the parameters of a simple fit to an ellipse. MOSAF is customized to the Spacewatch imaging system and is integrated into the image creation pipeline. MOSSUR uses the object catalogs created by MOSAF to search for moving objects and creates graphical displays for review and validation by the observer at the telescope. The catalogs are searched for motions with rates between 0.05 and 2.5 degrees per day. This system has been operational since early 2003 and will be described in a paper on the derived statistics of NEOs.

Mosaic Survey Pattern: Since 2003 April when it went into full operation, the mosaic of CCDs on the 0.9-m telescope has covered an average of 1400 deg² of sky each lunation. A survey pattern is usually concentrated near opposition, with three exceptions: in the times of the year when the ecliptic is most vertical in the evening and morning skies, in cases of targeted searches, and in the summer.

Figures 3(a), (b), and (c) illustrate the sky coverage with the mosaic of CCDs through May 2005. Since 2004 September we have been revisiting the same regions during each lunation to allow linking of main belt asteroids for statistical studies of that population. The overprinting of region symbols in Fig. 3(c) illustrates the effect. The region centers are actually moved between revisits to follow the motion of typical main belt asteroids. Simulations show that 4-8 day intervals of revisits are infrequent enough to refresh the content of NEOs in the images, although admittedly this practice makes sky coverage by Spacewatch look three times smaller on the Minor Planet Center's (MPC's) sky coverage plots than it really is.

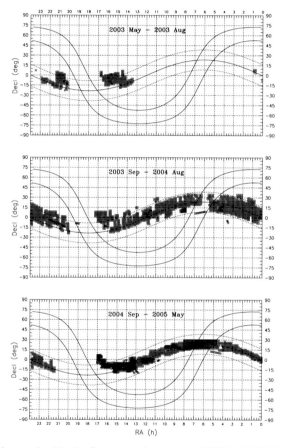

Figure 3. Regions observed with the Spacewatch mosaic of CCDs, 2003 March – 2005 May. The ecliptic and Milky Way are also shown. Coverage during the latest (2005 Sep – 2006 June) season looks similar to 3(c).

Spacewatch 1.8-meter Telescope: The 1.8-meter Spacewatch telescope is, as far we can tell, the largest telescope in the world dedicated exclusively to observations of comets and asteroids. It was built by and for Spacewatch (McMillan *et al.* 1998; McMillan *et al.* 2000; Perry *et al.* 1998). It has reached an apparent magnitude of V=23 for high priority recovery of asteroids. The field of view is 0.6×0.6 degrees on a $2K \times 2K$ CCD. Routine operation of the telescope by solo observers began in 2001 October and improvements to the efficiency of its operation have continued steadily. The drift scanning technique (Gehrels *et al.* 1986) provides smooth background and flatfield response as well as straightforward astrometry. The exposure time for each pass with the 1.8-m telescope is $136s/\cos\delta$. A typical scan to follow up an NEO covers from 0.3 to 1.0 deg^2. This telescope is dedicated to followup of faint NEOs, with emphasis on PHAs and objects on the MPC's NEO Confirmation Page and the JPL and NEODyS impact risk pages. About half of the followup targets are detected automatically and half are measured by hand, usually because they were too faint or too trailed for the software. The 1.8-m is also committed to lightcurves of Trans-Neptunian Objects (TNOs) under a grant to Jim Scotti from NASA's Planetary Astronomy Program.

3. RECENT RESULTS

Followup Observations: Spacewatch equipment and methods are best suited to followup observations nowadays. Followup of NEOs helps to consolidate their orbits as their brightness fades after discovery. Faint followup is also frequently required for recoveries of NEOs during later apparitions. We elaborate on recoveries in a later section.

About two-thirds of followup observations of PHAs by Spacewatch are deliberately targeted and one third occur incidentally during surveying. Discovery observations are counted here with the same weight as targeted and incidental followup observations. We quote most of our statistics for the last three years of operation in order to refer to the equipment we are presently using. We count one set of three observations of position as one "detection" by Spacewatch.

Figure 4 illustrates how Spacewatch's rate of detection of NEOs improved after the introduction of the 1.8-meter telescope in late 2001 and the mosaic of four CCDs and new optics of the 0.9-meter telescope in early 2003. Figure 5 shows a similar improvement for PHAs.

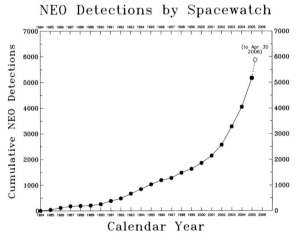

Figure 4. Cumulative count of detections of NEOs by Spacewatch vs. time. All discoveries, incidental detections, and targeted followup detections by either telescope are counted.

To show that Spacewatch samples a wide range of absolute magnitudes of NEOs, Figure 6 presents a histogram of the H values of NEOs detected with the 0.9-m telescope and mosaic. The results from the 1.8-m telescope are similar.

Spacewatch has recovered objects less than 60 degrees from the Sun. Figure 7 shows the distribution of pointings with the Spacewatch 1.8-m telescope during one lunation. Figure 8 shows the distribution of observations of PHAs with respect to opposition for both Spacewatch and the rest of the NEO community. The capability of ground-based telescopes to reach small elongations for special cases should not be overlooked.

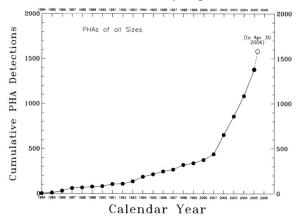

Figure 5. Cumulative count of detections of PHAs by Spacewatch vs. time.

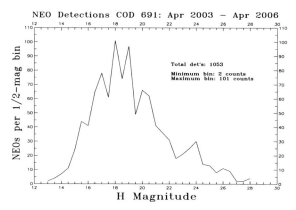

Figure 6. Histogram of absolute magnitudes H of NEOs detected by the Spacewatch 0.9-m telescope in the last three years.

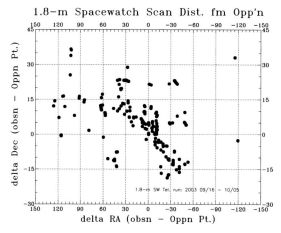

Figure 7. Distribution on the sky of observations made with the Spacewatch 1.8-m telescope during one lunation, illustrating how far from opposition we can observe.

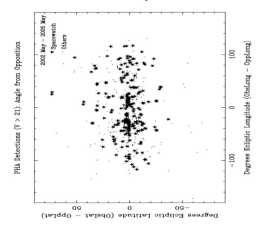

Figure 8. Distribution with respect to the opposition point of detections of PHAs by the asteroid followup community, with Spacewatch observations emphasized.

Another factor to consider is how our numbers of detections compare with those made by other stations. Some use four or five exposures instead of three, and sometimes an NEO might receive a long series to measure a lightcurve. Updating orbits is served better by distributing the observing effort over multiple objects and separate nights. For the statistics presented here, we count only one detection per object per station per night.

Figure 9 illustrates the substantial contribution of Spacewatch relative to the other stations that are active in followup of large PHAs when such objects are faint. Of course Spacewatch also follows up PHAs with absolute magnitudes down to the Minor Planet Center's (MPC's) defining limit of H=22; Figure 10 compares our contributions on those objects relative to the rest of the faint followup community. Spacewatch stands out similarly when the limiting magnitude is increased to V=21.5 and when numbers of different objects are counted instead of all detections on separate nights. Objects that appear on JPL's impact risk website are also priorities for Spacewatch followup. Spacewatch followup observations contributed to the removal of ~80 objects from that list in the last 3 years.

Recovery Observations: PHAs with uncertain ephemerides are targeted by Spacewatch. Some objects become uncertain due to the infrequency of favorable apparitions and/or interference by the Moon or galactic plane. If the object is faint during a return apparition, which is usually the case, recovery is labor intensive and time critical. A. S. Descour and others developed software tools and an observing regimen to aid recoveries.

Table 1 lists some examples. Of particular note is the Spacewatch recovery of 1990 SM, a lost H=16 PHA that had not been seen since the discovery apparition, 15 years before. The object had had windows of opportunity for recovery nearly every year since discovery, more than six of which reached V brighter than 18, but they were very short and near the galactic plane, so the surveys had not recovered 1990 SM incidentally.

Image Data Archive: This archive has yielded arc extensions for more than a dozen virtual impactors, 2004 MN_4 being the most important example. Images have also been provided for other objects of interest. About 3 Terabytes (TB) of data from the old (1990-2002) configuration of the 0.9-m telescope covers a sky area of 30,000 deg^2 including revisits on separate nights. Imagery to date from the Spacewatch 1.8-m telescope amounts to approximately 1.5 TB. Data from the mosaic of CCDs have been accumulated since 2003 Mar 23 and now consist of >5 TB. About 40,000 deg^2 have

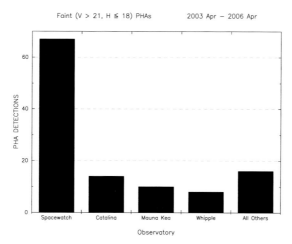

Figure 9. Numbers of detections of intrinsically bright PHAs while their apparent magnitudes were faint, sorted by observing station. "Catalina" combines station codes 703, G96, E12, and 413. NEAT (not shown separately) includes both 644 and 608, and Spacewatch includes both 691 and 291.

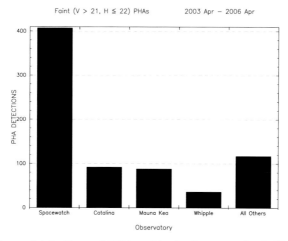

Figure 10. Numbers of detections of PHAs with absolute magnitude H ⩽ 22 while their apparent magnitudes were faint, sorted by observing station.

been covered by three passes with the mosaic. The URL http://fmo.lpl.arizona.edu/cgi-bin/mosaic_archive/point_history.cgi accesses the database of pointing centers and our protocol for requests for imagery.

Incidental Astrometry (IA): In the last three years, Spacewatch has sent more than a million astrometric detections of asteroids (∼3 million astrometry records) to the MPC. More than half of those detections have been linked by the MPC to previously known asteroids or with each other. At our request, the MPC has recently agreed to release all of Spacewatch's unlinked detections for the use of the asteroid community at large.

Precoveries: Observations made prior to discoveries can be found in archives of images as well as in previously reported incidental astrometry (IA). Here we give examples of both types of such "precoveries". Spacewatch has precovered high priority NEOs such

Table 1. Notable Examples of Spacewatch Recoveries of Uncertain PHAs.

Object	Unc. (deg)	H mag	V mag	MPEC	Arc Before	Arc After	Net O-C arcsec
2002 TW$_{55}$	1	18.0	21.7	2005E54	52d	831d	237
1990 SM	80	16.2	21.2	2005C26	24d	5225d	23022
1999 VT$_{25}$	3	21.4	21.5	2004U47	26d	1786d	7556
2000 EV$_{70}^{*}$	3	20.5	20.9	2004E11	46d	1193d	214
2001 US$_{16}$	2	20.2	20.7	2004B68	31d	802d	485
1998 VS*	4	22.3	21.3	2003Y18	32d	1831d	1581
2000 UL$_{11}$	2	20.1	21.9	2003S71	28d	1039d	3320
2003 BH		20.7	22.7	2005-J56	51d	844d	45
1998 VF$_{32}$	2	21.2	21.2	2005-W43	14d	2555d	5581
2001 YP$_3$	2	22.0	21.4	2005-X55	109d	1453d	21
2004 JQ$_1$	1	20.1	21.8	2006-C02	31d	600d	210
2004 RY$_{109}$	0	19.1	22.6	2006-C19	94d	510d	22
2005 TR$_{50}^{*}$	1	20.2	21.5	2006-F24	2d	164d	3660

$*$ Asteroids 1998 VS and 2005 TR$_{50}$ lost their PHA status due to the recoveries' updates of their orbits. Catalina's station G96 recovered 2005 TR$_{50}$ on the same night that Spacewatch did.

as 2004 MN$_4$, a PHA whose estimated probability of impact on Earth rose to a record high value of a few percent in 2004 December before we found images in the Spacewatch archives. We increased 2004 MN$_4$'s arc from 190 days to 255 days, enough to reduce the estimate of probability of impact to a much less alarming value. 2004 MN$_4$ now has the permanent designation and name of (99942) Apophis. Other precoveries within the last 3 years total about 130 NEOs and 4 comets.

Examples of precoveries extracted by the MPC from our IA include 2001 WG$_2$ (an Apollo of high eccentricity and inclination), 2001 XN$_{254}$ (a PHA with H=17.5), periodic comet P/2002 BV, Amor 2003 HB$_6$, comet P/2004 A1, Amors 2003 OB$_4$ and 2003 MT, Apollo 2003 YO$_1$, and Apollo PHA 2003 YK$_{118}$. The latter was found among our observations in 1993, predating the discovery by more than a decade. 1994 UG was retired from JPL's impact risk page as a result of Larsen's new astrometry of old imagery. During the last three years, at least 122 NEOs, 6 comets, and 3 outer solar system objects were found by the MPC in our IA.

4. ASSISTANCE to Pan-STARRS

The next few years will see the beginning of the operation of the Panoramic Survey Telescope and Rapid Response System (Pan-STARRS) of the University of Hawaii's Institute for Astronomy (Kaiser *et al.* 2005; Jedicke *et al.* 2005a; Jedicke *et al.* 2005b). Pan-STARRS' (PS's) revisits of areas surveyed during a lunation will allow PS to determine preliminary orbits of asteroids spanning 4-16 days. Spacewatch has been providing PS with copies of catalogs of detected point sources on which they are testing their linking software.

Additional followup observations with other telescopes would help PS make such linkages in their archives. We assume that PS will detect ~50% of the NEOs with H≤22 in 10 years of surveying (Jedicke 2006, personal communication). If there are ~50,000 NEOs with H≤22 (Stuart 2001), PS will detect ~25,000 NEOs to V≤24 with H≤22 in 10 years. We suppose that ~10% of those might lack a third night due to picket fence and other incompleteness effects, and another ~10% might have poorly determined orbits (Jedicke 2006, personal communication). So PS may need followup of ~5,000 NEOs in 10 years or ~500 NEOs per year. About 200 of those should be accessible to Spacewatch with V⩽23 and our weather and declination constraints. With ~1000 hrs/yr of clear observing time with 2 telescopes, and a current annual average of ~1000 separate-night detections of NEOs per year, Spacewatch should be able to target and follow up ~200 NEOs per year at least once. Absorbing this burden into the existing target list is feasible because some of the targets Spacewatch currently follows are lower-priority non-hazardous NEOs, and the brighter (V⩽21.5) PS detections can be absorbed into the survey pattern covered by the automated 0.9-meter telescope. Furthermore, enhancements to software at the 1.8-meter telescope can probably gain us ~20% in telescope time efficiency and ~0.2 mag in sensitivity.

Acknowledgements

The Spacewatch Team members, past and present, and others who contributed their talents and fine work toward these results include L. Acedo, L. Barr, M. Block, T. H. Bressi, R. A. Buchroeder, K. C. Cochran, A. S. Descour, T. Gehrels, R. Jedicke, J. A. Larsen, B. W. Lawrie, J. L. Montani, M. L. Perry, M. T. Read, J. V. Scotti, A. F. Tubbiolo, W. T. Verts, J. T. Williams, and M. Williams.

Spacewatch is funded by grants from NASA's NEO Observation and Planetary Astronomy Programs, The Brinson Foundation, the estates of Richard S.Vail and Robert L.Waland, and other private individuals. At *http://spacewatch.lpl.arizona.edu* there is a description of Spacewatch, the present and former team members are listed, as well as past contributors of funds.

Extensive use of the subscription-supported on-line data services of the Minor Planet Center was made for this report.

References

Bertin, E. & Arnouts, S. 1996, *A&AS* 117, 393
Bottke, W. F., Durda, D. D., Nesvorný, D., Jedicke, R., Morbidelli, A., Vokrouhlický, D. & Levison, H. F. 2005a, *Icarus* 179, 63
Bottke, W. F., Durda, D. D., Nesvorný, D., Jedicke, R., Morbidelli, A., Vokrouhlický, D. & Levison, H. F. 2005b, *Icarus* 175, 111
Bottke, W. F., Morbidelli, A., Jedicke, R., Petit, J.-M., Levison, H. F., Michel, P. & Metcalfe, T. 2002, *Icarus* 156, 399
Gehrels, T. 1991, *Space Sci. Rev.* 58, 347
Gehrels, T., Marsden, B. G., McMillan, R. S. & Scotti, J. V. 1986, *Astron. J.* 91, 1242
Jedicke, R. 1996, *Astron. J.* 111, 970
Jedicke, R. & Herron, J. D. 1997, *Icarus* 127, 494
Jedicke, R. & Metcalfe, T. S. 1998, *Icarus* 131, 245
Jedicke, R., *et al.* 2005a, *BAAS* 37, 637
Jedicke, R., *et al.* 2005b, *BAAS* 37, 1363
Kaiser, N., *et al.* 2005, *BAAS* 37, 1409
Larsen, J. A., *et al.* 2001, *Astron. J.* 121, 562
Larsen, J. A., *et al.* 2005, *BAAS* 37, No. 4, abstract 4.09: `http://www.aas.org/publications/baas/v37n4/aas207/1500.htm` .
Larsen, J. A., *et al.* 2006a, The search for distant objects in the solar system using Spacewatch, submitted
Larsen, J. A., *et al.* 2006b, The absolute magnitude distribution of near-Earth asteroids from the Spacewatch Survey (approximate title), in preparation

McMillan, R. S. 1999a, Paper presented at the Space Resources Utilization Roundtable at the Colorado School of Mines in Golden, CO: LPI Contrib. No. 988 and `http://www.mines.edu/research/srr/first_srr.html` .

McMillan, R. S. 1999b, *Space Manufacturing 12: Challenges and Opportunities in Space: Proceedings of the Fourteenth Space Studies Institute's Princeton Conference on Space Manufacturing, May 6-9, 1999,* B. Greber, Ed. (SSI Publishing, Princeton, NJ), 72-75.

McMillan, R. S., Block, M. & Descour, A. S. 2005, *The ALPO Minor Planet Bulletin* 32, 53: `http://www.minorplanetobserver.com/mpb/MPB%2032-3.pdf` .

McMillan, R. S., *et al.* 1998m *BAAS* 30, 1114: `http://www.aas.org/publications/baas/v30n3/dps98/176.htm` .

McMillan, R. S., Gehrels, T., Scotti, J. V. & Frecker, J. E. 1986, In *Instrumentation in Astronomy VI: Proc. S.P.I.E. 627* (D.L. Crawford, Ed.), 141

McMillan, R. S., Perry, M. L., Bressi, T. H., Montani, J. L., Tubbiolo, A. F. & Read, M. T. 2000, 0.9-m telescope. *BAAS* **32,** 1042: `http://www.aas.org/publications/baas/v32n3/dps2000/8.htm` .

McMillan, R. S. & Stoll, C. P. 1982, *Proc. SPIE 331, Instrumentation in Astronomy IV,* 104

Mink, D. J. 2002, *ASP-CS Astronomical Data Analysis Software and Systems XI,* 281, 169

Monet, D. B. A., *et al.* 1998, The USNO-A2.0 Catalogue. *VizieR Online Data Catalog: I,* 252.

Morbidelli, A., Jedicke, R., Bottke, W. F., Michel, P. & Tedesco, E. F. 2002, *Icarus* 158, 329

Ostro, S. J., *et al.* 2003, *Icarus* 166, 271

Pence, W. 1999, CFITSIO, v2.0: A new full-featured data interface. *ASP-CS Astronomical Data Analysis Software and Systems VIII,* 172, 487

Perry, M. L., Bressi, T. H., McMillan, R. S., Tubbiolo, A. F. & Barr, L. D. 1998, *Proc. SPIE 3351, Telescope Control Systems III,* 450-465: `http://bookstore.spie.org/index.cfm?fuseaction=detailpaper&cachedsearch=1&productid=308809&producttype=pdf&CFID=1982426&CFTOKEN=46151179`

Rabinowitz, D. L. 1991, *Astron. J.* 101, 1518

Rabinowitz, D. L. 1993, *Astrophys. J.* 407, 412

Roe, E. A, Larsen, J. A. & the Spacewatch Team 2005, *BAAS* 37, #4, abstract 4.07: `http://www.aas.org/publications/baas/v37n4/aas207/1308.htm` .

Scotti, J. V. 1994, in *Asteroids, Comets, and Meteors 1993,* A. Milani *et al.,* eds., Kluwer, 17

Stuart, J. S. 2001, *Science* 294, 1691

Near Earth Objects, our Celestial Neighbors: Opportunity and Risk
Proceedings IAU Symposium No. 236, 2006
A. Milani, G.B. Valsecchi & D. Vokrouhlický, eds.

© 2007 International Astronomical Union
doi:10.1017/S1743921307003419

The next decade of Solar System discovery with Pan-STARRS

†Robert Jedicke, E. A. Magnier, N. Kaiser and K. C. Chambers

Institute for Astronomy, University of Hawaii, Honolulu, HI, 96822
email: jedicke@ifa.hawaii.edu

Abstract. The Panoramic Survey Telescope and Rapid Response System (Pan-STARRS) at the University of Hawaii's Institute for Astronomy is a funded project to repeatedly survey the entire visible sky to faint limiting magnitudes ($m_R \sim 24$). It will be composed of four 1.8m diameter apertures each outfitted with fast readout orthogonal transfer Giga-pixel CCD cameras. A single aperture prototype telescopes has achieved first-light in the second half of 2006 with the full system becoming available a few years later. Roughly 60% of the surveying will be suitable for discovery of new solar system objects and it will cover the ecliptic, opposition and low solar-elongation regions. In a single lunation Pan-STARRS will detect about five times more solar system objects than the entire currently known sample. Within its first year Pan-STARRS will have detected 20,000 Kuiper Belt Objects and by the end of its ten year operational lifetime we expect to have found 10^7 Main Belt objects and achieve $\sim$90% observational completeness for all NEOs larger than $\sim$300m diameter. With these data in hand Pan-STARRS will revolutionize our knowledge of the contents and dynamical structure of the solar system.

Keywords. Survey; telescope; instrumentation; astrometry; data analysis; identification; population models

1. Pan-STARRS Overview

The conceptual design of the Pan-STARRS system derives from a desire to survey the sky as deeply, rapidly and inexpensively as possible. Relatively simple technological and economic arguments suggest that the minimum cost for a surveying system is achieved with telescope mirrors in the range of 1.5m to 2.5m diameter. Simply put, it is more cost effective to build more small telescopes to reach the same effective limiting magnitude and sky-coverage as a massive system due to the fact that duplicating a small system increases costs linearly while the cost of a single large system increases as a power of its light gathering ability. Another major benefit of a distributed aperture system is the reduced time to deployment over a monolithic system with equivalent etendue. Thus, Pan-STARRS opted to develop a design for a coordinated set of four small telescopes with the capability of a large synoptic survey system.

A prototype single Pan-STARRS telescope (PS1) on Haleakala is nearing completion and is expected to start survey operations in mid to late 2007. It is designed to act as a test-bed for the commissioning, testing, and calibration of the Pan-STARRS hardware and software in anticipation of the full (four aperture) Pan-STARRS array. PS1 uses a 1.8m primary mirror to image a roughly 7 square degree field on a 1.4 Gigapixel CCD camera. With 0.26 arcsecond pixels, the images from PS1 will be well-matched to the sub-arcsecond seeing of Haleakala. The large entendue of PS1 will make it the most

† for the Pan-STARRS team.

efficient survey telescope for the near future. The PS1 Survey Mission is expected to last 3.5 years, and will produce about 1.8 PB of raw image data.

The PS1 Survey Mission will consist of multiple survey components, including a very large area survey covering three quarters of the full sky (the 3π survey) and targeted fields observed frequently and repeatedly for both extensive temporal coverage and depth. A major driver for the observing strategies is to efficiently detect solar system objects including potentially hazardous asteroids. The telescope will use 4 filters very similar to the Sloan griz set and a y filter at the reddest end of the camera sensitivity to exploit the high quantum efficiency of the detectors at 1 micron.

Figure 1. The prototype Pan-STARRS telescope installed near the top of Haleakala, Island of Maui, Hawaii. The main optics are in place, the dome is operational and first light has already occurred. Installation and testing of the OTA camera (in the hole at the center of the primary mirror) is the next step.

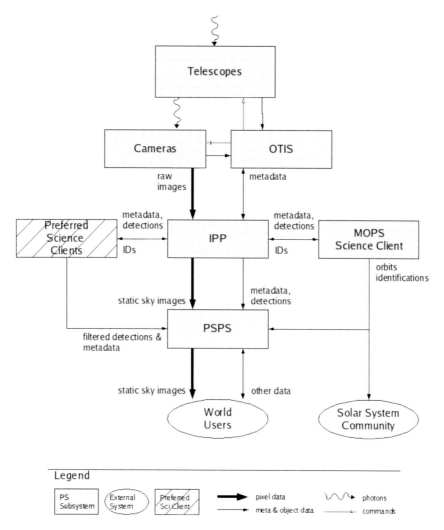

Figure 2. PS1 concept of operations.

The primary enabling technology for Pan-STARRS is the development at the Institute for Astronomy (IfA) of Orthogonal Transfer Array (OTA) CCDs Tonry *et al.* (2002). These CCDs boast an amazing set of characteristics including fast readout (a few seconds for a 4K×4K array), low read noise (about 5 e$^-$), superb response at long wavelengths (40% QE at 1μm) and low cost. Perhaps the most innovative features of the OTAs are their ability to move charge between pixels in both directions on the CCD and to read out sub-arrays on the CCD (called cells) as an image is being acquired. These two features in combination allow a poor mans first order correction to both atmospheric and mechanically induced image motion. The ∼20% reduction in the PSF width for point sources provides an important increase in limiting magnitude and source identification. On Mauna Kea, we expect to acquire images with 0.5″ width in median seeing of 0.6″.

Figure 2 shows the high level concept of operations for the PS1 system. We expect that it will be representative of PS4 operations as well but will modify the actual concept based upon our experience with PS1. Major elements represented in fig. 2 are already well developed, with sub-system completeness generally decreasing from top to bottom.

The telescope is controlled by the Observatory Telescope and Instrumentation System (OTIS). OTIS is responsible for monitoring the weather and sky conditions, deciding on an image-by-image basis which field to obtain next, issuing telescope and dome commands, etc. Metadata collected by OTIS are made available for image processing and science analysis

Each of the four apertures has its own 1.44Gpix camera. The camera system is responsible for implementing the OTA image correction. A few percent of the cells containing bright stars that would otherwise saturate or bleed into the system are read out at about 20Hz. Their centroids are calculated in real time and their positions are used to determine appropriate (x, y) offsets to the positions of the charge in all the other cells in the camera. Finally, the entire image is read out and passed to the Image Processing Pipeline (IPP).

The IPP is responsible for all the image reduction and preliminary analysis (see §2). It identifies known stars in the field to solve for the astrometric reference for the image, measures the flux from each object in the field, characterizes the shape of all detections, adds the current image to a static-sky image that gradually builds a deep representation of the non-moving objects on the sky, warps and convolves the existing static-sky image so that it may be subtracted from the current image and leave only transient objects in the difference image and, finally, detects and characterizes transient sources in the difference image. The set of all transient detections (stationary, moving and false) are passed to the Moving Object Processing System (MOPS, see §3) for asteroid and comet detection.

The Pan-STARRS published science products system (PSPS) is the database of detections and images produced by the IPP. It is the window into Pan-STARRS for the world and the repository of all information for internal access by Pan-STARRS users.

The Pan-STARRS concept of operations (Fig. 2) allows for many preferred science clients (PSC). These software sub-systems obtain their data from the IPP and perform analysis-specific tasks for a particular science goal. PS4 will have many PSCs including supernovae, extra-solar planet, weak gravitational lensing and high-z galaxy clients. The only existing PSC for PS1 and PS4 is the MOPS that is funded through the development and construction costs as a recognized major part of the core Pan-STARRS goal of detecting NEOs and other solar system objects.

2. Image Processing Pipeline

The Pan-STARRS PS1 Image Processing Pipeline (IPP) performs the image processing and data analysis tasks needed to enable the scientific use of the images obtained by the Pan-STARRS PS1 prototype telescope. The primary goals of the IPP are to process the science images from the Pan-STARRS telescopes and make the results available to other systems within Pan-STARRS. It is also responsible for combining all of the science images in a given filter into a single representation of the non-variable component of the night sky – the 'Static Sky'. To achieve these goals, the IPP also performs other analysis functions to generate the calibrations needed in the science image processing, and to occasionally use the derived data to generate improved astrometric and photometric reference catalogs. It also provides the infrastructure needed to store the incoming data and the resulting data products.

Like other large-scale surveys (e.g., the Sloan Digital Sky Survey or the 2 Micron All-Sky Survey), end users will have access to derived data products, not the raw image data stream. The IPP will perform the individual image calibrations, image combinations, and object measurements needed to characterize the astronomical sources detected in the images. During clear, dark nights, the PS1 telescope will produce images at a sustained rate of about one 3 GB image every 45 seconds for periods as long as 10 hours. The IPP needs to perform the data processing with a high enough throughput to keep up with the raw image data. The resulting data products and the extensive supporting metadata stream will be made available to the other components of the project including the Published Science Products System (PSPS) which will, in turn, make the data products available to users via a database and sophisticated query mechanism. Other science clients performing additional interpretation of the science data products will also receive subsets of the full IPP output data stream.

The IPP receives data from two Pan-STARRS subsystems: the Camera, from which it receives the large volume of image data, and OTIS (Observatory, Telescope and Infrastructure Subsystem), from which it receives metadata describing the images and the environmental conditions. The users of the IPP output are all systems internal to the Pan-STARRS project. They consist of: 1) The Preferred Science Clients, which receive specified data products on short timescales. 2) The Moving Object Processing System (MOPS) described in §3 that receives the detections of all transient objects. 3) The Published Science Products Subsystem (PSPS), that receives all data products of interest to the community external to the Pan-STARRS data processing systems and will act as the long-term archive and publishing clearinghouse.

The IPP performs several types of data analysis in a regular fashion. The most obvious of these is the science image analysis, from which the measurements of individual astronomical objects are actually derived. In preparation for this critical function, the IPP must also analyze the calibration images needed by the science image analysis. Downstream from the science image analysis, the IPP must perform data calibration on the collection of object detections, yielding improved calibrations and improved reference catalogs for astrometry and photometry. At 5-σ the astrometric floor is expected to be about $0.1''$ while the photometric floor at the same S/N should be about 0.1 mags (both improve with S/N).

The IPP science image analysis is separated into two major stages: the analysis of individual images and the analysis of groups of images taken of the same portion of the sky. The individual images are analyzed independently as they arrive from the telescope where standard image detrending steps (bias, flat, etc.) are performed. Objects are detected in the images using customized software developed by the IPP team (the image analysis software will eventually be publicly available). Stars and non-stellar objects are distinguished, but only limited effort is spent at this stage on characterizing the extended sources. Brighter stars are the used to perform astrometric and photometric calibrations. In the second major stage the images which correspond to the same portions of the sky are combined. Several image combinations may be performed. Sets of individual science images may be combined into a single, high-quality image which has been cleaned of cosmetic defects. Comparison between this image, or the individual images, and an archival reference image of the same location (the Static Sky image) may be performed. Image difference techniques are used to detect the variable, transient, and moving sources. Sources identified in difference images are passed to the Moving Object Processing System (described below) as a catalog list. Finally, the new images may be combined with the Static Sky image in order to improve its signal-to-noise.

Additional science image analysis is performed on the Static Sky images. It is in this stage that detailed analysis of the shapes of extended objects is performed. In the Static Sky analysis, the data from all five filters will be analyzed at the same time to improve the signal-to-noise for fainter sources. Common parameters such as the location of the galaxy may be fit to a single value in all filter images.

The IPP is also responsible for generating a high-quality photometric and astrometric (P&A) reference catalog from the collection of measurements of the astronomical sources. As the images are analyzed, the information about each object is supplied to the IPP object database software called DVO (the Desktop Virtual Observatory). Several programs interact with this database to iteratively improve the calibration of the individual images and to update the astrometric and photometric reference catalog.

The P&A analysis can be viewed as a very large least-squares problem, in which the astrometric or photometric parameters of the images are solved to minimize the residuals for individual objects, while the positions and magnitudes of individual objects are adjusted to minimize the residuals for individual images. It is a two step process where the first pass obtains the linear-only astrometric solution for each the CCDs independently. In the second pass the CCDs are treated together to solve for the camera distortion and high-order terms for each chip. Tests of this process with the CFHT's Megacam and CFH12K cameras have realized 10mas astrometric performance. Those cameras have pixel scales of $0.19''$ and $0.21''$ respectively, comparable to the pixel scale for the PS1 camera.

The P&A analysis will likely be limited to the objects that have been observed with sufficient signal-to-noise but the resulting image parameters can then be used to characterize all objects in the images. A natural consequence is that the objects that have significant residuals even after the iterations have run their course can be identified as photometric variables or objects with detectable proper-motion and/or parallax. Images which were obtained under less-than-ideal conditions will also be flagged and may be excluded from this analysis.

The IPP image analysis tasks are well suited to parallel processing. Not only are individual images processed independently, but most of the computational effort for each of the chips of the mosaic camera may be processed without reference to the other chips. The analysis of different patches of the sky may also be performed independently.

3. Moving Object Processing System

Pan-STARRS+MOPS (Moving Object Processing System) will be the worlds first integrated asteroid detection, linking, orbit determination and database system. Combining the processes provides Pan-STARRS a tremendous advantage in sky area coverage because it can spread the requisite number of detections over many nights with a concomitant increase in the number of NEO discoveries. Furthermore, by retaining control over all the processes Pan-STARRS can determine the efficiency of the entire sub-system as well as the efficiency of every step. This capability is critical to monitoring the MOPS performance and de-biasing the science data to account for observational selection effects.

Figure 3 provides a high level concept of operations for the MOPS. Input consists of image meta-data (e.g. boresight, limiting magnitude, filter) and all source detections with S/N>3 from the IPP's difference images as described above. This includes apparently stationary transients since they may be very distant slow moving objects, and also those trailed detections consistent with being images of nearby fast moving objects.

The combinatoric problem of properly linking detections of the same object observed on a few nights over a period of a couple weeks could be computationally expensive. On the ecliptic and to a limiting magnitude of r~24 we expect that there will be about 250 real moving objects per deg^2 or almost 2000 asteroids and comets per single Pan-STARRS field. Asteroid sky-plane density drops quickly off the ecliptic while the rate of false detections remains constant. At 5-σ we expect a maximum of a 1:1 ratio of false:real detections on the ecliptic (i.e. ~2000 false 5-σ detections per field) while at 3-σ we expect

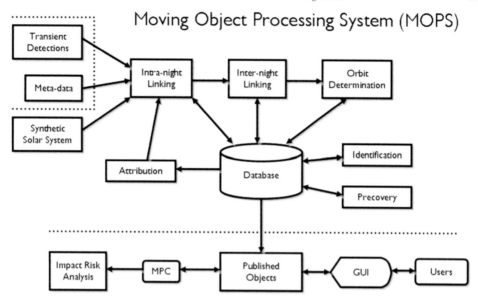

Figure 3. Moving Object Processing System concept of operations. Data in the upper left region surrounded by dotted lines is provided by the IPP. Systems below the horizontal dotted line are not part of the MOPS.

about 200,000 false detections per field or a false:real ratio of about 100:1! Even so, we expect that the 5-σ MOPS database(s) will require only about 2TB of disk storage including a full backup and other overhead requirements. The 3-σ data will probably not be stored in a relational DB but rather in compressed flat files on a distributed file network.

The current survey strategy for solar system objects is to obtain two images of each field on a night separated in time by a transient time interval (TTI) of about 15 to 30 minutes. Roughly the same fields are visited 3× per lunation. We envisage a PS1 survey strategy Chambers *et al.* (2006) that combines a photometric & astrometric survey, a medium-deep survey and a solar system survey in an efficient manner that will allow ~55% of surveying time to be utilized in a cadence suitable for linking solar system objects. Each of the three visits (two images) will be obtained in a different filter (g, r and i) with exposure times arranged to obtain the same limiting depth for a solar colored object (equivalent to V~23 in a 38s exposure in r). An additional ~5% of surveying time will be devoted to surveying in the 'sweet-spots' Chesley & Spahr (2004), the sky within about 10° of the ecliptic and with solar elongation from 60° to 90°. The sweet-spot regions will be covered in either r or i, whichever provides us the best limiting magnitude for NEO detections. Figure 3 shows a single lunation of surveying in a pattern similar to that envisioned for PS1. It is clear that the sweet-spots are particularly rich in NEOs due to looking along the Earths orbit where Potentially Hazardous Objects (PHOs with a minimum orbital impact distance with the Earth of <0.05AU) must pass.

It is the responsibility of the MOPS to identify candidate moving object 'tracklets' in the images acquired within a TTI on each night. A tracklet is composed of detections that are within an angular separation and position consistent with their elongation and orientation with respect to one another. The detections are those sources identified in difference images by the IPP as described above. The elongation of the detections is not used to identify moving object candidates, only to verify that the detections within a candidate tracklet have consistent motion vectors. Thus real tracklets are distinguished

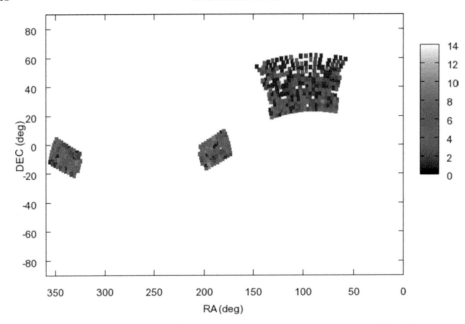

Figure 4. Sky-plane density of NEOs detected in a single lunation by PS4. The opposition region is the large area near 100° RA and the other areas represent the morning and evening sweet spots.

from spurious false tracklets merely by the spatial proximity and shape consistency of the detections within the tracklet *and* by the ability to link tracklets obtained on separate nights into tracks consistent with a heliocentric orbit (as described below).

Our efficiency (percentage of real tracklets that are properly identified) and accuracy (percentage of identified tracklets that are real) for creating tracklets Kubica *et al.* (2006) is essentially 100% as determined from a realistic solar system model and survey simulation. All known objects that might appear in each field are associated with identified tracklets when available (a process known as attribution). Tracklets that can not be associated with known objects are stored for future use.

Every time a new TTI image pair is acquired, the MOPS automatically searches the meta-data database of images acquired in the past 14 days to determine if there are 3 or more nights of images including the current image that might contain unknown moving objects that could be linked together. When a multi-night set of TTI image pairs are available the MOPS extracts all un-attributed tracklets from those images and attempts to link them together into candidate tracks that might be consistent with a newly identified solar system object. The combinatoric difficulty in creating tracks has been solved using a variable kd-tree algorithm Kubica *et al.* (2006). The formation of tracks does not incorporate any orbit information. By implementing a few passes through the data appropriate for different classes of solar system objects we have achieved >99% efficiency in creating tracks for objects in a realistic but synthetic survey. The accuracy of track formation is about 0.3% but this percentage is relatively unimportant as long as the incorrect tracks can be discarded through an orbit determination attempt.

Once tracks have been created each one must be tested for compatibility with a heliocentric orbit by attempting an initial orbit determination (IOD). Tracks for which the IOD provides a suitably low residual with respect to all the detections are then passed to a differential orbit determination routine (OD) that attempts to fit the orbital parameters to the observations and further reduce the residual while improving the orbit.

We are currently studying the IOD and OD efficiency at producing useful orbits. The orbits do not necessarily have to be 'correct' if they can be used to link the object to future (attribution) or past (precovery) detections. E.g. for slow moving objects we may choose to use an IOD or OD with fixed eccentricity to allow convergence in the orbital solution. Our preliminary indications are that we can achieve nearly 100% efficiency for orbit determination of correctly linked tracks with <1% contamination by orbits obtained with false linkages.

We have usable prototype code in place for moving object attribution and also precovery detection identification but have not yet tuned their performance. Similarly, the process of "orbit identification", in which the same object may be separately identified and linked in different apparitions without it being possible to attribute or precover mutual detections, but where their identity may be discovered through the similarity of their orbit elements, also exists in prototype form.

The MOPS incorporates efficiency determination software (EDS) directly into its architecture. The EDS requires a high-fidelity solar system model containing $> 10^7$ synthetic asteroids and comets that we have developed for this purpose. The model contains realistic orbit and size distributions for all the objects including tri-axial ellipsoid shapes, random pole orientations and a spin period distribution. Each time new detections are provided by the IPP the MOPS generates synthetic detections that should appear in the image according to the solar system model and the meta-data from the IPP (e.g. limiting magnitude, trailing effects, chip gaps). The synthetic detections are injected into the MOPS pipeline and analyzed at the same time and in exactly the same manner as the real detections. Since we know what synthetic detections went into the MOPS we can monitor their progress through the system in nearly real-time and determine if there is a problem with any MOPS sub-system. The ability to determine the overall efficiency of the system is critical to correcting the data for observational selection effects.

4. Pan-STARRS and the Solar system

The prototype Pan-STARRS telescope on Haleakala will yield the best measurement of all asteroid and comet populations to date by providing a sample that is much larger than the entire data set currently in hand and, more importantly, doing so with a single well-calibrated detection system. This will dramatically decrease the current errors in the orbit and size distributions of all populations (especially at small sizes).

Working radially outwards from the Sun we speculate below on the impact of PS1 on our knowledge of the solar system's small bodies.

Pan-STARRS has been designed to be capable of surveying at low altitude allowing detections of moving objects at small solar elongations especially in the range from $60°$ to $90°$. This region of the sky is important to the discovery of PHOs Chesley & Spahr (2004) and objects entirely Interior to the Earth's Orbit (IEO) so we expect that the number of known PHOs and IEOs will increase dramatically in the next few years. These observations will thereby allow a much better characterization of the Earth impact risk. The IEOs are particularly interesting because they provide a sensitive test on contemporary models Bottke *et al.* (2002) of transport of asteroids from the MB into NEO space.

Collapsing the error bars on the size and orbit distribution is also important because it allows a better tuning of the capabilities for future NEO surveys. For instance, it is possible that the impact risk is larger than currently estimated because there may be local enhancements in the orbit distribution of PHOs. The fact that Tunguska (a once in a 1000 year event) and Apophis (a once in a tens of thousands of years event) occured within a century of one another makes it reasonable to speculate that the orbit distribution of NEOs is not as smooth as current models Bottke *et al.* (2002), Stuart & Binzel (2004) would suggest. PS1 will resolve these issues.

PS1 will extend the completeness limit for objects in the entire MB (to its outer edge at about 3.3 AU) from the current value of about absolute magnitude 14.5 to 18 (complete for all objects >1 km diameter in the MB). This sample should resolve many open disputes about the size distribution of these objects Gladman *et al.* (2006), Ivezić *et al.* (2001). We expect that this will allow the identification of dozens of new young asteroid families and with the addition of five filter measurements we will be able to refine space weathering rate estimates Nesvorný *et al.* (2002). Simple extrapolation of the discovery rates of MB comets Hsieh & Jewitt (2006) suggest that we may identify hundreds of this new class of objects. Perhaps most interesting of all is the opportunity of identifying collisions between MB objects too small to otherwise be observed. A collision between a 100m and 10m asteroid might create a dust cloud that lasts sufficiently long to be identified as a 'transient moving object'. The collision rates in this size range Bottke *et al.* (2005) imply that PS1 might identify one such event per month but this estimate is almost entirely dependent upon how long the dust clouds remains at an optical depth greater than one.

The Trojan regions of all the planets will be surveyed to a consistent limiting magnitude and with known efficiency allowing for a detailed analysis of the stability of these regions and a resolution of the issue of different numbers of objects in the Jovian L4 and L5 regions. Similarly, we will obtain a far larger sample of Centaurs than currently exists allowing for detailed measurements of the transport of objects from the Trans-Neptunian region into giant-planet-crossing orbits.

The number of TNOs will increase by at least an order of magnitude. The impact that these new objects will have on our understanding of the structure of the Kuiper Belt will be incredible. If recent estimates of the number of small TNOs Chang *et al.* (2006) hold, they imply that collisions between TNOs will happen frequently and PS1 may be able to detect them as was argued for MB asteroid collisions above. If our solar system formed in a dense star forming region and interacted with another system it is possible that there exist retrograde orbit TNOs. The MOPS has been designed to identify objects moving in any direction so that if there are retrograde TNOs above the PS1 detection threshold they should be identified as such.

The MOPS is also being designed to identify extremely slow moving objects. Those objects that appear to be stationary in two images acquired on the same night only 15-30 minutes apart, yet move from night to night over the course of 7-10 days. We estimate that the MOPS is capable of detecting objects moving almost as slowly as Barnard's Star. Thus, if there exist large objects well beyond the orbit of (134340) Pluto we will find them. PS4 will be sensitive to an Earth sized object out to 620 AU Jewitt (2004).

New frontiers will open in the study of comets as the number of objects and the temporal coverage increases dramatically. Since PS1 will revisit the same location on the sky many times in different filters and identify comets when they are much further away than existing surveys, it will be possible to study the detailed morphological evolution of comets as a function of heliocentric distance and other physical or dynamical characteristics.

One of the many exciting possible new discoveries is that of identifying unambiguous interstellar asteroids or comets. While estimates and limits on the volume density of these objects vary greatly it has been estimated Jewitt (2004) that PS4 will either detect a few interstellar interlopers or set the best limit to date on their number density. The scientific potential of such a discovery are incredible as spectroscopic followup of the object could reveal the chemical and/or mineralogical composition of objects from another solar system.

Finally, after a number of years of surveying PS will obtain perhaps a hundred(s) of observations of some bright asteroids. At sufficiently high S/N the PS photometric error will be on the order of 0.01 mags, allowing the light curve inversion techniques of Kaasalainen (2004) to be applied in order to determine the spin period, pole orientation and even shapes of hundreds or thousands of asteroids. At the current time spin periods are known for about 1000 asteroids, pole orientations for just over 500 of them, and shapes for only a handful that have been close enough to obtain radar information or that were spacecraft targets.

5. Conclusion

The PS1 and MOPS will be the first integrated asteroid and comet discovery, linking, orbit determination and database system. We have developed new algorithms to handle the combinatoric problem of linking detections on a single night and between nights within a lunation. We have developed the first comprehensive model of the solar system including over 10^7 objects that might be discovered by Pan-STARRS during the course of its ten year operational lifetime. The MOPS is the first asteroid and comet linking system to embed an efficiency determination and monitoring system. By the end of 2007 the PS1 system should be discovering more asteroids each month than all other surveys in the world combined. When PS4 begins operations it will identify in a single month more asteroids than are currently known.

Acknowledgements

The authors thank the entire Pan-STARRS team for their dedicated effort to developing the entire system. The design and construction of the Panoramic Survey Telescope and Rapid Response System by the University of Hawaii Institute for Astronomy is funded by the United States Air Force Research Laboratory (AFRL, Albuquerque, NM) through grant number F29601-02-1-0268. The MOPS is currently being developed in association with the Large Synoptic Survey Telescope (LSST). Jeremy Kubica's work was funded in part by the LSST and by a grant from the Fannie and John Hertz Foundation. The LSST's research and development effort is funded in part by the National Science Foundation under Scientific Program Order No. 9 (AST-0551161) through Cooperative Agreement AST-0132798. Additional funding comes from private donations, in-kind support at Department of Energy laboratories and other LSSTC Institutional Members. The MOPS team also thanks Spacewatch (Robert McMillan), CINEOS (Andrea Boattini), NEODys (Andrea Milani), JPL (Steven Chesley), Mikko Kaasalainen and Joseph Ďurech for their contributions to the MOPS development effort.

References

Bottke, W. F., Morbidelli, A., Jedicke, R., Petit, J. M., Levison, H. F., Michel, P. & Metcalfe, T. S. 2002, *Icarus* 156, 399

Bottke, W. F., Durda, D. D., Nesvorný, D., Jedicke, R., Morbidelli, A., Vokrouhlický, D., & Levison, H. F. 2005, *Icarus* 179, 63

Chambers for the Pan-STARRS team, in the *Proceedings of the Air Force Maui Optical Surveillance meeting*, 2006.

Chang, H.-K., King, S.-K., Liang, J.-S., Wu, P.-S., Lin, L. C.-C., & Chiu, J.-L. 2006, *Nature* 442, 660

Chesley, S. R. & Spahr, T. B. 2004, in: M. J. S. Belton, T. H. Morgan, N. H. Samarasinha & D. K. Yeomans, *Mitigation of Hazardous Comets and Asteroids* (Cambridge University Press, Cambridge), p. 22

Gladman, B., *et al.* 2006, *Icarus*, submitted

Hsieh, H. H. & Jewitt, D. 2006, *Science* 312, 561

Ivezić, Ž., *et al.* 2001, *Astron. J.* 122, 2749

Jewitt, D. 2004 , *Earth, Moon & Planets* 92, 465

Kaasalainen, M. 2004, *Astron. Astrophys.* 422, L39

Kubica, J., *et al.* 2006, *Icarus*, submitted

Magnier for the Pan-STARRS team, in the *Proceedings of the Air Force Maui Optical Surveillance meeting*, 2006

Nesvorný, D., Bottke, W. F., Jr., Dones, L. & Levison, H. F. 2002, *Nature* 417, 720

Stuart, J. S. & Binzel, R. P. 2004, *Icarus* 170, 295

Tonry, J. L., Luppino, G. A., Kaiser, N., Burke, B., Jacoby, G. H. 2002, *Experimental Astronomy* 14, 17

Near Earth Objects, Our Celestial Neighbors: Opportunity And Risk
Proceedings IAU Symposium No. 236, 2006
A. Milani, G.B. Valsecchi & D. Vokrouhlický, eds.
© 2007 International Astronomical Union
doi:10.1017/S1743921307003420

LSST: Comprehensive NEO detection, characterization, and orbits

Željko Ivezić[1], J. Anthony Tyson[2], Mario Jurić[3], Jeremy Kubica[4], Andrew Connolly[5], Francesco Pierfederici[6], Alan W. Harris[7], Edward Bowell[8], and the LSST Collaboration[9]

[1]Department of Astronomy, University of Washington, Seattle, WA 98155, USA
email: ivezic@astro.washington.edu

[2]Department of Physics, University of California, Davis, CA 95616, USA [3]Princeton University Observatory, Princeton, NJ 08544, USA [4]Google Inc., 1600 Amphitheatre Parkway, Mountain View, CA 94043, USA [5]Department of Astronomy, University of Washington, Seattle, WA 98155, USA [6]LSST Corporation, 4703 E. Camp Lowell Drive, Suite 253, Tucson, AZ 85712, USA [7]Space Science Institute, 4603 Orange Knoll Ave., La Canada, CA 91011-3364, USA [8]Lowell Observatory, 1400 W. Mars Hill Rd., Flagstaff, AZ 86001, USA [9] www.lsst.org

Abstract. The Large Synoptic Survey Telescope (LSST) is currently by far the most ambitious proposed ground-based optical survey. With initial funding from the National Science Foundation (NSF), Department of Energy (DOE) laboratories, and private sponsors, the design and development efforts are well underway at many institutions, including top universities and national laboratories. Solar System mapping is one of the four key scientific design drivers, with emphasis on efficient Near-Earth Object (NEO) and Potentially Hazardous Asteroid (PHA) detection, orbit determination, and characterization. The LSST system will be sited at Cerro Pachon in northern Chile. In a continuous observing campaign of pairs of 15 s exposures of its 3,200 megapixel camera, LSST will cover the entire available sky every three nights in two photometric bands to a depth of V=25 per visit (two exposures), with exquisitely accurate astrometry and photometry. Over the proposed survey lifetime of 10 years, each sky location would be visited about 1000 times, with the total exposure time of 8 hours distributed over several broad photometric bandpasses. The baseline design satisfies strong constraints on the cadence of observations mandated by PHAs such as closely spaced pairs of observations to link different detections and short exposures to avoid trailing losses. Due to frequent repeat visits LSST will effectively provide its own follow-up to derive orbits for detected moving objects.

Detailed modeling of LSST operations, incorporating real historical weather and seeing data from Cerro Pachon, shows that LSST using its baseline design cadence could find 90% of the PHAs with diameters larger than 250 m, and 75% of those greater than 140 m within ten years. However, by optimizing sky coverage, the ongoing simulations suggest that the LSST system, with its first light in 2013, can reach the Congressional mandate of cataloging 90% of PHAs larger than 140m by 2020. In addition to detecting, tracking, and determining orbits for these PHAs, LSST will also provide valuable data on their physical and chemical characteristics (accurate color and variability measurements), constraining PHA properties relevant for risk mitigation strategies. In order to fulfill the Congressional mandate, a survey with an etendue of at least several hundred $m^2 deg^2$, and a sophisticated and robust data processing system is required. It is fortunate that the same hardware, software and cadence requirements are driven by science unrelated to NEOs: LSST reaches the threshold where different science drivers and different agencies (NSF, DOE and NASA) can work together to efficiently achieve seemingly disjoint, but deeply connected, goals.

Keywords. survey; telescope; instrumentation; completeness; astrometry; proper elements

1. Introduction

1.1. *The Challenges*

We are immersed in a swarm of Near Earth Asteroids (NEAs) whose orbits approach that of Earth. About 20% these, the potentially hazardous asteroids (PHAs), are in orbits that pass close enough to Earth's orbit (<0.05 AU) that perturbations with time scales of a century can lead to intersections and the possibility of collision. Beginning in 1998, NASA set as a goal the discovery within 10 years of 90% of the estimated 1000 NEAs with diameters greater than 1 km. It is expected that ongoing surveys will in fact discover about 80% of these large (>1 km) NEAs by 2008 (Jedicke *et al.* 2003). However, this mission has been recently extended by the US Congress. The following text became law as part of the NASA Authorization Act of 2005 passed by the Congress on December 22, 2005, and subsequently signed by the President:

"The U.S. Congress has declared that the general welfare and security of the United States require that the unique competence of NASA be directed to detecting, tracking, cataloguing, and characterizing near-Earth asteroids and comets in order to provide warning and mitigation of the potential hazard of such near-Earth objects to the Earth. The NASA Administrator shall plan, develop, and implement a Near-Earth Object Survey program to detect, track, catalogue, and characterize the physical characteristics of near-Earth objects equal to or greater than 140 meters in diameter in order to assess the threat of such near-Earth objects to the Earth. It shall be the goal of the Survey program to achieve 90% completion of its near-Earth object catalogue (based on statistically predicted populations of near-Earth objects) within 15 years after the date of enactment of this Act."

Ground-based optical surveys are the most efficient tool for comprehensive NEO detection, determination of their orbits and subsequent tracking. A survey capable of extending these tasks to NEOs with diameters as small as 140 m requires a large telescope, a large field of view (FOV) and a sophisticated data acquisition, processing and dissemination system.

A 140-meter object with a typical albedo (0.1), positioned in the main asteroid belt (at a heliocentric distance of 2.5 AU), and observed at opposition will have an apparent visual Johnson magnitude of V~25. In order to detect such a faint object with at least 5σ significance, in an exposure not longer than 30 seconds to prevent trailing losses (the maximum exposure time is even shorter for NEOs observed at low elongations around so-called "sweet spots"), a 10-meter class telescope is needed even at the best observing sites. A large FOV, of the order 10 square degrees, and short slew time are required to enable repeated observations of a significant sky fraction with a sufficient frequency. With such observing cadence and depth, observations will necessarily produce tens of terabytes of imaging data per night. In order to recognize NEOs, determine their orbits and disseminate the results to the interested communities in a timely manner, a powerful and fully automated data system is mandatory.

These considerations strongly suggest that in order to fulfill the Congressional mandate, a system with a high etendue (or throughput, defined as the product of the aperture area and field-of-view area), of at least several hundred $m^2 deg^2$, is required. The LSST system has nearly two orders of magnitude larger etendue than that of any existing facility (Tyson 2002), and is the only facility that can detect 140-meter objects in the main asteroid belt in less than a minute.

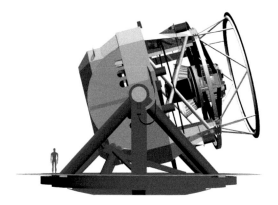

Figure 1. The left panel shows baseline design for LSST telescope, current as of April 2006. The telescope will have an 8.4-meter primary mirror, and a 10-square-degree field of view. The right panel shows LSST baseline optical design with its unique monolithic mirror: the primary and tertiary mirrors are coplanar and their surfaces will be polished into single substrate.

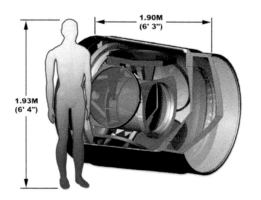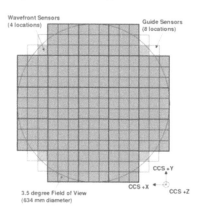

Figure 2. The left panel shows LSST camera with person to indicate scale size. The camera is positioned in the middle of the telescope and will include a filter mechanism and shuttering capability. The right panel shows LSST focal plane. Each cyan square represents one 4096x4096 pixel large sensors. Nine sensors are assembled together in a raft. There are 189 science sensors, each with 16.8 Mpix, for the total pixel count of 3.2 Gpix.

1.2. *The LSST Drivers*

Three recent committees comissioned by the National Academy of Sciences† concluded that a dedicated wide-field imaging telescope with an effective aperture of 6–8 meters is a high priority for US planetary science, astronomy, and physics over the next decade. The LSST system described here will be a large, wide-field ground based telescope designed to obtain sequential images covering the entire visible sky every few nights. The current baseline design allows us to do so in two photometric bands every three nights.

The survey will yield contiguous overlapping imaging of ∼20,000 square degrees of sky in at least five optical bands covering the wavelength range 320–1050 nm. Detailed simulations that include measured weather statistics and a variety of other effects which

† Astronomy and Astrophysics in the New Millennium, NAS 2001; Connecting Quarks with the Cosmos: Eleven Science Questions for the New Century, NAS 2003; New Frontiers in the Solar System: An Integrated Exploration Strategy, NAS 2003.

affect observations predict that each sky location can be visited about 100 times per year, with two 15 sec exposures per visit.

The range of scientific investigations which would be enabled by such a dramatic improvement in survey capability is extremely broad. The main science themes that drive the LSST system design are

(a) Constraining Dark Energy and Matter
(b) Taking an Inventory of the Solar System
(c) Exploring the Transient Optical Sky
(d) Mapping the Milky Way

In particular, the detection, characterization and orbital determination of Solar System objects impact the requirements on relative and absolute astrometric accuracy (10 milliarcsec and 50 milliarcsec, respectively, for sources not limited by photon statistics), and drive the detailed cadence design, discussed further below.

1.3. *The LSST Reference Design*

The LSST reference design†, with an 8.4 m diameter primary mirror, standard filters ($ugrizY$, 320 – 1050 nm), and current detector performance, reaches 24th V mag in 10 seconds‡. With an effective aperture of 6.5m and 9.6 square degree field of view, LSST has an etendue of 320 m^2deg^2. This large etendue is achieved in a novel three-mirror design (modified Paul-Baker) with a very fast f/1.25 beam, and a 3.2 gigapixel camera (with 0.2 arcsec large pixels). The baseline designs for telescope and camera are shown in Figs. 1 and 2. The LSST telescope will be sited at Cerro Pachon, Chile.

2. The LSST PHA Survey

2.1. *The PHA Survey Requirements*

The search for PHAs puts strong constraints on the cadence of observations, requiring closely spaced pairs of observations two or preferably three times per lunation in order to link observations unambiguously and derive orbits. Individual exposures should be shorter than about 20 sec each to minimize the effects of trailing for the majority of moving objects. Because of the faintness and the large number of PHAs and other asteroids that will be detected, LSST must provide the follow-up required to derive orbits rather than relying, as current surveys do, on separate telescopes. The observations should be preferentially obtained within ±15 degrees of the Ecliptic, with additional all-sky observations to increase the completeness at the small size limit (because the smaller asteroids must be closer to the Earth in order to be visible, and therefore are more nearly isotropically distributed).

The images should be well sampled to enable accurate astrometry, with absolute accuracy not worse than 0.1 arcsec. There are no special requirements on filters, although bands such as V and R that offer the greatest sensitivity are preferable. The images should reach a depth of at least 24 (5σ for point sources) in order to probe the $\sim$ 140 m size range at main-belt distances. Based on recent photometric measurements of asteroids by the Sloan Digital Sky Survey, the photometry should be better than 1-2% to allow for color-based taxonomic classification and light-curve measurements.

---

† More details about LSST system are available at http://www.lsst.org.
‡ An LSST exposure time calculator has been developed and is publicly available at http://tau.physics.ucdavis.edu/etc/servlets/LsstEtc.html.

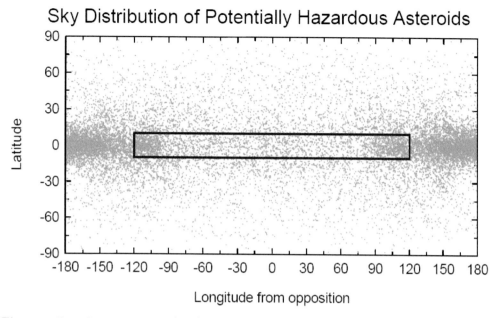

Figure 3. Distribution of PHAs (dots) on the sky in ecliptic coordinate system for objects with $V < H + 2$ (e.g. $V < 24$ for $H = 22$ objects). The box extending $\pm 10°$ in latitude and $\pm 120°$ in longitude from the opposition point represents a region targeted by most current surveys because it yields a good efficiency in surveying for PHAs. However, the baseline LSST PHA survey will not be limited to this region, and will greatly benefit from its frequent all-sky coverage with faint magnitude limits.

2.2. The PHA Survey Baseline Design

The baseline design cadence is based on two revisits closely separated in time (15-60 min) to enable a robust and simple method for linking main-belt asteroids (MBAs). Their sky surface density is about two orders of magnitude higher than the expected density of PHAs, and thus MBAs must be efficiently and robustly recognized in order to find PHAs. MBAs move about 3-18 arcmin in 24 hours, which is larger than their typical nearest neighbor angular separation at the depths probed by LSST (2.3 arcmin on the Ecliptic). Two visits closely separated in time ("re-visits") enable linking based on a simple search for the nearest moving neighbors, with a false matching rate of only a few percent.

The present planned observing strategy is to "visit" each field (9.6 sq. deg.) with two back-to-back exposures of ∼15 sec, reaching to at least V magnitude of ∼24.8. Two such visits will be spaced in time by about half an hour. Each position in the sky will be visited several times during a month, spaced by a few days. This cadence will result in orbital parameters for several million MBAs, with light curves and color measurements for a substantial fraction of each population. Compared to the current data available, this would represent a factor of 10 to 100 increase in the numbers of orbits, colors, and variability of the two classes of object. The large LSST MBA sample will enable detailed studies of the dynamical and chemical history of the Solar System.

2.3. The PHA Survey Simulations

The performance of the baseline design cadence is studied and quantified using detailed simulations (developed by A. Harris and E. Bowell), including real historical weather and seeing data from Cerro Pachon (LSST site). The survey is simulated by generating

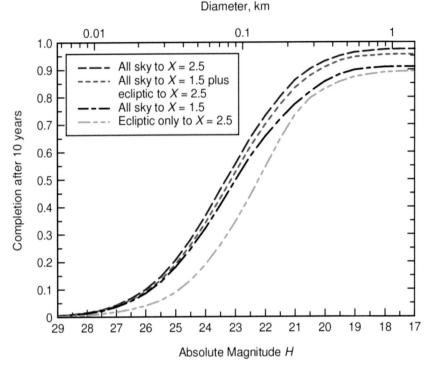

Figure 4. Completeness for PHAs for LSST baseline design cadence as function of absolute magnitude H (or, equivalently, object's size, as marked on top). Different curves correspond to different sky tiling strategies (X is airmass). LSST baseline design cadence results in 90% completeness for PHAs with diameters larger than 250 m, and 75% completeness for those greater than 140 m, after ten years. Ongoing simulations suggest that, with additional optimization of the observing cadence, LSST can achieve 90% completeness for PHAs with diameters larger than 140 m.

a set of 1000 synthetic orbital elements that match the distribution of discovered objects in the large size range where present surveys are essentially complete. Positions and magnitudes in the sky are computed at five day intervals for ten years. An example of the instantaneous distribution of the simulated sample on the sky is shown in Figure 3.

The resulting sample is then "filtered" according to assumed sky coverage and cadence pattern, limiting magnitude of survey instrument, visibility constraints, and magnitude loss according to observing conditions. It is assumed that detections on two days out of three days observed in a month suffice for orbit determination (justification for this assumption is discussed further below). This selection is repeated as a function of absolute magnitude, H, each time tabulating how many of the 1000 objects are "discovered". The completeness as a function of absolute magnitude (i.e. size) for the baseline design cadence is shown in Figure 4.

These simulations suggest that it is essential to cover the ecliptic band to as great an elongation as possible. We find that a band extending $\pm 10°$ in latitude and $\pm 120°$ in longitude from the opposition point (see the blue box in Figure 3) yields fairly good efficiency in surveying for PHAs. This is true because even objects in highly inclined orbits must pass through the ecliptic sometime. Nevertheless, adding coverage of the rest of the sky improves the survey efficiency at the small size limit markedly.

The LSST baseline design cadence can achieve, over 10 years, a completeness of 90% for objects larger than ∼250 m diameter, and 75% completeness for those greater than

Diameter, km

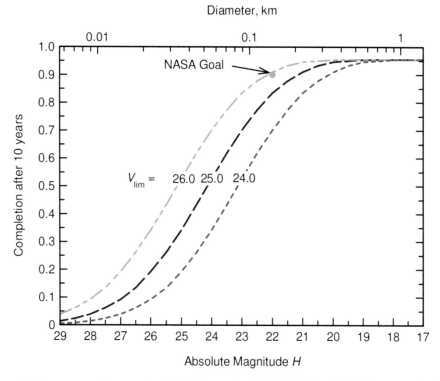

Figure 5. 10-year completeness for PHAs for various LSST single visit limiting magnitudes as function of absolute magnitude H (or, equivalently, object's size, as marked on top). Different curves correspond to different limiting magnitudes per visit. LSST baseline design cadence corresponds closely to the middle curve. If all the data in other bands were used in the search for PHAs, then the completeness rises by 10% reaching the goal of 90% completeness for PHAs with diameters larger than 140 m. For PHAs with large collision cross-section this actually corresponds to nearly 95% completeness, as described in the text.

140 m. However, ongoing simulations show that by using all available data, optimizing filter choice and operations, LSST would be capable of just reaching a completeness of 90% for PHAs larger than 140 m in ten years. For example, adjusting the depth of each visit affects completeness as shown in Figure 5. Simply including all the relevant data from other filter bands can also increase completeness. This could result in an increase of 10% for 140 m PHAs, moving up from the middle curve to the top curve. Moreover, note that none of the curves saturate at 100%. That remaining 5% represents PHAs with mainly low collision cross section – thus they pose no substantial hazard. In effect, the completeness for truly hazardous PHAs of >140m size is already close to 90% in the LSST baseline cadence. Due to its large etendue, LSST system enables sufficiently frequent sky coverage to assure multiple detections per lunation even at the 140 m limit.

Perhaps the most important insight from these simulations is the realization that a system with an etendue of at least several hundred $m^2 deg^2$ is mandatory for fulfilling the Congressional mandate to catalog 90% of PHAs larger than 140m.

2.4. *From Many Detections to Orbits*

The LSST PHA survey simulations described above assume that detections on two days out of three days observed in a month suffices for orbit determination. The linkage of individual detections and orbit determination will be a formidable task for any large area survey. The increase in data volume associated with LSST will make the extraction of tracks and orbits of asteroids from the underlying clutter a significant computational challenge.

The combinatorics involved in linking multiple observations of $\sim 10^6$ sources spread out over several nights will overwhelm naive linear and quadratic orbit prediction schemes. Tree-based algorithms for multihypothesis testing of asteroid tracks can help solve these challenges by providing the necessary 1000-fold speed-ups over current approaches while recovering 99% of the underlying objects.

In addition, observations of asteroids are often incomplete. Sources fall below the detection threshold in one or more of the series of observations. Weather results in incomplete sampling of the temporal data. Noise introduces photometric and astrometric uncertainties. Combined, these observational constraints can severely limit our ability to survey large areas by requiring that we revist a given pointing on the sky more often than necessary. The LSST's approach is designed to be robust to these effects.

LSST will use a three stage process to find new moving objects (Kubica 2005; Kubica *et al.* 2005). In the first stage intra-night associations are proposed by searching for detections forming linear "tracklets." By using loose bounds on the linear fit and the maximum rate of motion, many erroneous initial associations can be ruled out. In the second stage, inter-night associations are proposed by searching for sets of tracklets forming a quadratic trajectory. Again, the algorithm can efficiently filter out many incorrect associations while retaining most of the true associations. However, the use of a quadratic approximation means that a significant number of spurious associations still remains. In the third stage, initial orbit determination and differential corrections algorithms are used to further filter out erroneous associations by rejecting associations that do not correspond for a valid orbit. Each stage of this strategy thus significantly reduces the number false candidate associations that the later and more expensive algorithms need to test.

To implement this strategy, the LSST team has developed, in a collaboration with the Pan-STARRS project (Kaiser *et al.* 2002), a pipeline based on multiple kd-tree data structures (Barnard *et al.* 2006). These data structures provide an efficient way of indexing and searching large temporal data sets. Implementing a variable tree search we can link sources that move between a pair of observations, merge these tracklets into tracks spread out over tens of nights, accurately predict where a source will be in subsequent observations and provide a set of candidate asteroids ordered by the likelihood that they have valid asteroid tracks. Tested on simulated data, this pipeline recovers 99% of correct tracks for NEOs and MBAs, and requires less than a day of CPU time to analyze a night's worth of data. This represents a several thousand fold increase in speed over a naive linear search. It is noteworthy that comparable amounts of CPU time are spent on kd-tree based linking step (which is very hard to parallelize) and on posterior orbital calculations to weed out false linkages (which can be trivially parallelized).

Given the predicted astrometric accuracy of the survey (per exposure: 10 milliarcsec relative and 50 milliarcsec absolute, root-mean-square scatter per coordinate for sources not limited by photon statistics) and the efficiency of the tracking algorithms described above, LSST can lose all but two observations in one 24 hour period and still recover sufficient information to determine the tracklet (linking pairs of re-visits) required to initiate a search for an asteroid. Pairs of these tracklets can be observed as infrequently as once every eight nights and still result in a candidate asteroid being identified with an accuracy of 90% or greater.

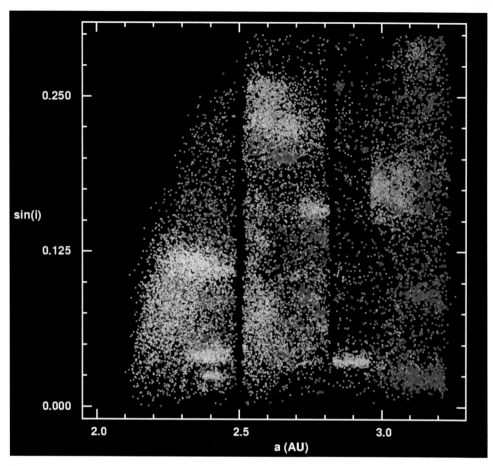

Figure 6. The distribution of ~30,000 asteroids from the SDSS Moving Object Catalog (Available from http://www.sdss.org/science/index.html) in the plane spanned by the proper semi–major axis and orbital inclination (approximately, the x axis is proportional to the distance from the Sun, and y axis is proportional to the distance from the orbital plane) computed by Milani & Knežević (1992). The dots are colored according to their *measured* SDSS colors. The clusters of points are Hirayama's dynamical families, proposed to represent remnants of larger bodies destroyed in collisions. SDSS data vividly demonstrate a strong correlation between dynamics and colors, in support of Hirayama's hypothesis. LSST will extend this map to a three times smaller size limit for more than ten times larger sample, and will also provide structural information by measuring light curves for most objects. LSST will make it possible to produce analogous plot for about the same number of trans-Neptunian objects as shown here for main-belt asteroids.

2.5. *PHA Characterization*

LSST will not only obtain orbits for PHAs, but will also provide valuable data on their physical and chemical characteristics, constraining the PHA properties relevant for risk mitigation strategies. LSST will measure accurate colors for a substantial fraction of detected moving objects, thereby allowing studies of their surface chemistry, its evolution with time, and of dynamical (collisional) evolution. As a recent example obtained by a modern large-area multi-color survey, Figure 6 shows a correlation between orbital elements and optical colors for MBAs measured by the Sloan Digital Sky Survey (Ivezić *et al.* 2001). The clusters of points visible in the figure are Hirayama's dynamical families, proposed to represent remnants of larger bodies destroyed in collisions. SDSS data vividly demonstrate a strong correlation between dynamics and colors, in support of Hirayama's hypothesis. LSST will extend this map to a three times smaller size limit, and will also provide structural information by measuring light curves for the majority of objects. The variability information carries important information about the physical state of an asteroid (e.g. solid body vs. a rubble pile; Pravec & Harris 2000), and these new data will constrain the size-strength relationship, which is a fundamental quantity that drives the collisional evolution of the asteroid belt.

Inevitably, some PHAs will be sufficiently interesting, based on provisional orbits from LSST data, that they should be continuously followed to ascertain their detailed light-curve (rubble pile or solid body). Through Las Cumbres Observatory, one of the LSSTC member institutions, we will have access to a worldwide array of 2m-class telescopes for this continuous tracking. These telescopes will be instrumented with identical detectors and filters to those on LSST in a dichroic camera capable of taking simultaneous images at multiple wavelengths. This color "movie" will aid characterization of these PHAs.

3. Conclusions

LSST is in a unique position to fulfill the Congressional mandate to to achieve 90% completion for 140 m large NEOs because such a survey requires a large telescope to achieve necessary depths (at least $V \sim 24.5$) with exposure times not longer than 15-20 seconds (to avoid trailing losses), a large field of view to be able to scan the sky at a required pace, and a sophisticated data acquisition, processing and dissemination system to handle billions of detectable stars and galaxies.

Detailed simulations suggest that the LSST baseline design cadence will achieve, over 10 years, a completeness of 90% for objects larger than $\sim$250 m diameter, and 70% completeness for those greater than 140 m. Ongoing simulations suggest that with further minor optimization of the baseline cadence LSST would be capable of reaching the Congressional target completeness of 90% for PHAs larger than 140 m. In addition, LSST will not only obtain orbits for PHAs, but will also provide valuable data on their physical and chemical characteristics, constraining the PHA properties related to risk mitigation.

In summary, LSST will, with its unprecedented power for discovering moving objects, make a giant leap forward in the studies of the dynamical and chemical history of the whole Solar system.

References

Barnard, K., Connolly, A., Denneau, L., *et al.* 2006, in: D. R. Silva & R. E. Doxsey (eds.), *Observatory Operations: Strategies, Processes, and Systems*, Procs. of the SPIE, 6270, p. 69
Harris, A. W. & Bowell, E. L. G. 2004, *BAAS* 36, 1530
Ivezić, Ž., Lupton, R. H., Jurić, M., *et al.* 2002, *Astron. J.* 124, 2943
Jedicke, R., Morbidelli, A., Spahr, T., Petit, J-M. & Bottke, W. F. 2003, *Icarus* 161, 17
Kaiser, N., Aussel, It., Burke, B. E., *et al.* 2002, in: J.A. Tyson & S. Wolff (eds.), *Survey and Other Telescope Technologies and Discoveries*, Proceedings of the SPIE, 4836, p. 154
Kubica, J., Masiero. J., Moore, A., Jedicke, R. & Connolly, A. 2005, in: Y. Weiss, B. Schölkopf & J. Platt (eds.), *Advances in Neural Information Processing Systems 18* (MIT Press, Cambridge), p. 691
Kubica, J. 2005, *Efficient Discovery of Spatial Associations and Structure with Application to Asteroid Tracking*, PhD Thesis, Robotics Institute, Dec 2005, Carnegie Mellon University, Pittsburgh, PA
Milani, A. & Knežević, Z. 1992, *Icarus* 98, 211
Pravec, P. & Harris, A. W., 2000, *Icarus* 148, 12
Tyson, J. A. 2002, in: J. A. Tyson & S. Wolff (eds.), *Survey and Other Telescope Technologies and Discoveries*, Proceedings of the SPIE, 4836, p. 10

Near Earth Objects, Our Celestial Neighbors: Opportunity And Risk
Proceedings IAU Symposium No. 236, 2006
A. Milani, G.B. Valsecchi and D. Vokrouhlický, eds.
© 2007 International Astronomical Union
doi:10.1017/S1743921307003432

Searching for NEOs using Lowell observatorys Discovery Channel Telescope (DCT)

Edward Bowell, Robert L. Millis, Edward W. Dunham, Bruce W. Koehn and Byron W. Smith

Lowell Observatory 1400 W. Mars Hill Road Flagstaff AZ 86001, USA email:elgb@lowell.edu

Abstract. We discuss the potential contribution of the Discovery Channel Telescope (or a clone) to a detection program aimed at discovering 90% of potentially hazardous objects (PHOs) larger than 140 m in diameter. Three options are described, each involving different levels of investment. We believe that LSST, Pan-STARRS, and DCT, working in a coordinated fashion, offer a cost-effective, low-risk way to accomplish the objectives of the extended NEO search program.

Keywords. survey; telescope; active optics; data management; data archiving; completeness

1. Introduction

Lowell Observatory's 4.2-m Discovery Channel Telescope is currently under construction at an outstanding new site in Northern Arizona (Fig.1). The product of a unique research and public education partnership between Lowell and Discovery Communications, Inc., the DCT offers uncommon versatility. Of particular relevance to NEO detection is the telescope's prime focus 2°-diameter wide-field imaging capability and its very rapid slew and settle time. In its alternative Ritchey-Chrétien, Nasmyth, and bent Cassegrain configurations, the DCT can bring an array of other instruments to bear on NEO characterization.

The 2360-m elevation Happy Jack site, on the Coconino National Forest 65 km SSE of Flagstaff, Arizona, was first identified following a years-long search, and was selected as the site for the DCT after an intensive site-testing program spanning all seasons. Median seeing, as measured in white light with a differential image motion system mounted a few meters above the ground, was found to be 0.84 arcsec, with the average of the best quartile seeing at 0.62 arcsec (Bida *et al.* 2004). The DCTs altitude axis will be about 12 m above the ground, so even better image quality can be expected. In November 2004, Lowell Observatory received a Special Use Permit from the U.S. Forest Service, allowing construction of the DCT.

The telescope, telescope enclosure, and the wide field camera designs were subjected to formal conceptual design review in July 2004. Since then, detailed design of the site infrastructure, telescope enclosure, and auxiliary building has been completed. The access road to the telescope site is finished, power and communications conduits have been brought to the site, and the buildings at the summit are well on the way to completion (see Fig. 2). The ULE meniscus primary mirror blank has been completed by Corning, and is undergoing figuring and polishing at the College of Optical Sciences at the University of Arizona. Meanwhile, designs of the telescope mount and wide-field corrector optics

Figure 1. Artists conception of the completed DCT installation.

have been further refined in response to issues identified during the conceptual design review. First light is scheduled for late 2009 or early 2010.

Figure 2. Aerial view of the DCT site, telescope enclosure, and auxiliary building (June 2006).

Telescope Facility. The DCT facility consists of a telescope building, auxiliary building, and equipment yard. The telescope building supports the dome, and includes the control room, computer room, instrument work room, and electrical equipment room. The heated work spaces below the telescope are insulated, and an intermediate airspace is flushed to prevent detrimental heat transfer to the observing chamber. The telescope mount is supported on an isolated foundation to avoid motion transfer from building wind loads. The auxiliary building is located away from the telescope building; it houses the mirror-coating facility and heat- and vibration-producing equipment, such as pumps and compressors. The equipment yard, next to the auxiliary building, includes glycol chillers, backup power generator, and propane, water, and sewage tanks.

Dome. The dome design is based on that of the similarly sized SOAR telescope dome. Using local wind statistics for computational fluid dynamics modeling of the telescope and enclosure, large ventilation openings around the equator of the dome have been included to minimize dome seeing effects. Incorporation of lessons learned from the SOAR dome is being considered in the detailed design of the DCT dome to improve its reliability and performance.

Mount. The telescope mount is a conventional altitude-over-azimuth design (Fig. 3), offering a $3°/s$ slew rate and a 6-s step and settle time for $2°$ offsets to adjacent fields. The design work is being performed by Vertex RSI, and draws on design experience with the SOAR 4.1-m and VISTA 4.0-m mounts by using similar or identical components for axial drives, bearings, and encoders. The mount structure is steel, with the exception of the tube truss elements, which are carbon composite to reduce thermal defocus sensitivity. Two top-end assemblies for the telescope will be available, enabling selection between the prime focus instrument and the secondary mirror for the Ritchey-Chrétien configuration. Finite-element analysis of the mount structure has been used to verify dynamic performance and deformations due to gravity and wind.

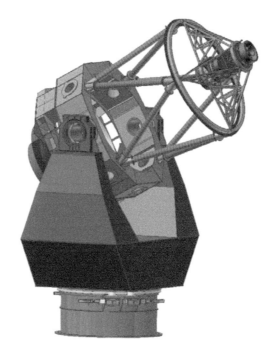

Figure 3. Rendering of the DCT mount in the prime focus configuration.

Active Optics. The active optics system provides active collimation and primary mirror figure control. In the prime focus configuration, collimation is achieved by a combination of actuation at the instrument support, and tip, tilt, and piston of the primary mirror using the figure control actuators. The primary mirror figure control system is based on the system developed for the SOAR telescope, with improvements to communications bandwidth necessary to support the step and settle requirements for the $2°$ offsets required by the NEO search mission. Optical and finite-element modeling has been performed to verify the correctability performance of the collimation and figure control systems.

Prime Focus Assembly. The DCT Prime Focus Assembly (PFA) consists of the Prime Focus Camera (PFC), the Wide Field Optical Corrector (WFOC), and the supporting spider structure. The PFC includes the CCD array, the dewar and cooling system, and the control and data-handling system to the point of initial data storage. Subsequent data handling, real-time analysis, archiving, etc., falls under the purview of the specific observational program for which the camera is used. The WFOC comprises the corrective optics, including an integral atmospheric dispersion compensator (ADC), filters and filter changer, shutter, and instrument rotator.

Key requirements for NEO search work have been factored into the PFAs design from the inception of the DCTs development. The requirements include image quality that degrades the median seeing by less than 10%; a single broad-band filter, encompassing the Johnson V and R passbands, for NEO and TNO searches; transmission from 330-110 nm; atmospheric dispersion compensation to a zenith distance of 75°; and CCD read time of 8 s or less, overlapped with a 6-s telescope move and settle for a 2° move.

The PFA has been the subject of several design studies carried out by EOST, Goodrich Corporation, and e2v. These companies delved into the optical design of the WFOC and its ADC (Blanco *et al.* 2002; MacFarlane and Dunham 2004, 2006), the mechanical aspects of the WFOC (Delp *et al.* 2004), and details of the PFC design (Dunham and Sebring 2004). Our current understanding of the PFA design derives from these studies, and is summarized here.

The DCT wide field optical corrector comprises five lenses (L1 through L5). L3 is an ellipsoid; L4 is the atmospheric dispersion compensator, which operates by means of a tilt and decenter mechanism; filters are placed between L4 and L5; and L5 is a 6th-order asphere. All elements are fused silica except for one Schott LLF6 or Ohara PBL6Y meniscus, the latter of which also provides the ADC function by means of a tilt and decenter mechanism.

The optical performance for a survey band broader than that envisaged for the NEO search is almost everywhere better than 0.5 arcsec FWHM, even at a zenith angle of 75°. Thus the design easily meets its image quality and ADC performance requirements.

The design has been scrubbed to improve the manufacturability and cost of the optics and to reduce their mounting tolerances. The opto-mechanical design described by Delp *et al.* (2004) has been simplified to take advantage of the change from a tumbling telescope top end to a swappable top end. A notable improvement in this regard is that the shutter is now commercially available through the University of Bonn. The design now rotates all components from the camera up to, but not including, the ADC element, which permits accurate flat fielding of frames. The ADC mechanism is a simple one-dimensional motion, with loose tolerances, using a flexure and cam to provide the necessary tilt and decenter motions. Focus and alignment of the PFA are accomplished by shimming and adjusting the primary mirror position and tilt through the active primary mirror support, as described above.

The PFC contains a CCD mosaic array mounted in a dewar and cooled to -100 C by four CryoTiger mechanical coolers. The last element of the WFOC serves as the dewar window. The focal plane is baselined as 40 e2v CCD44-82 CCDs. These are $2K \times 4k$ back-illuminated CCDs with 15 μm pixels and two output amplifiers. Thirty-six are used for science data, and 4 are set aside for guiding and wavefront curvature sensing, as shown in Fig. 4. The entire science array can be read out in 6 s, with read noise of about 6-8 e^-, using two NOAO Monsoon CCD controllers having synchronized clocks. A third controller will independently operate the guide and wavefront sensing CCDs. Focal plane data are delivered to the control and reduction computers in the computer room over a gigabit fiber network connection (Wiecha & Sebring 2004).

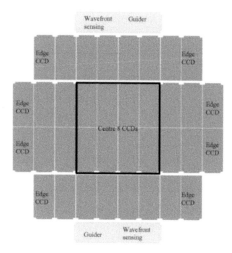

Figure 4. PFC focal plane layout. Pairs of CCDs at left and right are set aside for guiding and wavefront sensing.

We are taking advantage of lessons learned in similar large format cameras such as MMTs Megacam, VSTs OmegaCAM, CFHTs MegaCam, and the Kepler mission focal plane. The primary challenges in the PFC design are not related to the size of the array, per se, but to the fast focal ratio and the associated focal plane flatness requirement. A secondary issue is the basic logistical problem of cable routing. These items, the subject of the e2v study, were described by Dunham & Sebring (2004). The baseline approach is to use a lightweighted Invar mounting plate backed by a copper thermal spreader, all mounted using titanium flexures. This compromise, among gravitational sag, thermal distortion, and thermal conductivity, meets all our requirements. Cable management inside the dewar will be accomplished using custom flat flex cables and interface PC boards. The flex cables run from the science CCDs to interface boards located inside the dewar. The boards are not yet designed, but will include line filters and zener diode protection, and possibly preamplifiers as well. A second set of flex cables, potted into vacuum feedthroughs, will bring the signals outside the dewar. These cables will be terminated with more robust connectors to mate with the controller cables. Potting the cables into feedthroughs greatly reduces the required real estate on the dewar walls. This approach has been used in particle physics experiments and is in use with a Kepler test unit at Ball Aerospace.

2. NEO Detection Using the DCT

DCTs prime focus wide-field camera, with its 2.3 deg^2 FOV and 0.32 arcsec pixel size, will be a fine NEO search instrument. The tiling of the science CCDs on the sky plane has essentially no overlap and little area is lost to gaps between CCDs. The telescopes étendue of 39 m^2 deg^2 will exceed the total étendue of existing NEO search telescopes by almost an order of magnitude, which implies that DCT could, by itself, carry out NASA's current search for 90% of NEAs larger than 1 km in diameter in three or four years.

Our effort using DCT will concentrate on discovering and identifying potentially hazardous asteroids (PHAs: asteroids whose orbits pass within 0.05 AU of the Earths orbit), although other NEAs (aphelion distance $q \; \mathrm{i} \; 1.3$ AU) and comets, some of the latter also potentially hazardous, will be found in abundance. Note that, to a given diameter limit, there are about four times as many NEAs as PHAs. For PHA detection, using a VR-type filter (probably a long-pass Schott filter), the limiting magnitude of the camera/telescope system will depend on a number of factors, including the PHAs sky-plane motion and the chosen observing cadence. Nominally, we plan to make four exposures/night on a chosen region of the sky (two back-to-back 20-s exposures, repeated within, say, 30 minutes) and then to make similar observations on one or two additional nights during a lunation. Such a monthly cadence will provide the necessary self-followup, except for PHAs that "leak" out of the search region. In average seeing and moonless conditions, the cadence should lead to a limiting magnitude, for most PHA detected, of $VR_{lim} = 23.8$ (4.1σ above noise) on a single night. For PHAs detected on three nights, VR_{lim} should, on average, be near 24.1 mag.

Orbits for most PHAs observed on two or three nights/lunation will be accurate enough for meaningful Earth-impact calculations to be carried out by others within a few days to a few weeks of their discovery. Most PHAs larger than 140 m in diameter will be observed for more than one lunation, and the resulting ephemeris accuracy will be good enough, if necessary, for their recovery during a future apparition. Our observing cadence and planned rapid dissemination of moving-object astrometry will also facilitate expeditious physical observations by others. It is planned that all imaging data will be stored indefinitely, which will allow post facto searches to small S/N ratio.

To understand what DCTs 15-year search performance might be, we refer to Fig. 5 (from Harris & Bowell 2004), which pertains to a 10-year search to $V_{lim} = 24$ mag, a limiting magnitude close to what DCT will achieve.

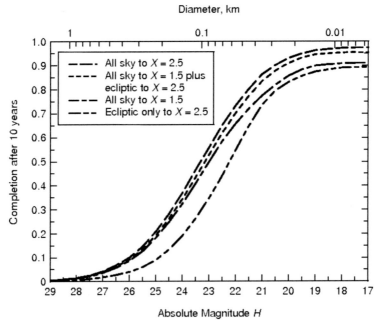

Figure 5. The estimated PHA completeness of various sky survey patterns as a function of PHA diameter/absolute magnitude.

The computations, using the method of Harris (1998), assumed that detections on two days out of three days observed in a month suffice for orbit determination. A moderate magnitude loss with air mass X of $dm_{lim} = -2.0 log X$ was applied. Clearly, an all-sky survey is physically impossible, so the three upper curves represent upper bounds. Harris and Bowell showed that, for deep, wide-field survey telescopes such as DCT, it is profitable to confine PHA searches largely to an ecliptic-centered zone (although faint comets are more isotropically distributed).

Using the 3-nights/lunation cadence described above, DCT could, if dedicated 100% to PHA searching, observe an annualized average of about $4,000 \ deg^2$ of "fresh" sky/lunation, which corresponds to an ecliptic zone about 17° wide covering the entire zodiac at solar elongations ¿ 60°. For a 2-night/lunation cadence, the areal coverage would increase by 50%. Thus DCTs predicted performance corresponds approximately to that of the lowest curve in Fig. 9, which represents a 20°-wide ecliptic zone covering solar elongations > 45°. Therefore, in a 10-year search, DCT could, by itself, discover about half of the PHAs larger than 140 m in diameter (for which, on average, $H \simeq 22.8$ mag). One can estimate the result of a longer search program by noting that the discovery rate of PHAs of a given size decays exponentially with time. Allowing for the lowest curves asymptote to 90% completeness for a search of infinite duration, we estimate that DCT could detect about two-thirds of PHAs larger than 140 m in diameter–about 15,000 PHAs–in 15 years.

We can estimate the number of PHAs of all sizes that DCT would detect in 15 years from Fig. 8 of Bowell & Muinonen (1994). Using an improved PHA population model, and ignoring trailing losses for very small, fast-moving objects, we find, very roughly, that DCT could detect about 10^6 PHAs, the median size of which would be just tens of meters. We estimate that, on average, about 40 PHAs per hour could be detected. In addition, we estimate that DCT could detect more than 5000 comets.

Slavish adherence to the ecliptic-zone observing regime is certainly not optimum for PHA detection. For example, after dusk and before dawn, it will likely be advantageous to observe in the so-called sweet spots (Stokes *et al.* 2003), on the ecliptic at small solar elongations, where the surface density of PHAs is enhanced. Note that DCTs atmospheric dispersion compensator provides good images to a zenith distance exceeding 75°. For 50% or 25% usage of DCT, the rate of PHA detection would, to first order, be cut proportionately, in which case the shape of the ecliptic-zone search region could be altered to reduce PHA "leakage". However, much more sophisticated numerical modeling than has hitherto been carried out is required to understand the effects of moonlight, interruptions due to weather and equipment failure, variable seeing, PHA rotational brightness modulation, and other impediments to optimum observing.

Finally, we note that DCT could usefully partner with other anticipated groundbased NEO search facilities. For example, LSST (southern hemisphere) and DCT plus Pan-STARRS (northern hemisphere) could, with careful coordination, observe the entire dark sky two or more times per lunation. Modeling the results of such a combined search effort has not yet been carried out, but one can be almost certain that NASAs mandate to catalog 90% of the impact risk within 15 years could be achieved using these three groundbased facilities alone.

3. Data Management and Archiving

The data processing and archiving stream will be based on the processing architecture of the currently operational Lowell Observatory Near-Earth-Object Search (LONEOS). A separate processor will be assigned to each CCD amplifier, just as at LONEOS, so an entire frame will be reduced in near real-time. With a planned exposure time of 20 s, a read time of 6 s, and a slew time of 6 s, the data rate will be 82 GB/hr and each processor will have 26 s to read and reduce a frame segment. LONEOS currently performs at approximately this rate using 2.4 GHz processors. The per-amplifier cost of computation will be similar to that at LONEOS. There we use dual-processor LINUX-based computers, one for each CCD.

We plan to use at least two reduction techniques. The first will be based on traditional source detection. Any NEO or PHA detected in this process will be immediately reported to the Minor Planet Center. The second technique, requiring far more computer processing, will be based on image subtraction, which is already in routine use at Lowell. All moving objects detected using these techniques will be made public as soon as possible, usually within a few hours of observation. Note that, although we have good methods linkage and orbit computation, they are not discussed here.

We plan to archive all frames. A typical night of 10 hours' observation will produce about 0.9 TB of data. Using the techniques developed at LONEOS, we will transfer the data to a RAID server and also keep a backup disks offline in a remote vault. We have developed software that can make these frames available to other researchers via the server. Higher density disks will become available as the project proceeds, so the number of archive computers will likely never exceed 20. Possibly, the archiving task could be outsourced.

To maintain data processing reliability, existing software will have to be slightly modified to permit continued data stream processing when one or more computers are unavailable. At LONEOS, the reductions are loosely synchronized to frame production, so if the reductions fall behind, the frames are statistically queued. By adding dynamic assignment of a reduction computer to a queue, the reduction stream will be able to tolerate multiple computer failures.

4. Schedule

As already noted, the construction of the DCT is well underway, and, if sufficient funds become available, will be completed by 2010. The PFA, which is not currently funded, will take about four years to construct. For a DCT clone, the time to first light is limited by the fabrication of the primary mirror, which it is thought would take about 4.5 yr. Thus "DCT2" could see first light by 2012 if funding is made available soon.

References

Bida, T. A., Dunham, E. W., Bright, L. P. & Corson, C. 2004, *Proc. SPIE* 5489, 196†

Blanco, D., Pentland, G., Smith, C., Dunham, E. & Millis, R. 2003, *Proc. SPIE* 4842, 85

Bowell, E. & Muinonen, K. 1994. in T. Gehrels (ed.), *Hazards Due to Comets and Asteroids* (Tucson: The University of Arizona Press), p. 149

Delp, C., Duffy, M., Neill, D. & Sebring, T. A. 2004, *Proc. SPIE* 5495, 216

Dunham, E. W. & Sebring, T. A. 2004, *Proc. SPIE* 5492, 1471

Epps, H. W. & DiVittorio, M. 2004, *Proc. SPIE* 4842, 355

Harris, A. W. 1998, Planet. Space Sci. 46, 283

Harris, A. W. & Bowell, E. 2004, *Bull. Amer. Astron. Soc.* 36, 1530

MacFarlane, M. J. & Dunham, E. W. 2004, *Proc. SPIE* 5489, 796

MacFarlane, M. J. & Dunham, E. W. 2006, *SPIE* paper 6267-05

Smith, B. W., Bida, T. A., Millis, R. L., Dunham, E. W., Wiecha, O. M., & Marshall, H. K. 2006, *SPIE* paper 6267-05

Stokes, G. H., *et al.* 2003. Study to determine the feasibility of extending the search for near-Earth objects to smaller limiting diameters. NASA.

Wiecha, O. M. & Sebring, T. A. 2004, *Proc. SPIE* 5496, 701

† Most of these papers are available from `http://www.lowell.edu/DCT/html/papers.html`

Near Earth Objects, our Celestial Neighbors: Opportunity and Risk
Proceedings IAU Symposium No. 236, 2006
A. Milani, G.B. Valsecchi & D. Vokrouhlický, eds.
© 2007 International Astronomical Union
doi:10.1017/S1743921307003444

NEO-related scientific and outreach activities at KLENOT

Jana Tichá, Miloš Tichý and Michal Kočer

Kleť Observatory & KLENOT Project, Zátkovo nábřeží 4, CZ-370 01 České Budějovice,
South Bohemia, Czech Republic
email: klet@klet.cz

Abstract. In the recent times, there has been a noticeable increase in interest about NEOs. In the light of results of recent NEO surveys the need for continuous follow-up astrometry to secure the orbits of the discovered bodies has risen.

The Kleť Observatory Near Earth and other Unusual Objects Observation Team and Telescope, project KLENOT, started in 2002. It is dedicated to confirmation, follow-up, and recovery of NEOs. For this task the 1.06-m KLENOT telescope, equipped with a high-efficiency CCD is used. Since it was set in service in March 2002, a significant number of results have been obtained. For instance, 10,000 positions have been determined, more than 400 confirmations made, including 16 recoveries and 104 Virtual Impactors measured.

An important part of NEO search is the discovery of comets. Therefore an inseparable component of NEO follow-up includes the detection of possible cometary features of newly discovered bodies and the confirmation of Near Earth Comets (NEC).

In the next decade, surveys will be characterized by several aspects. Pan-STARRS opens the question whether "classical" confirmation and early follow-up of newly discovered NEO candidates will be necessary by others, and we discuss the need for possible changes in existing system of follow-up process. The most important challenge for follow-up will likely become the NEOs which are in urgent need of astrometric positions over longer arcs including Virtual Impactors, radar, and mission targets. The observing strategies and obtained results of the KLENOT Project will be presented as well as future plans will be discussed.

One of the most important duties of NEOs scientists and research institutions is to maintain contact with the public. NEO related issues have outstanding educational value and outreach potential. Nowadays the Internet has proved itself to be an excellent mean of bringing NEO knowledge to a wider audience.

Keywords. Astrometry; follow up; telescope; outreach

1. Introduction

Mainly due to increased interest from the scientific community and fortunately also from the general public, a tremendous work has been done in the study of NEOs in the last decade. Although, at least according to our present knowledge, we are not living under immediate danger of collision with potentially hazardous object, we have to carefully consider our response to such threats. Therefore the orbit determination with high precision is imperative. The task is, of course, more difficult and complex then it can be seen at the first sight. In its complexity the astrometry of NEOs is important but often overlooked activity. This paper presents some of the methods and procedures which contribute to reliable orbit determination of NEOs. The importance of follow-up astrometry including recovery in the second opposition is particularly stressed.

2. Kleť Observatory

The Kleť Observatory is located in the Czech Republic in central Europe.

The IAU/MPC observatory codes are (046) Kleť Observatory, České Budějovice and (246) Kleť Observatory-KLENOT. The geographical position of the observatory is longitude $\lambda = 14°17'17''E$, latitude $\phi = +48°51'48''N$ and elevation $= 1069$ m. There are about 180 clear nights per year with best weather conditions in February, March, August, September and October.

3. KLENOT Project

3.1. KLENOT telescope

The KLENOT telescope was built using an existing dome and infrastructure of the Kleť Observatory. The original mounting was upgraded and the optoelectronical control system was added. A new control and computer room was built. All observing time belongs to the KLENOT team.

Optical system consists of 1.06-m f/3 main mirror (Sital glass, Carl Zeiss Jena) and four lenses primary focus corrector computed by Sincon, Turnov, Czech Republic and manufactured at the Optical Facility of Charles University, Prague, Czech Republic. It yields a 1.06-m f/2.7 optical system.

For the KLENOT telescope we use CCD camera Photometrics Series 300 with chip SITe 003B 1024 x 1024 pixels, pixel size 24 microns, liquid nitrogen cooling and Q.E. > 80 per cent in range 550–800 nm and > 60 per cent in range 370–880 nm.

Field of view is 33×33 arcminutes. Image scale is 1.9 arcseconds per pixel.

The limiting magnitude is $m_V = 22$ for 180-sec exposure time in standard observing conditions.

3.2. KLENOT Goals

- Confirmatory observations of newly-discovered, faint NEO candidates

Some of new search facilities produce discoveries fainter than $m_V = 20$ (for example 1.8-m Spacewatch II, 1.5-m Mt. Lemmon Survey) which need a larger telescope for confirmation and early follow-up. A 1-m class telescope is also very suitable for confirmation of very fast moving objects because larger aperture helps minimize trailing losses and our larger field of view enables us to search for NEO candidates having a larger ephemeris uncertainty.

- Follow-up astrometry of poorly observed NEOs

It is necessary to observe newly discovered NEOs in a longer arc during the discovery opposition often as they are getting fainter. Special attention is given to "Virtual Impactors" and PHAs, targets of future space missions or radar observations. On the other hand, it is necessary to find and use an optimal observing strategy to maximize orbit improvement of each asteroid.

- Recoveries of NEOs in the second opposition

For the determination of reliable orbits it is required to observe asteroids in more then one opposition. If the observed arc in a discovery apparition is long enough, the chance for a recovery in the next apparition is good. If the observed arc at single opposition is not sufficiently long, a larger uncertainty in its orbit determination requiring a search along the line of variations. For this purpose our larger field of view is a clear advantage.

- Follow-up astrometry of other unusual objects

We plan to make follow-up astrometry of other unusual objects, such as Centaurs and transneptunian objects, both at discovery and subsequent apparitions. To obtain positions of brighter transneptunians, we propose to use longer exposures with magnitude limit about $m_V = 22$. Considering the problem with acquiring adequate data for orbit computation of these objects, follow-up astrometry, at least for some of them, will be useful.

• Cometary features

The majority of new ground-based discoveries of comets come from large NEO surveys. The first step in finding these new members of the population of cometary bodies consists in confirmatory astrometric observations along with inspection for cometary features. A timely recognition of new comets can help in planning further observation campaigns.

• Search for new asteroids

Even though our primary goal is astrometric follow-up of NEOs and other unusual objects, all of our CCD images are processed not only for target objects, but also examined for possible new object(s). This can be achieved because the effective field of view, exposure time, and the limiting magnitude of $m_V = 22$ enable us to find many possible new object(s). Obtained images are processed with special attention to fast- and slow-moving objects.

4. Technology

A special software package has been developed for the KLENOT Project using a combination of programs running on Windows and Linux platforms. The system consists of observation planning tools, data-acquisition, camera control and data processing tools.

4.1. *Observations planning*

The SQL database stores information on minor planets updated on daily bases from text-based databases; the MPC Orbit Database (`MPCORB`), maintained by the Minor Planet Center, and from the Asteroid Orbital Elements Database (`ASTORB`), created and maintained by E. Bowell at the Lowell Observatory. The asteroids listed on Spaceguard system Priority List and objects listed as a Virtual Impactors by SENTRY (JPL) or by CLOMON2 (NEODyS) are flagged in the SQL database as well.

In addition the database holds orbital elements and other useful data of all solar system objects discovered at Kleť and also information on comets created and updated from several sources by Kleť. Positions, times, and observed objects on all of the processed plates and CCD images are also stored in the database.

For observation planning a web-based tool called `ephem` is used. The tool allows an observer to get ephemeris for one minor planet and/or for minor planets in a specified field at given time. The results can be sorted out to objects of given magnitude and/or type; i.e. to NEAs, PHAs, Virtual Impactors, Kleť discoveries, critical list objects, unusual or distant minor planets, Trojans, Spaceguard Priority List objects and comets. The output list also includes designation, position in the sky, magnitude, and other ephemeris data, as well as information on object type, ephemeris uncertainty, date of the last observation and length of orbital arc used in orbit computation.

There is also used another tool for observation planning. Program `KAC` –*Kleť Atlas Coeli* — shows stars and solar system objects together with a line showing their daily motion in a selected region on the sky. The size of the region corresponds to the FOV of the telescope and can be used to check the telescope's position during observation. The USNO-A2.0 star catalog is used as a source of positions and magnitudes of stars.

4.2. *Data processing*

Program `Astrometry` has been developed for astrometric measurement of CCD images. It uses from 20 to 200 reference stars from the USNO-A2.0 catalogue per measured object.

Images are reduced and all objects with realized condition for signal to noise ratio are detected on the image; objects are then identified with catalogue and coordinates of objects are determined.

The user then selects which object on the image the output should be measured and results are written directly in the MPC format. The time of observation and other information needed for the output are derived from the data stored in the header of the image file. Information about the processed CCD image (time, filename, frame number, equatorial coordinates of the center of the frame, telescope used, exposure time, position of objects on frame, etc.) are added into the SQL database of processed CCD images.

The residuals of the measured astrometric positions are checked before providing them to the community. The calculation of residuals is based on osculating elements of the object near the current epoch, so they are acceptable mainly for the evaluation of observations. In addition the Δ–T variation of the mean anomaly is determined. Checking of both residuals and the Δ–T variation in mean anomaly helps verify object identification.

Program `orbit` is used for preliminary orbit determination. It allows a computation of preliminary orbital elements from observations for one or two nights on the assumption that an object is in perihelion, has been also developed. From observations over several nights the orbit of a new minor planet can be determined using Lagrange-Gauss method improved by variation of geocentric distances and variation of elements. Orbital elements of new discovered objects are stored in K_KLET database. We are working now on an improvement of this software for orbit determination taking into account perturbations from a N-body system. For exact determination of planetary positions the system uses The Planetary and Lunar Ephemerides DE405 provided by JPL.

5. Results

We review here the KLENOT Project results obtained from March 2002 to July 2006 (between 2005 May – 2005 December the KLENOT Telescope was out of operation for maintenance).

5.1. *Astrometric measurements*

We have measured and sent to the Minor Planet Center 30,900 positions of 3706 objects, including 10,139 positions of 1233 NEAs, 104 of which were Virtual Impactors (VIs), and 1,922 positions of 130 comets.

The majority of measured NEAs were confirmatory observations and early follow-up observations of newly discovered objects presented on the NEO Confirmation Page (NEOCP) maintained by the Minor Planet Center. These observations were included in 437 Minor Planet Electronic Circulars. Especially notable are Virtual Impactors 2002 MN, 2003 QQ$_{47}$, (99942) Apophis = 2004 MN_4, 2004 VD$_{17}$, and 2006 BQ$_6$, as well as close Earth approachers 2003 SW$_{130}$, 2004 FH, and 2004 XP$_{14}$.

5.2. *Recoveries*

In the framework of the KLENOT Project we recovered 16 NEAs including 2 PHAs, as it is presented in Table 1.

5.3. *Discoveries*

In the framework of the KLENOT Project we discovered 2 NEAs, as it is presented in Table 2. Their closest Earth approaches in discovery apparition are given in the last column.

Table 1. Near-Earth asteroids recovered by KLENOT

designation	orbit type	disc. opp. arc [days]	last obs. in disc. opp.	recovery date	Δ-T [days]	reference MPEC
2001 FZ_6	Amor	76	2001 June 2	2003 Feb. 26.10	+0.0074	2003-D31
2001 FT_6	Aten	26	2001 Apr. 18	2003 Mar. 25.03	-0.15	2003-F37
1993 FS	Amor	86	1993 June 13	2003 Mar. 25.09	+.024	2003-F38
1999 TF_{211}	Apollo [PHA]	36	1999 Nov. 11	2003 Aug. 4.01	+1.7	2003-P10
2001 FB_7	Amor	118	2001 July 14	2003 Aug. 24.04	+0.0059	2003-Q24
2002 SR_{41}	Apollo	31	2002 Oct. 31	2003 Aug. 24.88	-0.0019	2003-Q31
2000 RS_{11}	Apollo [PHA]	290	2001 June 19	2003 Aug. 27.10	-0.0076	2003-R28
2002 RR_{25}	Aten	56	2002 Oct. 30	2003 Sept. 5.05	+0.0010	2003-R30
1999 VS_6	Apollo	181	2000 May 5	2003 Sept. 5.12	+0.00028	2003-R31
2001 SG_{276}	Amor	56	2001 Nov. 21	2003 Sept. 5.13	-0.025	2003-R34
2003 LP_6	Apollo	30	2003 July 12	2004 Feb 17.15	+0.005	2004-D02
2001 YF_1	Apollo	250	2002 Aug. 23	2004 July 16.98	+0.0046	2004-O20
2003 EN_{16}	Amor	145	2003 Apr. 27	2004 July 17.01	-0.032	2004-O24
2002 PQ_{142}	Apollo	41	2002 Sept. 21	2004 Sept. 4.91	-0.36	2004-R25
2003 MA	Amor	83	2003 Sept. 7	2005 Jan. 18.05	+0.00000	2005-B29
2001 GQ_2	Apollo	14	2001 Apr. 28	2005 Mar. 31.98	+0.009	2005-G09

Table 2. Near-Earth asteroids discovered by KLENOT

designation	orb.type	a [AU]	e	i [deg.]	H	observed arc	Earth approach [AU]
2002 LK	Apollo	1.10	0.15	25	24.2	2002 June 1-8	0.023
2003 UT_{55}	Aten	0.98	0.15	16	26.8	2003 Oct. 26-27	0.0074

5.4. *Comets*

A part of Near-Earth object population consists of comets. The first step of recognizing this comet fraction is an analysis of possible cometary features of newly discovered bodies. Since March 2002 we have confirmed 28 newly discovered comets (i.e. we found cometary features of objects with unusual motion presented on the NEO Confirmation page). A natural further step should be to pursue the behavior of such cometary bodies i.e. to obtain observation data of Near-Earth comets outbursts, fragmentation, splitting and so on. In the framework the KLENOT Project we detected nucleus duplicity of comet C/2004 S1 (Van Ness) and provided astrometric measurements of 17 fragments of comet 73P/Schwassmann-Wachmann 3 including an independent detection of its several new fragments during its 2006 close approach to the Earth.

6. Future

The role of the follow-up in connection to the new generation of all-sky deeper surveys should be discussed. These surveys plan to provide their own follow-up, so the role of "small" observatories providing confirmations and early follow-up of NEOs will be changed. If it will be the case the role of such observatories should be moved towards NEOs which are in urgent need of astrometric positions determination over longer arc including Virtual Impactors, radar and mission targets. These changes would depend also on possible improvements made by the Minor Planet Center (MPC), SENTRY and NEODys.

Our contribution to the changing role of follow-up observatories is included in our future plans. In order to improve our possibility to observe fainter objects we plan to prepare software for co-adding of KLENOT multi-TIFF images. Secondly, in order to improve precision of KLENOT astrometric measurements we plan to change the USNO A2.0 astrometric catalogue to the USNO B1.0 and UCAC3 astrometric catalogues. Finally, due to the improvements planned by the MPC we have been working on incorporating of the new MPC format for astrometric observations to the Kleť Software Package. These improvements will make us able to meet the new scientific and practical challenges.

7. Outreach Activities

Considering recent research results there is a growing interest in minor planets and comets among general public and media. Especially NEOs and the impact hazard attract much attention.

There are many interesting and well designed websites, but naturally all of them are in English. Considering that there is still an unsatisfactory level of knowledge of English language in our society, we have decided to design a NEOs website in Czech as a clear, comprehensive and up to date information about NEOs based on our long-time observing program at the Kleť Observatory and our experience in education programs.

Czech Public Service on NEOs – www.planetky.cz – was designed by the Kleť Observatory on February 2001. Up to now, we have published here 173 articles spread into six themes. This number includes 58 articles about Near-Earth Objects. There are following services for visitors on www.planetky.cz pages: full-text search procedure, sorting out articles into six themes, server statistics, external links to important NEO webpages all over the world, orbit diagrams of minor planets, electronic postcards, RSS Chanel, and the latest news from an additional Kleť server www.komety.cz about comets.

It is been indicated (August 1, 2006) that more then 122,000 visitors has viewed more then 730,000 pages on website www.planetky.cz so far. Our website serves to general public including journalists, students and educators at different levels. On the basis of language similarities this website has also been visited by many people from Central and East European countries like Slovakia, Poland, Ukraine, Russia and others.

The system used on server www.planetky.cz was created with PHP-j00k, a web portal system based on PHP-Nuke, Free Software released under the GNU/GPL license with changes made by the Kleť Observatory. Pages are displayed using PHP and mySQL.

The server www.planetky.cz has its own International Standard Serial Number ISSN 1214-6196.

Acknowledgements

The work of the Kleť Observatory and the KLENOT Project is funded by the South Bohemian Region.

References

Tichá, J., Tichý, M. & Kočer, M. 2002, *ESA SP-500: ACM 2002* 793
Tichý, M., Tichá, J. & Kočer, M. 2005, *International Comet Quarterly* 27, 87
Sekanina, Z., Tichý, M., Tichá, J. & Kočer, M. 2005, *International Comet Quarterly* 27, 135
Pascu, D., *et al.* 2002, *ASP Conf. Series* 272, 361
Chamberlin, A. B., Chesley, S. R., Chodas, P. W., Giorgini, J. D., Keesey, M. S., Wimberly, R. N., & Yeomans, D. K. 2001, *Bull. Am. Astron. Soc.* 33, 1116
Chesley, S. R. & Milani, A. 1999, *Bull. Am. Astron. Soc.* 31, 28

Near Earth Objects, our Celestial Neighbors: Oportunity and Risk
Proceedings IAU Symposium No. 236, 2006
A. Milani, G.B. Valsecchi & D. Vokrouhlický, eds.
© 2007 International Astronomical Union
doi:10.1017/S1743921307003456

Astrometry of small Solar System bodies at the Molėtai observatory

K. Černis, J. Zdanavičius, K. Zdanavičius and G. Tautvaišienė

Institute of Theoretical Physics and Astronomy, Goštauto 12, Vilnius LT-01108, Lithuania
email:cernis@itpa.lt

Abstract. We describe an observational project devoted to astrometric observations of Near-Earth Objects (NEO), main belt asteroids and comets at the Molėtai Observatory, Lithuania. Exposures are obtained with the two telescopes of the observatory: 0.35/0.50 m f/3.5 Maksutov telescope and the 1.65 m reflector with focal reducer f/3.1 and CCD camera. The results of more than 10 000 positions of asteroids and comets have been published in the Minor Planet Circulars and Minor Planet Electronic Circulars. During the 2001–2006 period 130 new asteroids were discovered. The latest discovery is the high-inclination asteroid 2006 SF$_{77}$ belonging to the NEO Aten group.

Keywords. Astrometry; follow up; discovery

1. Observations

Our project serves as a study of asteroids and comets for a better prediction of their orbits and thus their Earth-impact threat. The Molėtai observing program is centered on observing Main belt asteroids, newly discovered bright (15-18 mag) NEA and for search of new asteroids. The targets for observations were selected using public WEB tools for observers (IAU MPC, The NEO Confirmation Page). First astrometric observations of comets at the Molėtai Astronomical Observatory (longitude 25.57 E, latitude 55.32 N, altitude 210 m) have been started in 1998 using a 0.41 m f/6 Schmidt-Cassegrain telescope with a small Meade CCD camera (Černis & Janulis 1998). In the period of 2000–2001 the Tromso (Norway) CCD camera with a thinned TK1024 chip and two-stage thermo-electric cooling was used with the 0.35 m Maksutov Newtonian reflector for astrometric observations of asteroids and comets (Černis & Laugalys 2002).

The Maksutov telescope with Tromso CCD produced a scale of 4.1″ /pixel. First three new asteroids have been discovered in 2001: 2001 OM65, 2001 UM14 and 2001 UU175. The first asteroid, deep Mars crosser 2001 OM65, has been discovered with the 1.65 m telescope with focal reducer f/8. Other two asteroids where discovered with the Maksutov reflector.

Systematic astrometric observations of asteroids and comets began in 2002 when a new VersArray CCD camera (with liquid nitrogen cooling) has been purchased by the Institute of Theoretical Physics and Astronomy. With this camera the Maksutov telescope produces a scale of 3.4″/pixel (Zdanavičius 2003). The 1.65 m reflector with a new focal reducer gives much better astrometric precision, having a scale of 0.9″/pixel. All measurements were done using the Astrometrica software (Raab 2003). The catalogues USNO-A2.0, USNO-B1.0 and UCAC-2 were used for selection of the reference stars. The limiting magnitude for stars with the Maksutov telescope is about 20.5 R magnitude on unfiltered images with the exposure time about 360 s (field-of-view 76′ ×80′). It is a very useful instrument to follow-up astrometry of poorly observed bright NEOs, unusual

Table 1. Distribution of numbers of discovered asteroids $N_{disc.}$, numbers of astrometric observations $N_{obs.}$ and numbers of observed objects $N_{obj.}$ according to time.

Year	$N_{disc.}$	$N_{obs.}$	$N_{obj.}$	References (MPC No.)
2000	0	58	26	41639
2001	3	141	35	42977, 43111, 43450, 43833
2002	12	442	99	44289, 44718, 45048, 45452, 45855, 46218, 46511, 46858
2003	13	643	179	47507, 47996, 48621, 49430, 49886
2004	55	2233	503	50599, 51502, 52495, 52889
2005	36	5252	1230	53631, 54346, 54967, 55473
2006*	11	1252	399	56150, 56765, MPEC 2006-S57
Total	130	10021	2471	

Note: until 2006 October 1.

Table 2. Distribution of numbers of observed comets $N_{com.}$ and numbers of astrometric observations $N_{obs.}$ of comets according to time.

Year	$N_{com.}$	$N_{obs.}$	References (MPC and MPEC)
2000	7	48	MPC 42236, MPC 42959
2001	1	21	MPEC 2001-R57
2002	2	25	MPC 43426, MPEC 2002D-38, MPEC 2002G-38, MPEC 2002-G40
2005	6	52	MPC 54967, MPC 55473, MPEC 2005-U04, MPEC 2005-V88, MPEC 2005-V90, MPEC 2005-V91, MPEC 2005-V95
2006*	10	193	MPC 56735, MPEC 2006-H61, MPEC 2006-J10, MPEC 2006-J31, MPEC 2006-J54, MPEC 2006-K18, MPEC 2006-K55, MPEC 2006-L18, MPEC 2006-L48, MPEC 2006-S50
Total	26	339	

Note: until 2006 October 1.

objects and comets. About 3000 CCD images were obtained for the astrometric work of asteroids and comets during the four last years.

During sky survey in near ecliptic regions in 2002–2006 and during NEO asteroid follow-up astrometry 119 new asteroids have been discovered. Our site (IAU Code 152) in total has discovered 122 objects, 46 of the discoveries are involved in multiple-apparition, 46 of the discoveries are involved in one-opposition object orbits, 22 objects of the one opposition objects have calculated orbits with low accuracy. For 103 objects the orbit have been determined.

Eight additional asteroids have been discovered by one of the authors (K. Černis) using the NEAT CCD frames obtained from the Palomar Mountain Observatory (IAU Code 644) with 1.24 m Oschin Schmidt telescope in 2003 (Helin *et al.* 2003, 2004).

Among the discovered asteroids there are a few unusual objects: the Aten group PHA asteroid 2006 SF_{77} with a=0.92 AU, e= 0.33 and i= 33 deg, the Hilda group asteroid 2004 TB_{21} with a=3.98 AU, 2005 TW_{52} with e= 0.4, Mars crossers 2001 OM_{65} and 2005 TB_{50}. During the investigation of our CCD frames two NEO asteroids, the Apollo-type object 2004 EP20 (q= 0.58 AU) and the Amor-type asteroid 2004 DK_1 (q= 1.1 AU) have been discovered independently, but after the original discovery was published in the MPECs. The same happened for the unusual object 2005 EL_1 (q= 1.35 AU) and the Hungaria-type object 2005 SK_1 (a= 1.92 AU).

Table 1 shows the number of our discoveries, number of astrometric observations, number of observed objects and references for published data in the IAU Minor Planet Circulars. Among more than 10 thousands observations, 144 observations belong to NEO. This means about 1.5 % of total number of observations belong to NEO asteroids.

Table 2 shows the distribution of numbers of observed comets, astrometric observations of comets and references to the published data (IAU Minor Planet Circulars and Minor Planet Electronic Circulars). Among the 26 observed comets at the Molétai Observatory, the brightest one was the comet C/2002 C2 (Ikeya-Zhang) being an object of mag. 3, and the faintest was P/2005 T2 (Christensen), a mag. 20 object. In 2006 the comet 73P/Schwassmann-Wachmann 3 made a close approach to the Earth. During the period Apr 19 – May 16 we got 267 CCD images of 7 comet components (73P-C, 73P-B, 73P-AQ, 73P-G, 73P-R, 73P-M and 73P-N). Splitting of the comet component 73P-B into two parts (73P-B and 73P-AQ) has been confirmed in our images of April 23, 2006.

2. Results

The results of about 10 000 astrometric positions of more than 2400 asteroids and 26 comets, including the NEO asteroids (about 1.5%) (2000 LL, 2000 PH5, 2001 KP41, 1990 SB, 2001 LF, 2001 MZ7, 2001 MF1, 2001 KX67, 2002 EX11, 2004 EP20, 2005 TR, 2005 TP45, 2005 TG50, 2005 TD, 2005 TS45, 2006 SF77) and transneptunian objects (2002 UX25, 2003 UB313) have been already published in Minor Planet Circulars and MPEC circulars (Černis & Zdanavičius 2002, 2005; Černis, Zdanavičius, Zdanavičius 2004, 2005).

Almost all asteroids have absolute magnitudes in the range H= 15–18 mag. The asteroids 2002 FU10 with H= 14.2 mag and Aten-type object 2006 SF77 with H= 21.7 mag have the extreme values of H. Spatial distribution of absolute magnitudes H versus semi-axes for 102 our discovered objects is shown in Fig. 1(left). All asteroid data are taken from Minor Planet Center database (here and for other figures). The vertical axis of H is connected with a photometric diameter of the asteroid. We estimate that our objects are of diameters from 150 m to 7 km.

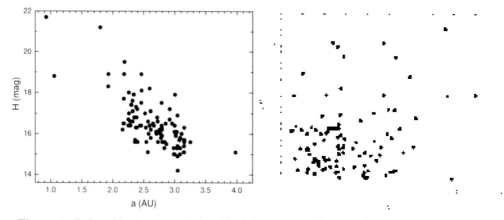

Figure 1. Left – Absolute magnitudes H of the discovered asteroids versus their semiaxes a. Right – Distribution of the asteroid orbit inclination (i) versus eccentricity (e).

Fig. 1(right) shows the distribution of i (orbital plane inclination) and orbital parameters e (eccentricity) of the discovered asteroids. We can see that most orbits have e< 0.3 and i< 20 deg. Extreme values of e (about 0.4) have four objects: NEO asteroids (2004 DK1, 2004 EP20), one main-belt asteroid 2005 TW52 and asteroid 2005 EL1. The largest values of orbit inclination (i= 25 − 28 deg) have: the asteroid 2005 EL1, main-belt asteroids 2002 TP303, 2003 FB123 and two Hungaria-type asteroids. The analysis of statistical properties of the discovered main-belt asteroids share the known characteristics of the main belt.

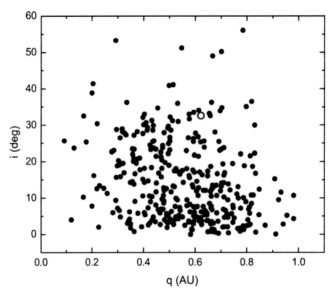

Figure 2. Distribution of Aten group asteroids in the perihelion distance vs. orbit inclination. Open circle – the asteroid 2006 SF77.

The asteroid 2006 SF77 was discovered on September 23, 2006 (Černis & Zdanavičius, 2006). This object belongs to Aten group asteroids whose orbits lie almost completely inside of Earth's. Only about 340 of such objects are known. At the discovery moment PHA 2006 SF77 was 0.081 AU from Earth. It orbits the Sun once in every 325 days. Astrometry of this Aten group asteroid of about 150 meters was carried out during 5 nights at the Molėtai observatory. The next close approache will happen in September 22, 2037 when this asteroid will pass at 0.061 AU from Earth. Fig. 2 shows the distribution of Aten group asteroids in the perihelion distance vs. orbit inclination diagram. We can note that the object 2006 SF77 belongs to the high inclinated Aten group asteroids (i= 33 deg).

References

Černis, K. & Janulis, R. 1998, *M.P.C.* 32377-32382

Černis, K. & Laugalys, V. 2002, *M.P.C.* 44686-44718

Černis, K. & Zdanavičius, J. 2002, *M.P.C.* 45855

Černis, K. & Zdanavičius, J. 2005, *M.P.C.* 54346

Černis, K. & Zdanavičius, J. 2006, *M.P.E.C.* 2006-S57

Černis, K., Zdanavičius, J. & Zdanavičius, K. 2003, *M.P.C.* 48621

Černis, K., Zdanavičius, J. & Zdanavičius, K. 2004, *M.P.C.* 52889

Černis, K., Zdanavičius, J. & Zdanavičius, K. 2005, *M.P.C.* 54967

Černis, K., Zdanavičius, J. & Zdanavičius, K. 2005, *M.P.C.* 55471

Helin, E., Pravdo, S., Lawrence, K., Černis, K., *et al.* 2003, *M.P.C.* 49471

Helin, E., Kervin, P., *et al.* 2003, *M.P.C.* 49909

Helin, E., Pravdo, S., Lawrence, K., *et al.* 2004, *M.P.C.* 50635

Raab, H. 2005, *Astrometrica*, http://www.astrometrica.at/ (electronic version)

Zdanavičius, J. & Zdanavičius, K. 2003, *Baltic Astronomy* 12, 642

Near Earth Objects, our Celestial Neighbors: Opportunity and Risk
Proceedings IAU Symposium No. 236, 2006   © 2007 International Astronomical Union
A. Milani, G.B. Valsecchi & D. Vokrouhlický, eds.   doi:10.1017/S1743921307003468

NEO search telescope in China

Yuehua Ma[1,2], Haibin Zhao[1,2] and Dazhi Yao[1,2]

[1]Purple Mountain Observatory, Chinese Academy of Sciences, Nanjing 210008, China
[2]National Astronomical Observatories, Chinese Academy of Sciences, Beijing 100012, China
email: yhma@pmo.ac.cn

Abstract. Chinese scientists have contributed many research works to the field of asteroid surveying and related topics. In the early 1960s, Purple Mountain Observatory (PMO) began observing asteroids and found over 130 new numbered asteroids during the following decades. The Schmidt CCD Asteroid Program (SCAP) of Beijing Astronomical Observatory (BAO) started in 1995 and found 575 asteroids in several years.

NEOs represent a great threat to the near-Earth environment by closely approaching Earth or even impacting with Earth. After the impact of comet SL9 on Jupiter, Chinese proposed to build a NEO Search Telescope to take part in the joint international NEO survey. It is a 1.0/1.2 m Schmidt telescope with a 4K×4K drift scanning CCD detector. Here we elaborate upon its applications to objects of scientific interest, its parameters, and its observing station.

Keywords. Survey; telescope; observation site

1. Introduction

Many Near Earth Object (NEO) Searching Plans are being carried out all over the world, e.g. LINEAR (Stokes, Evans, Viggh, *et al.* 2000), SPACEWATCH (Gehrels & Jedicke 1995) etc. China is one of the countries paying great attention to the space environment near Earth. In 1928, Asteroid (1125) China was found at Yerkes Observatory by C.Y. Chang who was the former director of Purple Mountain Observatory (PMO). In 1955, Asteroid Purple 1 was found at PMO by C.Y. Chang & J.X. Zhang, which was the first asteroid found by Chinese at home. Up to the mid-1980s, the PMO asteroid research group found over 130 new numbered asteroids with a 40 *cm* double-tube refracting telescope. Among them, Asteroids (2077) Jiangsu and (2078) Nanjing are Mars-crossing asteroids.

The Schmidt CCD Asteroid Program (SCAP) of Beijing Astronomical Observatory (BAO) was put into practice in 1995. The telescope used is a 60/90 *cm* Schmidt type with a 2K×2K CCD detector. In 1997, 2 NEOs and a Mars-crossing asteroid were found by SCAP. 575 asteroids were found by this scheme in the 1990s, which performance was among the top in the world at that time.

In 1994, Professor J.X. Zhang and his colleagues predicted the impact of comet Shoemaker-Levy 9 (SL9) on Jupiter. Although lacking the last key observational data, their prediction agreed, to a high degree of accuracy among all those in the world, with the actual one observed by the Galileo spacecraft.

In 1995, Chinese proposed to build a NEO Search Telescope to take part in the international NEO joint survey at the International Conference on Near Earth Objects held at the United Nations, New York and at the Planetary Defense Workshop held in Livermore, California. After a long period of preparation from 1999, the telescope and site construction were carried out at Xuyu Station of the PMO.

Figure 1.   The dome for the telescope   The 1.0/1.2 m Schmidt Telescope

2. 1.0/1.2 m Schmidt telescope

Figure 1 shows the main building for the telescope and the telescope mounted in the dome. The telescope is equipped with a 4096×4096 high quantum efficiency CCD camera (Figure 2) with a field of view (FoV) of $2° \times 2°$. It will soon start operations after the last course to test the optical system and mechanical and electrical controls. The technical features of the instrument and the good quality of the sky at the site supply optimal conditions for an effective NEO survey. The high quantum efficiency of the CCD allows achieving deep magnitudes with short exposure times, while the wide FoV of the Schmidt telescope makes this facility competitive for the detection of new objects. The main purposes of the telescope are to look for NEOs and to carry out a NEO follow-up and recovery program according to the suggestions provided by the Minor Planet Center (MPC) and Spaceguard Central Node (SCN). After that, we will assess the impact probability of NEOs with Earth and establish a pre-alarm system for the Near Earth environment.

2.1. Scientific background and objects

The interest to study NEOs has grown steadily within the astronomical community on the grounds of science and the potential hazard to mankind. Recent studies (Bottke, Morbidelli, Jedicke, *et al.* 2002; D'Abramo, Harris, Boattini, *et al.* 2001) estimate the population of NEOs to be 900 members with H<18 (corresponding to a diameter of about 1 km). The estimated number of the objects with diameters larger than 100 m varies between 30,000 and 300,000.

At the end of 2005, 748 Potentially Hazardous Objects (PHOs) have been discovered, which is still far from the so-called Spaceguard goal, consisting of the discovery of 90% NEOs with H<18 (more exactly, with D>1 km). In fact, even though we estimate to be around halfway in the number of objects discovered to this size, all simulations show that the discovery of the remaining part to this size will require a longer time than the one that has been necessary to reach the present stage. Among the NEO groups, Atens are rather difficult to discover (Boattini & Carusi 1997). This is due mostly to geometrical reasons; a large fraction of their orbits are inside that of the Earth, so most of the time these objects are observable only at small solar elongations. On the other hand, the main NEO surveys cover these regions of the sky only occasionally. Numerical simulations predict the existence of Inner Earth Objects (IEOs), asteroids or comets with orbits completely inside that of the Earth. But few of them have been discovered so far.

This is not surprising, since the observational requirements for IEOs are even more stringent than those for Atens, requiring observations carried out even closer to the sun (no more than 75-80 degrees of elongation). To discover a larger fraction of Atens, and to start discovering IEOs, the above constraints need to be implemented in the sky survey.

The scientific objectives of the telescope summarized explicitly are

(i) Survey NEOs,

(ii) Determine the orbits of asteroids and comets,

(iii) Predict possible collision events,

(iv) Research the dynamical evolution of the orbits of asteroids and comets,

(v) Inspect space debris.

2.2. *Main parameters and capability of the telescope*

The telescope is a 1.0/1.2 *m* Schmidt type with a 4K×4K drift scanning CCD camera. A special design allows fitting the whole camera inside the telescope tube, directly in the focal station of the Schmidt optical system. The camera is cooled by a CryoTiger PT30, which guarantees the camera working with the CCD chip stable at -110 °C independently of the position of telescope. The CCD camera has the function of drift scanning so that the readout rate is clocked to the sidereal drift rate across the CCD. With F=1.8 for the optical system of the Schmidt telescope, CCD camera delivers a 1.705 arcsec/pixel sampling that corresponds to a FoV about $2° \times 2°$. Its high quantum efficiency (up to 92% at the peak) and the extremely low dark current guaranteed by the low working temperature ($<0.007e^-$/pixel/sec at -100 °C) with a full capacity of 100,000 electrons make this system very valuable on moving objects when very fast exposures are required to avoid the trailing loss problem. The CCD camera is a Lockheed CCD 486 operated with a multiport readout channel at 700 kHz giving a readout time of 12 seconds. Read noise was about 6 electrons at 100 kHz and 15 electrons at 400 kHz under -100°C. The camera and controller are built by Spectral Instruments Inc. in Tucson, Arizona.

The telescope mechanical structure was built by Nanjing Astronomical Instrument Co. Ltd and PMO with new and more accurate encoders, better motors and an advanced control system. The telescope is now able to perform open loop tracking (without any guider) for periods of 1 minute within $1''$ stretching and to perform CCD guiding tracking for periods of 10 minutes within $1''$ stretching. Due to the Schmidt system, the telescope has a large FoV of $3.14° \times 3.14°$, and the center wavelength of the correcting lens is 656.3nm. The configuration of the hardware system is very suitable for our main scientific objective to search for NEOs. The detailed parameters are summarized as follows

(i) Diameter of the correcting lens: 104 *cm*,

(ii) Diameter of the primary mirror: 120 *cm*,

(iii) Focal ratio: F=f/D=1.8 (f=180 *cm*),

(iv) Effective field: $3.14°$ (linear diameter: 100 *mm*),

(v) Limiting magnitude of the telescope: 20.7,

(vi) Center wavelength of the correcting Lens: 656.3 *nm*,

(vii) Light power distribution: 80% of light in less than $2''$(linear diameter is less than 20 *μm*),

(viii) Distortion caused by optical designing and machining: less than 15 *μm*,

(ix) Bearing of the tube: bending to focal plane less than 0.02 *mm*,

(x) Tracking precision: $1''/4$ *min*,

(xi) Pointing Precision: less than $10''$.

Figure 2. 4K by 4K CCD camera

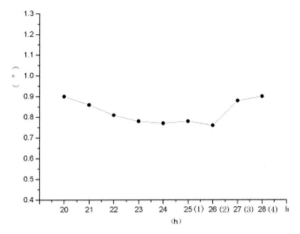

Figure 3. Seeing curve of the site

3. Site of the observational station

The NEO Search Telescope is mounted at Xuyu station. The site is 120 km north of Nanjing. Its position is at $N32°44'.2 \pm 0.5'$, $E118°27'.9 \pm 0.5'$ and the elevation above sea level is 180.9 m. The survey will cover sky north of declination $S20°$. By historical record, the observing nights from 1995-1999 averaged 214.6 per year, and from 1990-1999 207.5 per year. We observed from Aug 1999 to July 2000 for a whole year and got 194 observing nights. In that year the number of astronomical observing hours was 1652. Seeing Observation Results: From June to December in 1999, we made 1493 estimates of seeing; the average result is $\theta = 0.81''$.

Figure 3 is the seeing curve and it shows that seeing changes symmetrically about midnight.

Acknowledgements

This work was supported by Natural Science Foundation of China (No. 10273024, 10573037, 10503013) and the Minor Planetary Foundation of Purple Mountain Observatory. The paper was finished at Tuorla Observatory, Piikkio, Finland during the exchange program between Finnish Academy and NSFC was performed. We thank the referee R.S. McMillan for useful comments and for taking care of language editing.

We would like to acknowledge Chinese Ministry Of Sciences and Technology (MOST), Chinese Academy of Sciences (CAS) and Jiangsu Province of China for the financial support for the telescope construction.

References

Boattini, A. & Carusi, A. 1997, *Vistas in Astronomy* 41, 527

Bottke, W. F., Morbidelli, A., Jedicke, R., Petit, J.-M., Levison, H. F., Michel, P., Metcalfe, T. S. 2002, *Icarus* 156, 399

Carusi, A., Gehrels, T., Helin, E. F., Marsden, B. G., Russell, K. S., Shoemaker, C. S., Shoemaker, E. M., Steel, D. I. 1994, in: T. Gehrels (ed.), *Hazards Due to Comets and Astroids* (University of Arizona Press, Tucson), p. 127

D'Abramo, G., Harris, A. W., Boattini, A., Werner, S. C., Harris, A. W., Valsecchi, G. B. 2001, *Icarus* 153, 214

Gehrels, T. & Jedicke, R. 1995, *Earth, Moon and Planets* 72, 233

Stokes, G. H., Evans, J. B., Viggh, E. M., Shelly, F. C., & Pearce, E. C. 2000, *Icarus* 148, 21

Near Earth Objects, our Celestial Neighbors: Opportunity and Risk
Proceedings IAU Symposium No. 236, 2006   © 2007 International Astronomical Union
A. Milani, G.B. Valsecchi & D. Vokrouhlický, eds.   doi:10.1017/S174392130700347X

Kharkiv study of near-Earth asteroids

Yu. N. Krugly[1,2], N. M. Gaftonyuk[3], I. N. Belskaya[1], V. G. Chiorny[1], V. G. Shevchenko[1], F. P. Velichko[1], D. F. Lupishko[1], A. A. Konovalenko[4], I. S. Falkovich[4] and I. E. Molotov[5]

[1]Institute of Astronomy of Kharkiv National University, Sumska str. 35, Kharkiv 61022, Ukraine, email: krugly@astron.kharkov.ua

[2]Main Astronomical Observatory, NASU, Zabolotny str. 27, Kyiv 03680, Ukraine

[3]Crimean Astrophysical Observatory, Simeiz 98680, Ukraine

[4]Institute of Radio Astronomy, NASU, Chervonopraporna str. 4, Kharkiv 61002, Ukraine

[5]Keldysh Institute of Applied Mathematics, RAS, Miusskaya sq. 4, Moscow 125047, Russia

Abstract. The regular CCD observations of near-Earth asteroids (NEAs) in the Institute of Astronomy of Kharkiv National University were initiated in 1995 within the framework of asteroid hazard problem in collaboration with the DLR, Institute of Planetary Research (Berlin). The main aim of the study is a determination of rotation periods and shapes of NEAs as well as astrometry of newly discovered objects. We also carry out the absolute photometry of NEAs in *BVRI* bands in order to put constraints on surface properties and to estimate their diameters. The observations are carried out with 0.7-m telescope of the Institute of Astronomy (Kharkiv) and with 1-m telescope of the Crimean Astrophysical Observatory (Simeiz) in the standard Johnson-Cousins photometric system. Some observations were made as an optical support of radar observation of NEAs. We present the results of photometric observations of 21 NEAs obtained in 2004-2006 which include asteroid rotation properties, diameters and shapes.

Keywords. Photometry; asteroid, lightcurves; asteroid, rotation

1. Introduction

The regular CCD observations of NEAs were initiated in Institute of Astronomy of Kharkiv National University in 1995 within the framework of collaboration with the DLR, Institute of Planetary Research (Berlin). The main aim of the study is determination of rotation properties and estimation of shapes and sizes of NEAs as well as carrying out astrometry of newly discovered ones.

A possibility of ground-based photometric observations of NEAs during their close approaching the Earth occurs usually once per tens years. It makes difficult an accumulation of information about these bodies. Therefore, the observations of newly discovered NEAs were the most important part of the reported project. We tried to observe NEAs in a wide range of aspect and phase angles to constrain their rotation and shapes. The absolute photometry together with the measurements of color indices was used to estimate diameters and surface properties of the bodies.

We present the results of the CCD observations of 21 NEAs obtained in 2004-2006.

2. Observations

The observations were carried out in the standard Johnson-Cousins UBVRI photometric system (mainly in the R band as the most effective for CCD), using the 0.7-m telescope at the Chuguevskaya Station (70-km to the south-east of Kharkiv) and the 1-m telescope at the Crimean Astrophysical Observatory (Simeiz). Several CCD cameras (SBIG ST-6,

Apogee Alta U42, FLI IMG1024S and CCD47-10) were used in Newtonian focus of the 70-cm telescope (f/4) and in the Cassegrain focus of the 1-m telescope equipped with a focal reducer (f/5). The coordinated observations from the two sites (Kharkiv and Simeiz), made with the same routine of observations and reduction, promote the obtained results which are less dependent from weather conditions.

All the asteroids were observed during the maximum interval of their night visibility, usually during two-three consecutive nights, to constrain the rotation period. To achieve a good signal-to-noise ratio, the integration time for each image was chosen to range from 10 s to 360 s, depending on the asteroid magnitude, magnitudes of the comparison stars in the field of view, and the moving rate of the asteroid. Differential photometry of an asteroid in the CCD frame provides an uncertainty of measurements of the order of 0.01-0.03 mag (rms). To compensate, at least partially, the asteroid motion during the exposure, and thus to increase efficiency of the observations, a correction of the telescope tracking was applied. Reduction of the observations and the aperture photometry routine were made in a standard way with the AstPhot software developed by S. Mottola (Mottola *et al.* 1995). In the case of the asteroid fast motion the method of overlapping CCD fields was used (Krugly 2004). Special attention was paid to absolute magnitude calibration which was fulfilled during nights with the best photometric conditions. The method of observations and data reduction was described in more details in Krugly *et al.* (2002).

3. Results

Since 2004 the observations of 21 NEAs were carried out. Their results are presented in Table 1, which contains year of observations, the absolute magnitude, the diameter estimate, the rotation period, and the maximal observed lightcurve amplitude. Figures 1-5 illustrate a variety of rotation periods and amplitudes of the observed sample of near-Earth asteroids.

The most important results of the project can be shortly summarized as following:

• estimations of rotation periods for 13 NEAs. For (8567) 1996 HW_1, (66251) 1999 GJ_2, 2005 AB, and 2006 BQ_6 they have been determined for the first time; more accurate period definitions were obtained for (3200) Phaethon, (23187) 2000 PN_9, and (54509) 2000 PH_5. The observations of asteroid (54509) 2000 PH_5 can be also used for investigation of an influence of the Yarkovsky-O'Keefe-Radzievskii-Paddack (YORP) effect (Bottke *et al.* 2002).

• determinations of the absolute magnitudes and diameters of 15 NEAs.

• measurements of the B-V, V-R, and R-I colors for about a half of the observed asteroids.

• determinations of NEA lightcurve amplitudes which characterize the asteroid shape elongation. Some of NEAs, like (1685) Toro, (11405) 1999 CV_3, (54509) 2000 PH_5, and 2006 BQ_6) show amplitudes as large as 1.0-1.65 mag.

• photometric observations of the newly discovered asteroids 2005 AB, 2005 CV_{69}, and 2006 BQ_6 during their close approaches the Earth.

• observations at different aspects and phase angles in order to constrain shape, pole coordinates, and surface properties of the following NEAs: (1036) Ganymed, (1627) Ivar, (1862) Apollo, (1685) Toro, (1980) Tezcatlipoca, (3103) Eger, (3200) Phaethon, (4179) Toutatis, and (11405) 1999 CV_3. The asteroids Ganymed and Ivar were observed down to phase angles as small as 1.5 and 3.5 deg, respectively.

• photometric and astrometric observations in support of the radar observations of the NEAs (4179) Toutatis, 2000 PH_5, and 2004 XP_{14} during their close approaching the

Table 1. Results of NEA photometry

Asteroid	Orbit type	Year of observ.	H (mag)	D (km)	Rotation period (hrs)	Amplitude (mag)
1036 Ganymed	Amor	2006	9.58	39	10.314 ± 0.004	0.12
1627 Ivar	Amor	2005	13.06	8.4	4.7956 ± 0.0002	0.35
1685 Toro	Apollo	2004	(14.23)	3.5	10.19	>1.0
1862 Apollo[B]	Apollo	2005	16.1*	1.6	3.065 ± 0.005	0.26
1980 Tezcatlipoca	Amor	2006	13.78	6.1	7.246 ± 0.009	0.58
3103 Eger	Apollo	2006	(15.38)	2.5	5.706	0.7
3200 Phaethon	Apollo	2004	14.52*	6.7	3.6052 ± 0.0008	0.17
4179 Toutatis	Apollo	2004	15.38	2.8	176.4[1]	0.8
6611 1993 VW[B]	Apollo	2005	16.61	1.0	2.556[2]	0.06
8567 1996 HW1	Amor	2005	15.04*	3.5	8.75 ± 0.05	0.4
100085 1992 UY4	Apollo	2005	17.72	1.55	12.88 ± 0.06	>0.5
11405 1999 CV3	Apollo	2006	15.73	2.45	6.507 ± 0.006	1.1
13553 1992 JE	Amor	2005	16.0	3.4	>30	>0.7
23187 2000 PN9	Apollo	2006	(16.1)	2.1	2.5325 ± 0.0004	0.13
54509 2000 PH5	Apollo	2004	22.50	0.1	0.202875 ± 0.00003	1.0
66251 1999 GJ2	Amor	2005	17.1*	1.1	2.462 ± 0.001	0.10
1999 LF6	Apollo	2004	(18.2)	0.8	>12	0.3
2004 XP14	Apollo	2006	(19.4)	0.45	>24	~0.05
2005 AB[B]	Amor	2005	(17.5)	2.	3.346 ± 0.008	0.11
2005 CV69	Apollo	2005	18.1	1.3	-	~0.1
2006 BQ6[B]	Apollo	2006	19.45	0.44	4.414 ± 0.008	1.65

Note: H values in brackets were adopted from *the Minor Planet Circulars.* Diameters were calculated assuming an albedo 0.15, if the asteroid albedo and type were unknown.
[B] Supposed to be binary.
* The absolute magnitude was obtained from R-band observations and the V-R=0.43 was used to transform H_R to H.
[1] Period of the long-axis precession Hudson & Ostro (1995).
[2] Pravec (2005, private communication).

Earth. On 2006 July 3 the international radar experiment was arranged to investigate the NEA 2004 XP_14 with participation of radar and antenna facilities in USA, Ukraine, Russia, Italy and China. The object was sounded with RT-70 in Goldstone (USA) at 3.6 cm and with RT-70 in Evpatoria (Ukraine) at 6 cm wavelength. The outside of USA the echoes were detected with RT-64 in Kalyazin (Russia) at 6 cm and with RT-70 in Evpatoria at 3.6 cm (see Figure 6 which presents the first result of the radar experiment).

• observations of four binary NEAs: (1862) Apollo and 2006 BQ_6 identified as binaries from radar observations; (6611) 1993 VW and 2005 AB found out to be binary in the Photometric Survey for Asynchronous Binary Asteroids Project (Pravec 2005). Additionally, observations of (1862) Apollo helped to determine the YORP effect in the rotation of this asteroid (Kaasalainen *et al.*, 2006, to be submitted).

4. Conclusions and perspectives

The regular CCD observations of NEAs in Kharkiv have been launched since 1995. It was the beginning of follow-up program of NEAs study in Ukraine. The observations were carried out in the Institute of Astronomy of the Kharkiv National University in cooperation with the Crimean Astrophysical Observatory (Simeiz). Over 500 lightcurves of more than 100 NEAs have been obtained. Recently the larger and more sensitive CCD cameras have become available for both observation sites (Kharkiv and Simeiz), which noticeably improve our potential to investigate key properties of the NEA population.

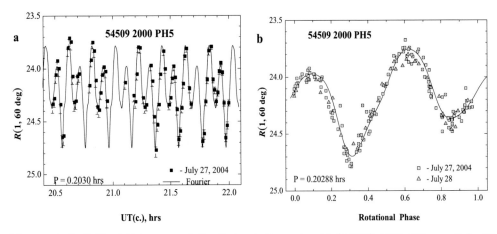

Figure 1. Individual lightcurve of the fast-rotating asteroid (54509) 2000 PH$_5$ on July 27, 2004 with the best 6th order Fourier series fitting (a). The composite lightcurve of the asteroid obtained from two consecutive nights in July 2004 (b).

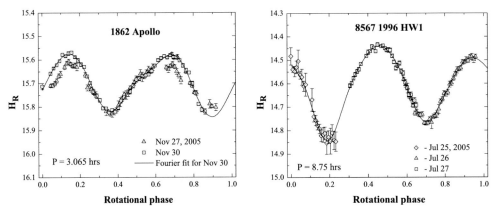

Figure 2. Composite lightcurves of NEAs: (1862) Apollo observed in November 2005 (left); (8567) 1996 HW$_1$ observed in July 2005 (right).

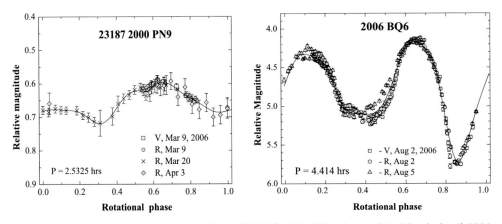

Figure 3. Composite lightcurves of NEAs: (23187) 2000 PN$_9$ observed in March-April 2006 (left); 2006 BQ$_6$ observed in August 2006 (right).

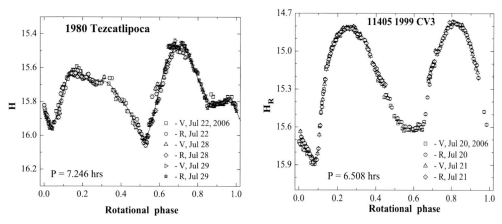

Figure 4. Composite lightcurves of NEAs: (1980) Tezcatlipoca observed in July 2006 (left); (11405) 1999 CV$_3$ observed in July 2006 (right).

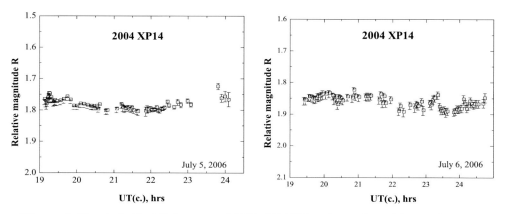

Figure 5. Two individual lightcurves of NEA 2004 XP$_{14}$ observed on July 5 and 6, 2006.

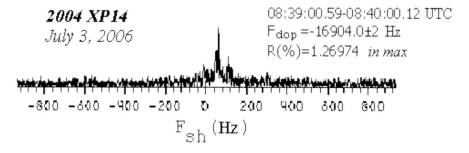

Figure 6. The measurement of Doppler shift of NEA 2004 XP14 echo received with Evpatoria RT-70. The echo-signal is cross-correlated with model of sounding signal of Goldstone RT-70; correlation coefficient R (y axes) is plotted vs. Doppler shift Fsh (x axes).

Last years we have provided an optical support of radar observations of the selected NEAs with RT-70 antenna in Evpatoria (Ukraine) (Konovalenko *et al.* 2005). Such collaboration seems to be very promising.

Acknowledgements

Yu.N.K. and D.F.L. are thankful to IAU for awarding grants to attend the XXVIth General Assembly and, personally, to Oddbjorn Engvold and Monique Orine for their helpful assistance. This work was partially supported by Ministry of Education and Science of Ukraine. Since June 2006 observations in Kharkiv were carried out with CCD camera obtained thanks to INTAS grant Ref. No 03-70-567.

References

Bottke, W. F., Vokrouhlický, D., Rubincam, D. P. & Brož, M. 2002, in: W. F. Bottke, A. Cellino, P. Paolicchi & R. P. Binzel (eds.), *Asteroids III* (Tucson: University of Arizona Press), p. 395

Hudson, R. S. & Ostro, S. J. 1995, *Science* 270, 84

Konovalenko, A. A., Falkovich, I. S., Lytvynenko, L. M., Nabatov, A. S., Petrenko, A. G., Fedorov, A. P., Kamelin, A. B., Malevinskyi, S. V., Molotov, I. E., Dement'ev, A. F., Lipatov, B. N., Nechaeva, M. B., Tukkari, J., Lu, Sh., Gorshenkov, Yu. N. & Agapov, V. M. 2005, *Radio Physics and Radio Astronomy* 10, Special issue, S20

Krugly, Yu. N., Belskaya, I. N., Shevchenko, V. G., Chiorny, V. G., Velichko, F. P., Erikson, A., Mottola, S., Hahn, G., Nathues, A., Neukum, G., Gaftonyuk, N. M. & Dotto, E. 2002, *Icarus* 158, 294

Krugly, Yu. N. 2004, *Solar System Research* 38 (3), 241

Mottola, S., De Angelis, G., Di Martino, M., Ericson, A., Hahn, G. & Neukum, G. 1995, *Icarus* 117, 62

Pravec P. 2005, in: B. D. Warner, D. Mais, D. A. Kenyon & J. Foote (eds.), *Proceedings for the 24th Annual Conference of the Society for Astronomical Science* (Published by Society for Astronomical Sciences, Inc.), p. 61

Near Earth Objects, our Celestial Neighbors: Opportunity and Risk
Proceedings IAU Symposium No. 236, 2006
A. Milani, G.B. Valsecchi & D. Vokrouhlický, eds.

© 2007 International Astronomical Union
doi:10.1017/S1743921307003481

The observations of Near Earth Objects by the automatic mirror astrograph ZA–320M at Pulkovo observatory

A. V. Devyatkin, A. P. Kulish, V. V. Kouprianov, D. L. Gorshanov, A. S. Bekhteva, O. V. Krakosevich, E. Yu. Aleshkina, F. M. Ibragimov, V. N. L'vov, R. I. Smekhacheva and S. D. Tsekmeister

Main (Pulkovo) Astronomical Observatory of Russian Academy of Science

Abstract. At Pulkovo Observatory, we conduct observations of various Solar System bodies: major planets, their satellites, comets, and asteroids, including Near Earth Objects. For these purposes, a robotic telescope was constructed on the base of the ZA–320 Mirror Astrograph ($D = 320$ mm, $F = 3200$ mm). It can perform CCD observations of Solar System bodies with the limiting magnitude of up to 19.0.

Independent ephemeris support is provided by the EPOS software package developed at Pulkovo Observatory; it includes tracing of catalogues of comets and asteroids, regular ephemeris calculations, and control of observations. CCD frame processing is done by the Apex automatic data reduction package developed at Pulkovo Observatory.

In 2001-2006, more than 12000 observations of minor Solar System bodies were collected, including more than 6000 positions of 656 NEAs, about 1200 observations of 27 comets, and about 2000 observations of major planets satellites. The mean accuracy of obtained positions is $0''.09 - 0''.40$. Results of observations are regularly submitted to the Minor Planet Center.

Currently, ZA–320M is the 16-th of more than 680 telescopes in the worldwide rating of those that observe NEAs (by the number of observations).

In the near future, our group is planning to start observations with another two robotic telescopes: MTM–500 ($D = 500$ mm, $F = 4000$ mm Maksutov) and 1-meter telescope ($D = 1000$ mm, $F = 1200$ mm) of the Pulkovo mountain station at Northern Caucasus (Kislovodsk, 2100 m above sea level). These two instruments will allow to increase the number of observations, their accuracy, and limiting magnitude (up to 20.5 mag).

Keywords. Astrometry; follow up; photometry; telescope

1. Introduction

The ZA–320 Mirror Astrograph ($D = 320$ mm, $F = 3200$ mm) has been working at Pulkovo Observatory (Saint-Petersburg, Russia) since 1997. The telescope is used for various kinds of astrometric and astrophysical CCD observations. The following list presents celestial objects observed by our telescope.

I. Asteroids:

(*a*) Near Earth Asteroids
(*b*) Recently discovered asteroids
(*c*) Other asteroids from the MPC Critical List
(*d*) Double asteroids
(*e*) Asteroids – extinct comets (3200 Phaethon)
(*f*) Close approaches of asteroids
(*g*) Visual approaches of asteroids
(*h*) Asteroids – space mission targets (25143 Itokawa)

(*i*) Stellar occultations by asteroids (111 Ate, 2000)

II. Comets

III. Satellites of giant planets:

(*a*) Jupiter: Himalia, Elara, Pasiphae

(*b*) Saturn: Titan, Hyperion, Iapetus, Phoebe

IV. Planets (and dwarf planets):

(*a*) Jupiter: Himalia, Elara, Pasiphae

(*b*) Saturn: Titan, Hyperion, Iapetus, Phoebe

V. Objects of the geosynchronous Earth orbit

VI. Lagrange points of Earth

VII. Photometric observations:

(*a*) Asteroids, comets, satellites of giant planets

(*b*) Star with "eclipsing" planet – HD209458

(*c*) Mutual phenomena in systems of satellites of Jupiter and Saturn

(*d*) SNe

(*e*) Variable stars

(*f*) Open clusters

Still the primary goal of the telescope includes astrometric observations of asteroids and comets, including Near Earth Objects.

Up to 2005 we used a small-size CCD camera SBIG ST–6 (375×242 pix, $9'.5 \times 7'.5$ field of view). Since 2005, we use FLI IMG1001E CCD camera (1024×1024 pix, $28' \times 28'$).

The ZA–320M telescope control system (TCS) includes capabilities for remotely controlling the telescope, as well as for the fully automatic operation without any human intervention. TCS is able to control pointing, dome and slit, filter wheel, CCD camera, focuser, autoguider, and to perform observation scheduling via the built-in ephemeris interface. Another software package is used for automatic processing of raw observation results.

The TCS software consists of the following main modules:

- telescope control subsystem,
- image acquisition and CCD camera control subsystem,
- ephemeris support,
- reduction of CCD-frames.

2. Telescope control subsystem

The TCS software for the ZA–320M robotic telescope has the distributed modular design and thus can be used to control various types of telescopes. In the following we briefly describe the major TSC modules.

High precision clock interface. Provides accurate high-resolution UTC time for all TCS modules. Various time sources, like GPS receivers and synchronization signal from atomic standards, are supported. Implemented as a standalone package (*AccuTime*) including the software development kit (SDK) for C/C++, Ada95, Delphi, and Python programming languages, with easy integration into other software packages and customization to various time sources.

Distributed datalogging service. Provides a centralized datalogging facility based on the client-server approach, with a simple application programming interface (API) allowing all TCS components running on any of the local control units (LCUs) of the ZA–320M network to register events in the global journal.

TCS hardware layer. Implements a number of high-level commands, like pointing to the specified target, tracking control, dome control, etc., for the actual TCS hardware protocols.

Operator interface. Provides the graphical user interface (GUI) for planning observations, watching the current TCS state, and manual control of the telescope operation. Automatic object scheduler can choose the sequence of observed objects to achieve the optimal observation conditions for every object, taking into account their priorities and possible constraints and requirements. TCS can operate in the manual, semi-automatic, and fully automatic modes.

Image acquisition subsystem. ZA–320M image acquisition subsystem is a stand-alone software package CameraControl designed for controlling various types of CCD cameras via the graphical user interface (GUI). It is fully integrated into the whole TCS package and is able to operate as a pure image acquisition module. Like all other TCS components, CameraControl is based on the modular design. Dedicated modules implement a generic set of high-level commands, like cooler, shutter, and exposure control, and image downloading, for the particular CCD camera.

Apart from the built-in FITS format support using the CFITSIO library, various image storage formats are supported via extension module API. Basic image examination facilities, including brightness and contrast control, histogram equalization, magnifying glass, 2D and 3D profile visualization, area statistics, equatorial grid etc., are included into the GUI.

CameraControl TCS integration is based on the built-in Python interpreter with a full set of commands for operating the camera and querying its state, and on the integrated TCP server for remote camera control. Other TCS modules, like observation scheduler, working as clients, can thus obtain the full control over the image acquisition process, though the capability of performing exposures manually is still left intact.

3. Ephemeris support – the EPOS software

EPOS (Ephemeris Program for Objects of the Solar system) is the effective application for study and ephemeris support of observations of the Solar systems objects. This program was developed at Pulkovo Observatory. The orbital elements of asteroids and comets as well as the observatories' coordinates obtained via Internet are used by EPOS. The applications operation is also based on the data of modern numerical ephemerides (DE200/LE200, DE405/LE405, DE406/LE406) and star catalogs (Hipparcos, Tycho-2, USNO, etc.) that are distributed on the CDs. EPOS is intended for use under Windows 98/XP and has bilingual (English and Russian) interface. EPOS (version 5.x) includes the following components:

The main program. It transfers the control to one of the programs that solve the restricted set of tasks. This program also imports the observatories' data and controls the use of various numerical ephemerides.

"Catalogs of objects". This program stores in the internal database the orbital elements and other parameters of the minor bodies of the Solar system (now more than 300000 asteroids and more than 500 comets). This helps to browse, edit and export these data, to select the objects with respect to various conditions and to import the most recent data from the well known catalogs (B.Marsden, E.Bowell). One can get the histograms and distributions for various parameters. It is possible to work with numerous catalogs of objects (for example, NEAs, trans-Neptunians, Centaurs, Trojans and Hildas, etc.).

"Ephemerides". This program calculates the Solar system objects' ephemerides of various type and accuracy. The calculated values can be observable and geometric

coordinates, and the osculating orbital elements. The observable are astrometric or apparent coordinates along with their first two derivatives or Pulkovo parameters of apparent motion and a set of other values. The advantage of this program is the ability to create the list of necessary parameters located in the required order and properly formatted. The geometric are the rectangular and spherical coordinates of any object referred to the centers of the Sun, Moon or major planets. The ephemerides may be calculated with various accuracy, from elliptic motion to the high precision numerical integration of equations of motion taking into account the perturbations from all major planets and some minor bodies. One can use the values of orbital elements or rectangular coordinates that are placed to the special input files with higher formal accuracy. Various combinations of elements and some group operations are possible in the ephemeris calculation.

"*O − C: Comparison of Observations with Calculations*". This program compares observed coordinates and velocities with the calculated ones. This helps to evaluate the observational accuracy, to reveal the large errors, to identify the objects. Thus an observer can control the results of observations before sending them to Minor Planet Center.

"*Frame*". The program visualizes the apparent motion of many objects on the star background and their diurnal motion as well. One can get the list of objects that are observable at the specified moment in the specified sky area or even in the space box. This helps in identification of objects and in the study of the structure of the Solar system.

"*Orbits*". This program visualizes the perturbed motion of many objects and groups of objects along their heliocentric orbits in the space. It is possible to vary the observers position, the scale, the velocity and direction of motion, to switch on/off the numbers and names of the objects, their orbits, apsidal and nodal lines and the ecliptic plane. The picture may be saved in BMP or EMF files. The useful quality of the program is the opportunity to accumulate the images of the lines that helps to follow the orbital evolution in time.

"*Hazardous Objects*". By this program one can get the current list of PHA – potentially hazardous objects for the Earth and other major planets, and the list of close approaches of asteroids to the specified major planet within the specified time span.

"*What to Observe*". The program generates the list of objects observable at the specified place in the specified night. One can limit the values of visual magnitude, the objects altitude and elongation from the Sun.

So with the EPOS software package one can calculate the accurate ephemerides for many objects and get the quick illustration of their motion as well. This helps to support the existing observational programs, to develop the new ones and to put the ephemeris data at the disposal of the interested observers and publishers.

EPOS was used in research of the dynamical structures of the Solar system, in preparation and analysis of Pulkovo observations of asteroids and comets, 6-meter BTA observations of trans-Neptunian objects, in support of NEA international radar observations, in the accuracy analysis of asteroidal observations of many observatories throughout the world and in other tasks.

4. Processing of CCD frames – Apex software package for data reduction

Observational data are being processed using the *Apex* image processing software series – the former semi-automatic package *Apex 1.0* and the recent fully automatic pipeline based on the *Apex II* platform, both developed at Pulkovo observatory.

Apex II is a general-purpose software platform for astronomical image processing. Its architecture and design concepts are similar to those of the major image processing packages including IRAF, MIDAS, and IDL. Like them, *Apex II* consists of several components:

- *core* – high-level interpreted dynamic (scripting) programming language;
- *standard library* of general-purpose utility functions and algorithms specific to the area of astronomical image processing;
- object-oriented *graphical user interface* (GUI) subsystem with interactive image examination and data visualization capabilities, built on top of the core language and library;
- a set of *user functions and scripts* which utilize the above components to perform particular image processing tasks.

This structure has proven to be most flexible and versatile. It allows implementing the full range of image processing applications – from interactive command-line driven tools with interactive examination of intermediate processing results to fully automated pipelines for processing large data volumes to stand-alone GUI applications for specific image reduction tasks.

Unlike the image processing packages mentioned above, *Apex II* is not based upon a dedicated interpreted programming language, but rather upon the widely-used general-purpose object-oriented scripting language Python. This choice is motivated primarily by the clarity, power, and flexibility of the language, existence of implementations for all major hardware and software platforms, and the extensive standard library for most routine tasks like input/output, data visualization, matrix algebra, curve and surface fitting, n-dimensional image processing etc. Despite the widespread opinion about the low performance of scripting languages, pure Python scripts in *Apex II* are often faster than similar programs written in conventional compiled programming languages. This is mostly due to the high level of vectorization of mathematical operations and to effective optimization of underlying C/Fortran libraries.

All these advantages currently attract attention of the leading scientific software developers. The evidence for this are Python interfaces to the two major astronomical image processing systems, PyRAF and, recently, PyMIDAS.

The *standard Apex II library* is built primarily on top of the two Python packages, Numerical Python and Scientific Python (NumPy/SciPy). The first of them implements the basic functionality for working with multidimensional arrays, including vectorization and matrix algebra. The second one provides implementation of most of the algorithms commonly used in scientific applications: Fourier transform, integration, solving PDEs, interpolation, optimization and nonlinear regression, signal and image processing, special functions etc. Based on these algorithms, as well as on the built-in Python functions, the *Apex II* library implements various higher level tasks specific to the field of astronomical image processing, like timescale conversions, calibration and filtering of CCD images, automatic object detection, PSF fitting, astrometric and photometric reduction, catalog access and so forth.

The *graphical subsystem* (still under active development) is based on wxWidgets/ wxPython, the cross-platform GUI toolkit, and on matplotlib, the scientific data visualization package modeled after MATLAB. These packages can be used to display individual CCD frames or catalog fields, plot various data obtained during image processing, as well as create standalone GUI applications intended for processing of specific kinds of astronomical images.

Thus *Apex II* is primarily a general-purpose software platform for development of reduction systems for various astronomical data.

Currently, the *Apex II* platform is used primarily for automated astrometric and photometric reduction of large volumes of observational data for various Solar system objects.

The dedicated image processing pipeline is capable of producing reports in the standard Minor Planet Center format, as well as estimating the accuracy of observations with the help of the EPOS software package. Astrometric reduction is performed in the system of the USNO–A2.0, USNO–B1.0, UCAC–2 catalogs, accounting for chromatic refraction.

The main *Apex II* image processing pipeline for Solar system objects includes the following steps:

• automatic selection (and, if necessary, creation) of required calibration frames for each frame being processed; bias/dark/flat correction;

• estimation and subtraction of sky background;

• optimal image filtering for increasing the objects signal-to-noise ratio (SNR);

• object extraction by thresholding with logical filtering intended to increase the probability of detection of faint objects and to reduce the number of false detections;

• deblending of overlapping objects;

• finding the objects centroids by PSF fitting;

• flux measurement using the aperture, PSF, or optimal photometry techniques;

• elimination of the remaining false detections (including cosmic rays) using the various criteria;

• reference astrometric catalog matching, with the possibility to find the target field in the catalog area of the given size;

• astrometric reduction – obtaining the LSPC solution using the selected plate model;

• reference photometric catalog matching and differential photometry in the instrumental band;

• post-identification of all unknown objects within the frame using all available ephemeris databases and other catalogs;

• generation of the standard MPC report for all Solar system objects within the frame.

This pipeline allows to perform reduction of large volumes of raw observational data without sacrificing accuracy. The possible manual operations here reduce to the final examination of series of measurements to eliminate the possible processing errors that are hard to detect by any formal criteria.

5. Results: Astrometry

The "Pulkovo Program for Research of Near Earth Objects" started in 2002 using the ZA–320M telescope. During 2001–2005, more than 12000 observations of minor Solar System bodies were performed, including more than 6000 observation of 656 NEAs and about 1200 observations of 27 comets. These observations were processed in the framework of the USNO–A2.0, USNO–B1.0, and UCAC–2 catalogues. The mean accuracy of the obtained positions is $0.''09 - 0.''40$.

The results of the observations are regularly submitted to the Minor Planet Center. According to the Center information, the ZA–320M telescope is rated 16-th by the number of observations in the world-wide list of more than 680 telescopes that observe NEAs.

In the following table, we highlight several examples of our astrometric results. Here N is the number of observations for the object, $\overline{(O-C)}_\alpha \cos\delta$ and $\overline{(O-C)}_\delta$ are mean values of $(O-C)$ difference for the whole observational period, for right ascension and declination, respectively; $\sigma_\alpha \cos\delta$ and $\sigma\delta$ are accuracies of a single observation for right ascension and declination, respectively (estimation by the $(O-C)$); Δm is the magnitude range of the object during observational period.

Object	N	$\overline{(O-C)_\alpha \cos\delta}$	$\sigma_\alpha \cos\delta$	$\overline{(O-C)_\delta}$	$\sigma\delta$	Δm
9 Metis	46	+0.″38	±0.″18	+0.″42	±0.″24	8.6–10.9
433 Eros	65	+0.14	0.27	+0.43	0.28	11.4–12.6
1866 Sisyphus	24	0.00	0.24	+0.19	0.29	15.1–17.7
3122 Florence	18	−0.12	0.23	+0.19	0.25	15.9–17.7
3199 Nefertiti	27	+0.14	0.40	+0.18	0.35	15.5–17.9
3200 Phaethon	18	0.00	0.40	+0.03	0.37	16.3–18.4
4179 Toutatis	10	−0.17	0.42	+0.39	0.24	14.8–16.8
2004 FX31	22	+0.05	0.46	−0.02	0.50	17.3–18.9
2004 JA	33	−0.01	0.21	−0.09	0.17	15.6–16.8
2P Encke	7	+0.33	0.51	+0.72	0.24	14.5–16.3
1036 Ganymed	44	−0.21	0.20	+0.02	0.12	13.3–13.9
3800 Karayusuf	41	−0.35	0.12	−0.31	0.08	16.0–17.0
423 Diotima	70	−0.10	0.16	−0.06	0.10	11.1–12.9
5164 Mullo	34	−0.08	0.10	−0.03	0.07	16.4–17.4
85709	45	−0.28	0.12	−0.27	0.07	15.5–16.6
C/2003 WT42	24	+0.89	0.09	−0.32	0.06	15.5–16.3

6. Results: Photometry

Photometric measurements of minor bodies of the Solar System are made along with their astrometric measurements. Photometric observations are made in the Johnson BVR bands, as well as in the wide instrumental band of the telescope (350–1000 nm) – in the cases of weak objects and when very short exposure time is needed. For photometric processing of asteroids and comets, we use USNO–A2.0, USNO–B1.0, and UCAC–2 catalogues.

We obtain photometry only for asteroids that demonstrate fast magnitude changes. For example, NEA 1999 HF1 has shown variability with characteristic time less then half an hour. A number of series of its observations were made on several nights. They have shown that its magnitude changes with the quasi-period of about 20 minutes. The amplitude of magnitude changes from night to night lies in the range of $0.^m1$–$0.^m4$. Mean accuracy of photometry in these observations was about $0.^m06$. The following figure shows an example of the 1999 HF1 observation series.

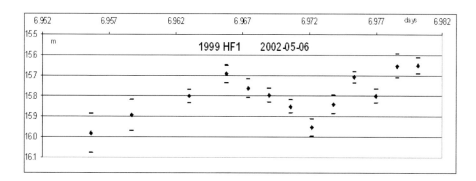

7. Conclusions

The experience gained in the course of construction of the ZA–320M robotic telescope is used now for modernization of the two other telescopes: Maksutovs MTM–500 ($D = 500$ mm, $F = 4000$ mm) and 1-meter telescope ($D = 1000$ mm, $F = 1200$ mm). We are planning to mount MTM–500 on the mountain observational site of Pulkovo Observatory at Northern Caucasus (Kislovodsk, 2100 m above sea level) in 2007. The telescope will be used for astrometric and photometric observations of asteroids, comets and other Solar System bodies. We are planning that these instruments will produce a significant increase in the number of observations, their accuracy, and limiting magnitude (up to 20.5 mag).

References

Devyatkin, A. V. & Gorshanov, D. L. 2001, *IBVS* 5072
Devyatkin, A. V., Gorshanov, D. L., Gritsuk, A. N., Mel'nikov, A. V., Sidorov, M. Yu., & Shevchenko, I. I. 2002, *Solar System Research* 3, 248
Devyatkin, A. V., L'vov, V. N., Kornilov, E. V., Gorshanov, D. L., Kouprianov, V. V.,Sidorov, & M. Yu. 2002, *Izvestia GAO* 216, 120 (in Russian)
Devyatkin, A. V., L'vov, V. N., Gorshanov, D. L., Kouprianov, V. V., Aleshkina, E. Yu., Bekhteva, A. S., Baturina, G. D., Kornilov, E. V., & Sidorov, M. Yu. 2004, *Izvestia GAO* 217, 236 (in Russian)
Devyatkin, A. V., Aleshkina, E. Yu., Barshevich, K. V., Baturina, G. D., Bekhteva, A. S., Gorshanov, D. L., Krakosevich, O. V., Kouprianov, V. V., L'vov, V. N., Smekhacheva, R. I., Sochilina, A. S., & Tsekmejster, S. D. *Proc. of Sixth US/Russian Space Surveilence Workshop* (August 22-25, 2005, Pulkovo, St.-Petersburg, Russia), 2005, p. 220
Frolov, V. N., Jilinski, E. G., Ananjevskaja, Y. K., Bronnikova, N. M., Poljakov, E. V., & Gorshanov, D. L. 2002, *A&A* 369, 115
Frolov, V. N., Ananjevskaja, J. K., Jilinski, E. G., Gorshanov, D. L., & Bronnikova, N. M. 2006, *A&A* 451, 901
Kanaev, I. I., Devyatkin, A. V., Kulish, A. P., Rafalskiy, V. B., Vinogradov, V. S., Kouprianov, V. V., & Kornilov, E. V. 2004, *Izvestia GAO* 217, 505 (in Russian)
Kanaev, I. I., Sochilina, A. S., L'vov, V. N., Devyatkin, A. V., Kouprianov, V. V., Gorshanov, D. L., Sidorov, M. Yu., Guseva, I. S., & Molotov, I. E. 2003, *Proc. of Fifth US-Russian Space Surveillance Workshop, (24-27 September, 2003, Pulkovo, St.-Petersburg, Russia)* p. 6
Minor Planet Electronic Circ. No. 2003-M40, 2003-H21, 47449, 47506, 47994, 48317, 48619, 49222, 49389, 49426, 50321

Part 7

Current and Future Missions to NEOs

Near Earth Objects, our Celestial Neighbors: Opportunity and Risk
Proceedings IAU Symposium No. 236, 2006
A. Milani, G.B. Valsecchi & D. Vokrouhlický, eds.
© 2007 International Astronomical Union
doi:10.1017/S174392130700350X

The nature of asteroid Itokawa revealed by Hayabusa

M. Yoshikawa, A. Fujiwara, J. Kawaguchi
and
Hayabusa Mission & Science Team

JAXA, 3-1-1 Yoshinodai, Sagamihara, Kanagawa, 229-8510, Japan,
email: makoto@isas.jaxa.jp

Abstract. The spacecraft Hayabusa, which was launched in 2003, arrived at its destination, asteroid (25143) Itokawa in September 2005. The appearance of Itokawa, a small S-type near Earth asteroids, was totally unexpected. The surface is covered with a lot of boulders and there are only a few craters on it. It looks like a contact binary asteroid. The surface composition is quite similar to LL-chondrite. The estimated density is 1.9 ± 0.13 (g/cm^3), so the macro-porosity is about 40%. This means that Itokawa is a rubble pile object. In Itokawa, we may see such things that are very close to building blocks of asteroids. In this paper, we review the mission and the first scientific results.

Keywords. space missions; rendezvous; asteroid surface; asteroid landing; sample return; instrumentation; mass determination

1. Introduction

Hayabusa mission, which was originally called MUSES-C mission, is the asteroid sample return mission of Japan (The literal meaning of Hayabusa is "falcon."). The spacecraft was launched in May 2003, and it arrived at its target asteroid, (25143) Itokawa, in September 2005. Before the arrival, it was already known by ground-based observation that the size of Itokawa is rather small, about 500 m in length. This is correct. What Hayabusa saw was certainly a very tiny object as expected. However the appearance of its surface was completely unexpected. There are only a few craters on its surface but it is covered with a lot of boulders. Figure 1 shows the artistic images of Hayabusa mission before and after arriving at the asteroid. As this figure shows, the concept of small asteroid has been largely changed by Hayabusa mission.

Hayabusa stayed around Itokawa for about three months. In the first two months, Hayabusa carried out detailed scientific observations. And in the third month, November 2005, it tried to approach closely to Itokawa several times and to touch down twice. Although the sampling sequence was not executed as planed, Hayabusa became the first spacecraft that lifted off from a solar system bodies except the Earth and the Moon. After the second touchdown, some troubles occurred in the spacecraft and the communication was lost for about one and half months. Fortunately, the communication was recovered at the end of January 2006, and since then daily operations are going on. But the Earth arrival date has been delayed and now Hayabusa is planed to come back to the Earth in June 2010.

The asteroid Itokawa was known as 1998 SF$_{36}$, discovered by LINEAR (Lincoln Near-Earth Asteroid Research) team. The perihelion distance is 0.95 AU and the aphelion distance is 1.70 AU, so it is an Apollo-type NEO (Near Earth Object). The orbital inclination is about 1.7 degrees, so the orbital plane of 1998 SF36 is almost the same

Figure 1. Artistic images of Hayabusa mission. Before arriving at Itokawa, we assumed a lot of craters on the surface (left). However, actually there were only a few craters but a lot of boulders on it (right). The illustrations were made by A. Ikeshita.

as those of the Earth and Mars. Because of this orbital character, 1998 SF36 is a good target for a space mission.

After 1998 SF_{36} was selected as the target of Hayabusa mission, a lot of observations were carried out. Therefore, it was given the number 25143 in June 2001, and in 2003 it was named Itokawa after Prof. Hideo Itokawa, who is the father of Japanese rocketry. The first launch experiment of Japanese rocket, which was called "Pencil Rocket", was done by Prof. Itokawa in 1955, exactly 50 years before Hayabusa's arrival at Itokawa. In this paper, we review the first scientific results of Hayabusa (See special issue of Science journal on 2 June 2006: Fujiwara *et al.* (2006); M. Abe *et al.* (2006); Okada *et al.* (2006); Saito *et al.* (2006); S. Abe *et al.* (2006); Demura *et al.* (2006); Yano *et al.* (2006)).

2. Mission summary

In this section, we briefly summarize the Hayabusa mission. Hayabusa was launched on 9 May 2003 by M-V rocket from Uchinoura Space Center, Japan. The launch was successful and no correction maneuvers were needed. About one month later, the ion engines were started. The orbital control is basically done by the ion engines.

For the first year, Hayabusa was orbiting near the orbit of the Earth. And about one year later, on 19 May 2004, Hayabusa passed the position of 3,700 km from the Earth and performed the Earth swingby, and its orbit was changed to the orbit that is similar to the orbit of Itokawa. About one and half months before the Earth swingby, the ion engines were stopped and accurate navigation (or targeting) was done. The error of navigation at the closest point to the Earth was about 1km in position. This was accurate enough so we did not have to carry out the correction maneuver after the swingby. Then we stared the ion engines again.

In July 2005, Hayabusa came behind the Sun from the Earth. This is the solar conjunction. We cannot keep good communication with the spacecraft at solar conjunction, because the noise by the Sun becomes very large. Therefore, we stopped the ion engines during this period. The error of orbit determination became large and at the end of the solar conjunction (at the end of July 2005), the position error of Hayabusa was almost 2000km. However, just after the solar conjunction, the optical navigation camera on board was able to take images of Itokawa. Then we started the optical navigation, so the accuracy of the orbit of Hayabusa was improved very rapidly. At the end of August 2005, the orbit error in position was about 1km. Also at the end of August, we stopped the operation of ion engines. The ion engines fulfilled their duty. In Figure 2, the orbit history up to the arrival is summarized schematically.

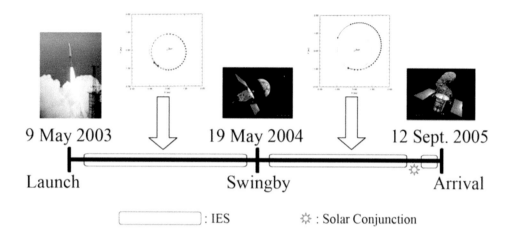

Figure 2. Orbit history from the launch to the asteroid arrival. IES means the periods when the ion engines were operated.

The approach phase was started from the end of August 2005. The distance from Itokawa became smaller and smaller everyday, and the image of Itokawa became larger and larger. We stopped the motion of Hayabusa relative to Itokawa for the first time on 12 September 2005, so we say that the arrival date is 12 September 2005. The arrival point of Hayabusa was about 20 km away from the surface of Itokawa. We call this position as "Gate position."

At first, Hayabusa was moving around this gate position, and gradually it went down toward the asteroid. At the end of September 2005, Hayabusa arrived at the position of about 7 km from the asteroid. We call this position as "Home position." Hayabusa did not revolve around the asteroid, but it moved along the line that connects the Earth and the asteroid. In October 2005, Hayabusa was located around the Home position and made some "tour" to observe Itokawa from various angles. At the end of October, Hayabusa approached to less than 4 km from the surface of Itokawa. The orbital operation is schematically summarized in Figure 3. The actual orbit and the orbital operation in detail are shown in Figure 4.

By the end of October, we almost finished the basic observations of Itokawa. We had information about the surface of Itokawa and we constructed the shape model of Itokawa. Moreover, we have selected the candidate places for touchdown. Thus, we were ready for touchdown.

In November 2005, Hayabusa tried the touchdown. The first touchdown rehearsal was executed on November 4, 2005. In this first descent to Itokawa, we encountered several problems that we did not expect, so we carried out two more descent operations on November 9 and 12. Then we had confidence to carry out touchdown, and the first touchdown was executed on November 20. In this first touchdown, totally unexpected things happened again. We learned later that Hayabusa stayed on the surface of Itokawa more than 30 minutes. However, the sampling sequence was not performed in this first touchdown, so we tried a second touchdown on November 26. All of these descent and touchdown operations are summarized in Table 1.

Table 1. Descent and Touchdown Operations

Operation	Date	Comments
Rehearsal #1	Nov. 4, 2005	
Nav & Guidance Practice	Nov. 9, 2005	Target Marker Release #1 (rehearsal)
Rehearsal #2	Nov. 12, 2005	MINERVA Lander Release
Touchdown #1	Nov. 20, 2005	Target Marker Release #2
		Two Touchdowns + One Landing
Touchdown #2	Nov. 26, 2005	One Touchdown

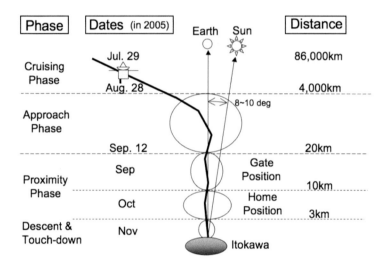

Figure 3. Schematic figure of Hayabusa operation near Itokawa.

The mission sequence was perfect in the second touchdown, so all of us thought that Hayabusa was successful in getting the surface material. However, after the liftoff from Itokawa, a fuel leak occurred. We were forced to operate Hayabusa in a very difficult status and we were not able to confirm that the bullet for collecting samples was fired at the moment of the second touchdown. Anyway, at present (November 2006) we are operating Hayabusa, and we think we can make it return to the Earth in June 2010, which is three years behind the schedule. We hope that some of the surface materials are inside the capsule of Hayabusa.

This is the short history of Hayabusa mission. We have learned a lot of engineering matters as well as we have discovered many scientifically new things. In the following sessions, we review the scientific results.

3. Features of Itokawa

3.1. *Science instruments*

Hayabusa has four science instruments, AMICA, NIRS, XRS, and LIDAR. They are shown in Figure 5.

The Asteroid Multi-band Imaging Camera (AMICA) is also called the telescopic optical navigation camera (ONC-T), and it is used both for navigation and scientific observations. AMICA has both a wide band-pass filter and seven narrow band filters, the central wavelengths of which are nearly equivalent to those of the Eight Color Asteroid Survey

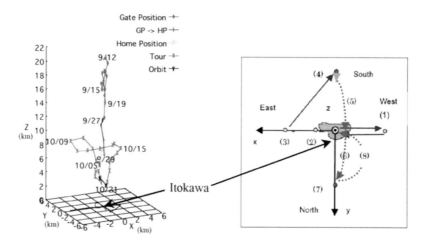

Figure 4. The orbit of Hayabusa near Itokawa. The left figure shows the actual orbit of Hayabusa near Itokawa. The Earth is located in the +Z direction. The numbers in the figure are month/date in 2005. The right figure shows the path of what we call "tour." The spacecraft undertook a tour near the asteroid after achieving the home position observation. The view is from the Sun. Arrows indicate the sequential path of the spacecraft. Dots show the hovering sites. Each spacecraft position indicated by the number in the figure corresponds to the dates as follows: (1) 8 to 10 October, westward, high phase angle; (2) 12 and 13 October, zero solar phase angle; (3) 15 October, east side high, phase-angle; (4) 17 and 18 October, south pole; (5) 20 October, south pole, low altitude (~4 km); (6) 22 October, north pole, low altitude (~4 km); (7) 23 and 24 October, north pole; (8) 27 and 28 October, low altitude observation (~3 km). Dashed lines include low-altitude observation.

(ECAS) system as follows: 380 (ul), 430 (b), 550 (v), 700 (w), 860 (x), 960 (p), and 1010 nm (zs). AMICA imaged the entire surface of Itokawa with a solar phase angle of ~10 degrees at the home position. Because the angular resolution is 0.0057 deg / pixel (99.3 micro-rad/pixel), the nominal spatial resolution is 70 cm/pixel at the home position. Four position-angle glass polarizers were mounted on an edge of the 1024 pixel by 1024 pixel CCD chip.

The Near-Infrared Spectrometer (NIRS) has a 64-channel InGaAs photodiode array detector and a grism (a diffraction grating combined with a prism). The dispersion per pixel is 23.6 nm. Spectra were collected from 0.76 to 2.1 mm. The NIRS field of view (0.1deg × 0.1deg) was aligned with the fields of view of LIDAR and AMICA.

The X-ray Fluorescence Spectrometer (XRS) is an advanced type spectrometer with a light-weighted (1.5 kg) sensor unit based on a CCD X-ray detector. This is the first time a CCD has been used for such a purpose on a planetary mission. The CCD has an energy resolution of 160 eV at 5.9 keV when cooled, which is much higher than that of the proportional counters used in previous planetary missions. In addition, the XRS has a standard sample plate (SSP) for concurrently calibrating the X-ray fluorescence when it is excited by the Sun. The SSP is a glassy plate whose composition is intermediate between those of chondrites and basalts. By comparing X-ray spectra from the asteroid and from the SSP, quantitative elemental analysis can be achieved, although the intensities and spectral profiles of solar X-rays change over time.

The Light Detection and Ranging Instrument (LIDAR) measures distance by determining the time of flight for laser light to travel from the spacecraft to the asteroid and return. The LIDAR averages the topography within the LIDAR footprint on the surface of the asteroid, which approximates 5 by 12 m at a 7-km altitude for normal

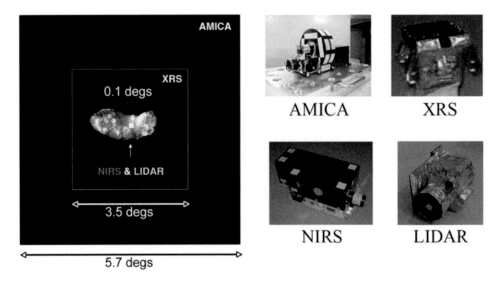

Figure 5. Science Instruments of Hayabusa and their fields of view. Left image shows the comparison of the field of view of each instrument.

incidence. The accuracy of LIDAR ranging obtained from ground calibration was 1 m from a distance of 50 m and 10 m from 50 km.

3.2. Basic parameters of Itokawa

There are a lot of ground based observations of Itokawa, so we knew several characteristics of Itokawa before Hayabusa arrived at Itokawa. For example, the rotational period of Itokawa is 12.1324 hours, the spin axis is almost perpendicular to the ecliptic and retrograde, and its shape is elongated, about 0.5 km long. By Hayabusa, the length of the principal axes were determined as X=535 m, Y=294 m, Z=209 m. The orientation of the spin axis in the ecliptic coordinate is $[\beta,\lambda]=[128.5, -89.66]$, and the nutation was not detected.

Other important parameters are the mass, volume, and density. These were estimated as follows: mass = $(3.51\pm0.105) \times 10^{10}$ kg, volume = $(1.84\pm|,0.092) \times 10^7$ m^3, density = 1.90 ± 0.13 g/cm^3. We will discuss about the mass estimation in section 3.5.

3.3. Surface features of Itokawa

As already mentioned, the most distinctive feature of Itokawa is the large number of boulders. Most of its surface is covered by many small and large boulders. Figure 6 shows the images of Itokawa taken from four directions. We can see some smooth areas, but such regions occupy only a small fraction in the total surface. Thus, the surface of Itokawa is clearly divided into two parts; one is rough terrains, that is boulder rich region, and the other is smooth terrains. There are several large boulders, and one of them is shown in Figure 7, which is the enlargement of the region marked in Figure 6. The sharply sticking boulder in this image is what we call "Pencil Boulder." Another large boulder is seen at the right side of Itokawa in the lower images of Figure 6. Actually this is the largest boulder on Itokawa and we call it "Yoshinodai Boulder."

Some parts of the surface of Itokawa seem to be polygonal planes (one of the examples is indicated by an arrow in Figure 6). We call such planes as "facet." We think that facets were created by collisions or they are parts of the original surface of the parent body. Facets may be the part of basic building blocks of the asteroid.

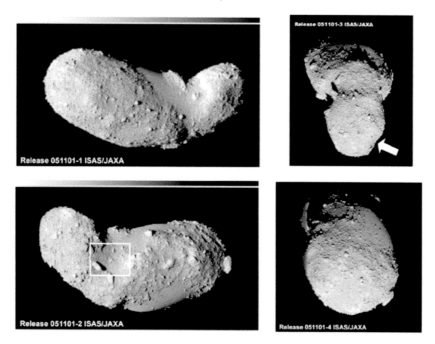

Figure 6. Images of Itokawa. The arrow in upper right figure shows facet. The region marked in lower left figure is shown in detail in Figure 8.

Figure 7. A part of Itokawa marked in Figure 6. The very sharp boulder in the lower left corner is called "Pencil boulder" by the mission team.

At first sight, we do not notice craters, but crates actually exist. Figure 8 shows some of the examples of craters. They are very small and maybe they were formed by impacts of small meteorites. Moreover, we found much smaller features, which might be created by impacts of small meteorites.

In Figure 9, we show two close-up views of the surface of Itokawa. One is the close-up view of one of the smooth terrains, "MUSES-Sea" (IAU officially approved name for this

Figure 8. Some of the craters on the Itokawa. Mission team call each crater as follows ; A : Komaba crater (D=27m), B : Kamisunagawa crater (D=10m), C : Fuchinobe crater (D=36m).

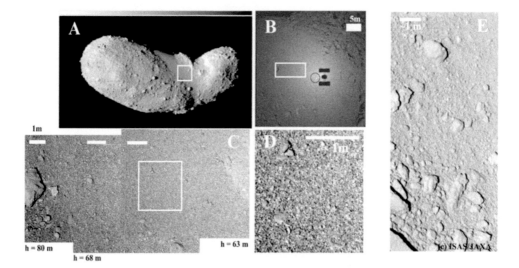

Figure 9. Images of surface close up. Images A, B, C, and D are the sequential images, and each image shows the region within the rectangular section in the previous image. The final image D shows the region where touchdown was performed. Image E is the boundary region between rough and smooth terrains.

region is "MUSES-C"). In this region, the touchdown of Hayabusa was performed. The final image, whose spatial resolution is 6-8 mm, shows that this area is covered by cm-sized gravel and it looks like pavement. Another close-up image in Figure 9 is the boundary between smooth terrain and rough terrain, which is the border of "MUSES-Sea." The resolution is about 20 mm. We can see a clear transition of rock size distribution from the rough terrain to smooth terrain.

Local topography is also measured by LIDAR. For example, in Figure 10, we show one of the results by LIDAR, where "Tsukuba Boulder" was measured. We know the roughness of the surface and in this region it is about 2.2 m. The roughness of "MUSES-Sea" region is about 0.6 m.

Figure 11 shows the distribution of the surface potential and the surface slope. The lowest potential regions coincide with the smooth terrains, suggesting the mobility of fine materials due to external forces after the formation of the asteroid such as seismic shaking and perturbation by planetary encounters. Evolution of the smooth terrains likely

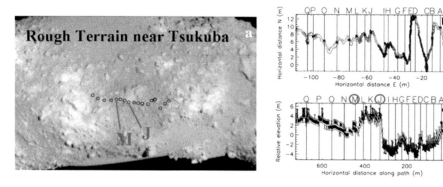

Figure 10. Local topography measured by LIDAR. Figure (a) shows the measured points and (b) is the horizontal distance between each point and (c) is relative elevation.

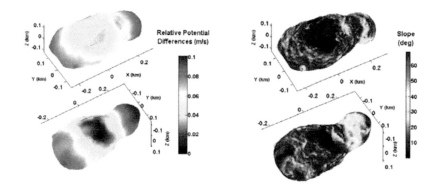

Figure 11. Potential and slope. Left figure shows the potential difference and right figure shows the slope.

involve processes for grain-size sorting and dynamical interactions between regolith and boulders, although the transport/deposition mechanisms must be carefully investigated further.

3.4. *Surface composition of Itokawa*

The surface composition of Itokawa can be investigated by NIRS and XRS. Figure 12 shows the comparison of NIRS data with the reflectance spectrum of ordinary chondrites. The features of these spectra agree well, so this indicates that the surface of Itokawa is olivine and pyroxene assemblage. NIRS took the spectrum of various part of the surface, and the spectrum was almost the same. So we can also say that the mineralogical material is almost the same all over the surface of Itokawa. Figure 13 also shows the data obtained by NIRS, and it shows the correlation of the band strength ratios for average Itokawa spectra. This figure shows that the surface of Itokawa is especially olivine-rich surface, compared with other S-type asteroids, and it supports that the surface materials of Itokawa are quite similar to LL-chondrites.

Consistent results were obtained by the analysis of XRS data. One of the XRS data is shown in Figure 14, where both spectra of the standard on-board sample and of Itokawa are plotted. X-rays from Itokawa (right) have larger Mg/Si and smaller Al/Si than those of X-rays from the standard sample (left), which indicates that Itokawa is similar to

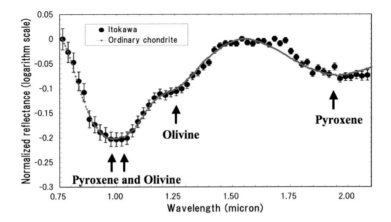

Figure 12. Near infrared reflectance spectrum. Dots show the observed data of Itokawa and the line is a spectrum of ordinary chondrite.

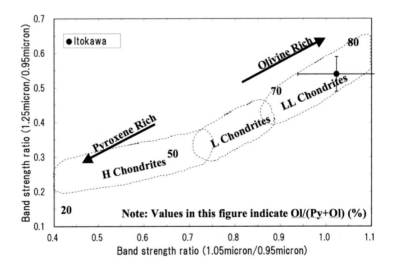

Figure 13. Natural log absorption strengths at 1.05 and 1.25 mm relative to that at 0.95 mm of average Itokawa spectra (filled circle).

ordinary chondrites in composition. Major elemental ratios obtained by XRS are Mg/Si = 0.78±0.09 and Al/Si = 0.07±0.03. This result also shows that Itokawa seems similar to ordinary chondrites, especially LL- or L-chondrites but some primitive achondrites with a small degree of melting cannot be ruled out. No substantial regional variation is found, indicating homogeneity in composition.

Thus, from the observations by NIRS and XRS, we can say that the surface composition of Itokawa is quite similar to LL-chondrites and the surface of Itokawa is homogeneous from the point of view of the composition.

However, we detected that there is heterogeneity in the distribution of the color and albedo. Figure 15 shows the image of Itokawa in enhanced composite color. We can see that some parts are brighter than the basic brightness of Itokawa. Some bright regions correspond to the borders of facets or land-slide region. So we think we can explain this by space weathering. By the effect of space weathering, the surface of asteroid becomes dark. But if meteorites collide with Itokawa or the asteroid encounters planets, it will be

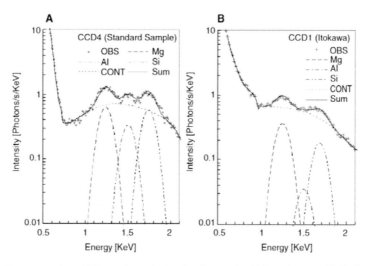

Figure 14. X-ray spectra of the onboard standard sample (A) and asteroid Itokawa (B) were simultaneously observed by the XRS at 9:27 UTC on 19 November 2005. The observed spectra (OBS) are fitted by Gaussian profiles to K- lines of major elements (Mg, Al, and Si) and by a background continuum component (CONT).

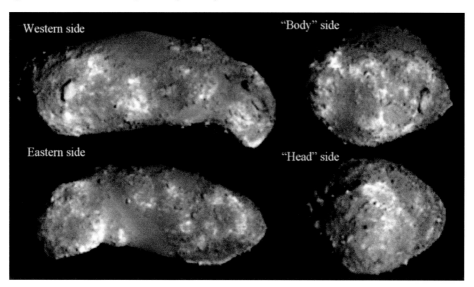

Figure 15. Color composite images constructed from b-, v-, and w-band data. The contrast adjustment was done in each image to enhance the color variation.

shaken and sub-surface fresh materials, whose color is not dark, appear on the surface. In fact, generally, the brighter area is bluer, while the darker area is redder, which supports this hypothesis of space weathering.

3.5. *Mass of Itokawa*

The mass of an asteroid is quite important because it will be an important clue to know the structure of the asteroid, when the mass is converted to the density. The mass itself is also important for the spacecraft navigation. Therefore, four independent teams tried to estimate the mass of Itokawa by using different data and different methods. In this section, we summarize the results of these mass estimations.

Figure 16 shows the schematic image of the mass estimation. As we mentioned earlier, basically Hayabusa was moving along the line that connects the Earth and Itokawa. When Hayabusa was near Itokawa, the Sun and Earth were located almost in the same direction from Hayabusa. The forces that affect Hayabusa are the gravity from Itokawa, the solar radiation pressure, and the gravity from other celestial bodies, such as the Sun and planets. The gravitational accelerations from the Sun and the planets are well known, so we should estimate the solar radiation pressure and the gravitational attraction from Itokawa separately. In this way we can determine the mass of Itokawa. The data used for mass estimation are the range and Doppler from ground stations, the distance between Hayabusa and the surface of Itokawa measured by LIDAR, and the direction of Itokawa from Hayabusa obtained by the navigation cameras.

Figure 17 shows that the variation of the distance between Hayabusa and Itokawa. The motion of Hayabusa, which was mentioned in the previous section, can be understood clearly in this figure. In Figure 18, the periods when the mass estimations were carried out are marked by circles of A to D. Here we briefly summarize what were done in each phase.

The first mass estimation was done when Hayabusa moved around the gate position and the home position (A). Hayabusa had acceleration toward the asteroids, partly due to the gravitational attraction from Itokawa, but mainly due to the solar radiation pressure. When Hayabusa was around the gate position (about 20 km from Itokawa), the effect of the solar radiation pressure was about 20 times larger than that of the gravitational attraction by Itokawa. Using the range and Doppler data, the mass of Itokawa was estimated. The estimation error is about 15%. This error is rather large because in this period, the effect of the gravitational attraction by Itokawa was much smaller than that of the solar radiation pressure.

The mass estimation in the period "A" was successful, so we expected that we would perform a more precise mass estimation when the spacecraft approached the asteroid

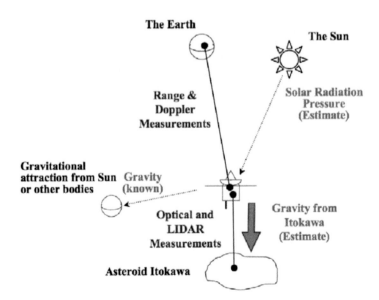

Figure 16. Schematic image of the estimation of the mass of Itokawa. The gravity of Itokawa and the solar radiation pressure act in similar direction, so we should estimate these effects separately.

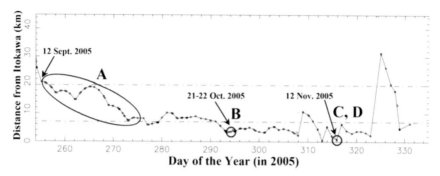

Figure 17. Variation of distance between Hayabusa and Itokawa and the periods when the mass estimations ware carried out. The circles marked by "A" to "D" indicate the periods when the mass of Itokawa was estimated. The distance data in November 2005 (after DOY of 307) does not represent actual changes especially when Hayabusa was far from Itokawa, because the data points are not dense enough. The distance data were based on the range measurement from the Earth.

much closer. However at the beginning of October 2005 (just after the period "A" in Figure 17), one of the reaction wheels had a trouble and stopped functioning. Hayabusa has three reaction wheels, and one of them broke down before arriving at the asteroid. The attitude control did not have serious problems with two reaction wheels. However, by the break down of the second reaction wheel, we had to use the chemical thrusters to control the attitude of Hayabusa. When chemical thrusters are used, small acceleration is generated in its orbital motion. This makes very difficult to estimate the mass of Itokawa. In order to get the proper orbital data to estimate the mass, we intentionally stopped the attitude control for two days, and let Hayabusa move without the artificial force. This period is October 21 and 22, which is shown by "B" in Figure 17. Using the data of LIDAR and the navigation camera in addition to the range and Doppler, the mass of Itokawa was estimated with an error of 5%.

When Hayabusa came quite near to Itokawa, the gravitational attraction by Itokawa became much stronger than the solar radiation pressure. As we mentioned in Chapter 2, Hayabusa made approaches to Itokawa five times in November 2005. By using the data (mainly LIDAR and navigation cameras) of the third descent (Nov. 12), two independent mass estimations were carried out. They are "C" and "D" in Figure 17. Each mass estimation was done by using the data of a short period. The parts of the orbit used for the mass estimation here are so close to the asteroid that the mass distribution (shape of the asteroid) was taken into account assuming that the density of asteroid is homogeneous. (In the analysis of "A" and "B", the asteroid is treated as a point mass.) The error of the mass estimation of "C" and "D" is 5 or 6%.

The mass estimations of Itokawa are summarized in Table 2, where the estimated values are shown in GM (G is the gravitational constant and M is the mass of Itokawa). In order to compare the results shown in Table 2, the error range of each estimate is plotted in Figure 18. From this figure, we can say that the estimated mass is consistent within each other.

Taking the weighted mean of these results, the final value of GM is $(2.34 \pm 0.07) \times 10^{-9}$ (km^3/s^2), which is converted to the mass of Itokawa as $(3.51 \pm 0.105) \times 10^{10}$ (kg). According to the shape modeling team, the volume of Itokawa was estimated as $(1.84 \pm 0.092) \times 10^7$ (m^3). Thus the density of Itokawa is calculated as 1.9 ± 0.13 (g/cm^3). This value of density is quite important to understand the nature of Itokawa. These

Table 2. Results of Mass Estimation of Asteroid Itokawa

Phase	Period	Data Type*	Distance from Itokawa	Model of Itokawa	GM $10^{-9}\mathrm{km}^3/\mathrm{s}^2$	Error
A	Sep. 12 – Oct. 2	R, Dop	20 – 7 km	point mass	2.34	15%
B	Oct. 21 – 22	R, Dop, Opt, LI	3.5 km	point mass	2.29	5%
C	Nov. 12	LI, Opt	1427 – 825 m	polyhedron	2.39	5%
D	Nov. 12	Opt, LI	800 – 100 m	polyhedron	2.36	6%

*R: Range, Dop: Doppler, LI: LIDAR, Opt: Optical images

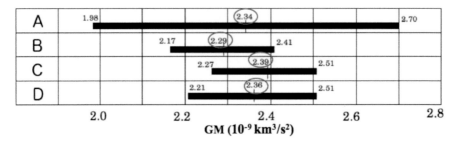

Figure 18. Estimated GM values with the error range. The values of GM and error in Table 2 are shown graphically. The dark lines indicate the range of estimated GM for each phase of the orbit.

results of the mass and the density were taken as the first reference for the basic physical characteristics of Itokawa, which was mentioned in chapter 3.2.

4. Summary

Now we know that the density of Itokawa is 1.9 g/cm^3, and the surface material is probably LL-chondrite. A macro-porosity equal to 40% was calculated. This means that there is large vacant space inside of Itokawa. Therefore, we concluded that Itokawa is a rubble pile object. A possible formation scenario is as follows (Figure 19) : (a) The parent body was disrupted by impact. (b) A portion of fragments coagulated each other forming two objects, which were forming a contact binary. (c) These two bodies were merged into one and became Itokawa.

For the first time, we saw a very small asteroid from its vicinity. This tiny asteroid was very different from what was expected. We can say that we saw more basic and original elements that created larger asteroids and planets.

The Hayabusa mission is still going on. Although the probability that some surface materials are inside the capsule is small, we have not gave up. The operation of the spacecraft is rather difficult because we can use only the ion engines, but we will try to bring it back to the Earth in June 2010. From the point of the engineering of Hayabusa, the only thing that we have not yet tried is the capsule reentry.

At the same time, we are planning post-Hayabusa missions. One is what we call "Haybusa-2." Hayabusa-2 is almost the copy of Hayabusa, but the target asteroid is a C-type NEO. The target asteroid is tentatively 1999 JU3. Since Hayabusa-2 is a copy mission of Hayabusa, we can save the time for development. The possible launch window is November 2010 (backup window is November 2011). It will arrive at the asteroid in 2013 and it will come back to the Earth in 2015 or 2016. The other mission is what we call "Hayabusa-Mk2," where we will develop a totally new spacecraft. Especially the

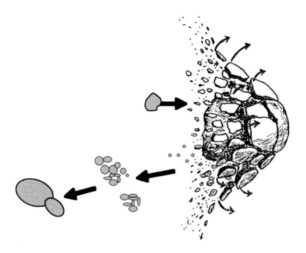

Figure 19. Possible origin of Itokawa.

sampling method will be significantly changed. The target object is not selected yet, but it will be much more primitive object.

Space missions to small solar system bodies are quite important to understand the origin and evolution of the solar system. Also it will be important to know the origin of life. We hope that many exciting missions will be carried out with international collaborations.

Acknowledgements

We think that the Hayabusa mission was successful because many people joined forces to overcome many difficulties. Especially here we would like to express our appreciation to the navigators in Jet Propulsion Laboratory and the operators of Deep Space Network for their excellent assistance. Also we acknowledge LINEAR team, who kindly accepted the proposal of the name of Itokawa.

References

Abe, M., Takagi, Y., Kitazato, K., Abe, S., Hiroi, T., Vilas, F., Clark, B. E., Abell, P. A., Lederer, S. M., Jarvis, K. S., Nimura, T., Ueda, Y., & Fujiwara, A. 2006, *Science* 312, 1334

Abe, S., Mukai, T., Hirata, N., Barnouin-Jha, O. S., Cheng, A. F., Demura, H., Gaskell, R. W., Hashimoto, T., Hiraoka, K., Honda, T., Kubota, T., Matsuoka, M., Mizuno, T., Nakamura, R., Scheeres, D. J., & Yoshikawa, M. 2006, *Science* 312, 1344

Demura, H., Kobayashi, S., Nemoto, E., Matsumoto, N., Furuya, M., Yukishita, A., Muranaka, N., Morita, H., Shirakawa, K., Maruya, M., Ohyama, H., Uo, M., Kubota, T., Hashimoto, T., Kawaguchi, J., Fujiwara, A., Saito, J., Sasaki, S., Miyamoto, H., & Hirata, N. 2006, *Science* 312, 1347

Fujiwara, A., Kawaguchi, J., Yeomans, D. K., Abe, M., Mukai, T., Okada, T., Saito, J., Yano,H., Yoshikawa, M., Scheeres, D. J., Barnouin-Jha, O. S., Cheng, A. F., Demura, H., Gaskell, R. W., Hirata, N., Ikeda, H., Kominato, T., Miyamoto, H., Nakamura, A. M., Nakamura, R., Sasaki, S., & Uesugi, K. 2006, *Science* 312, 1330

Okada, T., Shirai, K., Yamamoto, Y., Arai, T., Ogawa, K., Hosono, K., & Kato, M. 2006, *Science* 312, 1338

Saito, J., Miyamoto, H., Nakamura, R., Ishiguro, M., Michikami, T., Nakamura, A. M., Demura, H., Sasaki, S., Hirata, N., Honda, C.,. Yamamoto, A, Yokota, Y., Fuse, T., Yoshida, F., Tholen, D. J., Gaskell, R. W., Hashimoto, T., Kubota, T., Higuchi, Y., Nakamura,

T.,Smith, P., Hiraoka, K., Honda, T., Kobayashi, S., Furuya, M., Matsumoto, N., Nemoto, E., Yukishita, A., Kitazato, K., Dermawan, B., Sogame, A., Terazono, J., Shinohara, C., & Akiyama H. 2006, *Science* 312, 1341

Yano, H., Kubota, T., Miyamoto, H., Okada, T., Scheeres, D. J., Takagi, Y., Yoshida, K., Abe, M., Abe, S., Barnouin-Jha, O. S., Fujiwara, A., Hasegawa, S., Hashimoto, T., Ishiguro, M., Kato, M., Kawaguchi, J., Mukai, T., Saito, J., Sasaki, S., & Yoshikawa, M. 2006, *Science* 312, 1350

Near Earth Objects, our Celestial Neighbors: Opportunity and Risk
Proceedings IAU Symposium No. 236, 2006
A. Milani, G.B. Valsecchi & D. Vokrouhlický, eds.
© 2007 International Astronomical Union
doi:10.1017/S1743921307003511

Keplerian consequences of an impact on an asteroid and their relevance for a deflection demonstration mission

Andreas Rathke[1] and Dario Izzo[2]

[1]Astrium GmbH, Dept. AED41, 88039 Friedrichshafen, Germany
e-mail: andreas.rathke@astrium.eads.net

[2]European Space Agency, Advanced Concepts Team, EUI-ACT, ESTEC, Keplerlaan 1,
2201 AZ Noordwijk, The Netherlands
e-mail: dario.izzo@esa.int

Abstract. We investigate upon the change of an asteroid orbit caused by an impact. We find that, given the assumption of two dimensional motion, the asteroid displacement may be described by an analytic and explicit expression that is the vectorial sum of a radial component and a component along the asteroid velocity. The new formulation bridges the gap between the study of short-term effects, using numerical methods and the analytic study of secular changes of the asteroid orbit. The relation of the method to the established formulations is described and the known results are derived as limiting cases.

The application of the new method for the performance evaluation of an asteroid deflection demonstration mission is illustrated. In such a mission the measurement of the change of the asteroid orbit by an impact will be conducted by radio-ranging to a spacecraft orbiting the deflected asteroid. Hence the measurement will primarily be sensitive to the deflection projected onto the Earth-asteroid line of sight. We discuss how the new formulation of the deflection can conveniently be employed for the estimation of the measurement accuracy and the optimal planning of a deflection demonstration mission.

Keywords. Celestial Mechanics; Space vehicles

In memoriam Floh

1. Introduction

One of the most promising strategies for mitigating the threat of an asteroid on collision course with the Earth is the impulsive deflection of the asteroid by a kinetic impactor. The key advantages of this method compared to deflection by applying long-duration thrust to the asteroid or its explosive destruction are simplicity and technical maturity. In particular the Deep Impact space mission [see A'Hearn *et al.* (2005) for an overview] has demonstrated that a high velocity impact onto a small celestial body is indeed technologically feasible.

From the astrodynamics point of view the impulsive deflection of an asteroid has already received considerable attention and can be regarded as well understood. It has been demonstrated by Carusi *et al.* (2002) that the most promising strategy is to apply a small velocity change to the asteroid several orbital periods before the collision with Earth or before the passage through the keyhole of a resonant return [see Valsecchi *et al.* (2003) for an exposition of the keyhole concept] by a tangential impact. The tangential impact leads to a change of the orbital period of the asteroid and hence to a secular change of the orbit. This secular change will accumulate over time and — even for a minuscule change of asteroid velocity — lead to a large accumulated miss distance. This method has meanwhile been implemented into a full end-to-end trajectory optimisation procedure that maximises the secular change of an asteroid's orbit achieved by a spacecraft departing from Earth and reaching the asteroid either by chemical or electric propulsion in Izzo (2005), Izzo et al. (2006a) and Izzo (2006b).

It is also known from the numerical study of Conway (2001) that, if the time between the deflection and the potential collision with the Earth is small, the optimal impact direction is no longer tangential to the asteroid orbit. This has recently also been verified by Kahle et al. (2006) for a deflection taking place shortly before the passage through a keyhole.

Up to now the short term regime (i.e. deflections taking place two or less orbits before the dangerous close encounter) has not been described by analytical methods. In the present paper we fill this gap by deriving an analytical description of the response of an asteroid to a small impulsive velocity change. Our result is a solution for the relative orbital dynamics between the perturbed and unperturbed orbit to linear order in the change of velocity. In its present form the solution is restricted to a velocity change in the orbital plane of the asteroid but it can be used advantageously both for short term and for long term deflections. In particular the well known secular orbital change is recovered in the limit of a long term deflection.

The study of the linearised equations of relative orbital motion has already received long-standing attention in the context of spacecraft rendezvous manoeuvres and formation flying. In this realm the relative motion is typically treated in a local-horizontal local-vertical (LHLV) coordinate system. The equations of relative motion in these coordinates are the so-called Tschauner-Hempel equations [cf. Tschauner & Hempel (1964)], a system of three linear differential equations in the true-anomaly difference with periodic coefficients. For a circular orbit the Tschauner-Hempel equations reduce to three linear differential equations with fixed coefficients, the Clohessy-Wiltshire equations [cf. Clohessy & Wiltshire (1960)]. If rewritten in LHLV coordinates, our solution is an explicit algebraic solution of the Tschauner-Hempel equations. For the case of a circular orbit, we establish the equivalence of our solution to the well know analytic solution of the Clohessy-Wiltshire equations.

In order to illustrate the usefulness of our solution we apply it to the performance evaluation for a space mission that aims at demonstrating the capability to deflect an asteroid. A deflection demonstration mission that is currently being studied by ESA under the name Don Quijote [cf. Carnelli et al. (2006)] is taken as the example case.

The layout of this paper is the following: We start by formulating the deflection of an asteroid in terms of the perturbation of the asteroid state vector in Sec. 2. For this formulation of the problem we first obtain the solution for the radial perturbation in Sec. 2.1 and then for the tangential perturbation in Sec. 2.2. The combined result, yielding the total deflection, is then discussed in Sec. 2.3. In Sec. 3 we apply the solution to the performance evaluation for a deflection demonstration mission. We close with a summary of our key results and give an outlook on further extensions and applications of the new method in Sec. 4.

2. Impulsive deflection as a perturbation of the state vector

For the present study we restrict our analysis to the case of a velocity change in the orbital plane of the asteroid. While the generalisation to out of plane deflection is straightforward it is of little practical relevance for a space mission that performs an asteroid deflection rehearsal. In the following we will determine the deflection of the asteroid with respect to the unperturbed orbit of the asteroid. For the unperturbed orbit we choose a description in terms of the propagation of the initial state vector $[\mathbf{r}_0, \mathbf{v}_0]$ at the time t_0 by means of the Lagrange coefficients. The radius vector at other times is then given by

$$\mathbf{r} = F\mathbf{r}_0 + G\mathbf{v}_0. \tag{2.1}$$

Here F and G are the Lagrange coefficients,

$$F = 1 - \frac{r}{p}(1 - \cos\theta) \quad \text{and} \quad G = \frac{rr_0}{\sqrt{\mu p}}\sin\theta, \tag{2.2}$$

where

$$r = \frac{pr_0}{r_0 + (p - r_0)\cos\theta - \sqrt{p}\sigma_0\sin\theta}, \tag{2.3}$$

and p is the orbit parameter, θ is the true anomaly difference between $\mathbf{r}_0$ and $\mathbf{r}$, $\sigma_0 \equiv \mathbf{r}_0 \cdot \mathbf{v}_0/\sqrt{\mu} = r_0 v_0 \cos\gamma_0/\sqrt{\mu}$, γ_0 denotes the flight path angle at t_0 and μ is the gravitational parameter of the Sun.

If the orbit is perturbed by a velocity change $\mathbf{\Delta V}$ at the instant t_0 then the perturbation on the radius (that from now on will be called the deflection) is be given by

$$\mathbf{\Delta r} = \Delta F\mathbf{r}_0 + \Delta G\mathbf{v}_0 + G\mathbf{\Delta V}. \tag{2.4}$$

Here ΔF and ΔG denote the perturbed Lagrange coefficients.

In order to facilitate the calculation it is helpful to choose a non-orthogonal coordinate system for $\mathbf{\Delta r}$. We decompose $\mathbf{\Delta r}$ into a component along the unperturbed radius vector and a component along the unperturbed velocity,

$$\mathbf{\Delta r} = \Delta q\mathbf{i}_r + \Delta s\mathbf{i}_v, \tag{2.5}$$

where $\mathbf{i}_r$ is the unit vector in radial direction and $\mathbf{i}_v$ is the unit vector in velocity direction. Using this system of skew axis the contravariant component of the perturbation in the radial direction is equal to the perturbation of the absolute value of the radius vector under the constraint of unperturbed difference in true anomaly θ between $\mathbf{r}_0$ and $\mathbf{r}$,

$$\Delta q = \Delta r|_{\Delta\theta=0}. \tag{2.6}$$

Hence one can easily perturb Eq. (2.3) for the determination of Δq. A further advantage of the use of these skew axis is that the secular change of orbit will only affect the component along the velocity vector. Hence one retains the intuitive interpretation of the secular term that would for instance be lost in the LHLV coordinate system.

2.1. *The radial part*

The equation for the absolute value of the radius in terms of the initial radius, initial velocity and true-anomaly difference is given by the polar form of the equation of orbit Eq. (2.3). The change of the radius Δq is determined to linear order by variation of parameters under the constraint that θ is unperturbed,

$$\Delta q = \frac{r^2}{\sqrt{p}r_0}\sin\theta\Delta\sigma - \frac{r^2\cos\theta}{pr_0}\Delta p + \frac{r^2\sigma_0}{2p^{3/2}r_0}\sin\theta\Delta p + \frac{r}{p}\Delta p. \tag{2.7}$$

The variations of p and σ_0 by a $\mathbf{\Delta V}$ transfer are in turn given by

$$\Delta p = 2\sqrt{\frac{p}{\mu}}|\mathbf{r}_0 \times \mathbf{\Delta V}| = 2\sqrt{\frac{p}{\mu}}r_0\Delta V\sin\phi, \tag{2.8}$$

$$\Delta\sigma = \frac{\mathbf{r}_0 \cdot \mathbf{\Delta V}}{\sqrt{\mu}} = \frac{r_0\Delta V}{\sqrt{\mu}}\cos\phi, \tag{2.9}$$

where ϕ is the angle between $\mathbf{r}_0$ and $\mathbf{\Delta V}$. Inserting these expressions we arrive at

$$\Delta q = r \frac{2r_0}{\sqrt{\mu p}} \sin\phi \, \Delta V$$
$$+ r^2 \frac{r_0 v_0}{\mu p} \left[\sin\gamma_0 \cos\phi \sin\theta - 2\sin\gamma_0 \sin\phi \cos\theta + \cos\gamma_0 \sin\phi \sin\theta \right] \Delta V , \quad (2.10)$$

where we have used $\sqrt{\mu p} = r_0 v_0 \sin\gamma_0$ and $\mathbf{r}_0 \cdot \mathbf{v}_0 = r_0 v_0 \cos\gamma_0$. In Eq. (2.10) the terms with $\sin\phi$ correspond to a change of orbital angular momentum while the term proportional to $\cos\phi$ corresponds to a change of orbital energy.

2.2. The tangential part

In order to obtain the tangential part of the deflection one has to consider the full variation of the state vector. Singling out the components along the velocity vector $\mathbf{v}$ one obtains

$$\Delta s\mathbf{i}_v = \mathbf{v} \Bigg\{ -\frac{3a}{\mu}[t - t_0](\mathbf{v}_0 \cdot \mathbf{\Delta V}) - \frac{3a^2}{\mu^{3/2}}(\sigma - \sigma_0)(\mathbf{v}_0 \cdot \mathbf{\Delta V})$$
$$-\frac{3a^2 r \sigma_0}{\mu^{3/2}p}(1 - \cos\theta)(\mathbf{v}_0 \cdot \mathbf{\Delta V}) + \frac{a r_0 r}{\mu^{3/2}p}(\sigma - \sigma_0)(1 - \cos\theta)(\mathbf{v}_0 \cdot \mathbf{\Delta V})$$
$$+\frac{3a^2 r}{\mu^{3/2}\sqrt{p}} \sin\theta(\mathbf{v}_0 \cdot \mathbf{\Delta V}) + \frac{r_0^2 r^2}{(\mu p)^{3/2}}(1 - \cos\theta)\sin\theta(\mathbf{v}_0 \cdot \mathbf{\Delta V})$$
$$-\frac{r_0 r}{\mu p}(1 - \cos\theta)(\mathbf{r}_0 \cdot \mathbf{\Delta V}) - \frac{r_0 r^2 v_0}{(\mu p)^{3/2}}(1 - \cos\theta)\sin(\theta - \gamma_0)(\mathbf{r}_0 \cdot \mathbf{\Delta V})$$
$$-\frac{r^2 r_0}{\mu p}\sin\theta \sin(\theta - \phi)\Delta V \Bigg\} , \quad (2.11)$$

where we have introduced $\sigma = \mathbf{r} \cdot \mathbf{v}/\sqrt{\mu}$ and a denotes the semimajor axis.

Several of the terms of Eq. 2.11 have a direct interpretation. Most notably the first term proportional to the time after impact, $t - t_0$, is the only secular perturbation of the orbit. This term had already been obtained in Izzo (2005) and compared in magnitude to the other terms for different orbits and times. The next three terms are effects of the eccentricity of the unperturbed orbit and vanish for a circular orbit. Again all terms with the exception of the last term can be associated with either a change of orbital energy (term 1–6) or angular momentum (term 7 and 8). The last term is the projection of the third term in Eq. (2.4).

2.3. The total deflection

The total deflection is obtained by inserting Eqs. (2.10) and (2.11) into Eq. (2.5). The total deflection is a an exact solution to linearised equations of motion for the relative motion between the perturbed and unperturbed orbit. Higher order terms will only lead to corrections of order $O(\Delta V/v_0)^2$. For the case of a real asteroid deflection ΔV is foreseen to be of the order 10^{-5} m/s while the orbital velocity will be of the order 10 km/s. Hence corrections to Eq. (2.5) will be suppressed by a factor 10^{-9} and higher order terms can be neglected without introducing significant errors.†

† The situation is analogous to that of spacecraft rendezvous in Earth orbit. Also there the relative motion is usually only treated to linear order for all practical applications.

While the new expression is fully analytical and explicit, it depends on both the true anomaly difference, θ, and time after the impulsive velocity change, $t - t_0$. Hence the application of the solution requires solving Kepler's equation for the unperturbed orbit.

In conclusion, the new deflection equation gives a full description of the change of a Keplerian orbit after an impulsive velocity change. After transformation into a LHLV coordinate system the solution becomes an analytic and explicit solution of the Tschauner-Hempel equations for the relative dynamics on elliptic orbits.

Indeed similar formulations in terms of perturbation matrices have already been obtained previously [see e. g. Battin (1994) p. 463ff. for an introduction to the topic]. The new formulation has however a significant advantage over the known formulations. In order compute the deflection in terms of perturbation matrices one has to numerically solve the variational equation for the secular part of the deflection. In our new formulation the secular term is given explicit and analytic in terms of the time after impact and the non-secular terms are given explicit and analytic in terms of the true anomaly difference.

Our new formulation encompasses the established treatments of long-term impulsive asteroid deflection in Carusi *et al.* (2002) and Izzo (2006b). These results of these studies are recovered in the limit of long times after the deflective impact. The particular power of the new method lies however in the regime of mid-term deflection, that is the case where the deflection has to be achieved a few orbital periods before the collision with the Earth. On the one hand, in this regime the radial term is not yet negligible and needs to be taken into account. On the other hand the propagation time after deflection is already so long that an analytical treatment of the secular term has significant advantages over numerical schemes.

It is also straightforward to show that our solution reduces to the solution of the Clohessy-Wiltshire equations [cf. Clohessy & Wiltshire (1960)] for the case of a circular orbit because for this case the LHLV coordinate system and the radial-tangential coordinate system coincide. The equivalence of the two solutions is readily established by noting that for a circular orbit we have $\sigma = \sigma_0 \equiv 0$ and exploiting the simplifications of the Lagrange coefficients that arise for circular orbits

$$F_t = -G, \quad \text{and} \quad G_t = F, \tag{2.12}$$

where F_t and G_t denote the Lagrange coefficients for the equation for the velocity vector $\mathbf{v} = F_t \mathbf{r}_0 + G_t \mathbf{v}_0$. From Eq. (2.12) we have the relation

$$(1 - F)\mathbf{r} + G\mathbf{v} = -(1 - F)\mathbf{r}_0 + G\mathbf{v}_0, \tag{2.13}$$

by which we can easily transform Eq. (2.11) into the tangential component of the solution of the Clohessy-Wiltshire equations for an impulsive orbit change.

3. Application to deflection demonstration missions

The perturbative solution is particularly suited for the assessment and optimisation of the performance of a deflection demonstration space mission. The basic principle of a deflection demonstration mission is to impact an asteroid that poses no threat to the Earth and measure the achieved deflection. The measurement of the achieved deflection is of crucial importance because it yields a measurement of the momentum carried away by ejecta. It is expected that the momentum transfer to the asteroid from the shedding of ejecta will considerably exceed the momentum transfered by the impact itself. Hence, the determination of the momentum transfer caused by the impact achieved by measuring the orbital change of the asteroid gives more reliable estimates for future deflection missions.

A deflection demonstration that follows this principle is currently being studied by ESA under the name Don Quijote [see Carnelli *et al.* (2006) for an overview]. The Don Quijote mission is envisaged to comprise two spacecraft, the Orbiter and the Impactor. The Orbiter will be launched first. It will travel to the asteroid and conduct a precision determination of the asteroid orbit and carry out a precision measurement of its mass, gravity field, topography, rotational motion and composition. After the successful arrival of the Orbiter at the asteroid the Impactor will be launched from Earth. It will travel to the asteroid and collide with it in a way to achieve the largest possible change of the orbital energy of the asteroid. The minimum goal is to change the semimajor axis of the asteroid orbit by 100 m. The Orbiter will observe the collision of the Impactor from a safe position and analyse the ejecta. Then it will return to closer proximity of the asteroid and conduct a second orbit determination campaign in order to measure the deflection that has been achieved.

In a deflection demonstration mission the smallness of the orbital changes makes it necessary to measure it via radio-tracking of a transponder orbiting the asteroid (or placed on it). Consequently the mission layout has to take into account the peculiarities of radio-tracking measurements. In particular the measurement will be sensitive primarily to special components of the deflection in the *geocentric frame* because the measurement will be carried out from a groundstation on Earth.

The methods of choice for the deflection measurement are ranging and differential very large baseline interferometry (ΔVLBI) [see Thornton & Border (2000) for a description of these techniques and their performance]. In particular Doppler tracking can immediately be excluded as a suitable method because the velocity change of the asteroid will be in the order of 1 to 10×10^{-5} m/s whereas the absolute accuracy of present-day Doppler spacecraft tracking is limited to approximately 10^{-3} m/s post-processing accuracy.

Radio ranging, in which the runtime of a signal to the spacecraft and back to the ground station is measured, is sensitive primarily to the distance change between the asteroid and the Earth. Hence this method will be sensitive to the deflection along the line-of-sight between the asteroid and the Earth. As a rule of thumb we can expect a post processing accuracy for the ranging of about 10 m, which will be dominated by the uncertainty in the position of the asteroid centre of mass which respect to the Orbiter.

ΔVLBI measures the angular position of the Orbiter on the sky relative to a known astronomical radio source such as a quasar by triangulation with two ground stations. Hence it is mainly sensitive to the asteroid's change in angular position, i.e. the component of the deflection orthogonal to the Earth line of sight. With present-day equipment a post-processing accuracy of 50 nrad is achieved. For upcoming missions with new equipment we can however assume a post-processing accuracy of 5 nrad, which we will assume in our analysis [cf. Thornton & Border (2000) p. 65]. As a consequence, the deflection needs to be assessed in terms of the projections onto the measurable quantities in the geocentric frame.

While the above performance values give a rule of thumb, the precise expected performance for the orbit determination of the asteroid from the radio measurements can only be determined from a complete simulation. Important parameters, that will influence the measurement performance, are the Yarkowsky effect on the asteroid, and the correct estimation of it, the determination of the asteroid centre of mass position with respect to the Orbiter, the knowledge of non-gravitational disturbances on the Orbiter, and effects on the radio signal propagation.

Hence a determination of the measurement performance from simple astrodynamical considerations remains elusive. Nevertheless the orbital change of the asteroid in the measurement frame provides an important figure of merit. Such a figure of merit is

Table 1. Asteroid and Impact parameters for the two Don-Quijote strawman targets.

Asteroid	(10302) 1989 ML	2002 AT$_4$
Asteroid parameters		
a (AU)	1.27	1.87
perihelion (AU)	1.10	1.03
e	0.14	0.45
P (days)	524	931
Impact parameters		
Impact date	6 Feb. 2018	9 April 2017
ϕ (deg)	-29	-70
ΔV (m/s)	10^{-5}	6×10^{-5}
Δa (m)	96.4	2232

in particular necessary for the design a suitable Impactor trajectory. Due to the large number of possible trajectories, it is not feasible, in practice, to evaluate the suitability of each of them by a full radio-tracking simulation. The simple geometrical quantities provide a reliable criterion if a particular Impactor trajectory is suitable to achieve a measurable deflection. The projections of the deflection onto the Earth line of sight $x_\parallel$ and orthogonal to it $x_\perp$ are given by

$$x_\parallel = \Delta q \left(\mathbf{i}_r \cdot \mathbf{i}_{\oplus A} \right) + \Delta s \left(\mathbf{i}_v \cdot \mathbf{i}_{\oplus A} \right), \tag{3.1}$$

$$x_\perp = \Delta q \left| \mathbf{i}_r \times \mathbf{i}_{\oplus A} \right| + \Delta s \left| \mathbf{i}_v \times \mathbf{i}_{\oplus A} \right|, \tag{3.2}$$

where $\mathbf{i}_r \equiv \mathbf{r}/r$, $\mathbf{i}_v \equiv \mathbf{v}/v$ and $\mathbf{i}_{\oplus A}$ is the unit vector along the Earth asteroid line of sight.

In a deflection demonstration mission the measurement time after the deflection will typically be limited in order to limit the overall mission duration. For example, for the Don Quijote study ESA aims at completing the deflection measurement within $1/2$ year after the impact. Noting that the period of a Near Earth Asteroid will be one year or even significantly more it is immediately obvious that such a measurement time lies in the last-minute regime of asteroid deflection. Hence it becomes necessary to build the performance meter for the deflection on the full (albeit linearised) orbital change taking into account both, secular and non-secular terms.

For the Don Quijote study ESA chose not to aim at a deflection that is optimally measurable but instead has chosen a scenario that is close to a real case of long-term deflection: An Impactor trajectory is considered optimal if it achieves a change of semi-major axis of $\Delta a = 100\,\text{m}$ at minimal mission cost and complexity. This corresponds to an optimisation of the secular deflection. In order to determine if this is a viable strategy it has to be considered if the trajectories optimised in this way lead to a deflection that is measurable in the geocentric frame.

We approach this problem by considering trajectories to both Don Quijote targets that have been optimised for a change of semimajor axis and evaluate the measurable deflection for them. The trajectories that we consider have been obtained in the framework of an ongoing industrial phase A study of the Don Quijote mission by the company Deimos Space. They assume a launch with the Russian Dnepr launcher and Earth escape with a LISA-Pathfinder propulsion module. Furthermore the Sun-asteroid-Impactor angle is limited to 70 deg during the terminal approach to ensure sufficient illumination for visual terminal navigation.

Some key parameters of the target asteroid orbit and the parameters characterising the impact are given in Table 1. The achievable velocity increments ΔV are based on rough mass estimates for the asteroids and the assumption of a totally inelastic collision.

Figures 1 to 4 display the result of our analysis. Figure 1 displays the deflection for the asteroid (10302) 1989 ML. The maximal orbital change along the direction of the Earth line-of-sight reaches approximately 550 m. At the end of the measurement period the measurable deflection rapidly decays because the Earth has 'overtaken' the asteroid and the tangential and and radial deflection partially compensate each other in the projection along the Earth line-of-sight. The maximal measurable deflection is reached near the opposition of the asteroid. Briefly before that the Earth line-of-sight is aligned with the vector of the deflection. At this instance the measurable deflection and the total deflection coincide. The change of the relative position of Earth causes a strong modulation of the measurable deflection component. Clearly the secular term is neither representative for the total deflection nor for the tangential deflection. Still in this scenario, that has been optimised for the secular term, the measurable deflection is one order of magnitude larger than the expected ranging accuracy in the second half of the measurement period. This shows that a reliable measurement of the deflection is feasible in this scenario. The angular deflection of 1989 ML seen from the Earth is displayed in Figure 2. The angular deflection remains below the capabilities of ΔVLBI for the first 300 days after the impact. A precise measurement of the deflection is not possible with this method.

Figure 3 shows the deflection achieved for 2002 AT$_4$. The deflection is nearly one order of magnitude larger than for 1989 ML. For this asteroid the observation period comprises less than half of orbital period and is hence truly in the 'last minute' regime of deflection. Also for this object the secular term is not representative of the total deflection or the tangential deflection. Despite of the much larger deflection, the angular deflection is even smaller than for 1989 ML. The reasons for this are the larger geocentric distance of the asteroid during the measurement period and the higher eccentricity of the orbit by which the tangential deflection is quite well aligned with the Earth direction for a considerable part of the orbit after the impact.

The two examples show that for the limited measurement duration of a deflection demonstration the secular and non-secular contributions to the deflection are equally important. The modulation of the measureable deflection by the relative motion of the Earth has a strong influence on the magnitude of the measureable deflection. In particular, for certain Earth-asteroid configurations the contributions of the radial and tangential deflection to the apparent deflection can compensate each other. The optimal measurement situation is only achieved if the apparent deflection is optimised.

The two examples also demonstrate that the secular term (i.e. the semimajor axis change) is by no means representative of the measurable deflection in a deflection demonstration mission.

4. Conclusions

In the present study we have analysed the orbital change of a Keplerian orbit by an impulsive velocity change in the orbital plane. A solution for the relative dynamics with respect to the unperturbed orbit was obtained up to linear order. The solution is algebraic, analytic and explicit albeit depending on both the time after the velocity change and the relative true anomaly difference to the point of velocity change. It is well suited for an implementation into a delfection optimisation procedure.

The solution was formulated in a non-orthogonal coordinate system with its principle axis along the unperturbed radius vector and the unperturbed velocity. This coordinate system is advantageous because the secular change of the orbit is oriented tangentially to the orbit and hence secular changes are limited to one of the coordinate vectors. From the

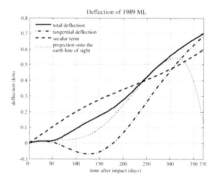

Figure 1. Deflection of 1989 ML and components thereof

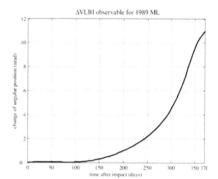

Figure 2. Angular deflection of 1989 ML as seen from Earth

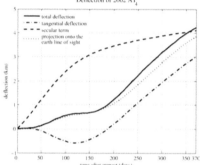

Figure 3. Deflection of 2002 AT$_4$ and components thereof

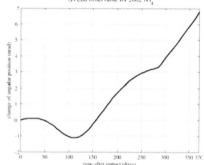

Figure 4. Angular deflection of 2002 AT$_4$ as seen from Earth

new solution an explicit analytic solution of the Tschauner-Hempel equations is easily obtained by projecting the solution into the LHLV coordinate system.

As an application of the new solution we studied the orbital deflection of an asteroid after the impact of a spacecraft. For a velocity change which is small compared to the orbital velocity of the asteroid the solution describes accurately the relative dynamics of the perturbed orbit compared to the unperturbed one. The new formulation is in particular relevant for the description of the perturbation in the first few orbital revolutions after the impact. For long times after the impact the secular perturbation of orbit becomes dominant because it exhibits a linear growth with time whereas all the other terms show a periodicity shorter or equal to one orbital period. A scenario in which the non-secular terms are still important is that of mid-term deflection in which a collision with Earth in the next few orbital periods needs to be mitigated. The effect of the non-secular terms is even more significant for a deflection demonstration mission because for practical reasons one will want to measure the achieved deflection on a timescale that will hardly exceed one year.

The equal importance of all contributions in the deflection measurement of a demonstration mission has been illustrated using two example cases which are based on the Don Quijote study currently being carried out by ESA. For each of the two strawman targets of the Don Quijote mission, 1989 ML and 2002 AT$_4$, a trajectory optimised for a maximal secular change of the asteroid orbit was considered. It was found that the secular term is not representative for the perturbation of the asteroid orbit that is achieved. The reasons are two-fold: Firstly, the non-secular terms are of the same magnitude. Secondly the measurement of the deflection is primarily sensitive to the deflection along the Earth-asteroid line of sight. This singles out a projection of the total deflection that does not coincide with the deflection tangential to the asteroid orbit. While the two effects do not exclude the measurement of the deflection in the example cases, they delay the reliable detectability of the deflection — in the case of 1989 ML by several month.

In conclusion, a deflection demonstration mission can either be designed as closely as possible to a real case deflection by optimising the secular change of the asteroid orbit or it can be optimised to achieve a large measurable deflection by maximising the deflection along the Earth-asteroid line-of-sight. Choosing the latter option has the potential of considerable cost savings because a smaller velocity change of the asteroid orbit will be sufficient to verify the ability of carrying out a real-case deflection.

Trajectory planning that optimises the measurable deflection is readily accomplished by implementation of the new analytic deflection description into an optimiser. Due to the analyticity of the new expressions the optimisation procedure will only marginally exceed the computational requirements of a trajectory optimisation that maximises the secular deflection. An implementation of an optimisation procedure for the measurable deflection in a deflection demonstration mission is currently underway.

Acknowledgements

The authors are grateful to Juan L. Cano from Deimos Space for providing the example Impactor transfer trajectories and to Diego Escorial Olmos for useful discussions on the ranging measurements.

References

A'Hearn, M.F., *et al.* 2005, *Science* 310, 258

Battin, R.H. 1994, *An Introduction to the Mathematics and Methods of Astrodynamics* AIAA Education Series (AIAA, Washington, DC)

Carnelli, I., Galvez, A. & Izzo, D. 2006 "Don Quijote: A NEO deflection precursor mission" submitted to NASA Workshop: Near-Earth Object Detection, Characterization, and Threat Mitigation, http://www.esa.int/gsp/ACT/doc/ACT-RPR-4200-IC-NASANEOWS-DonQuijote.pdf

Carusi, A., Valsecchi, G.B., D'Abramo, G. & Boattini, A. 2002, *Icarus* 159, 417

Clohessy, W.H. & Wiltshire, R.S. 1960, *Journal of the Aerospace Sciences* 27, 653

Conway, B.A. 2001, *Journal of Guidance, Control, and Dynamics* 24, 1035

Izzo D. 2005, in: *Proceedings of the 2005 AAS/AIAA Space Flight Mechanics Conference* (Univelt Inc.), Paper AAS 05-150, Vol. 121, p. 611

Izzo D., Bourdoux A., Walker R. & Ongaro F. 2006, *Acta Astronautica* 59, 294

Izzo, D., "Optimisation of interplanetary trajectories for impulsive and continuous asteroid deflection," 2006, to appear in *Journal of Guidance Control and Dynamics*

Kahle, R., Hahn, G. & Kührt, E. 2006, *Icarus* 182, 482

Thornton, C.L. & Border, J.S. 2000, *Radiometric tracking techniques for deep-space navigation* Deep-space communications and navigation series 00-11, Jet Propulsion Laboratory, http://descanso.jpl.nasa.gov/Monograph/series1/Descanso1_all.pdf

Tschauner, J. & Hempel, P. 1965, *Astronautica Acta* 11, 104

Valsecchi, G.B., Milani, A., Gronchi, G.F. & Chesley, S.R. 2003, *Astron Astrophys.*, 408, 1179

Near Earth Objects, our Celestial Neighbors: Opportunity and Risk
Proceedings IAU Symposium No. 236, 2006
A. Milani, G.B. Valsecchi & D. Vokrouhlický, eds.
© 2007 International Astronomical Union
doi:10.1017/S1743921307003523

A family of low-cost transfer orbits for rendezvous-type missions to NEAs

Ştefan Berinde

Babeş-Bolyai University, Cluj-Napoca, Romania
email: sberinde@math.ubbcluj.ro

Abstract. We develop a family of transfer orbits suitable for the peculiarity of NEAs orbits. They involve one or more carefully planned deep-space maneuvers and one or more close encounters with the Earth. Basically, it is a gravity assist technique, but only Earth is used as the driving planet. In this manner, the orbit of the spacecraft stays near the Earth, as the orbits of NEAs. The idea is to obtain a large relative velocity in respect to the Earth, using these impulsive maneuvers. Next, successive rotations of this relative velocity vector due to resonant encounters with the Earth will, finally, shape the orbit of the spacecraft, in order to match the inclination and the orientation of apsidal line of the asteroid's orbit.

With a velocity budget less than 9 km/s most of NEAs orbits are reachable, even highly inclined ones, or orbits with relatively large nodal distances. An example is given: a rendezvous mission to asteroid (1036) Ganymed.

Keywords. space missions; rendezvous; resonant returns

1. Introduction

The population of discovered near-Earth asteroids (NEAs) is continuously growing. They mimic various (and mostly unknown) physical characteristics (internal structure, rotation, companions, etc.) which qualify them for in situ exploration. From the dynamical point of view, many of them have orbits allowing close encounters with the Earth in the near or distant future. Mitigation of an impact hazard might require in situ exploration of these objects. Among the types of possible interplanetary missions: rendezvous, nodal flyby or resonant flyby missions (Perozzi *et al.* 2001), the first one offers the greatest scientific return, since the spacecraft will stay longer near the object to study or interact with it. Rendezvous missions require a transfer orbit in such a way that the spacecraft will end on the same orbit as the target asteroid. This alteration of the orbit is usually very expensive in terms of velocity budget. Even if NEAs are "near" the Earth, many of them are not easily reachable on rendezvous missions.

In this paper we develop a family of transfer orbits for rendezvous missions, equally applicable to most of the NEAs, with a total velocity budget at a reasonable level (<9 km/s). The orbits involve deep-space maneuvers (DSMs), and one or more close encounters with the Earth. Being rather complex, the major side effect is the large transfer time, sometimes above 10 years. We use the piecewise two-body approximation for this approach. Next section will summarize this theory. In section 3 we introduce the concept of *true intersection orbits*. In section 4 we analyze the velocity requirements to accomplish a rendezvous-type mission to NEAs using simple transfers (two-impulse transfers). In section 5 we introduce a technique to increase the velocity of a spacecraft in respect to the Earth using deep-space maneuvers and in section 6 we show how to build highly inclined orbits using resonant encounters with our planet. In section 7 we give a complete example of how the theory works and, finally, we draw some conclusions.

2. Piecewise two-body approximation

Piecewise two-body approximation is basically a two "two-body" problem approach for the motion of an infinitesimal body in interplanetary space (Carusi *et al.* 1990). It can be viewed as a simplified version of patched-conics technique used in the field of Astrodynamics (Battin 1987). We recall some of its principles: Earth (the perturbing planet in this case) moves in circular orbit around the Sun, the spacecraft (the infinitesimal body) has a geocentric hyperbolic orbit near the planet and an elliptic heliocentric orbit in interplanetary space. The geocentric phase is in fact an instantaneous event (close encounter), intended to change the orientation of the unperturbed geocentric relative velocity vector of the spacecraft (Figure 1a).

We adopt the following units for distance and time, such that the radius of the Earth's orbit equals 1 and its heliocentric velocity is also 1. It follows that heliocentric gravitational parameter is $\mu_\odot = 1$ and Earth's orbital period is $T_\oplus = 2\pi$.

The heliocentric orbit is given as a set of well known keplerian elements (a, e, i). The other two ones ω and Ω are not independent variables (since the orbits must intersect each other at a known location). The geocentric orbit is fully described by the unperturbed geocentric relative velocity (in short, *relative velocity*) u, which is an invariant of the encounter, and the orientation of this velocity vector, given by two angles (θ, ϕ). We do not reproduce here the formulas relating these variables, just say that the heliocentric orbit can be propagated analytically through a close encounter, as follows

$$(a, e, i) \longrightarrow (u, \theta, \phi) \longrightarrow (u, \theta', \phi') \longrightarrow (a', e', i') \qquad (2.1)$$

The link between the two sets of angles (θ, ϕ) and (θ', ϕ') involves two additional angles: the deflection angle γ and the inclination of geocentric orbit ψ in respect to a given reference plane. The deflection angle gives a measure of how much the encounter affects the orbit of the spacecraft, and its expression is

$$\sin\frac{\gamma}{2} = \left[1 + \frac{r_{min}}{r_{LEO}}\left(\frac{u}{v_{LEO}}\right)^2\right]^{-1}, \qquad r_{min} \geqslant r_{LEO} \qquad (2.2)$$

It depends on the minimum distance of the encounter r_{min}, but also on the physical properties of the planet, which are embedded in the values of the circular velocity in low-Earth orbit (LEO), $v_{LEO} = 0.26$ (7.8 km/s), and in the radius of LEO orbit r_{LEO}. We set this value to be at an altitude of 200 km above the Earth's surface. This is also the minimum allowed altitude for the encounter. So, the maximum value for the deflection angle, γ_{max}, is obtained for $r_{min} = r_{LEO}$.

3. True intersection orbits

It is useful for our approach to introduce the following concept. We define the *true intersection orbit* related with the orbit of a NEA, that orbit which geometrically intersects the Earth's one and is obtained from the original orbit using a minimum single-impulse transfer (Figure 1b).

The velocity increment dv_0 required to change an orbit to a true intersection one can be computed numerically. However, for the sake of simplicity we use the following analytical approximation. We consider that the velocity dv_0 is applied along the heliocentric velocity vector at predefined locations on original orbit (aphelion, perihelion or a nodal point). In this manner, the nodal distances changes till one of them equals 1 (true intersection). The orbital inclination is not altered nor the longitude of ascending node.

It's easy to compute explicit formulas for this velocity increment applied at aphelion dv_Q, perihelion dv_q and at the opposite orbital node dv_n. Since there are two nodal

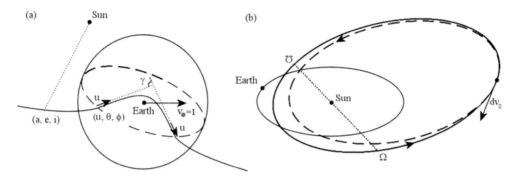

Figure 1. (a) Geometry of piecewise two-body approximation; (b) The true intersection orbit (dotted) of an original non-intersecting one (solid).

distances aimed to be changed, the minimum velocity increment is computed as $dv_0 = \min\{dv_Q^1, dv_Q^2, dv_q^1, dv_q^2, dv_n^1, dv_n^2\}$, where superscript selects the nodal distance involved in computation. Using a simple plot we conclude that most of the discovered NEAs require $dv_0 < 0.1$ (3 km/s).

If (Q_0, q_0) denotes the aphelion and perihelion distances of the true intersection orbit, the geocentric relative velocity at the intersection point is computed from

$$u_0^2 = 3 - \frac{2}{Q_0 + q_0} - \sqrt{\frac{8Q_0 q_0}{Q_0 + q_0}} \cos i \tag{3.1}$$

In this manner, each orbit of a NEA has an associated geocentric relative velocity u_0.

4. Velocity requirements for rendezvous-type missions

The most expensive task in a rendezvous mission is to match the *orbital inclination* of the target asteroid. Also, the complexity of the mission depends on how distant the orbit is, in respect to the Earth's one. For orbits not very close to the ecliptic, this aspect is quantified by the *closest nodal distance* of the asteroid's orbit. In Figure 2a we depict the distribution of discovered NEAs function of these two orbital parameters. There is an apparent clustering of orbits having an orbital node near the Earth's orbit and low orbital inclination, an effect of observational bias. Anyway, many NEAs have "wild" orbits, even if, by definition, they are "near" the Earth.

It becomes clear that a rendezvous mission to such bodies involves a high velocity budget. We will give a measure of this budget below, using a simple (and naive) tranfer scenario: two-impulse orbital transfer. First impulse dv_{node} is performed in Earth's orbit, in order to obtain a Hohmann transfer to one of the orbital nodes of the NEA's orbit (most favourable one). Its expression is

$$dv_{node} = \sqrt{\frac{2d}{(1+d)}} - 1, \tag{4.1}$$

where d is the nodal distance of the selected node. From here, a second impulse, given by

$$u_{node}^2 = \frac{2(2+d)}{d(1+d)} - \frac{1}{a} - \frac{2}{d} \sqrt{\frac{2a(1-e^2)}{d(1+d)}} \cos i, \tag{4.2}$$

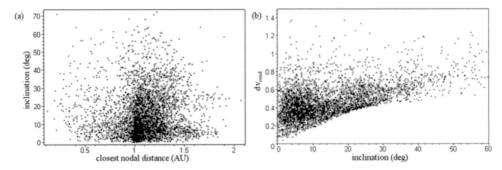

Figure 2. (a) Distribution of discovered NEAs in the plane of closest nodal distance (in respect to Earth's orbit) versus orbital inclination; (b) Velocity budget required to accomplish a rendezvous mission to each discovered NEA versus their orbital inclination (see text). Values on the vertical axis should be multiplied roughly by 30 to get the velocity expressed in km/s.

will put the spacecraft on the target orbit (a, e, i). The total velocity budget is now $dv_{rend} = dv_{node} + u_{node}$ (Berinde 2005). This approach was actually used by Perozzi *et al.* (2001), in order to estimate the accessibility of the whole NEA population, but with additional assumptions corresponding to "best case" mission profiles. Figure 2b depicts this velocity budget for each discovered NEA, function of the orbital inclination. Most of the NEAs require velocities much larger than 0.3 (9 km/s), which we take it as the upper limit for a low-cost interplanetary mission.

5. Increasing relative velocity with deep-space maneuvers

In this section we describe a technique in which a spacecraft can gain high relative velocities in respect to the Earth, using several small velocity impulses along the transfer orbit (deep-space maneuvers). At first, we give the following portrait of an orbit intersecting the Earth's one and located entirely in the ecliptic plane. Let u denotes the relative velocity at the intersection point and angle θ describes the orientation of this velocity vector against the heliocentric velocity vector of the Earth. The pair (u, θ) gives a complete description of the orbit. All other orbital quantities are computable, like semi-major axis a, orbital parameter p, aphelion distance Q and true anomaly of the intersection point f. They are given bellow

$$\begin{cases} \dfrac{1}{a} = 1 - 2u \cos \theta - u^2, & p = (1 + u \cos \theta)^2 \\[2mm] Q = a \left(1 + \sqrt{1 - \dfrac{p}{a}}\right), & \cos f = \dfrac{p - 1}{1 - p/Q} \end{cases} \tag{5.1}$$

As in Figure 3a, let the orbit of a spacecraft, with parameters (u_1, θ_1), have an intersection point A with the Earth's orbit. At the aphelion point B a small velocity impulse dv is applied in order to reduce the heliocentric velocity by this amount. Other two intersection points with the Earth's orbit appear, C (post-perihelion) and C' (pre-perihelion). Let (u_2, θ_2) be the new parameters of the orbit computed at one of these intersection points. We obtain the following relations

$$\begin{cases} dv = \sqrt{u_2^2 + y} - \sqrt{u_1^2 + y}, & \text{where } y = Q^2 + \dfrac{2}{Q} - 3, \\[2mm] Q dv = u_1 \cos \theta_1 - u_2 \cos \theta_2, \end{cases} \tag{5.2}$$

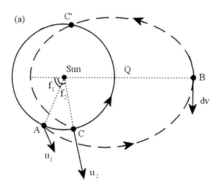
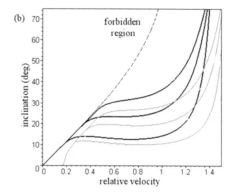

Figure 3. (a) Increasing relative velocity in respect to the Earth, using a deep-space maneuver; (b) Maximum value of the orbital inclination after one, two and three successive encounters with the Earth, on resonant orbits with semi-major axis $a_{1:1}$ (black), $a_{2:1}$ (dark gray) and $a_{3:1}$ (light gray).

which enable us to express the parameters (u_2, θ_2) function of (u_1, θ_1) and dv. More than that, it can be shown that $u_2 > u_1 + dv$. The difference $du = u_2 - (u_1 + dv)$ is significant, as can be seen later. It is an increasing function of Q and dv, but decreasing in u_1.

We must ask for Earth and the spacecraft to depart simultaneously from point A and arrive also simultaneously in point C (or C'). We need to compute the travel times for each of these bodies. Let (a_1, f_1, Q) be the set of elements derived from (u_1, θ_1), and (a_2, f_2, Q) the set of elements derived from (u_2, θ_2) as shown in equations (5.1). If Earth passes m times before the next encounter through the conjunction point with point B, its travel time reads

$$dt_\oplus = -f_1 + f_2 + 2\pi m \qquad (5.3)$$

If the spacecraft passes n times before the next encounter through the aphelion point B, its travel time reads

$$dt = dt_1(a_1, f_1, Q) + dt_2(a_2, f_2, Q) + 2\pi(n-1)a_2^{3/2}, \qquad (5.4)$$

where dt_1 and dt_2 are computable, but their expressions are rather lengthy to be reproduced here (Battin 1987).

The motion of these two bodies is synchronized if $dt = dt_\oplus$. If the pair (u_1, θ_1) is given, there exist at most one value of the velocity impulse dv for which the motion is synchronized. We name such a transfer orbit as of type "$\pm \frac{m}{n} \pm$". The sign "+" stands for a post-perihelion intersection point and "−" for a pre-perihelion intersection point. If there exist only one intersection point, we omit this sign. For example, a transfer orbit of type "$+\frac{2}{1}+$" will start at point A and will end at point C as depicted in Figure 3a. A transfer orbit of type "$+\frac{1}{1}-$" does not exist at all.

The process can be continued further. At the encounter point C, the angle θ_2 is altered due to a close encounter to a new value θ_2', freely available in the interval $[\theta_2 - \gamma_{max}, \theta_2 + \gamma_{max}]$. The previous procedure is restarted with the new pair (u_2, θ_2'). We obtain, eventually, a transfer orbit of type "$\pm \frac{m}{n} \pm \frac{m'}{n'} \pm$", and so on.

In practice, in order to maximize the efficiency of the process, we choose

$$\begin{cases} \theta_1 = 0 \\ \theta_2' = \max\{0, \theta_2 - \gamma_{max}\} \\ \theta_3' = \max\{0, \theta_3 - \gamma_{max}\} \\ \ldots \end{cases} \qquad (5.5)$$

Table 1. Families of solutions used to increase the relative velocity in respect to the Earth. Values on the second and third column should be multiplied roughly by 30 to get the velocity expressed in km/s (see text for details)

transfer type	sum of all impulses	final relative velocity	travel time (years)
1/1+	0.07 − 0.30	0.14 − 0.62	1.27 − 1.35
2/1−	0.17 − 0.30	0.21 − 0.69	1.93 − 1.70
2/1+	0.17 − 0.30	0.17 − 0.69	2.00 − 2.27
1/1 + 2/1−	0.09 − 0.29	0.20 − 0.87	2.91 − 3.21
1/1 + 2/1+	0.08 − 0.29	0.17 − 0.93	3.28 − 3.58
1/1 + 2/1 − 2/1+	0.17 − 0.26	0.62 − 0.93	5.47 − 5.55
1/1 + 2/1 + 2/1+	0.09 − 0.26	0.20 − 0.93	5.33 − 5.68
1/1 + 2/1 − 3/1−	0.10 − 0.22	0.29 − 0.91	5.85 − 6.10
1/1 + 2/1 + 3/1−	0.10 − 0.24	0.27 − 0.97	6.02 − 6.29
1/1 + 2/1 − 3/1+	0.09 − 0.23	0.24 − 1.00	6.20 − 6.50
1/1 + 2/1 + 3/1+	0.09 − 0.23	0.23 − 1.00	6.35 − 6.70

These assumptions produce maximum values for the aphelion distances on each transfer orbit and a minimum value for u_1. In this way, we get a maximized value for the final relative velocity in respect to the Earth. For each type of transfer orbit and for a given value of the final relative velocity, the starting value of u_1 is unique (if exist). Table 1 summarizes all practical types of transfer orbits, together with the range of sum of all impulses performed, the range of final relative velocities and the range of travel times until the last encounter. We have three categories of solution families: one-encounter solutions, two-encounter solutions and three-encounter solutions. The last ones are more efficient, but they require larger travel times (see table). We conclude here that a final relative velocity of at most 1.0 (30 km/s) can be reached using a total velocity budget less than 0.23 (7 km/s). We have limited our search to a reasonable range of orbital elements of transfer orbits, as follows $0.3 = q_{min} \leqslant q \leqslant 1 \leqslant Q \leqslant Q_{max} = 4.0$ (AU).

6. Increasing orbital inclination with resonant encounters

After completing the procedure from previous section, the final relative velocity vector is still on the ecliptic plane. Now, using one or more encounters with the Earth, this vector can be rotated, in order to match the inclination of the target orbit and, also, the orientation of its apsidal line. This process is performed without additional energy consumption, provided that this relative velocity is large enough.

If a single encounter cannot perform the desired inclination change, the spacecraft must be set on a resonant orbit with the Earth for an additional encounter at the same intersection point. We consider only the following mean motion resonances 1:1, 2:1, 3:1 and corresponding semi-major axis $a_{1:1}$, $a_{2:1}$, $a_{3:1}$. For each of these semi-major axis we depict on Figure 3b the maximum value of the orbital inclination after one, two and three successive encounters with the Earth. Low-order resonant orbits are more efficient and have shorter travel times, but they are not always accesible from the initial orbit. Sometimes is possible to change a resonant semi-major axis into another resonant semi-major axis after an additional encounter, in order to increase the efficiency and reduce total travel time. For a given relative velocity u, the maximum possible orbital inclination is $i = \arcsin u$, also depicted in the figure (dotted curve).

7. Putting all together: A random example

Let suppose we have the orbit of an NEA and we want to target it for a rendezvous mission. We do not take here into account the phasing requirements between the spacecraft and the asteroid. We will show only how to build a low-cost transfer orbit.

• Step 1. Consider the true intersection orbit related with the original orbit of the asteroid. We record the longitude Ω (or $\mho$) of its intersection point with the Earth's orbit, the relative velocity u_0 at this point, the semi-major axis a_0 and the orbital inclination i (the same as the original orbit).

• Step 2. Compute the synchronized transfer orbits used to increase the relative velocity exactly at a value of u_0. There is at most one solution for each type of transfer orbit. We choose that one which offers a good balance between the sum of required velocity impulses and travel time. The last encounter with the Earth should occur at the same longitude as recorded in step 1. This restricts the launch opportunity once every year.

• Step 3. Since the final relative velocity is exactly u_0, a proper orientation of this velocity vector will put the spacecraft on the true intersection orbit computed in step 1. We aim for an orbital inclination i and a final semi-major axis a_0. One or more encounters with the Earth on resonant orbits might be required to perform these orbital changes.

• Step 4. From the true intersection orbit a final impulse dv_0 is required to get into the asteroid's orbit.

This is a general purpose technique, equally applicable to most of the NEAs orbits. Only orbits with very high relative velocities, $u_0 > 1.0$ (30 km/s), or orbits with very large aphelion distances are not suitable for this technique. Also, there is no point to apply this method for orbits with low relative velocities $u_0 < 0.14$ (4 km/s), since a low-cost two-impulse transfer orbit does exist in this case.

To show the generality of our method, we pick up a random NEA as an example. It is random from the point of view of its orbital characteristics, but not from the physical point of view. Because we choose the largest known NEA – asteroid (1036) Ganymed. Some of its orbital elements are summarized in table 2. For sure, it is not an easy target.

Table 2. Some orbital parameters of the asteroid (1036) Ganymed

a	e	Q	q	d_1	d_2	i	u_0
2.67	0.53	4.09	1.24	1.40	2.90	26.7°	0.67

Among the types of transfer orbits found as solutions, we consider the "$\frac{1}{1} + \frac{2}{1}+$" one, which offers a good compromise between total velocity budget and completion time. We summarize all important quantities in table 3. The mission is completed in 11.4 years with a velocity budget of 7.3 km/s. A schematic view of orbits is given in Figure 4.

8. Conclusions

In this paper we introduce a low-cost rendezvous technique applicable to most of the NEAs orbits. It involves one or more carefully planned deep-space maneuvers and one or more close encounters with the Earth. Basically, it is a gravity assist technique, but only Earth is used as the driving planet. In this manner, the orbit of the spacecraft stays near the Earth, as the orbits of NEAs. With velocity budgets less than 9 km/s most of NEAs orbits are reachable, even highly inclined ones, or orbits with relatively large nodal distances. A random example is given: a rendezvous mission to asteroid (1036) Ganymed.

Besides of cost advantages offered by this technique, we should discuss here its limitations. One is the completion time, which may easily go beyond 10 years, depending

Table 3. Quantities related with the transfer orbit found for the asteroid (1036) Ganymed. The mission is completed in 11.4 years with a velocity budget of 7.3 km/s.

Event	minimum enc. altitude (km)	geocentric relative velocity	orbital incl. (deg)	sum of velocity impulses	travel time (years)
launch	–	0.05 (1.3 km/s)	0	0.05 (1.3 km/s)	0
enc. #1	200	0.21 (6.4 km/s)	0	0.11 (3.1 km/s)	1.3
enc. #2	200	0.67 (20.2 km/s)	10.2	0.20 (6.0 km/s)	3.5
enc. #3	200	0.67 (20.2 km/s)	19.2	0.20 (6.0 km/s)	5.5
enc. #4	200	0.67 (20.2 km/s)	26.3	0.20 (6.0 km/s)	7.5
enc. #5	1300	0.67 (20.2 km/s)	26.7	0.20 (6.0 km/s)	9.5
rendezvous	–	–	26.7	0.24 (7.3 km/s)	11.4

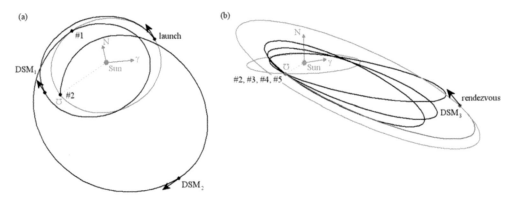

Figure 4. Schematic view of the transfer orbit found for the asteroid (1036) Ganymed. All velocity impulses are marked with arrows (the interplanetary insertion and three deep-space maneuvers). (a) Orbit of the Earth and powered transfer orbits used to increase the relative velocity of the spacecraft; (b) Orbit of the Earth, orbit of the asteroid, its true intersection orbit and resonant transfer orbits used to increase orbital inclination

on the target orbit. Second, the method was developed in the framework of a simplified theory of motion: piecewise two-body approximation. Because of that, it should be considered a first approximation transfer solution, which needs numerical improvement. Third, we did not take into account the phasing requirement between the spacecraft and the target asteroid. They should arrive (almost) simultaneously at the rendezvous point. But this is not hard to be accomplished, since the total travel time of the spacecraft can be slightly altered by changing some independent variables involved in computation or using another type of transfer orbit. Also, there is a launch opportunity every year, when the asteroid will have a different location on its orbit.

References

Battin, R. H. 1987, *An introduction to the mathematics and methods of astrodynamics* (AIAA Education Series)

Berinde, S. 2005 in: Z. Knežević & A. Milani (eds.), *Dynamics of Populations of Planetary Systems*, Proceedings of the IAU Colloquium 197 (Cambridge University Press), p. 265

Carusi, A., Valsecchi, G. B. & Greenberg, R. 1990, *Celest. Mech. Dyn. Astron.* 49, 111

Perozzi, E., Rossi, A. & Valsecchi G. B. 2001, *Planet. Space Sci.* 49, 3

Near Earth Objects, our Celestial Neighbors: Opportunity and Risk
Proceedings IAU Symposium No. 236, 2006
A. Milani, G.B. Valsecchi & D.Vokrouhlický, eds.
© 2007 International Astronomical Union
doi:10.1017/S1743921307003535

Asteroid mass determination with the Gaia mission

Serge Mouret[1], D. Hestroffer[1] and F. Mignard[2]

[1]IMCCE/CNRS, UMR 8028, Paris observatory, Denfert-Rochereau 77, 75014 Paris, France
email: mouret@imcce.fr, Hestroffer@imcce.fr

[2]OCA/Cassiopée/CNRS, UMR 6202, Nice observatory, Le Mont Gros, BP 4229, 06304 Nice
Cedex 4, France
email: francois.mignard@obs-nice.fr

Abstract. The main-belt asteroids are generally considered as the principal source for the near-Earth asteroids (NEAs). The ESA astrometric mission Gaia, due for launch in late 2011, will observe a very large number of asteroids (∼300,000 brighter than V = 20), the majority lying in the main-belt and with an unprecedented positional precision (at the sub-milliarcsecond level). Such high precision astrometry will enable to derive direct measures of the masses of the largest asteroids of which their precise determination will be of utmost significance for the knowledge of physical properties of asteroids. The method for computing the masses is based on the analysis of orbital perturbations during close encounters between massive asteroids (perturbers) and several smaller minor planets (targets). From given criteria of close approaches selection, we give the list of asteroids for which the mass can be determined, and the expected precision of these masses at mission completion.

Keywords. astrometry; orbit determination; space missions; mass determination

1. Introduction

Two main sources for the near-Earth asteroids (NEAs) are generally recognized: the main-belt asteroids (MBAs the principal source) and nuclei of extinct comets. Gaia is an astrometric cornerstone mission of the European Space Agency. With a launch due in late 2011, Gaia will have a much more ambitious aim than its precursor Hipparcos: obtain a "3D census" of our galaxy with astrometric, photometric and spectroscopic observations. It will pinpoint its sources with an unprecedented positional precision (at the sub-milliarcsecond level for single observation) which will allow it to observe about 300,000 asteroids brighter than V = 20, the majority being in the main-belt. The high precision astrometry will enable us to considerably improve the orbits of almost all observed asteroids, yielding the masses of the largest one from mutual perturbations. The objective of this work is the determination of the masses of a small subset of minor planets, modeling the signature of the gravitational perturbation affecting smaller objects during close encounters. Not only does the method consider multiple close encounters with different targets, but it also simultaneously treats all perturbers together with their target asteroids. Here we explain our interest in the knowledge of the masses which when associated to the asteroids dimensions will probably bring precious information about the physical characteristics of main-belt asteroids and consequently of NEAs (albeit Gaia will probably not derive directly masses of NEA). We give a list of asteroids for which the mass can be determined, and the expected precision of these masses at mission completion from realistic simulations of the Gaia observations (such as geometry, time sequence and magnitude).

2. On the importance of masses

Knowing the mass and size of an asteroid (the latter being from IRAS but also from Gaia) allows to determine its bulk density, which includes any space occupied by its pores. The calculation of the porosity, which enables us to estimate the part of void which makes up the asteroid, is of a great interest. It directly depends on the ratio between the bulk density of the asteroid and the grain density of an analog meteorite of similar composition. Knowing that during a collision, the internal structure of an asteroid acts upon the stress-wave propagation, which in turn, modifies its constitution, the porosity becomes a precious indicator of the collisional history of asteroids and their internal structure (Britt et al. 2002).

In addition, such measure of the bulk density will shed light on the relationship, if any, between the spectroscopic taxonomic class and the density, or possibly the porosity. Such progress could provide information on the origin and on the formation process of the solar system. If a well defined relationship can be evidenced, we could make reliable density estimates for objects of a known taxonomic characterization.

In the same way, the comparison between these densities for objects belonging to different taxonomic classes and the density of meteorites may strengthen the idea about the genetic relationship between different classes of meteorites and their supposed asteroid source (Zappalà & Cellino 2002).

3. The mass determination method

The effect of a close encounter as well as is to modify the trajectory of the perturbed bodies or equivalently to change the initial conditions of its motion at some reference time. The observed minus calculated positions (**O-C**) for each observation of minor planets, which is expressed in Gaia longitude λ projected over a given Great Circle, can be linearized and solved by the least squares technique,

$$\mathbf{O} - \mathbf{C} = \mathbf{A}\left(\begin{array}{c} \delta\mathbf{u_0} \\ \delta\mathbf{m_p} \end{array}\right) \Rightarrow \left(\begin{array}{c} \delta\mathbf{u_0} \\ \delta\mathbf{m_p} \end{array}\right) = (\mathbf{A}^t\mathbf{A})^{-1}\mathbf{A}^t(\mathbf{O} - \mathbf{C}) \qquad (3.1)$$

where $\delta\mathbf{u^0}=(\delta\mathbf{u_1^0}, .., \delta\mathbf{u_n^0})^t$ is the correction to initial conditions of the all n asteroids with $\delta\mathbf{u_k^0} = (\delta x^0, \delta y^0, \delta z^0, \delta\dot{x}^0, \delta\dot{y}^0, \delta\dot{z}^0)$ the corrections on the position and velocity of asteroid k at the initial time, and $\delta\mathbf{m_p}=(\delta m_1, .., \delta m_p)^t$ are the corrections on masses of the p perturbers.

The matrix $\mathbf{A}$ in equation (3.1) represents the partial derivatives of the longitudes λ of the minor planets with respect to their initial parameters,

$$[\mathbf{A}]_{i,j} = \sum_{q=1}^{3} \frac{\partial\lambda_i}{\partial x_q}\frac{\partial x_q}{\partial C_j^0} \quad \text{where} \quad \begin{cases} \mathbf{x}_{q=1,..,3} = (x,y,z) \\[2mm] \mathbf{C} = (\mathbf{u}_1^0, .., \mathbf{u}_n^0, m_1, .., m_p) \\[2mm] \mathbf{u_k^0} = (x^0, y^0, z^0, \dot{x}^0, \dot{y}^0, \dot{z}^0) \text{ for the } \mathbf{k}^{th} \text{ asteroid} \\[2mm] \lambda \text{ is the vector made up of every observations} \\ \quad \text{of all asteroids.} \end{cases}$$

$$(3.2)$$

The expression $\partial\lambda_i/\partial x_q$ is determined analytically, while the right-hand part of the decomposition $\partial x_q/\partial C_j$ is evaluated numerically. The latter is obtained by expressing the variations of the rectangular heliocentric coordinates $x_{q=1,..,3}$ (in an inertial frame) of each asteroid with respect to the adjustable parameters by integrating the variational equations simultaneously with the equations of motion (Herget 1968).

From equation (3.1), it is possible to give the precision $\sigma(\delta\mathbf{m_p})$ with which we will be able to determine the masses of perturbers $\mathbf{m_p}$ knowing the precision of asteroid positions. If the normal matrix $\mathbf{A}^t\mathbf{A}$ is invertible, we obtain a relation between the variance-covariance matrix of the foreseen parameters $\text{cov}(\delta\mathbf{u^0},\delta\mathbf{m_p})^t$ and the variance-covariance matrix $\text{cov}(\mathbf{O}-\mathbf{C})$. Each line of the matrix $\mathbf{A}$ corresponding to an asteroid k for an observation i is weighted according to the error on the position $\sigma_{k,i}$ (depending on the magnitude) by a constant σ_0 and so, the measures of the weighted $(\mathbf{O}-\mathbf{C})$ have the same variance σ_0^2. In addition, we make the assumption that the measures on the positions are independent, consequently $\text{cov}(\mathbf{O}-\mathbf{C}) = \sigma_0^2\mathbf{E}$ and we obtain ($\mathbf{E}$ is a unit matrix):

$$\text{cov}\left(\delta\mathbf{u^0},\delta\mathbf{m_p}\right)^t = \sigma_0^2(\mathbf{A}^t\mathbf{A})^{-1} \tag{3.3}$$

4. The selection of target asteroids

An important step for the mass determination is the selection of suitable target asteroids from which the gravitational signature of their perturbations by a larger asteroid is detectable. We restricted ourselves at this stage to the first 20,000 numbered asteroids for being potential targets, and considered that a potential perturber has a mass greater than $1\times10^{-13}\ M_\odot$ and belongs to the first 2,000 numbered asteroids. A close approach is considered meaningful if the impact parameter b (the minimal distance between the two asteroid trajectories in the case where we do not take into account their mutual perturbations) is smaller than 0.2 AU and the deflection angle θ greater than 5 mas estimated by:

$$\tan\frac{\theta}{2} = \frac{G(m+M)}{v^2b} \tag{4.1}$$

where G is the gravitational constant, M the mass of the perturber, m the mass of the target asteroid and v the relative velocity of the encounter.

We selected the target asteroids from a systematic exploration of all the close approaches between perturbers and targets from 2011 to 2017 (see Table 1), so that the whole Gaia mission is covered.

Table 1. Results of close approaches simulations

	Number of		
perturbers	149	perturber and target	
target asteroids	2508	simultaneously	54

5. Results

Calculations have been performed by taking into account realistic simulation of the Gaia observations (such as geometry, time sequence and magnitude). First the number of perturbers with respect to the precision with which we will be able to determine their masses is given in Table 2. The results are expressed in terms of the relative precision $\sigma(m)/m$ where m is the reference mass used in the simulation. Then, we give the ranking of the best relative precisions for the mass determination (Table 3). We can also see the number of targets for each mass determination, the number of close approaches for which one of the two characteristic periods (before or after the closest approach) will be out of the range of Gaia observations for the corresponding target asteroids, the formal precisions $\sigma(m)$, the standard masses m taken for the perturbers and the relative precision.

Table 2. Number of the determined masses with respect to the relative precision

Number of perturbers			
Total	149		
$\sigma(m)/m < 0.1\%$	2	$\sigma(m)/m < 20\%$	61
$\sigma(m)/m < 1\%$	7	$\sigma(m)/m < 30\%$	75
$\sigma(m)/m < 10\%$	46	$\sigma(m)/m < 50\%$	81

Table 3. The formal precisions on the best mass determination

Asteroid n° IAU name	number of target asteroids	close approaches not observed	sigma $\sigma(m)$ [$M_\odot$]	mass m [$M_\odot$]	$\sigma(m)/m$
4 Vesta	562	214	4.22×10^{-14}	1.30×10^{-10}	0.033 %
1 Ceres	336	109	1.68×10^{-13}	4.50×10^{-10}	0.037 %
14 Irene	113	31	4.56×10^{-14}	2.60×10^{-11}	0.175 %
27 Euterpe	137	39	2.54×10^{-14}	1.00×10^{-11}	0.254 %
10 Hygiea	211	78	1.32×10^{-13}	4.70×10^{-11}	0.282 %
52 Europa	134	73	9.46×10^{-14}	2.40×10^{-11}	0.394 %
2 Pallas	23	2	5.78×10^{-13}	1.30×10^{-10}	0.445 %
16 Psyche	165	70	1.41×10^{-13}	1.40×10^{-11}	1.010 %
46 Hestia	21	5	1.22×10^{-14}	1.20×10^{-12}	1.020 %
88 Thisbe	21	4	5.41×10^{-14}	5.00×10^{-12}	1.080 %

6. Conclusions

This simulation taking into account 149 perturbers and some 2508 target asteroids gives very encouraging results since about 50 masses could be estimated with a precision better than 10%. Keeping in mind that Gaia will actually observe about 300,000 asteroids, so that the number of close approaches is significantly increased, one can expect more mass determinations at that precision level. It is now planned to include these 300,000 asteroids in the simulations as well as more perturbers. Besides, we noticed that an important percentage (>30%) of targets asteroids, for half of the perturbers, was found to be not observed by Gaia, either before or after the close approach time. So, a great part of information from the encounter between the perturber and target asteroids is lost and, a study of what could bring ground-based observations before and after the space mission seems to be essential.

References

Britt, D.T., Yeomans, D., Housen, K. & Consolmagno, G. 2002, in: W.F. Bottke, A. Cellino, P. Paolicchi & R.P. Binzel (eds.), *Asteroids III*, (University of Arizona Press, Tucson) p. 485
Zappalá, V. & Cellino, A. 2002, in: O. Bienaymé & C. Turon (eds.), *GAIA: a european space project*, (EAS Publications Series) vol. 2, p. 343
Herget, P. 1968, *AJ* 99, 225

Part 8

Impact Monitoring and Risk Estimates

Near Earth Objects, our Celestial Neighbors: Opportunity and Risk
Proceedings IAU Symposium No. 236, 2006
A. Milani, G.B. Valsecchi & D. Vokrouhlický, eds.
© 2007 International Astronomical Union
doi:10.1017/S1743921307003559

The cometary impactor flux at the Earth

Paul R. Weissman†

Science Division, Jet Propulsion Laboratory, Pasadena, CA 91109 USA
email: paul.r.weissman@jpl.nasa.gov

Abstract. Comets account for a small but very significant fraction of impactors on the Earth. Although the total number of Earth-crossing comets is modest as compared with asteroids, the more eccentric and inclined orbits of the comets result in much higher encounter velocities with the planet. Additionally, some Earth-crossing comets are significantly larger than any current near-Earth asteroids (NEAs); comets 1P/Halley and C/1995 O1 Hale-Bopp are good examples of this. Thus, the most energetic impacts on the Earth likely result from comets and not NEAs. The mean impact probability for long-period comets is 2.4×10^{-9} per comet per perihelion passage, assuming the perihelion distribution of Everhart (1967), with a most probable encounter velocity of 53.5 km sec^{-1}. There are 21 known Earth-crossing Jupiter-family comets with a mean impact probability of 1.6×10^{-9} per comet per year and a most probable encounter velocity of 17.0 km sec^{-1}. For the 16 known Earth-crossing Halley-type comets the mean impact probability is 1.2×10^{-10} per year with a most probable encounter velocity of 51.3 km sec^{-1}. The poor knowledge of the size distribution of cometary nuclei makes it difficult to estimate actual impact energies at this time, though that situation is slowly improving, in particular for the Jupiter-family comets.

Keywords. Comets, orbit; impacts; Earth

1. Introduction

Comets pose a number of challenging problems when attempting to assess the impact hazard at the Earth. Long-period comets enter the planetary region from the Oort cloud (Weissman 1996, 1997) at random times and from random directions, making detection and prediction of potential impacts more difficult, and providing perilously short warning times. Determination of the sizes and masses of cometary nuclei is difficult, requiring specialized observations (Weissman & Lowry 2003; Meech *et al.* 2004; Lamy *et al.* 2004), some of which are not possible without spacecraft flyby or rendezvous missions. Even those comets classified as periodic can have intervals of up to 200 years between perihelion passages, and thus the time to catalog all of the periodic comets with telescopic surveys can be quite long. The problem is also complicated by observational selection effects whereby many comets are missed if they pass perihelion (where they are brightest) on the opposite side of the Earth's orbit.

The number of Earth-crossing comets (ECCs) per year is considerably less than near-Earth asteroids (NEAs). However, the higher eccentricities and inclinations of cometary orbits as compared with NEAs lead to far higher encounter velocities with the Earth. Thus, cometary impactors are potentially far more energetic than asteroids for bodies of comparable mass.

Moreover, observed ECCs include comets that are far larger than any currently known NEA. For example, comet Hale-Bopp, a long-period comet that passed perihelion inside the Earth's orbit in 1997, had an estimated nucleus diameter of 27-42 km (Weaver

† Present address: Jet Propulsion Laboratory, 4800 Oak Grove Drive, Mail stop 183-301, Pasadena, CA 91109 USA

et al. 1997). Taking a median value of 35 km and assuming a mean bulk density of 0.6 g cm^{-3} (Lowry & Weissman 2003; Weissman *et al.* 2004) results in an estimated mass of 1.3×10^{19} g. The impact probability for Hale-Bopp on the Earth is 2.54×10^{-9} per perihelion passage, fairly typical for a long-period comet. Because of the comet's high orbital eccentricity, 0.9951, and inclination, 89.43 degrees, the impact velocity would be 52.5 km sec^{-1}. The resulting impact energy is 1.9×10^{32} ergs, equivalent to to 4.4×10^{9} megatons. This is $\sim$44 times the estimated energy of the Cretaceous-Tertiary extinction impactor 65 Myr ago, the event that wiped out the dinosaurs. A Hale-Bopp like impact would probably be a planet sterilizing event.

This paper will review our current knowledge of the impactor flux at the Earth from active comets and the estimated frequency of impacts. Dormant comets are not included in the flux statistics presented here. Since dormant comets are difficult to differentiate from NEAs, they are generally included in the discovery and flux statistics for NEAs. A discussion on dormant comets is included later in this paper.

2. A cometary primer

Cometary orbits are classified by their orbital period and by their Tisserand parameter (Levison 1996). Long-period comets (LPCs) have orbital periods > 200 years. The choice of 200 years is somewhat arbitrary and largely reflects the lack of good orbital solutions for comets in the past. Short-period comets have orbital periods < 200 years, and are subdivided into two dynamical groups: Jupiter-family comets (JFCs) with periods < 20 years, and Halley-type comets (HTCs) with periods 20 < P < 200 years.

More formally, the JFCs have Tisserand parameters relative to Jupiter > 2 and the HTCs and LPCs have Tisserand parameters < 2. The Tisserand parameter is a pseudo-constant of the motion in the restricted 3-body problem, Sun-Jupiter-comet, and was devised to recognize returning comets even after they might have been perturbed by a close approach with Jupiter. Because of its large mass, Jupiter is the major perturber of comets in the solar system, with Saturn a distant second.

Some dynamicists define a third class of short-period comets, Encke-type, with Tisserand parameters > 3. These are short-period comets that have evolved to orbits totally interior to Jupiter's orbit. For purposes of discussion here, we will include the Encke-type comets with the JFCs.

Comets are transient objects in the planetary region. They come from two cometary "reservoirs" located at large distances from the Sun. The Oort cloud (Oort 1950) is a roughly spherical cloud of several times 10^{12} comets with semi-major axes between $\sim$3,000 and 100,000 AU. Oort cloud comets fill the gravitational sphere of influence of the Sun out to $\sim$200,000 AU, or approximately 1 parsec. Their orbits have been randomized by perturbations from the galactic tide and from random passing stars. These perturbations occasionally act to throw comets back into the planetary region where they can be observed (Dones *et al.* 2004).

It is generally agreed that Oort cloud comets are icy planetesimals thrown out of the giant planets region following the formation of the giant planets (Dones *et al.* 2004) Although the clearing of the planetary zones ejected most comets to interstellar space, a small fraction were captured to these very distant bound orbits. The Oort cloud is the source of the long-period comets. LPCs have random orbital inclinations and high eccentricities, approaching 1.

The second cometary reservoir is the Kuiper belt, located beyond the orbit of Neptune (Fernández 1980; Duncan *et al.* 1988). The Kuiper belt consists of two dynamically distinct populations: the "classical Kuiper belt" in low inclination, low eccentricity orbits beyond Neptune (plus some higher eccentricity and inclined objects captured into mean-motion resonances with Neptune), and the "scattered disk" in higher eccentricity, higher inclination orbits (though still relatively modest compared with the Oort cloud) that generally have perihelia near Neptune. Each of these sub-populations is estimated to contain $\sim 10^9$ comets > 1 km in radius.

The Kuiper belt is the source of the Jupiter-family comets. More specifically, because it is dyanmically interacting with Neptune, the scattered disk is considered the primary source for the JFCs (Duncan & Levison 1997). JFCs have dynamically evolved from the scattered disk to orbits with perihelia in the terrestrial planets region. The Centaurs, objects in orbits between Jupiter and Neptune, are believed to be that transiting population. Most JFCs are Jupiter-crossing with orbits with moderate eccentricities, typically ~ 0.6, and moderate inclinations, ~ 5–25 degrees. Both of these are considerably higher than the corresponding values for NEA orbits.

The source of the Halley-type comets is still debated and could be the Oort cloud or the Kuiper belt, or some combination of the two (Levison *et al.* 2001). The HTC orbits have substantially higher eccentricities and inclinations than the JFCs, but are not yet fully randomized like the LPCs.

Cometary nuclei are typically 1–10 km in radius, irregularly shaped, and are composed of an icy-conglomerate mixture of ices (primarily water ice), organics and silicate dust, in roughly equal proportions. Bulk densities of cometary nuclei have been estimated to be between 0.3 and 1.0 g cm^{-3} (Weissman *et al.* 2004) with a most likely value of 0.6 g cm^{-3}. Like many asteroids, comets are believed to be weakly-bonded "rubble piles" of smaller icy planetesimals (Weissman 1986). Cometary nuclei are often observed to spontaneously split and shed one or many smaller pieces; random disruption is likely the principal physical loss mechanism for cometary nuclei (Weissman 1979).

Because of their lower densities and lower material strengths as compared with NEAs, cometary nuclei smaller than ~ 300 meters are not believed to survive atmospheric entry at the Earth, and likely explode relatively high in the atmosphere (Chyba *et al.* 1993). Thus, comets likely do not contribute significantly to craters less than a few kilometers in diameter on the Earth.

3. Cometary impact rates and encounter velocities

The impact probability, p, for an Earth-crossing comet, per perihelion passage, can be calculated using Öpik's classic equation (Öpik 1951)

$$p = \frac{s^2 \, U}{\pi \, \sin \, i \, |U_x|} \tag{3.1}$$

where s is the capture radius of the target planet (including gravitational focusing), U is the encounter velocity of the comet with the Earth, i is the inclination of the comet's orbit relative to the ecliptic, and U_x is the component of the encounter velocity in the radial direction. The encounter velocity U (in units of the Earth orbital velocity) is given by

$$U = \sqrt{3 - 1/a - 2\sqrt{a(1 - e^2)} \cos i} \tag{3.2}$$

where a, e, and i are the semi-major axis, eccentricity, and inclination of the comet's orbit. The encounter velocity U is also known as the hyperbolic excess velocity or the velocity at infinity, and represents the velocity of the comet relative to the Earth as it crosses the Earth's orbit, but not yet perturbed by the Earth's gravity. The impact velocity, V_i, the velocity at which the comet impacts the Earth's surface (not accounting for atmosphereic effects) is given by

$$V_i = (U^2 + V_e{}^2)^{1/2} \tag{3.3}$$

where V_e is the escape velocity from the Earth's surface, equal to 11.18 km sec^{-1}.

The Öpik equation has singularities for comets with perihelion distances close to 1 AU and/or with inclinations close to 0 or 180 degrees. Also, the equation assumes a circular orbit for the Earth, whereas the Earth's orbital eccentricity is 0.0167. As a result, more detailed means of calculating the impact probability have been devised that eliminate these shortcomings (e.g. Kessler 1981). These more complex methods for finding the impact probability have been used in the estimates below.

Impact probabilities and velocities for long- and short-period comets have been calculated by a number of researchers: Weissman 1982; Weissman 1989; Marsden & Steel 1994; Shoemaker et al. 1994. These papers are all in good general agreement among themselves and with the results presented herein.

3.1. Long-period comets

For long-period comets, a random sample of say 10^7 hypothetical comets can be used to estimate the mean impact probability and encounter velocity, assuming a random distribution of orbital inclinations and a uniform perihelion distribution for the LPCs passing interior to 1 AU. The resulting mean impact probability is 2.2×10^{-9} per comet per perihelion passage, with a mean encounter velocity of 50.5 km sec^{-1}. If one weights the individual encounter velocity estimates by the impact probability for each comet, the most probable encounter velocity is 53.5 km sec^{-1}. This results because high impact probabilities and high encounter velocities are correlated for retrograde comets, those in orbits going around the Sun in the opposite direction to the Earth's revolution.

Everhart (1967) corrected the observed orbital element distributions for LPCs for observational selection effects and found that the perihelion distribution was not uniform inside the Earth's orbit; it increased linearly from a relative value of 0.45 at $q \sim 0.01$ AU to 1.0 at $q = 1$ AU. Using this perihelion distribution, the mean impact probability rises to 2.4×10^{-9} per comet per perihelion passage. The mean and most probable encounter velocities remain almost identical with that for the uniform perihelion distribution at 50.5 and 53.6 km sec^{-1}.

3.2. Jupiter-family comets

There are 21 known Earth-crossing Jupiter-family comets, as of August 2006. Of these, 9 have been observed on more than one apparition, 8 are single apparition objects (7 of which were discovered by the automated telescopic surveys searching for NEOs), and 4 have been either destroyed or lost, or are no longer in an Earth-crossing orbit. Interestingly, the closest observed approach to the Earth by an active comet was comet D/1770 L1 Lexell in 1770. This comet was thrown into an Earth-crossing orbit by Jupiter in 1767, approached the Earth to 0.0146 AU (about 2 million km) in 1770, and after two orbits around the Sun again encountered Jupiter (with closest approach at ~3.5 Jupiter radii) in 1779 and was removed from the terrestrial planets region.

The Earth-crossing JFCs are listed in Table 1. The mean impact probability is 8.4×10^{-9} per perihelion passage or 1.6×10^{-9} per year. The mean encounter velocity is 19.9 km sec^{-1} and the most probable encounter velocity is 17.0 km sec^{-1}. The lower most probable encounter velocity results because of the relatively low inclination distribution of the JFCs, generally < 30 degrees, which results in higher impact probabilities being correlated with lower encounter velocities.

Table 1. Impact probabilities and encounter velocities for Earth-crossing JFCs

Name	q AU	e	i deg	Period yrs	Impact P per orbit	Impact P per year	U km/s
Multiple apparitions							
2P/Encke	0.339	0.847	11.76	3.30	3.34e-09	1.01e-09	29.5
26P/Grigg-Skjellerup	0.997	0.664	21.09	5.11	8.55e-09	1.67e-09	15.3
45P/Honda-Mrkos-Pajdusakova	0.530	0.825	4.25	5.27	9.32e-09	1.77e-09	24.7
72P/Denning-Fukiwara	0.780	0.820	8.64	9.02	5.58e-09	6.18e-10	18.8
73P/Schwassmann-Wachmann 3	0.939	0.694	11.39	5.38	7.83e-09	1.46e-09	13.1
96P/Machholz 1	0.124	0.959	60.18	5.26	8.73e-10	1.66e-10	43.7
141P/Machholz 2	0.753	0.750	12.80	5.23	3.81e-09	7.29e-10	19.2
169P/NEAT	0.605	0.768	11.32	4.21	3.76e-09	8.92e-10	22.5
D/1819 W1 Blanpain	0.892	0.670	9.11	4.44	7.51e-09	1.69e-09	13.7
Single apparitions							
P/2001 J1 NEAT	0.937	0.758	10.16	7.62	8.28e-09	1.09e-09	13.6
P/2001 Q2 Petriew	0.946	0.696	13.94	5.49	6.89e-09	1.26e-09	13.8
P/2001 WF2 LONEOS	0.976	0.667	16.92	5.02	9.68e-09	1.93e-09	13.8
P/2003 K2 Christensen	0.549	0.829	10.14	5.75	3.99e-09	6.93e-10	24.9
P/2004 CB LINEAR	0.912	0.689	19.15	5.02	4.17e-09	8.30e-10	16.5
P/2004 R1 McNaught	0.988	0.682	4.89	5.48	3.35e-08	6.12e-09	10.0
P/2004 X1 LINEAR	0.782	0.727	5.14	4.85	9.63e-09	1.99e-09	17.0
P/2005 JQ5 Catalina	0.826	0.694	5.70	4.43	9.59e-09	2.16e-09	15.3
Destroyed, lost or no longer Earth-crossing							
3D/Biela	0.879	0.751	13.22	6.63	4.83e-09	7.27e-10	16.1
5D/Brorsen	0.590	0.810	29.38	5.47	1.63e-09	2.98e-10	27.6
D/1766 G1 Helfenzrieder	0.406	0.848	7.87	4.37	4.93e-09	1.13e-09	28.1
D/1770 L1 Lexell	0.674	0.786	1.55	5.59	2.76e-08	4.94e-09	20.5

3.3. Halley-type comets

There are 16 known Earth-crossing Halley-type comets, including 9 seen on more than one apparition, 6 seen on only one apparition (only 1 of which was discovered by the automated NEO surveys), and 1 that is lost. The prototype for this class, comet 1P/Halley, has been observed on every apparition since 240 B.C. and was last seen in 1986. It is a particularly large and active comet, with nucleus dimensions of $\sim 16 \times 8$ km. It was the first cometary nucleus encountered by flyby spacecraft.

The Halley-type comets are listed in Table 2. The mean impact probability is 4.8×10^{-9} per perihelion passage or 1.2×10^{-10} per year. The much lower impact probability per year results from the considerably longer orbital periods of the HTCs. The mean encounter velocity is 46.0 km sec^{-1} and the most probable encounter velocity is 51.3 km sec^{-1}. The higher most probable encounter velocity results because of the large number of retrograde HTCs with relatively high inclinations, which correlate with higher impact probabilities for those retrograde comets.

4. Cometary fluxes and total impact rates

There is considerable uncertainty as to the total flux of Earth-crossing comets. Everhart (1967) estimated that $\sim$11 LPCs brighter than $H_{10} = 11$ passed within 1 AU of the Sun per year, yet only 2 or 3 of these were actually observed. Everhart showed that

Table 2. Impact probabilities and encounter velocities for Earth-crossing HTCs

Name	q AU	e	i deg	Period yrs	Impact P per orbit	Impact P per year	U km/s
Multiple apparitions							
1P/Halley	0.586	0.967	162.26	74.83	4.90e-09	6.54e-11	66.4
8P/Tuttle	0.997	0.824	54.71	13.48	5.36e-09	3.98e-10	33.4
12P/Pons-Brooks	0.774	0.955	74.18	71.33	1.47e-09	2.06e-11	44.7
23P/Brorsen-Metcalf	0.479	0.972	19.33	70.76	2.06e-09	2.91e-11	31.0
27P/Crommelin	0.735	0.919	29.10	27.33	1.81e-09	6.61e-11	26.6
35P/Herschel-Rigollet	0.748	0.974	64.21	154.31	1.36e-09	8.81e-12	41.0
55P/Tempel-Tuttle	0.976	0.906	162.49	33.46	2.53e-08	7.55e-10	69.8
109P/Swift-Tuttle	0.960	0.963	113.45	132.16	4.92e-09	3.72e-11	59.9
122P/de Vico	0.659	0.963	85.38	75.17	1.25e-09	1.67e-11	49.3
Single apparition							
D/1917 F1 Mellish	0.190	0.993	32.68	141.41	1.26e-09	8.91e-12	41.0
D/1937 D1 Wilk	0.619	0.981	26.02	185.95	1.72e-09	9.25e-12	29.2
D/1989 A3 Bradfield	0.420	0.978	83.07	83.41	9.54e-10	1.14e-11	49.0
P/1991 L3 Levy	0.987	0.929	19.16	51.83	1.25e-08	2.41e-10	16.9
C/2001 OG108	0.994	0.925	80.25	48.25	4.37e-09	9.05e-11	46.4
P/2005 T4 SWAN	0.649	0.930	160.04	28.23	4.87e-09	1.72e-10	66.4
Lost							
D/1827 M1 Pons-Gambart	0.807	0.946	136.46	57.77	3.13e-09	5.42e-11	64.8

observational selection effects led to many comets being missed, for example when they arrived at perihelion on the opposite side of the Sun from the Earth.

H_{10} is the magnitude of the active comet, with coma, as viewed from a distance of 1 AU from the Earth at zero phase, when the comet is 1 AU from the Sun (somewhat equivalent to the absolute magnitude for an asteroid). The estimate of H_{10} assumes that the cometary brightness varies as $r^{-4}d^{-2}$ where r and d are the heliocentric and geocentric distances, respectively. Although this provides an acceptable first-order estimate, brightening rates versus heliocentric distance vary considerably from one comet to the next. The rates appear to correlate with age, though age itself is often difficult to determine. Dynamically new LPCs exhibit brightening exponents as low as r^{-2} on the inbound legs of their initial orbits, whereas JFCs often display very steep lightcurves, brightening as r^{-6} or even higher. Unfortunately, there is no good conversion from H_{10} to nucleus dimensions, as the active fraction of the surfaces of cometary nuclei can also vary considerably from one comet to the next.

Levison (personal communication) estimates that there are 9×10^6 comets in the Oort cloud with perihelia $\leqslant 1$ AU. Given a typical orbital period of 4 Myr for an LPC entering the planetary system for the first time from the Oort cloud, this translates into 2.2 dynamically new LPCs per year. Since the average LPC makes 5 passages through the planetary system (Weissman 1979), the total flux of LPCs is ~ 11 ECC per year, in surprisingly good agreement with the Everhart estimate.

Discovery statistics for LPCs show a steady increase in the rate of discoveries over time. Since 1900 there have been 414 discoveries of LPCs with perihelia < 3 AU, not including Sun-grazing comets with q < 0.08 AU. The first 207 of these were found from 1900–1978, an interval of 79 years, while the next 207 were found from 1979–2006, an interval of only 27.6 years. Further dividing the second interval, one finds that 105 LPCs were found from 1979–1997 (19 years) versus 102 LPCs found from 1998–2006 (8.6 years).

The increase in discovery rate is clearly associated with the automated telescopic surveys searching for NEAs. From 1900–1978 the discovery rate of long-period ECCs was 2.6 per year. From 1979-1997 it was 2.3 long-period ECCs per year, statistically the same. But from 1998–2006 the rate increased to 4.3 long-period ECCs per year. Although this rate is still short of the Everhart and Levison estimates, it is conceivable that the more sensitive surveys planned for future years may eventually yield rates comparable to those values. Discoveries of JFCs and HTCs have also been aided by the automated surveys.

Given a rate of 11 long-period ECCs per year, the total impact rate is 2.6×10^{-8} per year, or one impact every 38 Myr. Comet showers caused by random stars passing directly through the Oort cloud will double the mean impact rate over time, though the showers come in bursts lasting only 2-3 Myr. We are currently not in a cometary shower.

Unfortunately, it is not currently possible to estimate the sizes of the nuclei associated with this flux. There exist very few reliable estimates of the radii of LPCs. Even the estimate for the nucleus of comet Hale-Bopp in Section 1 is highly uncertain.

Estimates of the total flux of Earth-crossing JFCs are also somewhat indefinite. New Earth-crossing JFCs continue to be discovered by the automated NEO surveys and by amateur observers. As a rough guess, we might assume that the current discovery of Earth-crossing active JFCs is 50% complete. Then the average impact rate is 6.7×10^{-8} per year, or one impact every 15 Myr.

A similar calculation can be made for the HTCs. Discovery completeness for this population is lower because of their longer orbital periods. If we assume that completeness is 33%, then the impact rate is 5.8×10^{-9} per year or one impact every 176 Myr. If we assume that completeness is only 10%, then the average impact rate is 1.9×10^{-8} per year or one impact every 52 Myr.

Again, the problem is matching this impactor flux with the sizes of the impactors and predicting the energy of the impacts. Progress has been made in estimating the sizes of JFCs and HTCs through a variety of observational techniques. Because these comets are periodic, their returns can be predicted very accurately, and they can be observed at distances of 4–5 AU from the Sun on the inbound legs of their orbits, where they are most likely to be inactive. Brightness estimates can then be converted to nucleus radius estimates by assuming a typical cometary albedo of 0.04. Another successful method for estimating the sizes of cometary nuclei is to use the high angular resolution of the Hubble Space Telescope to image active comets when they are close to the Earth, and then model and subtract the coma signal from the total brightness. Again, a typical cometary albedo of 0.04 needs to be assumed. Lastly, four cometary nuclei (Halley, Borrelly, Wild 2, and Tempel 1) have been directly imaged by spacecraft flybys. Over 65 JFCs and HTCs have had their radii estimated in this way (Weissman & Lowry 2003, Meech *et al.* 2004, Lamy *et al.* 2004).

This include six of the JFCs listed in Table 1, whose radius estimates range from 0.17 to 4.4 km, the smallest being comet Blanpain and the largest comet Encke. In many cases, radius estimates by different observers agree well, but in other cases, such as Encke, there is considerable disagreement on the size of the nucleus. The radius estimates for JFCs (and HTCs) are shown in Table 3.

Radius estimates also exist for four of the HTCs in Table 2. These range from 1.78 to 15.0 km; the smallest corresponds to comet Tempel-Tuttle and the largest to comet Swift-Tuttle. Size estimates for HTCs are typically much larger than for JFCs. However, these four Earth-crossing HTCs are the only HTCs for which reliable radius estimates exist, versus ~60 measured JFCs. There are strong observational biases here, as there are far fewer opportunities to measure the radii of HTCs because of their longer orbital

Table 3. Nucleus radius estimates for periodic ECCs

Name	Impact P per year	U km/s	Radius estimates km
Jupiter-family comets			
2P/Encke	1.01e-09	29.5	3.4, 2.4, 4.0, 4.4
26P/Grigg-Skjellerup	1.67e-09	15.3	1.4, 1.44, 2.90
45P/H-M-P	1.77e-09	24.7	0.34, 1.34
73P/S-W 3	1.46e-09	13.1	0.6, 0.4
96P/Machholz 1	1.66e-10	43.7	3.3
D/1819 W1 Blanpain	1.69e-09	13.7	0.17
Halley-type comets			
1P/Halley	6.54e-11	66.4	5.73, 5.25
8P/Tuttle	3.98e-10	33.4	7.3
55P/Tempel-Tuttle	7.55e-10	69.8	1.78, 1.80, 1.85
109P/Swift-Tuttle	3.72e-11	59.9	11.8, 13.73, 15.0

periods. Also, the various techniques for estimating radii are most successful for nuclei that are relatively large and/or relatively inactive.

If one combines the impact rate estimates for LPCs, JFCs, and HTCs above, then the total impact rate is 1.1×10^{-7} per year, or one impact every 9.1 Myr. If comet showers are included, these numbers increase to 1.4×10^{-7} per year, or one impact every 7.2 Myr. These are only a few per cent of the rates for NEAs. However, as noted earlier, the higher encounter velocities and larger nuclei result in comets contributing significantly to the more energetic, though rare, impacts on the Earth.

5. Dormant comets

The most common loss mechanism for cometary nuclei is dynamical ejection from the solar system. Random disruption, or splitting, is likely the most common physical loss mechanism, whereby the weakly bonded nucleus fragments separate and rapidly sublimate away. A less common but possible mechanism is that the cometary nuclei evolve to a dormant state, in which a lag deposit of non-volatile material remains on the nucleus surface and slowly cuts off cometary activity. In the case of comet Halley, ∼70–80% of the sunlit surface appeared to be inactive. For most JFCs, the fraction of active surface area is typically only a few per cent.

In recent years there has been increasing evidence for dormant cometary nuclei among the NEA population. Bottke *et al.* (2002) showed that the orbital distribution of NEAs could best be explained if ∼6% of the NEA population had evolved from Jupiter-family comet orbits. Weissman *et al.* (2002) examined the list of NEAs identified by Bottke *et al.* as likely dormant comets, and showed that a significant fraction were either low albedo or of a primitive taxonomic type, both consistent with a cometary origin. Fernández *et al.* (2005) showed that two-thirds of NEAs with Tisserand parameters < 3, suggestive of a cometary origin, had low albedos, less than 0.075. Some estimates have placed the fraction of dormant comets among the NEAs as high as 15%.

An unusual object, 1996 PW, was discovered in 1996. This asteroidal-appearing object was in a highly eccentric orbit, similar to that of a long-period comet, with an orbital period of ∼4,000 years. Weissman & Levison (1997) showed that this object was equally likely to be a dormant long-period comet or an asteroid that had been ejected to the Oort cloud and was now returning. Color measurements suggested that 1996 PW was a D-type asteroid, consistent with an outer main belt asteroid or a primitive nucleus surface. No evidence of cometary activity was detected.

The actual fraction of comets that evolve to a dormant state is highly uncertain. The process is often described as "cometary fading," suggesting that active nuclei slowly become less luminous and eventually incapable of producing any visible coma. Cometary fading is often invoked by dynamical modelers as a convenient mechanism to account for the difference between predicted and observed numbers of comets in a particular dynamical class (e.g., Bailey 1984; Emel'yanenko & Bailey 1996). However, the differences may also be the result of incorrect initial assumptions and/or poor modeling of other physical loss mechanisms or of the dynamics itself, as well as observational selection effects and incompleteness in the observed sample of comets and asteroids. Indeed, Levison *et al.* (2002) argued that cometary fading had to be a relatively minor end-state because of the relatively low numbers of dormant objects discovered in orbits that could have evolved from long-period comet orbits. Further studies of the fraction of dormant comets among the NEAs will be quite valuable in providing an improved understanding of "cometary fading" and its role in other cometary evolution problems.

6. Summary and discussion

Comets make up a small but important fraction of near-Earth objects. Although Earth-crossing comets are far less numerous than NEAs, their high orbital inclinations and eccentricities result in very high hyperbolic encounter velocities with the Earth, up to 72 km sec^{-1}. Additionally, some Earth-crossing comets are considerably larger than any known NEA. Thus, comets likely provide the largest and most energetic impactors on the Earth.

The total cometary impact rate on the Earth, based on observed comets and crudely correcting for observational incompleteness, is 1.4×10^{-7} per year, or one impact every 7.2 Myr. About 48% of this total is due to Jupiter-family comets, 14% due to Halley-type comets, and 38% due to long-period comets, though half of the LPCs arrive in brief cometary showers of 2-3 Myr duration. We are currently not in a cometary shower. Most probable encounter velocities range between 17.0 km sec^{-1} for JFCs, 51.3 km sec^{-1} for HTCs, and 53.6 km sec^{-1} for LPCs.

It is difficult to compare these cometary impact rates with those for NEAs because the sizes of the cometary nuclei are not well known. However, progress is being made, particularly in the case of the JFCs. This is fortunate since JFCs appear to provide about half of the cometary impactors on the Earth. Unfortunately, the situation is considerably poorer in the case of the HTCs and LPCs. In addition, the long orbital periods of the HTCs mean that it will take at least several hundred years to survey this population, though technology improvements over that time may be able to speed the process. Lastly, the problem of the short warning time for LPCs on potential Earth-impacting orbits needs to be dealt with. LPCs are typically discovered at distances < 5 AU from the Sun, ⩽ 1 year prior to perihelion passage. There are exceptions, such as comet Hale-Bopp, discovered at 7.14 AU, but still < 2 years from perihelion. The next generation surveys are likely to improve the warning time but it is also likely that new technologies are required to make truly significant improvements.

The automated telescopic surveys for NEOs have been very successful at increasing our knowledge of the populations of comets in Earth-crossing orbits. This can be expected to improve further with the next generation of surveys, including PAN-STARRS and LSST.

Acknowledgements

I thank Hans Rickman for useful comments on an earlier draft of this paper. This work was supported by the NASA Planetary Astronomy and Planetary Geology & Geophysics Programs, and was performed at the Jet Propulsion Laboratory under a contract with NASA.

References

Bailey, M.E. 1984, *MNRAS* 211, 347
Bottke, W.F., Morbidelli, A., Jedicke, R., *et al.* 2002, *Icarus* 156, 399
Chyba, C.F., Thomas, P.J. & Zahnle, K.J. 1993, *Nature* 361, 40
Dones, L., Weissman, P.R., Levison, H.F. & Duncan, M.J. 2004, in: M.C. Festou, H.U. Keller & H.A. Weaver (eds.), *Comets II* (University of Arizona Press, Tucson), p. 153
Duncan, M.J., Quinn, T. & Tremaine, S. 1988, *Astrophys. J.* 328, 69
Duncan, M.J. & Levison, H.F. 1997, *Science* 276, 1670
Emel'Yanenko, V.V. & Bailey, M.E. 1996, *Earth, Moon & Planets* 72, 35
Everhart, E. 1967, *Astron. J.*, 72, 1002
Fernández, J.A. 1980, *MNRAS* 192, 481
Fernández, Y.R., Jewitt, D.C. & Sheppard, S.S. 2005, *Astron. J.* 130, 308
Kessler, D.J. 1981, *Icarus* 48, 39
Lamy, P.L., Toth, I., Fernández, Y.R. & Weaver, H.A. 2004, in: M.C. Festou, H.U. Keller & H.A. Weaver (eds.), *Comets II* (University of Arizona Press, Tucson), p. 223
Levison, H.F. 1996, ASP Conf. Ser. *Completing the Inventory of the Solar System*, 107, p. 173
Levison, H.F., Dones, L. & Duncan, M.J. 2001, *Astron. J.* 121, 2253
Levison, H.F., Morbidelli, A., Dones, L., *et al.* 2002, *Science* 296, 2212
Lowry, S.C. & Weissman, P.R. 2003, *Icarus* 164, 492
Marsden, B.G., & Steel, D.I. 1994, in: T. Gehrels, M.S. Matthews & A.M. Schumann (eds.), *Hazards Due to Comets and Asteroids* (University of Arizona Press, Tucson), p. 221
Meech, K.J., Hainaut, O.R. & Marsden, B.G. 2004, *Icarus* 170, 463
Oort, J.H. 1950, *Bull. Astron. Inst. Neth.* 11, 91
Öpik, E.J. 1951, *Proc. R. Irish Acad.* 54, 165
Shoemaker, E.M., Weissman, P.R. & Shoemaker, C.S. 1994, in: T. Gehrels, M.S. Matthews & A.M. Schumann (eds.), *Hazards Due to Comets and Asteroids* (University of Arizona Press, Tucson), p. 313
Weaver, H.A., Feldman, P.D., A'Hearn, M.F. & Arpigny, C. 1997, *Science* 275, 1900
Weissman, P.R. 1979, In *Dynamics of the Solar System*, p. 277
Weissman, P.R. 1982, In *Geol. Soc. Amer. Special Paper 190: Geological Implications of Impacts of Large Asteroids and Comets on the Earth*, p. 15
Weissman, P.R. 1989, In *Geol. Soc. Amer. Special Paper 247: Global Catastrophes in Earth History*, p. 211
Weissman, P.R. 1986, *Nature* 320, 242
Weissman, P.R. 1996, *ASP Conf. Ser.: Completing the Inventory of the Solar System* 107, p .265
Weissman, P.R. 1997, In *Near-Earth Objects*, Annals NY Acad. Sci. 822, 67
Weissman, P.R. & Levison, H.F. 1997, *Astrophys. J. Lett.* 488 133
Weissman, P.R., Bottke, W.F. & Levison, H.F., 2002, in: W.F. Bottke, A. Cellino, P. Paolicchi & R.P Binzel (eds.), *Asteroids III* (University of Arizona Press, Tucson), p. 669
Weissman, P.R. & Lowry, S.C. 2003, *LPSC* 34, #2003
Weissman, P.R., Asphaug, E. & Lowry, S.C. 2004, in: M.C. Festou, H.U. Keller & H.A. Weaver (eds.), *Comets II* (University of Arizona Press, Tucson), p. 337

Near Earth Objects, our Celestial Neighbors: Opportunity and Risk
Proceedings IAU Symposium No. 236, 2006
A. Milani, G.B. Valsecchi & D. Vokrouhlický, eds.
© 2007 International Astronomical Union
doi:10.1017/S1743921307003560

Albedo and size of (99942) Apophis from polarimetric observations†

Alberto Cellino[1], Marco Delbò[1,2] and Edward F. Tedesco[3]

[1]INAF – Osservatorio Astronomico di Torino,
strada Osservatorio 20, 10025 Pino Torinese, Italy
email: cellino@inaf.oato.it

[2]Observatoire de la Côte d'Azur, BP 229, Nice, France
email: delbo@obs-nice.fr

[3]University of New Hampshire, USA
email: Ed.Tedesco@unh.edu

Abstract. We have obtained the first accurate determination of the albedo of (99942) Apophis, by means of polarimetric observations carried out at the VLT. The observations allowed us to obtain the slope of the polarization – phase curve of this object, from which an albedo estimate of 0.33 ± 0.04 could be obtained. From our observations we also obtained a new estimate of the absolute magnitude: $H = 19.7 \pm 0.2$ (assuming G=0.25, which applies to S- and Q-type asteroids). Based on these results, we derive for the size of Apophis a value of 270 ± 30 meters. The accuracy of this size estimate is mostly related to uncertainties in H, whereas the obtained albedo value should be considered more robust. Our observations convincingly show that polarimetry is an effective and efficient tool to obtain accurate albedos and sizes for small and faint potentially hazardous asteroids.

Keywords. Asteroids, polarization

1. Introduction

The near-Earth object (99942) Apophis will make an extremely close approach to the Earth in 2029, possibly followed by resonant returns starting in 2036. The computation of the orbital evolution of near-Earth asteroids (NEAs), including potentially hazardous asteroids (PHAs) like Apophis, is currently limited by insufficient knowledge of some physical properties of the objects, required to compute the orbital drift produced by the Yarkovsky effect (Bottke *et al.* 2002). Moreover, knowledge of the size of the objects is needed also to estimate the amount of kinetic energy delivered to the Earth in the case of a collision with our planet. For these reasons, the exploitation of observing techniques which can provide reliable and accurate estimates of the basic physical properties of NEAs, including their sizes, is a high-priority task.

We present here the first determination of the albedo of (99942) Apophis, obtained by means of polarimetric observations carried out at the Very Large Telescope of the European Southern Observatory at Cerro Paranal (Chile). The observations allowed us to obtain the slope of the polarization – phase curve of Apophis, from which we derived the albedo of the object. Our observations also allowed us to obtain a new estimate of the absolute magnitude of Apophis. Based on these results, we use the well known relation relating albedo, absolute magnitude and size, to derive the equivalent diameter of this asteroid.

† Based on observations obtained at the European Southern Observatory (ESO), DDT request 276.C-5030

We note that a detailed description of our VLT observations and a thorough discussion of the obtained results is given in a separate paper (Delbò, Cellino & Tedesco 2006). Here, we only briefly summarize the results of our investigation, and briefly mention the potential role that polarimetry can play in the future for the purposes of physical characterization of NEAs and PHAs.

2. Asteroid Polarimetry

It is known that polarimetry is among the best available tools that may be adopted to derive asteroid albedos. Here, we briefly summarize the fundamental background of this technique. The light that we receive from asteroids at visible wavelengths consists of sunlight scattered by the solid surface of the bodies, and is therefore in a state of partial linear polarization. The degree of linear polarization varies for changing illumination conditions in a characteristic way, and can be described by plotting it as a function of the phase angle, namely the angle between the directions of the Earth and the Sun as seen from the asteroid. More precisely, the state of linear polarization is a vector, characterized by a module, (given by the square root of the sum of the squares of the Stokes parameters P_x and P_y which characterize the state of linear polarization of the received light) and a direction, represented by a Position Angle (given by $\arctan(P_y/P_x)$) which gives the orientation of the plane of polarization with respect to a reference direction. In the case of polarimetric measurements of atmosphereless solar system bodies like the asteroids, it is found that the plane of linear polarization is in general either parallel or perpendicular to the plane of scattering, defined as the plane containing the Sun, the target and the observer at the epoch of observations. For this reason, the state of linear polarization of the asteroids is commonly described by means of the parameter $P_r = \frac{(I_\perp - I_\parallel)}{(I_\perp + I_\parallel)}$. In the above definition $I_\perp$ and $I_\parallel$ indicate the intensity of the asteroid light having the plane of polarization perpendicular and parallel to the scattering plane, respectively (Dollfus & Zellner 1979, Dollfus *et al.* 1989).

When one plots the variation of P_r as a function of the phase angle, a well defined curve is usually obtained. A typical example is shown in figure 1. Phase-polarization curves are always characterized by the presence of a range of phase angles, usually between $0°$ and $20°$, for which P_r is negative and reaches a minimum at a phase angle around $10°$. This general trend characterizes, with some minor differences depending on the taxonomic class, all asteroids observed so far. Beyond a phase angle of about $20°$ (inversion angle), the polarization usually changes sign, and becomes positive. Around and beyond the inversion angle, the trend of Pr for increasing phase angle is essentially linear, and a well defined relation is known to exist between the slope of the linear part of the polarization curve and the albedo of the surfaces. This is the so-called slope-albedo law, and takes the form $\log p_V = C_1 \log h + C_2$, where p_V is the albedo, and h the polarization slope. The most recent derivation of the C_1 and C_2 coefficients has been published by Cellino *et al.* (1999).

3. Observations and results

In our 2006 observing campaign we observed (99942) Apophis in V light on four nights, on February 16, March 7, March 28 and April 2, respectively. The phase angles at the four epochs of observation were about $64°$, $68°$, $79°$ and $83°$, respectively. The P_r parameter was measured on each night, and from this we could derive a value of the slope equal to 0.069 ± 0.004. The results of our polarimetric measurements are graphically shown in the quoted Delbò, Cellino & Tedesco (2006) paper.

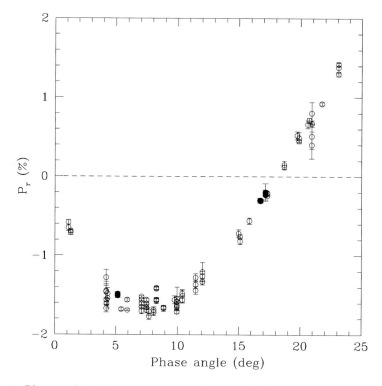

Figure 1. Phase–polarization curve of the asteroid (1) Ceres. Different symbols indicate measurements taken from different data-sets available in the literature.

Using the Cellino *et al.* (1999) calibration, and taking into account the nominal uncertainties both in the calibration coefficients and our measured polarimetric slope (the latter being a function of the error bars of our single polarimetric measurements), we derived an albedo value of 0.33 ± 0.04.

For each of our observations, we derived also some accurate estimates of the apparent magnitude of our target. From these photometric measurements, we could obtain four points of the phase – magnitude curve of (99942) Apophis during the 2006 apparition. These measurements can be used to extrapolate the magnitude of the object at zero phase angle, to derive its absolute magnitude, namely the magnitude at zero phase angle, as would be measured by an observer looking at the asteroid when it is at a distance of 1 AU from both the Sun and the observer. It is usual to indicate by means of the H symbol the absolute magnitude in V light, and to use another parameter, called G, to describe the non-linear variation of the magnitude as a function of phase angle, including the so-called "opposition effect", namely a non linear brightness surge at phase angles of a few degrees (Bowell *et al.* 1989). The (H, G) system has been officially adopted by the IAU Commission 20 in 1985. As mentioned above, our observations cover only four nights and were obtained at quite large phase angles, then they can hardly be used to derive a simultaneous solution for both the H and G parameters. For this reason, we computed best-fit values for the H parameter only, making some *a priori* assumptions on the value of G. By choosing for G the value of 0.25, which is the value commonly found for S- and Q-type asteroids, we derive for H a value of 19.7 ± 0.1. Assuming for G the typical default value of 0.15, gives correspondingly $H = 19.5 \pm 0.1$. Given the uncertainties in the choice of G, we assume the following value for the absolute magnitude of (99942)

Apophis: $H = 19.7 \pm 0.2$. Note that this is our choice based on the data we have at the moment of submitting this paper within the deadline for publication. We cannot rule out that in the final version of the Delbò, Cellino & Tedesco (2006) paper, which will present the definitive results of this analysis, we might adopt a slightly different value.

From a knowledge of the albedo and the absolute magnitude, we can finally derive an equivalent size of 270 ± 30 meters for Apophis. This value, is slightly but not negligibly smaller than the value previously assumed and reported in the NASA web page http://neo.jpl.nasa.gov/risk, namely 320 meters.

We note that our investigation has led to an albedo and size measurement for the smallest asteroid ever observed using the polarimetric technique. In particular, the most robust result of our investigation is certainly the albedo of (99942) Apophis, since this is the parameter that can be derived most directly from the polarimetric measurements. The size is more uncertain, due to the quoted uncertainties in our derived value of the absolute magnitude, which suffers from a poor knowledge of the G parameter. A more accurate derivation of the size will be possible in the future, when new observations of this asteroid will produce a more accurate determination of its absolute magnitude.

The present results show that the availability of large instruments equipped with modern CCD polarimeters can be a major breakthrough in the field of the physical characterization of NEAs and PHAs.

Acknowledgements

We thank the the staff and the Science Archive Operation of the ESO for their support to our observations. The work of Marco Delbò was partially supported by the European Space Agency (ESA) and that of E.F. Tedesco by the National Aeronautics and Space Administration (NASA) under grant NNG04GK46G, issued through the Office od Space Science Research and Analysis Programme.

References

Bottke, W.F., Vokrouhlický, D., Rubincam, D.P., & Brož, M. 2002 in: W.F. Bottke, A. Cellino, P. Paolicchi, R.P. Binzel, (eds.), *Asteroids III* (University of Arizona Press), p. 395
Bowell, E., Hapke, B., Domingue, D., Lumme, K., Peltoniemi, J., & Harris, A.W. 1989 in: R.P. Binzel, T. Gehrels, M.S. Matthews, (eds.), *Asteroids II* (University of Arizona Press), p. 524
Cellino, A., Gil-Hutton, R., Tedesco E.F., Di Martino, M., & Brunini, A. 1999, *Icarus* 138, 129
Delbò, M., Cellino, A. & Tedesco E.F. 2006, *Icarus*, submitted
Dollfus, A. & Zellner, B. 1979 in: T. Gehrels, (ed.), *Asteroids* (University of Arizona Press), p. 170
Dollfus, A., Wolff, M., Geake, J.E., Lupishko, D.F., & Dougherty, L. 1989 in: R.P. Binzel, T. Gehrels, M.S. Matthews, (eds.), *Asteroids II* (University of Arizona Press), p. 594

Near Earth Objects, our Celestial Neighbors: Opportunity and Risk
Proceedings IAU Symposium No. 236, 2006
A. Milani, G.B. Valsecchi & D. Vokrouhlický, eds.
© 2007 International Astronomical Union
doi:10.1017/S1743921307003572

Admissible regions for too short arcs: nodal distances and elongations

Stéphane Valk and Anne Lemaitre

Department of Mathematics, University of Namur, Belgium
email: stephane.valk@fundp.ac.be, anne.lemaitre@fundp.ac.be

Abstract. This study is based on the definition of the admissible region introduced by Milani *et al.* (2004); in the search for potential Earth impactors, this theory allows to take into account the partial data of the TSA (*Too Short Arcs*) from which it is impossible to deduce a full orbit. Only a set of 4 variables (two angles and their instantaneous time derivatives), called an *attributable*, is known; a few suitable boundary conditions allow to restrict the motions to a specific bounded 2-dimensional region. In this work, a new inner boundary of this region is introduced, based on the geocentric hyperbolic motion of the immediate impactors; the nodal distances (crossings of the virtual asteroidal orbits with the Earth's orbit) are drawn for two different test attributables, associated with a determination of circular and linear orbits. This could reduce the search for impactors (by propagation of the orbits) to a one-dimensional set. A few comments about elongations and complementary curves complete this paper.

Keywords. orbit determination; space debris; impact risk; MOID

1. Introduction

When a non identified object is observed, the first reaction of the scientific community is to try to determine its orbit. Unfortunately, for the data collected on very short periods of time, the arc of observation is not large enough to give any estimation of the curvature; the determination of the orbit is impossible, using traditional methods of orbital determination, such as Gauss method. If we intend to build a complete catalog of such objects, the conclusion is easy: this object is rejected, and the observers hope to be luckier a few months or years later, to re-observe the same body, on a larger timescale. However, in many cases, the right ascension, the declination and their instantaneous time derivatives are measured.

For the last few years, associations like Space Guard or the specialists of the Near-Earth Asteroids (Minor Planet Center† or NEODYS group‡) have a completely different point of view concerning these unexpected observed objects. The main question is not only the improvement of their orbit, but also the potential hazard that they represent for the Earth: could this unknown body becomes dangerous for us, in a delay of one or two hundreds years?

Virtually, this too short arc (TSA) corresponds to an infinity of orbits. We assume, because it is true in many cases, that its right ascension α and declination δ are know, as well as their time derivatives; on the opposite, there is no data concerning either their distance to the Earth or the time derivative of this distance. Consequently, on a set of six variables $(\alpha, \delta, \dot{\alpha}, \dot{\delta}, r, \dot{r})$, the first four are determined with a specific accuracy, while

---

† http://cfa-www.harvard.edu/cfa/ps/mpc.html
‡ http://newton.dm.unipi.it/cgi-bin/neodys/neoibo

the last two are completely arbitrary. This means that the object lies in a 2 dimensional subspace of a general 6 dimensional space.

This idea was introduced by Milani *et al.* (2004) and pushed further on in Milani *et al.* (2005a) and Milani *et al.* (2005b). This incomplete set of data (2 angles and 2 time derivatives) is called an *"attributable"* by the authors mentioned above and this denomination is conserved here. Thanks to reasonable hypotheses (the fact that the object belongs to the Solar System, or that it is not a satellite of the Earth), Milani *et al.* (2004) proved that this region, in the plane $(r, \dot{r})$ could be closed and formed of one or two connected sets. Curves of constant values of the osculating keplerian elements can be drawn on this region.

Unfortunately, if a second observation is not available, the admissible region is still very large; one of the challenges is to follow the propagation of this admissible region, by means of linear and non linear techniques, in order to compare its evolution with a potential new arc.

Our purpose here is to concentrate on some aspects of the initial admissible region. Firstly, we recalculate one of its boundaries, for the short distances to the Earth, introducing the hyperbolic shape of the orbit instead of its linear approximation ; secondly, we introduce, on the admissible zone, and in complement of the keplerian elements information, the nodal distances, corresponding to the intersections of the Earth and potential Earth's impactors orbits. We present different situations, where the singularity in inclination is inside or outside the admissible region, following the chosen attributable. These curves could be very interesting in the context of propagation of the motions, reducing the dimension of the admissible region (dimension one instead of two). Thirdly we introduce the concept of elongations on the graphics.

2. The admissible region

Let $\vec{P}_A$ and $\vec{V}_A$ be the heliocentric position and velocity vectors of a celestial body $\mathcal{A}$ at a reference time t. At the same time, the position vector $\vec{P}_\oplus$ and the velocity vector $\vec{V}_\oplus$ of the Earth are well known.

The heliocentric energy per unit mass of $\mathcal{A}$ is given by

$$E_\odot = \frac{1}{2}\|\vec{V}_A\|^2 - k_\odot^2 \frac{1}{\|\vec{P}_A\|} \tag{2.1}$$

and its geocentric energy takes the form

$$E_\oplus = \frac{1}{2}\|\vec{V}_A - \vec{V}_\oplus\|^2 - k_\oplus^2 \frac{1}{\|\vec{P}_A - \vec{P}_\oplus\|} \tag{2.2}$$

where $m_\oplus$ and $m_\odot$ are the masses of the Earth and of the Sun respectively. Gauss' constant is defined by: $k_\odot = \sqrt{Gm_\odot} = 0.01720209895$ and $k_\oplus^2 = k_\odot^2 \frac{m_\oplus}{m_\odot}$. The solar mass is taken as the mass unit, the mean semi major axis of the Earth orbit is the distance unit (AU) and the average day is the time unit.

An attributable is defined as a fourth dimension vector

$$\vec{A} = (\alpha, \delta, \dot{\alpha}, \dot{\delta}) \ \in \ [-\pi, \pi[\ \times \ ] -\frac{\pi}{2}, \frac{\pi}{2}[\ \times \ \mathbb{R}^2 \tag{2.3}$$

computed at the time t and to which an apparent magnitude M can be associated. We use the classical geocentric equatorial coordinates (α, δ) with α, the right ascension, and δ, the declination. The position vector $\vec{P}_A$ can be expressed as

$$\vec{P}_A = \vec{P}_\oplus + r\,\vec{u} \tag{2.4}$$

where r is the geocentric distance of the body $\mathcal{A}$ and $\vec{u}$ is the unit vector in the direction of the observation

$$\vec{u} = (\cos\alpha\cos\delta, \sin\alpha\cos\delta, \sin\delta). \tag{2.5}$$

The first time derivative of equation (2.4) gives the velocity vector

$$\vec{V}_{\mathcal{A}} = \vec{V}_{\oplus} + \dot{r}\,\vec{u} + r\,\dot{\alpha}\,\vec{u}_{\alpha} + r\,\dot{\delta}\,\vec{u}_{\delta}, \tag{2.6}$$

where

$$\vec{u}_{\alpha} = (-\sin\alpha\cos\delta, \cos\alpha\cos\delta, 0) \tag{2.7}$$
$$\vec{u}_{\delta} = (-\cos\alpha\sin\delta, -\sin\alpha\sin\delta, \cos\delta) \tag{2.8}$$

The geocentric position and velocity vectors can be computed as functions of r and $\dot{r}$

$$\|\vec{P}_{\mathcal{A}} - \vec{P}_{\oplus}\|^2 = r^2 \tag{2.9}$$
$$\|\vec{V}_{\mathcal{A}} - \vec{V}_{\oplus}\|^2 = \dot{r}^2 + r^2\,\dot{\alpha}^2\,\cos^2\delta + r^2\,\dot{\delta}^2 = \dot{r}^2 + r^2\eta^2 \tag{2.10}$$

where

$$\eta = \sqrt{\dot{\alpha}^2\cos^2\delta + \dot{\delta}^2} \tag{2.11}$$

is the proper motion. The energies are given by

$$E_{\oplus} = \frac{1}{2}\left[\dot{r}^2 + r^2\,\eta^2 - \frac{2k_{\oplus}^2}{r}\right] \tag{2.12}$$

and

$$E_{\odot} = \frac{1}{2}\left[\dot{r}^2 + c_1\,\dot{r} + W(r) - \frac{2k_{\odot}^2}{\sqrt{S(r)}}\right] \tag{2.13}$$

where the quantities $W(r)$ and $S(r)$ are functions of the geocentric distance r (see Milani *et al.* (2004) for details).

To determine the admissible regions, let us recall the conditions chosen by Milani *et al.* (2004):

(*a*) $\mathcal{A}$ is not a satellite of the Earth:

$$\mathcal{D}_1 = \{(r, \dot{r}) : E_{\oplus} \geqslant 0\}$$

(*b*) $\mathcal{A}$ is not influenced by the Earth's gravity field (of radius R_{SI}):

$$\mathcal{D}_2 = \{(r, \dot{r}) : r \geqslant R_{SI}\}$$

(*c*) $\mathcal{A}$ is on an elliptic orbit around the Sun:

$$\mathcal{D}_3 = (\{r, \dot{r}) : E_{\odot} \leqslant 0\}$$

(*d*) $\mathcal{A}$ is obviously outside the Earth's globe (of radius $R_{\oplus}$):

$$\mathcal{D}_4 = \{(r, \dot{r}) : r \geqslant R_{\oplus}\}$$

The admissible region is defined as

$$\mathcal{D} = \{\mathcal{D}_1 \cup \mathcal{D}_2\} \cap \mathcal{D}_3 \cap \mathcal{D}_4 \tag{2.14}$$

A schematic example of such a region is given in Figure 1.

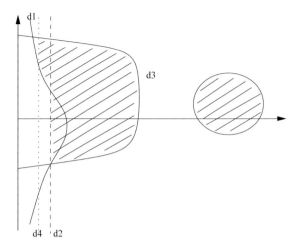

Figure 1. The topology of an admissible region with two connected components in the plane $(r, \dot{r})$; the curves d_i, for $i = 1, 2, 3, 4$ are the boundaries of the regions $\mathcal{D}_i$.

3. Immediate impact trajectory and inner boundary

Our first purpose is to exclude, from the admissible region $\mathcal{D}$, all the objects which are on a collision course with the Earth within a short time span; as mentioned by Milani *et al.* (2004), this gives an additional boundary for the left part of the admissible region, between d_4 and d_2 where the curves d_i are the boundaries of the regions $\mathcal{D}_i$. It is based on the assumption that the trajectory of the body $\mathcal{A}$ is rectilinear and can be written as

$$\frac{\eta\, r^2}{|\dot{r}|} \geqslant R_\oplus \tag{3.1}$$

where $R_\oplus = 4.26352 \times 10^{-5}$ AU. However, when the geocentric speed $\dot{r}$ is low, the hypothesis of linearity is not suitable anymore and the boundary description can be easy improved by using a two body (keplerian) formalism: the object is assumed to move within the sphere of influence of the Earth where its orbit is only controlled by the Earth's gravity field. In this context, our new condition can be expressed as

$$q_\oplus(r, \dot{r}) - R_\oplus > 0 \tag{3.2}$$

where $q_\oplus$ is the perigee corresponding to the geocentric orbit of the near-Earth object $\mathcal{A}$. This latest expression has no simple analytic form as a function of $(r, \dot{r})$, nevertheless it can be computed numerically. As it is obvious that $q_\oplus = a_\oplus\,(1 - e_\oplus)$, we compute the semi-major axis $a_\oplus$ (negative for hyperbolic orbits) and the eccentricity $e_\oplus$ by

$$a_\oplus = -\frac{k_\oplus^2}{2\, E_\oplus} \qquad\qquad e_\oplus = \sqrt{1 - \frac{C_\oplus^2}{a_\oplus\, k_\oplus^2}}$$

where

$$C_\oplus = \|\vec{C}_\oplus\| = \|(\vec{P}_\mathcal{A} - \vec{P}_\oplus) \times (\vec{V}_\mathcal{A} - \vec{V}_\oplus)\| = r^2\, \eta$$

is the norm of the angular momentum $\vec{C}_\oplus$ of the geocentric orbit. The linear and keplerian boundaries are represented in Figure 2 in a descriptive way and in the plane $(r, \dot{r})$ on Figure 3. The two conditions are very similar, even if, as expected, for low radial geocentric speed $\dot{r}$, the differences are significant. Nevertheless, as already pointed out by Milani *et al.* (2004), our new condition is only useful for the discovery of very small objects or of objects with a very small apparent magnitude. In other words, this condition

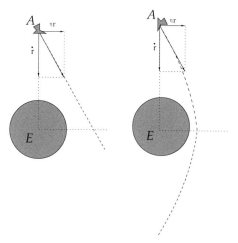

Figure 2. Immediate impact trajectory in the linear and keplerian cases (schematic representation).

allows to discriminate between the population of asteroids and that of future shooting stars.

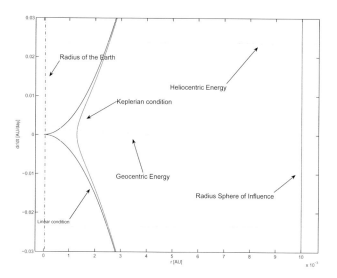

Figure 3. Immediate impact trajectory in the linear and keplerian cases; represented in the $(r, \dot{r})$ plane where r is given in astronomical unit.

A particular region appears clearly in Figure 3, inside the curve "geocentric energy" and to the right of the curve "keplerian condition": this is a region of bodies orbiting around the Earth. Of course, most of the artificial satellites have very precise orbits and do not require such a study. However, it is not always the case for Space Debris where orbital uncertainty is a common fact especially in the case of uncatalogued geostationary Space Debris. A detailed analysis of this confined area, associated to a suitable model of propagation, would give an interesting tool to follow this population and to measure the risk for future missions.

4. Nodal distances

Our second purpose is to define the subset of this admissible region associated with the objects which impact the Earth. The necessary conditions for an impact between the Earth (considered on a circular heliocentric orbit) and an hypothetic object can be easily formulated: the object and the Earth should be exactly at one of the nodes of the orbit at the same time

$$\Omega - \lambda_E = \frac{\pi}{2} \mp \frac{\pi}{2} \tag{4.1}$$

$$\omega + f = \frac{\pi}{2} \mp \frac{\pi}{2} \tag{4.2}$$

where λ_E is the longitude of the Earth, on its geocentric circular orbit, measured in the ecliptic plane, Ω, ω and f are respectively the longitude of the ascending node, the argument of the pericenter and the true anomaly of the heliocentric orbit of the body $\mathcal{A}$, in an ecliptic reference frame. The upper sign corresponds to an impact at the ascending node, and the lower sign to an impact at the descending node. For a collision at the ascending node, the so called *ascending nodal distance* must vanish

$$d_+ = \frac{a(1 - e^2)}{1 + e\cos\omega} - a_E = 0 \tag{4.3}$$

where a_E is the Earth's semi-major axis, a this of the body $\mathcal{A}$ and e its eccentricity. We have a very similar condition for a collision at the descending node, for the *descending nodal distance*

$$d_- = \frac{a(1 - e^2)}{1 - e\cos\omega} - a_E = 0 \tag{4.4}$$

For any set of values of $(r, \dot{r})$ in the admissible zone, we compute numerically the orbital elements using the usual transformations

$$a = \frac{k_\odot^2}{2\,E_\odot} \qquad e = \sqrt{1 - \frac{C^2}{a\,k_\odot^2}} \qquad \cos I = \left(\frac{C_z}{C}\right)$$

where $\vec{C} = (C_x, C_y, C_z)$ is the angular momentum of the body on its heliocentric orbit and C is its norm. $E_\odot$ is defined by the Equation (2.13). A smart way for calculating ω is to use a scalar product between the line of nodes and the Laplace vector defined by

$$\vec{q} = \vec{V}_A \times \vec{C} - k_\odot^2 \frac{\vec{P}_A}{\|\vec{P}_A\|} \tag{4.5}$$

We present two very different cases; in the first one (corresponding to the attributable $\alpha = 2.018$, $\delta = 0.204$, $\dot{\alpha} = -0.00623$ and $\dot{\delta} = 0.000302$), the level curve $I = 90°$ divides the admissible region into two parts; the level curves seem to converge towards a point located outside the admissible region. In the second one (corresponding to the attributable $\alpha = 2.018$, $\delta = -1.204$, $\dot{\alpha} = -0.0623$ and $\dot{\delta} = 0.00302$), the *convergence* point of the inclination curves is inside the admissible region. This point corresponds to a singularity: the orbit is so elliptic than it becomes a straight line; it means that the inclination is not defined anymore, the orbital plane being reduced to a line.

For both cases, the level curves of the ascending (solid) and descending (bold) nodal distances are plotted, giving a clear idea about the location of virtual impactors in the admissible zone (Figure 4). In the first case, on the left part of Figure 4, the ascending and descending nodal distances curves have no intersection, except for the case $r = 0$ and $\dot{r} = 0$ (i.e. the orbit of the Earth) which is obviously common to both conditions. On the opposite, on the right part of Figure 4, they cross several times, inside the admissible region, for different types of non circular and non coplanar orbits.

The location of the nodal distances in the admissible region is crucial to determine the potential hazard of this attributable. Indeed, it is easy to sample the curves $d_+ = 0$ or $d_- = 0$; to each point of this subset corresponds a set of six orbital elements, i.e. an orbit and an instantaneous position on this orbit. By propagating the motions of the body (on a keplerian orbit, for the simplest case) and of the Earth, we can rapidly check whether a close encounter is scheduled or not for the next few tens of years. By close encounter, we mean that the body enters the sphere of influence of the Earth. A that moment, another analysis has to be developed, using specific variables and formulae (see for example Öpik (1976) and Valsecchi *et al.* (2003)) to make the final model of approach and detect a significant probability of impact.

In a less restrictive use, this new information (the nodal distances location) may also be of great interest to improve the choice of the metric function to enhance some important subsets of the admissible region for future propagation (Milani *et al.* (2004)).

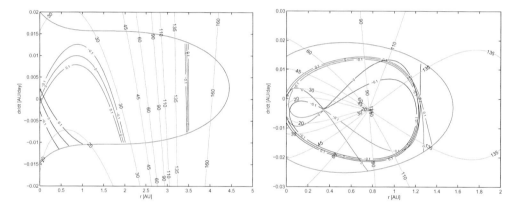

Figure 4. Level curves of the inclination ($I = 20, 30, 45, 60, 90, 110, 135, 160$ degrees), of the ascending ($d_+ = -0.01, 0, 0.01$ [solid]) and descending ($d_- = -0.01, 0, 0.01$ [bold]) nodal distances for the attributable ($\alpha = 2.018$, $\delta = 0.204$, $\dot{\alpha} = -0.00623$, $\dot{\delta} = 0.000302$) on the left and ($\alpha = 2.018$, $\delta = -1.204$, $\dot{\alpha} = -0.0623$, $\dot{\delta} = 0.00302$) on the right. The unit is the astronomical unit.

5. Circular and linear orbits

Let us draw the contour levels of the eccentricity in the admissible zone, for the two selected attributables (Figure 5). There are two apparent centers of circular orbits: the first one coincides with the Earth itself ($r = 0$ and $\dot{r} = 0$) which is assumed to be on a circular orbit; the second one is more interesting and can be characterized as a solution of the two following equations

$$\dot{r} = -\frac{A\,r}{r + B} \tag{5.1}$$

$$p_7 + 2\left[p_2\,\dot{r} + p_4\,r\,\dot{\alpha} + p_6\,r\,\dot{\delta}\right] + \dot{r}^2 + r^2\,\dot{\alpha}^2\,\cos^2\delta + r^2\dot{\delta}^2 = \frac{k^2}{\sqrt{p_0 + 2\,p_1\,r + r^2}} \tag{5.2}$$

where

$$A = p_2 + p_3\,\dot{\alpha} + p_5\,\dot{\delta}$$
$$B = p_1$$

as well as

$$p_0 = \langle \vec{P}_{\oplus}, \vec{P}_{\oplus} \rangle \quad p_7 = \langle \vec{V}_{\oplus}, \vec{V}_{\oplus} \rangle$$
$$p_1 = \langle \vec{P}_{\oplus}, \vec{u} \rangle \quad p_3 = \langle \vec{P}_{\oplus}, \vec{u}_\alpha \rangle \quad p_5 = \langle \vec{P}_{\oplus}, \vec{u}_\delta \rangle$$
$$p_2 = \langle \vec{V}_{\oplus}, \vec{u} \rangle \quad p_4 = \langle \vec{V}_{\oplus}, \vec{u}_\alpha \rangle \quad p_6 = \langle \vec{V}_{\oplus}, \vec{u}_\delta \rangle$$

These equations were obtained by combining two conditions characterizing circular orbits. First, the position vector $\vec{P}_A$ must be perpendicular to the velocity vector $\vec{V}_A$

$$\langle \vec{P}_A, \vec{V}_A \rangle = 0 \tag{5.3}$$

secondly, the orbital heliocentric velocity must correspond to

$$\| \vec{V}_A \|^2 = \frac{k_\odot^2}{a}$$

that is

$$\| \vec{V}_A \|^2 \, \| \vec{P}_A \| = k_\odot^2 \tag{5.4}$$

Let us notice that all the values of the eccentricities lie between 0 and 1, the values outside the admissible region correspond to hyperbolic orbits. The curve $a = 1$ (more visible on the right diagram) corresponds to the positions of Earth's Trojans. We plot

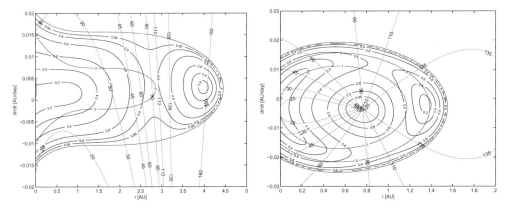

Figure 5. Values of the eccentricity ($e = 0.2, 0.4, 0.6, 0.8, 0.9$) and of the inclination ($I = 20, 30, 45, 60, 90, 100, 135, 160$ degrees) for the two test attributables: ($\alpha = 2.018, \delta = 0.204, \dot{\alpha} = -0.00623, \dot{\delta} = 0.000302$) on the left and ($\alpha = 2.018, \delta = -1.204, \dot{\alpha} = -0.0623, \dot{\delta} = 0.00302$) on the right. Two level curves of the semi-major axis are also drawn corresponding to $a = 1$ AU and $a = 2$ AU [bold].

the curves corresponding to conditions (5.3) and (5.4) in Figure 6. The condition (5.3) describes the objects which are exactly at perihelion or aphelion dividing the admissible region into two distinct parts. The intersections of the curves (5.3) and (5.4) give the circular orbits in the admissible region. Beside the obvious case ($\dot{r} = 0, r = 0$), two potential circular orbits appear in the case of the first attributable and only one for the second one. The supplementary solution hidden in (Figure 5) appears clearly as shown in (Figure 6, left). Let us remark that the virtual impactors detected by vanishing the nodal distances (except in the trivial case) correspond to non-circular orbits, which is obvious, the Earth moving on a circular orbit itself.

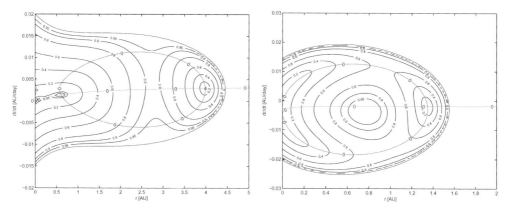

Figure 6. Condition (5.3) and (5.4) for the two test attributables: ($\alpha = 2.018$, $\delta = 0.204$, $\dot{\alpha} = -0.00623$, $\dot{\delta} = 0.000302$) on the left and ($\alpha = 2.018$, $\delta = -1.204$, $\dot{\alpha} = -0.0623$, $\dot{\delta} = 0.00302$) on the right.

6. Elongations and related angular distance

Let us remind that the attributable consists of two angles and their time derivatives; usually they are connected to α, the right ascension and δ, the declination, the geocentric equatorial coordinates, but they could also be replaced by λ and β, the ecliptic longitude and latitude of the object, or deduced from each other thanks to the relations.

$$\cos \beta \cos \lambda = \cos \delta \cos \alpha$$
$$\cos \beta \sin \lambda = \sin \epsilon \sin \delta + \cos \epsilon \cos \delta \sin \alpha$$
$$\sin \beta = \cos \epsilon \sin \delta - \sin \epsilon \cos \delta \sin \alpha$$

where ϵ is the obliquity, i.e the angle between the ecliptic and the equatorial plane. A quantity directly linked to the attributable is the *elongation*, denoted by ϕ, the angular distance between the Sun and the body $\mathcal{A}$ as viewed from the Earth. The elongation is given by the expression

$$\cos \phi = -x_E \cos \lambda \cos \beta - y_E \sin \lambda \cos \beta \tag{6.1}$$

where $(x_E, y_E, 0)$ is the heliocentric position of the Earth on its circular ecliptic orbit. For the first attributable, the elongation is $\phi = 166.87°$. This value suggests that the observations have been performed close to the opposition ($\phi = 180°$). The elongation value corresponding to the second attributable is $\phi = 91.48°$. In this particularly case, the observations would have been acquired near quadrature.

Let us notice that all the virtual asteroids corresponding to the same attributable have the same elongation. On the contrary the opposite angle θ, between the object $\mathcal{A}$ and the Earth, as viewed from the Sun, for a fixed elongation, is a function of r and its level contours are vertical lines in the admissible region, as shown in (Figure 6). Let us remark that the two selected attributables have quite different proper motions. Indeed, we have $\eta_2 \approx 4 \eta_1$ with $\eta_1 = 6.1 \times 10^{-3}$ rad/day where η_1 and η_2 are the proper motions of the first and second attributable respectively. On the other hand, the declination δ of the second attributable differs significantly from the first one giving to this last case a more theoretical and singular aspect. As a consequence, the internal structure of the admissible region associated to the second attributable shows several uncommon properties such as the inclination singularity for example.

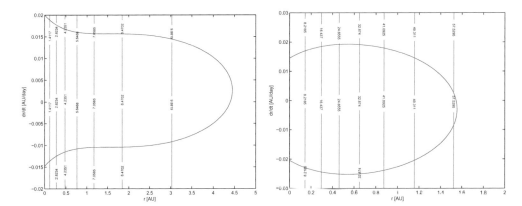

Figure 7. The vertical lines correspond to the values of opposite angular distance θ (in degrees), for the two selected attributables.

7. Conclusion

The topology of the admissible region is clearly dependent on the selected attributable, as shown by our two test attributables, corresponding to two very different observations: at the opposition or at the quadrature.

We have showed that the number and the positions of the potential circular orbits, the location, the shape and the length of the ascending and descending nodal distances, the behaviour of the inclination level curves, are very different from one situation to the other one and contain many informations about any virtual body compatible with the partial set of observations.

In the search for potential Earth's impactors, we propose to complete this preliminary study in two directions: the first idea would be to sample the curves of zero nodal distances, and to propagate this set of points for several years. This procedure reduces in a significant way the size of the admissible region and allows to use specific propagation methods, adapted to close encounters. The second project would be to compute the minimal orbital intersection distance (MOID) to the Earth using Öpik (1976) formalism instead of the nodal distances; however, this computation requires to check the values of Tisserand parameter, for any concerned orbit; the validity of the theory requires Tisserand parameters smaller than 3, which is not always the case, in particular for some of the orbits generated by our two test attributables.

References

Milani, A., Gronchi, J.F., De Michieli Vitturi, M. & Knežević, Z. 2004, *Celest. Mech. Dyn. Astron.* 90, 57

Milani, A. Gronchi, G.F., Knežević, Z., Sansaturio, M.E. & Arratia, O. 2005, *Icarus* 179, 350

Milani, A. & Knežević, Z. 2005, *Celest. Mech. Dyn. Astron.* 92

Öpik, E.J. 1976, *Interplanetary Encounters*, Elsevier, New York

Valsecchi, G.B., Milani, A., Gronchi, G.F. & Chelsey, S.R. 2003, *Astron. Astrophys.* 408, 1179

Part 9

IAU and Governments Role in the NEO Problem

Near Earth Objects, our Celestial Neighbors: Opportunity and Risk
Proceedings IAU Symposium No. 236, 2006   © 2007 International Astronomical Union
A. Milani, G.B. Valsecchi & D. Vokrouhlický, eds.   doi:10.1017/S1743921307003596

The IAU Role

Oddbjørn Engvold

Institute of Theoretical Astrophysics, University of Oslo, Oslo, Norway
email: oengvold@astro.uio.no

1. Introductory Comments

One knows currently close to 850 Near Earth Asteroids (NEAs) with diameters 1 km and larger, and one estimates that there may be of the order of 100 000 NEAs with diameters exceeding 140 m. Land and water impacts of NEAs with diameters between 100 m and 500 m will cause major damages.

Governments and international organizations are becoming increasingly concerned with natural hazards and disasters. The International Council for Science (ICSU) is developing a new program on "Natural and human-induced hazards and disasters" with participation from a large number of its scientific Unions. An adequate survey of Near Earth Objects (NEOs) is of notable interest and importance in such an initiative

A rapid development of NEOs survey programs illustrates the increased interest for detecting, tracking, characterizing and cataloging this family of bodies in our solar system. The first of four 2-m telescopes for such a survey program, the University of Hawaii Panoramic Survey Telescope and Rapid Response System (Pan-STARRS), will be operative in 2007. The planned Large Synoptic Survey Telescope (LSST) in Cerro Pachon, Chile, will further enhance the discovery rate of NEOs. One may also note that NASA has recently modified its charter to stress its increased responsibility for discovery and characterization of NEOs.

The discoveries of NEAs that will be "interesting" to the public and media may increase from about one per year to very possibly one per week. Based on preliminary orbit calculations, as much as a dozen of these may initially appear to pose a potential threat.

The IAU has been and will continue to be the primary international scientific organization with expertise on NEOs.

2. IAU's Responsibilities and Initiatives

The greatly increased discovery rate of NEAs will inevitably lead to an increased interest in characterization and studies of them. Calculation of precise orbits and determination of impact probabilities for potentially hazardous NEAs will in the very near future require a matching attention and augmentation in support. The IAU has an obligation to encourage its National Members to support and safeguard these important scientific activities.

The IAU has for several decades assumed responsibility for the IAU Minor Planet Center hosted at Smithsonian Astrophysical Observatory (SAO) with the important task of recording and maintaining the inventory of small bodies in the solar system, which constitutes an important service to our community at large. A new Memorandum of Agreement between the IAU and SAO was signed in June 2006, which will ensure continued operation of the MPC with funding from NASA. The IAU is pleased to note SAO's willingness and intention to revise and upgrade the operations of the MPC in order to meet the need to cope with up to a hundred-fold increase in incoming data.

One byproduct of the increasingly improved NEO surveys will be the finding of many new faint comets which concerns the IAU through their scientific value and interest.

With its responsibility for safeguarding and coordination of the science of astronomy in all its aspects IAU is the obvious organization and authority to be undertaking a sober and quality controlled information, on the matter of potential NEA threats to the Earth, to the public, to media and to governments. The IAU Executive Committee has created a small committee on "Impact Threats on the Earth" to enable the IAU President and General Secretary to respond quickly to media and the public, as well as to governmental and international organizations, on incidents of Near-Earth asteroids and comets.

The current members of this Advisory Committee are: David Morrison (Chair), Richard Binzel, Andrea Carusi, Andrea Milani, Donald Yeomans and the Director of IAU Minor Planet Center. The Committee will conduct its duties in accordance with the following rules and restrictions:

- Committee members will keep each other informed and will share information.
- Non-urgent communications to the IAU will be based on committee consensus.
- Urgent communications should represent a committee consensus, but can be initiated by a minimum of 2 (two) committee members .
- The committee will establish a public web page to be used for both general and urgent information. The site will cross-link to the official IAU web site and to the primary CEO web site (JPL, Pisa, MPC, etc.).
- Urgent statements on behalf of the IAU can be posted by a minimum of 2 committee members. All urgent messages will also be sent by e-mail to the IAU GS, President and Press Officer.
- The NEO web page will provide a historical record of IAU statements, including, for example, both original statements and subsequent modifications.
- The Committee will not issue IAU press releases or hold press conferences, but it will assist IAU in such press statements, if requested.
- Interaction with governments on issues of impact threats is under the sole control of the IAU President and General Secretary.

Acknowledgements

I thank David Morrison for helpful input to my preparation of this brief intervention.

Discussion

HARRIS: We had a small committee (the Technical Review Committee) that backfired on us because it was taking the data about possible collisions and was evaluating them essentially in secret, in order to advise the IAU General Secretary and the public at large. I would rather see things to operate as Steve Chesley said yesterday in his talk, with the impact monitoring groups posting their results on the web, when the results agree, leaving to the public to judge. I still see a very valuable use for the new committee, that is to provide expert evaluation about the problem to the IAU and to the general public.

VALSECCHI: You were speaking of the Technical Review Committee; that committee was established because at the beginning of impact monitoring there was only one group doing the job. The function of the committee was superseded by the existence of a second, independent impact monitoring system, witn the ensuing cross-checking. Concerning secrecy, let me say that when you have a confirmed news, and you do not disseminate it, then that is secrecy. However, when you are not yet convinced of the correctness of what you have found, and are trying to understand whether it is reliable, then that is not *secrecy*, that is *checking*. We (i.e., the community involved) were convinced from the beginning that secrecy backfires; at the same time, we wanted to be sure that what we had found was *news* worthy of being communicated. In the case of (99942) Apophis the information was released after three days of checking, since the initial data were of such bad quality that we could not trust the results that were coming out of the impact monitoring systems; as soon as we cleared all our doubts, the information was immediately released.

MILANI: I would like to comment on the recurring issues of verification and secrecy. First of all, we do wait for confirmation of all the most important (Palermo Scale $>$ -2) possible impacts. To understand how, you have to take into account time zones: there is a difference of 9 hours in local time between the location of the two current Impact Monitoring Systems, CLOMON2 in Europe and Sentry in California. Verification typically takes a short time, but in critical cases requires that the two teams are in their offices in the same moment, which is possible only in some afternoon hours (European time). As soon we are convinced the data have been verified by cross checking between the two systems we post immediately, within minutes. This is the situation now, and it is a very significant improvement with respect the time (before 2002) when only one impact monitoring system was available. Indeed, with duplication we have solved the problem of verification. This gives an important guarantee of reliability and also has relieved a lot of stress, which used to affect the previously unique team.

Another important point is that we do operate only on observational data which are public. Moreover, the algorithms and software we use for this are public and available on the web. The OrbFit software system is open source, free software, and contains the same subroutines we use for impact monitoring†. You can download it, compile on your own computer and run it on the same data we have, and reproduce the same results; there are people who have actually done this, as well as comparing with their own software‡. Thus there is secret neiter on the data, nor on the methods, nor on the results.

Once we have something which is out, there is typically a short time before the press notices it. Our sites, NEODyS ans Sentry, are monitored every 5 minutes to look for new and/or changed announcements on possible impact; e.g., such continous control is performed by the MPC, for obvious reasons, and by Bill Allen¶ who keeps a record so accurate of all the changes in our sites that we actually use his web site when we need a record of our previous postings. The response by the media may be slow or even null in some cases, e.g., at the worst moment of the (99942) Apophis crisis (26 December 2004) there was the tsunami and the 1/37 impact probability was essentially not reported by the media. In other cases, however, depending upon the hour in some time zone and upon the internal logic of the media, there can be a significant response in few hours.

This is the situation which has to be the main concern of the *Impact threat* committee: the information on a possible impact is not secret, it is out on the web and nobody has noticed it yet. During that short time span we need to know what to say to better convey the information to the general public, and in extreme cases to the authorities if they need to act for mitigation, as it has not happened yet. My understanding is that this is the main function of the new advisory group.

ENGVOLD: They have to keep the SG informed, but they are the best placed persons to decide how to pass the information to the press.

---

† http:newton.dm.unipi.it/orbfit/
‡ For example, I. Wlodarczyk in Poland.
¶ http://www.hohmanntransfer.com

MARSDEN: If we discover something, probably very small, which is going to hit in a matter of days, that will become obvious at the MPC. We hope it does not happen, but if it happens, it has to be handled in an appropriate way.

HARRIS: Current surveys have the capability to find objects with diameter few tens of meters, the next generation surveys could find meteoroids of few meters, with an impact rate on Earth's atmosphere of the order of one per year. Thus there is a significant probability that they would discover one of these while it is arriving.

MORRISON: The cases that will most likely cause difficulty in communicating with the public are those of impacts decades away in time, that may raise to a probability level of a few percent and then go away, rather than the rare cases in which a very small impactor is discovered hours or days before it impacts.

VALSECCHI: Questions related to the *announcement dilemma* are possibly not the hardest problem our community faces; rather, we have not yet been able to convince responsible people outside the astronomical community that what we have put in place (i.e., NEO discovery, impact monitoring and space missions to test mitigation techniques) is just the first link in the chain of actions to prevent consequences of impacts, and that all the other links, involving civil protection, policy issues, social issues, legal issues etc., have still to be put in place.

ANDERSEN: The IAU General Secretary, even if not a NEO-specialist, will make sure that the IAU has appropriate structures in place to deal with the NEO issue; it is clear that many, if not most, countries, have not the foggiest idea of what to do in case the announcement of a real impact were made, and that is where you need the IAU GS, who can speak in the name of the world astronomers, providing mankind the relevant information. It is then to mankind to find out what to do.

Near Earth Objects, our Celestial Neighbors: Opportunity and Risk
Proceedings IAU Symposium No. 236, 2006
A. Milani, G.B. Valsecchi, & D. Vokrouhlický, eds.
© 2007 International Astronomical Union
doi:10.1017/S1743921307003602

The UK Near Earth Object Information Centre (NEOIC)

Iwan P. Williams

Astronomy Unit, Queen Mary, University of London, London, E1 4NS, UK
email: i.p.williams@qmul.ac.uk

Abstract. Following the report of the 'task force', the UK Government decided to accept some of it's recommendations. In particular, it accepted two that recommended the setting up of a British National Centre for Near Earth Objects. The final outcome was the setting up of a Near Earth Object Information Centre to inform the general public of the dangers or otherwise from impact on the Earth of Near Earth Objects. The Centre has now been running for several years and in this publication we examine the current workings of the Centre and discuss some of its successes and failures.

Keywords. Impact risk; outreach activities

1. Historical Introduction

The story start in 1694 when Edmund Halley suggested that some global catastrophes could have been caused by cometary impacts. In particular, he suggested that one such impact may have been the cause of the event that lead to the biblical flood legend while an other impact could have formed the Caspian Sea. Such views were not however popular in Halley's time, when it was generally believed that the heavens, being God's creation, were perfect. Collisions between heavenly bodies simply could not take place. A hundred years later, Chladni (1794) proposed that meteorites could be of extra-terrestrial origin, but the idea that heavenly bodies could collide was still not generally accepted as is illustrated by the following. In December 1807 a huge fireball was seen by many people over a large section of New England and a meteorite was observed crashing to Earth near Weston, Connecticut. Two employees of what was then called Yale College, Sillman, a professor of Chemistry, and Kingsley, the college librarian, collected many samples of this Weston meteorite, but US President Thomas Jefferson is attributed with the probably apocryphal remark 'it is easier to believe that two Yankee Professors would lie than that stones would fall from the sky'.

However, evidence for collisions was mounting rapidly. For example Benzenberg and Brandes (1800) had observed the same 22 meteor trails from two different sites. They showed from parallax measurements that their average height was about 90 km. This was far too high for them to be a normal atmospheric phenomena and so they must have an extra-terrestrial cause. The spectacular Leonid meteor storms of 1833 convinced most people that small particles could regularly crash into the Earth. However, to believe that larger object, large enough to cause physical damage, could collide was a different matter. In the 1890's Barringer suggested that what is now called meteor crater in Arizona was of impact origin and in the 1930's, the Odessa crater in Texas was shown to be an impact crater.

Through the discoveries of an increasing number of asteroids on Earth approaching orbits, the idea that larger bodies could collide with the Earth slowly became acceptable and in the 1970's, Eleanor Helin together with Eugene and Carolyn Shoemaker started a systematic photographic survey of the sky with the primary aim of discovering Near-Earth Objects. At about the same time, Arthur C Clarke coined the term Spaceguard to describe the general activity of guarding the Earth against impacts from Near Earth Objects. In 1979, the film 'Meteor' was released, perhaps bringing the subject to the notice of the general public for the first time. The awareness of the public regarding the topic was increased further when Alvarez *et al.* (1980) suggested that the extinction of the dinosaurs could be explained through a massive asteroidal impact.

The 1980's saw a general increase in the acceptance of the idea that asteroids and comets could, and indeed did, collide with the Earth. This, in turn, led to an increase in the number of survey facilities being established in order to search for potential collision candidates. In 1991, the US Congress House Committee on Science and Technology direct NASA to study the feasibility of having a programme that would increase detection rates of asteroids on Earth-crossing orbits. This lead to the publication in 1992 of the Spaceguard Survey report which recommended a search programme and international collaboration to find objects greater than 1 km through the provision of six ground based telescopes suitably placed around the world. Half the cost of these telescopes was expected to come from international partners and half from the US.

In 1994, the above report was modified so that NASA was requested to report within one year, with help from the US Department of Defence and the Space Agencies of other countries, on the setting up of a programme to identify and catalogue all comets and asteroids with a diameter greater than 1 km that are on an orbit that crosses the Earth's orbit. In that year, the fragments from the break-up comet Shoemaker-Levy 9 were discovered and many of the fragment were observed as they collided with Jupiter, leaving visible scars on the planet. This provided a considerable upturn in the interest of the public in the matter of asteroid and comet collisions with the Earth. At the IAU General Assembly in the same year, a recommendation that some (unspecified) International Authority should take responsibility for NEO investigations was passed.

Over the next five years there are many conferences held on the subject and many resolutions passed. However, little action takes placed other than in the US, where new surveys facilities continue to be established and the IAU which continues to keep catalogues of Near Earth Object orbits through its Minor Planet Center.

In 1999, the UK Science Minister, Lord Sainsbury, in a speech to the House of Lords stated that Britain must cooperate internationally in order to tackle the threat presented by Near Earth Objects. One year later he created a Task Force that was required to investigate the topic of Potentially Hazardous Near Earth Objects and to make recommendations to the UK Government on how it should best contribute to an international effort on Near Earth Objects. In particular, the Task Force was required to

(*a*) confirm the nature of the hazard and assess the potential levels of risk;

(*b*) identify the current contribution of the UK to international efforts;

(*c*) advise the Government on what further action to take in the light of their findings with regard to points 1 and 2 above;

(*d*) advise the Government on how to communicate the issues to the public;

(*e*) send their report to the Director General of the British National Space Centre (BNSC) by the middle of the year 2000.

This paper deals with progress made in the UK from that date until now.

2. The Task Force Report

The Task Force was chaired by Dr Harry Atkinson with Sir Crispin Tickell and Professor David Williams as members. The report was published in September 2000 and can be found at `http://www.nearearthobjects.co.uk` . It provided an excellent overview of the subject at the time of publication and, in addition, made 14 recommendations to the UK Government. These recommendation are not reproduced in full here but a summary of each is given.

2.1. *Survey and Discovery of Near Earth Objects*

Recommendation 1

The UK Government should seek partners, preferably in Europe, to build a new advanced 3 meter survey telescope located somewhere in the Southern Hemisphere. This telescope should be dedicated to work on Near Earth objects.

Recommendation 2

Arrangements should be made so that any observational data obtained for other purposes by wide-field facilities such as the new VISTA telescope would be searched for NEO's on a nightly basis.

Recommendation 3

The UK Government should draw the attention of ESA to the role that GAIA and other space missions such as BepiColombo and NASA's SIRTIF could play in surveying the sky for NEO's

2.2. *Accurate orbit determination*

Recommendation 4

The 1m Johanes Kapteyn Telescope located on La Palma should be dedicated to follow-up astrometric observations of NEO's

2.3. *Composition and gross properties*

Recommendation 5

Negotiations should take place with all the partners that the UK share suitable telescopes with to establish an arrangement such that small amounts of time is provided for spectroscopic follow-up of NEO's

Recommendation 6

The UK Government, together with like-minded countries, should explore the case for mounting a number of coordinated space rendezvous missions based on using microsatellites.

2.4. *Coordination of astronomical observations*

Recommendation 7

The UK Government together with other Governments, the IAU and other interested parties should seek ways of putting the governance and funding of the IAU Minor Planet Center on a robust international footing.

2.5. *Studies of impact on the environmental and consequential social effects*

Recommendation 8

The UK Government should help to promote multi-disciplinary studies of the consequences on Earth of impacts from NEO's.

2.6. *Mitigation possibilities*

Recommendation 9

The UK Government, with other Governments, should set in hand studies to look at ways of deflecting any incoming objects and of mitigating the consequences of impact.

2.7. *Organization*

Recommendation 10

The UK Government together with other Governments and the IAU should urgently seek to establish a forum for an open discussion of the scientific aspects of NEOs

Recommendation 11

The UK Government should discuss with like minded Governments in Europe how Europe could best contribute to the international effort to cope with potential NEO impacts. ESA, ESO, the EU and the European Science foundation should work together on a strategy for this purpose.

Recommendation 12

The UK Government should appoint a single department to take the lead in the coordination and conduct of Government policy regarding NEOs.

2.8. *British National Centre for Near Earth Objects*

Recommendation 13

A British Centre for Near Earth Objects should be set up. It's mission would be to promote and coordinate work on the subject in Britain as well as providing an advisory service to the Government, the public and the media. It should also help to facilitate British involvement in international activities.

Recommendation 14

One of the most important functions of a British Centre for NEOs would be to provide a service for the public that would give balanced information in clear direct and comprehensible language as the need might arise.

3. The British National Centre for Near Earth Objects

It is a matter of some debate as to how many of the excellent recommendations listed above have actually been implemented by the UK Government. Some were dealt with in a different way from that envisaged in the report, for example recommendation 7 suggested one way of funding the MPC. However, at this meeting we have been informed of new arrangements for the MPC. Other recommendations were overtaken by events, for example recommendation 4 envisaged a role for the Jacobus Kapteyn telescope that can not be fullfilled because of the changed situation in UK optical astronomy, while for some more, other governments were perhaps more reluctant to take part. However at least two were implemented, recommendations 13 and 14 and we give an account here of the implementation of these two recommendations.

3.1. *The formation and structure of the Near Earth Object Information Centre*

The centre came into operation in April 2002, only 18 months after the publication of the Task Force report, a commendably short interval of time. The main centre is based at the National Space Centre located in Leicester. In passing it is important to clarify the distinction between the British National Space Centre (BNSC) which is based in London and the National Space Centre, based in Leicester. The former is a Quasi-Government agency that provides funding for Space Research and directly advises government on policy. The latter is a quasi-private enterprise that is more akin to the Science Museum and derives much of its financial support through its own activities. The Near-Earth Object Information Centre derives its funding primarily through a grant from BNSC, but is based at the National Space Centre in Leicester, which is more appropriate since much of its function is related to public contact. The main current functions of NEOIC are to provide (i) an up to date web site with all relevant information about NEO's, (ii) an exhibition which displays in an user-friendly way the full range of topics normally included under the heading of NEO studies and (iii) public lectures and talks, both at the centre and elsewhere, primarily at schools. The raw data is taken from the MPC, the JPL Impact table and the NEODyS risk page. To assist in these task the NEOIC has a team of experts that act as advisors. At present the team consists of Professor Martin Barstow, Dr John Davies, Professor Alan Fitzsimmons, Dr Mathew Genge, Professor Monica Grady, Professor Carl Murray and Professor Iwan Williams. In addition NEOIC has the responsibility of advising BNSC (and thus indirectly the UK Government) on any matters that deserve attention.

The NEOIC has also expanded to provide three Regional Centres, all also based at institutions with a primary function to provide education and information to the public. These are W5 in Belfast, the National History Museum in London and the Royal Observatory in Edinburgh. Each regional centre has its own NEO exhibition based on the main exhibition based at Leicester.

3.2. *The Activities of NEOIC*

In order to perform effectively its outreach tasks, it is necessary that all the information that NEOIC has is up to date. To facilitate this, the Minor Planet Electronic Circulars (MPEC) are monitored daily with all the new data incorporated into the data base. In addition, JPL's Impact Table and NEODyS Risk Page are both monitored daily for changes. All major sources of NEO news stories are also monitored.

With this up to date information, NEOIC basically performs a number of tasks.

1. NEOIC's own web page is updated to reflect any news gained through any of the above activities. It very occasionally highlights a particularly important piece of news, though experience has shown that this can be a dangerous practice, especially when the published orbit is based on a very short observing arc. The average number of users daily has remained fairly constant at about 300, though a peak of 20 000 was achieved during an unfortunate incident where NEOIC was premature in announcing a potential impact.

2. It provides regular reports to BNSC an in addition alerts them if a potential future threat has been identified. This is potentially an important role since occasionally a member of Parliament will ask the Science Minister what the UK Government is doing about the possible asteroid collision in 2XXX and the minister needs reliable and up to date information to reply sensibly to such questions.

3. It responds to media activities both through answering journalist questions by e-mail or telephone or by giving interviews. Here the Expert team play an important role since questions, especially from journalists, are often referred to them.

4. By far the main task, both in terms of time and public impact, is what might be called 'outreach activities '. These can be subdivided into three main categories:

a) Exhibitions: these are mainly unstaffed but interactive, and would more often than not be repeated at the Regional Centres. Typical examples are

(i) A computer simulation to 'find' an NEO, with illustration of how the predicted orbit improves with more observational data.

(ii) An interactive map showing that impacts have already happened. Impact sites are shown on a map, which when the mouse is clicked on a dot, a picture of the crater is shown (and possible also the likely size of impactor)

(iii) A view of the Solar System depicting the real motion of known Apollo, Amor or Aten asteroids and certain named comets.

Other 'hands on' activities are also often included, including the handling of real meteorites of different classes, images of comet Shoemaker-Levy 9 and so forth.

b) External Visits: these involve visiting Schools and Clubs to give an organized presentation via a lecture or talk and demonstrations of the topic. There is nothing particularly novel about this activity, but it is perhaps the back-bone of outreach activities. The main advantage of organizing such an activity through NEOIC is that presentational material can be prepared to a high standard.

c) Workshops: these can take place either at NEOIC or at an external venue. The difference between these and the external visits is to stress the hands-on aspect of the activity. Meteorite fragment are handled by the audience so that they can really see and feel the difference between a stony and an iron one, they help to make a comet nucleus model, or play with 'Newton's Cradle 'to appreciate the law of momentum conservation.

4. Conclusions

It is interesting after four years of operation to ask two questions, has it achieved its laid down objectives, and is it worthwhile?

The objectives outlined in recommendations 13 and 14 of the task force report were

(i) To promote and coordinate work on the subject in the UK;

(ii) To provide an advisory service to the UK Government;

(iii) To facilitate British involvement in international activities;

(iv) To provide a public service that gives balanced information;

(v) To provide an advisory service to the public.

In terms of giving the public information, NEOIC has exceeded the demands in terms of exhibitions and outreach activities, it advises the government and gives balanced information to the public. Through these it has also helped to promote and coordinate work in the field, primarily through the expert team. It has done very little to promote and facilitate British involvement in international activities primarily because of a lack of funds.

So, is it worthwhile? The answer has to be yes, the public are now much better informed at a relatively low cast. It also has the side benefit of getting young children interested in astronomy generally and indeed in finding out that science can be fun. It also provides a valuable service in ensuring that UK Government Ministers, particularly the Science Minister has accurate information. It has also, after a few errors, proved to be a reliable source of information for the Press-which is to the good of the subject in the long run.

Of course, it would be nice if it had more money so that it could fund both UK and international cooperation, but moving forward is always better than standing still.

Acknowledgements

The author would like to thank Kevin Yates the NEOIC project manager for providing lots of information without which this article could not have been written.

References

Alvarez, L.W. Alvarez, W. Asaro, F. & Michel, H.V., *Science* 208, 1095
Benzenberg, J.F. & Brandes H.W. 1800, *Annalen der Physik* 6, 224
Chladni, E.F.F. 1794, Hartknock (pub) Riga, 1794

Near Earth Objects, our Celestial Neighbors: Opportunity and Risk
Proceedings IAU Symposium No. 236, 2006
A. Milani, G.B. Valsecchi and D. Vokrouhlický, eds.
© 2007 International Astronomical Union
doi:10.1017/S1743921307003614

Near Earth Object impact simulation tool for supporting the NEO mitigation decision making process

Nick J. Bailey, Graham G. Swinerd,
Andrew D. Morley and Hugh G. Lewis

Department Aerospace Engineering, School ofEngineering Sciences,
University of Southampton, Southampton, UK

Abstract. This paper describes the development of a computer simulation tool, NEOSim, capable of modelling small NEO impacts and their effect on the global population. The development of the tool draws upon existing models for the atmospheric passage and impact processes. Simulation of the land and ocean impact effects, combined with a population density model, leads to casualty estimation at both a regional and global level. Casualty predictions are based upon the intensity of each impact effect on the local population density, with consideration given to the population inside or outside local infrastructure. Two case studies are presented. The first evaluates the potential threat to the UK, and highlights coastal locations as being at greatest risk. Locations around Cornwall demonstrate an increase in casualties above the local average. The second case study concerns the potential impact of asteroid (99942) Apophis in 2036. Propagation of the possible orbits along the line of variance leads to an extensive path of risk on the Earth. Deflection of the asteroid, by a variety of means, will move the projected impact site along this path. Results generated by NEOSim for the path indicate that South American countries such as Colombia and Venezuela are at a greatest risk with estimated casualty figures in excess of 10 million. Applications of this software to the NEO threat are discussed, along with the next stage of NEO impact simulation development.

Keywords. Impact risk

1. Introduction

Near Earth Objects have been the subject of continued study for many years with NASA's Spaceguard Survey providing the most comprehensive search and activity to date. Objects greater than 1 km in diameter are the focus of these surveys, while small sub-kilometre bodies remain largely uncatalogued. This NEO natural hazard was brought to public attention by the recent discovery of (99942) Apophis, a relatively small 320 m diameter asteroid, as it passed by Earth in December 2004. Following this discovery, the Earth witnessed the largest natural disaster on record when a 9.0 earthquake off the coast of Sumatra generated a tsunami wave which inundated coastlines around the Indian Ocean and parts of Africa. The calculated death toll for this event is 229,866 (United Nations 2006 and was a relatively 'small' tsunami compared to ocean impact generated tsunamis. Both these events demonstrate the significant threat posed by the sub-kilometre NEO impact hazard.

The work outlined in this paper began in the middle of 2004 and has focused on developing a computer simulation tool capable of modelling both land and ocean impact scenarios for sub-kilometre bodies. The program is called NEOSim and generates a casualty estimate by assessing the interaction of the impact generated effects with human populations. Two case studies have been investigated which deal with the local and global threat. The first scenario looks at the threat posed to a small region by objects landing in the vicinity. The UK was chosen as the test case, but this methodology can be applied to any location on the globe to assess the local risk. The second scenario investigates potential impact events of the asteroid Apophis along the predicted line of risk for the potential Earth encounter in 2036. This line stretches from Kazakhstan, across the north Pacific and Central America to the Cape Verde Islands in the Atlantic. Analysis shows the potential consequences for a space mission that seeks to mitigate the impact threat by altering the object's orbit.

2. NEOSim Methodology

A study of the literature revealed that there was a deficiency in tools that enabled the study of NEO impacts, including both land and ocean impacts. Three distinct phases of an impact were identified to be incorporated into the NEOSim program: atmospheric entry; impact energy transmission; and casualty prediction. These are briefly discussed here.

2.1. *Atmospheric entry*

NEOs impact the Earth at hypersonic velocities (typically in excess of 12 km/s) and, during atmospheric entry, generate complex hypersonic flows around the object. The dominant feature is the high stagnation pressure at the leading edge of the object and the expansion of the bow shock around the body. Behind the shock front the high pressure generates high temperatures and these combined effects lead to a mass loss through ablation of the surface material. If the stagnation pressure difference exceeds the internal strength of the object, which is determined by the physical composition of the NEO, the object will rupture leading to fragmentation. NEOSim incorporates three models for studying the effect of fragmentation.

- Single object model for robust, high strength objects
- Catastrophic fragmentation for weak objects
- Progressive fragmentation for pre-fractured objects

The single object model assumes the NEO remains intact throughout the atmospheric passage with mass loss only through ablation while the catastrophic fragmentation model, developed by Chyba *et al.* (1993) and Lyne *et al.* (1996) commonly known as the 'Pancake' model, deals with very low strength objects that disintegrate in the atmosphere. This model is thought to best represent the Tunguska event of 1908. the progressive fragmentation model was developed by Borovička *et al.* (1998), Baldwin & Sheaffer (1971) and Foschini (1998) to model objects that are pre-fragmented in orbit with relatively high internal strength. These objects break apart in the atmosphere but don't entirely disintegrate resulting in multiple ground impacts. A pre-defined break-up altitude is implemented according to Klinkrad *et al.* (2004), set at 30 km.

2.2. *Surface Impact Effects*

During impact the object's kinetic energy excavates a crater before being transmitted through a number of mechanisms identified by Collins *et al.* (2004). For land impact these mechanism include ejecta distribution, a seismic shock wave, a surface blast wave and thermal radiation generated by an expanding fireball. These impact generated effects can be likened to those associated with nuclear detonations. However, little data from nuclear testing is publicly available. Instead, the models by Glasstone & Dolan. (1977) and Collins *et al.* (2005) provide a good approximation and have been implemented in the NEOSim software.

Ocean impacts account for approximately two thirds of all NEO impact events due to the proportion of Earth covered by water. Ocean impacts are characterised by the excavation of a transient cavity (or crater) through the deposition of the object's kinetic energy at the impact site. This cavity is naturally unstable and immediately in-fills from the surrounding ocean. This in-filling water oscillates vertically generating the tsunami wave train. NEOSim implements the models by Chesley & Ward (2003) to calculate the tsunami's shoaling characteristics (wave run-up and run-in distances) on surrounding shorelines.

A special case exists where the transient cavity depth is greater than the ocean depth. In this 'bottoming out' scenario both a tsunami and some land impact generated effects are assumed to affect the surrounding region.

2.3. *Casualty Prediction*

Human population density data was obtained from NASA Visible Earth†. Although the data set is global, the low resolution limits the accuracy of the casualty estimations. This data is utilised by a dedicated casualty prediction algorithm that has been developed to assess each impact scenario. Following the simulation of an impact, the population in the area affected is used to calculate the casualty figure. The number of casualties at each point is depended on the local population and the severity of each effect. Increasing the distance from the impact site is the main factor in reducing the severity of each land impact effect. Several casualty prediction variables can be manipulated by the user for fine tuning of the scenario, including how many casualties are generated by the increasing effect intensity. Consideration is given to the percentage of people inside or outside local infrastructure, as this will offer some protection, assuming the buildings are not destroyed. The sum of casualties generated by each effect is output as a casualty distribution map.

For ocean impacts, casualty prediction is implemented in a similar manner, with casualties generated along the inundated littoral. The effectiveness of the wave in producing casualties is based upon the run-up height of the wave at shoaling and the run-in distance. The longer travel time of a tsunami provides populations with more time to evacuate, provided that there is a warning system. This casualty-reducing factor is incorporated, based on the wave travel time.

2.4. *Data Output*

NEOSim's primary output is the total casualty figure, which is provided to the user via a pop-up window. The user is then presented with the array of outputs provided by NEOSim. These outputs overlay data concerning each impact generated effect (both for land and ocean impacts) onto a world map. Furthermore, the casualty data is provided as an overlay onto the casualty density map with shading from dark to bright to denote low to high casualty densities. Figure 1 provides an example of the NEOSim outputs with a test impact into the North Sea (denoted by the small cross). The figure is a composition, with the left side showing the impact effects, in this example a tsunami, and on the right side the casualty density map. Verification of the absolute casualty figure for any impact is difficult due to the lack of first hand impact experience and the number of factors involved in casualty generation. Thus, NEOSim focuses on the relative casualty density over the area affected to provide information on the locations at greatest risk and to determine which populations should be evacuated in the event of an impact.

† NASA Visible Earth, Population Density Map, [Online] http://visibleearth.nasa.gov/view_rec.php?=116 [15 March 2005].

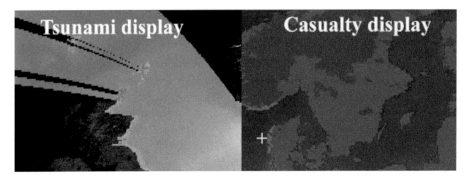

Figure 1. Figure shows the output provided by NEOSim. The left side shows the impact generated effects distribution, here the tsunami wave. The right provides the casualty density map with light regions representing more casualties.

Table 1. Characteristics of the test NEO used in the Case Study One simulation.

Parameter	Initial Value
Radius	150 m
Impactor Velocity	25 km/s
Altitude	106 km
Density 2600 kg/m^3	
Yield Strength	10000000 N/m^2
Heat of Ablation	8000000 J/kg

3. Case Study 1 – United Kingdom NEO Impact Risk

3.1. *Methodology*

In order to study the effect of NEO impacts around a region on the Earth rather than simply a single impact, a multi-run application was built into NEOSim. This multi-run tool allows the user to select a region defined by a latitude and longitude bin over a particular location of interest. This region is then divided into a number of cells (defined by the user) into which a single object is impacted. The casualty figure output from each cell impact is used to shade the output map. Mapping the results in this manner avoids the problem of validation of any specific casualty figure by providing the relative risk of an impact into each cell. For this case study the UK was chosen as the target with the latitude bin from 40° to 70° north and longitude bin from 30° west to 15° east. Resolution of the grid cells (the number of cells that fit within in the latitude longitude bins) is restricted by the processing time required for the simulation. Increasing the resolution and therefore number of cells improves the quality of the data but requires increasingly long runtime. For the present case study a 400 cell grid was chosen as a compromise between runtime and quality. The characteristics of the incident NEO are defined in Table 1.

3.2. *Results*

Figure 2 presents the results for the UK case study. The results show that the UK population is at greatest risk from objects impacting in the Atlantic Ocean to the west and south west. The peak amplitude of the tsunami is dependent on the depth of water at the point of impact with deep transient craters resulting in large amplitude oscillation as the cavity in-fills. Impacts into shallow oceans bottom out preventing the large oscillations of water, generating only small tsunami waves. This explains the low expected casualties

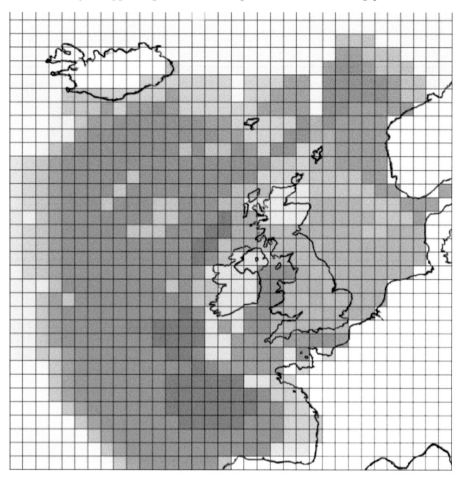

Figure 2. Shading denotes the number of casualties by the impact in the centre of each cell with dark shading representing more casualties.

from impacts into the North Sea where the water is only 30 m deep at maximum removing the potential to generate large waves. This provides some protection to communities on the eastern seaboard against tsunami inundation.

Populations highlighted as being at greatest risk are those along the westward coastline of Ireland and the south-western regions of the UK. In particular Cornwall and Devon, with a historically high dependence on maritime fishing industries, are at very high risk from such deep ocean impacts. NEOSim's tsunami model does not include wave diffraction which will reduce the effectiveness of the tsunami's propagation along the English Channel. Furthermore, the Bristol Channel, which experiences the second greatest tidal range in the world along with a large bore phenomenon, is likely to amplify the tsunami's amplitude as it travels up the constricting passage. Thus low-lying regions along the Severn Estuary are likely to experience catastrophic inundation from any deep Atlantic Ocean impact to the south west. Central coastlines of England and Wales, and in particular the major port of Liverpool, are shown at a reduced risk. This is explained by the protection afforded by Ireland which receives the majority of the inundation from impact tsunamis.

3.3. *Conclusions*

The threat to the UK from small, Earth-impacting asteroids is predominantly from Atlantic Ocean impacts to the west and south west. Atlantic impacts to the south west of the UK represent the greatest hazard with populations along the south west coastlines of the UK, including Cornwall, Devon and south Wales at greatest risk. A number of major cities are situated along the Bristol Channel including Cardiff, Newport, Bristol and Gloucester. These large cities will be at an increased risk due to the funnelling effect of the channel. It is expected that communities along the South coast of England will also be at increased risk due to the funnelling of the tsunami waves as they move up the Channel. Impacts in the shallow waters of the North Sea and Irish Sea present a much reduced casualty potential due to the small tsunami generated. These regions help to lower the risk to the UK by increasing the Earth's area where impacts can occur with little consequences. Land impacts in the UK will be very rare due to the small total land area (approximately 0.1% of the Earth's surface area). The consequences, while potentially severe even for these small NEOs, tend to be localised. The typically high population density of England in particular increases the risk of many casualties resulting from a land impact. NEOSim estimates casualty figures for England land impacts to be from 3 to 8 million, which is comparable to London's population of 7.5 million†. The cell that was created over London actually shows a larger casualty signal.

While this study focuses solely on the threat to the UK, in reality it is impossible to ignore the shared threat to other countries in the region. Particularly important in this study is the consequence for Ireland. When considering simply the UK, Ireland acts as a barrier to block the majority of tsunami waves from reaching the shores of England and Wales, effectively lowering the risk. However, it would be impossible not to view such impacts as threatening when, living in as we do in the global community, the consequences for neighbouring countries would be significant.

The high threat to the UK from Atlantic Ocean impacts calls for increased research into mitigating the threat. Furthermore, considering the very small risk from tectonic and volcanic activity and even catastrophic weather (such as hurricanes), the NEO risk is significant, potentially the UK's single greatest natural hazard (apart from, perhaps, the long-term effects of global warming).

4. Case study 2 – Apophis Path of Risk

The probability of impact for the asteroid 99942 Apophis has been reduced to 1 in 30,000 (equivalent probability of $3.3e^{-5}$) and thus no longer presents a significant risk. However, the example of Apophis has been used to demonstrate the potential of studying an object's path of impact risk. The speed of the simulation enables studies of newly discovered risk paths to be performed quickly to provide an assessment of the potential consequences of the impact.

4.1. *Methodology*

In order to study the effect of the potential impact of Apophis, the line of risk was required in latitude/longitude coordinates. Work by the B612 Foundation [footnote 1: website URL] has generated a path of risk for the predicted impact in 2036 using orbit intersections. The developed path is shown in Figure 3, which was processed to generate the latitude and longitude coordinates for every point along the path. A fixed longitude division was used to produce equal cell widths. NEOSim's multi-run tool was extended

† Wikipeda London [Online] http://en.wikipedia.org/wiki/London [15 August 2006].

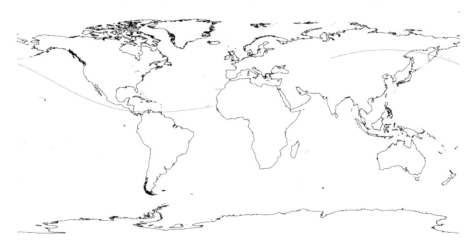

Figure 3. The curved line is the calculated path of risk for the potential impact of asteroid 99942 Apophis in 2036. This line is determined using orbit integrations.

Table 2. Assumed physical characteristics of the asteroid Apophis used as the test impactor. Data from http://neo.jpl.nasa.gov/risk/a99942.html.

Parameter	Initial value
Radius	320 m
Impactor Velocity	12.65 km/s
Object Density	2600 kg/m^3
Type	Monolithic

to model a series of impact cells along the path. A single 1-sigma cell was added above and below the path to represent the uncertainty in the lateral direction.

The estimated characteristics of Apophis, given in Table 2, were used to define the impacting object in NEOSim. This object was impacted into the centre of each cell along the path. The casualty figure for each impact was used to shade each cell as in the first case study.

4.2. Results

The western and eastern limbs of the modeled risk path are shown in Figures 4 and 5 respectively. For publishing reasons the shading has been converted to a grey scale with black representing most casualties and white least. The western limb of the path stretches from the Pacific Ocean, through Kamchatka and into Central Russia. The region of greatest casualty generation is found at the path's most westerly point with impacts into Kazakhstan. The second most significant region of risk is associated with impacts into the Sea of Okhotsk, north of Japan. Here impacts generate a tsunami that inundates the populated coastal region of Hokkaido, Japan, generating many casualties. Land impacts over northern parts of Russia appear to produce relatively few casualties.

The eastern limb of the path show ocean impacts into the Pacific far from land generating relatively few casualties compared to those close to populated land masses such around Central America mainland or the Hawaiian archipelago. Impacts around the coast of Central America (up to a range of 1200 kilometres, [750 miles]) and in the Gulf of Mexico represent the greatest threat to human life, indicated by that dark shading in

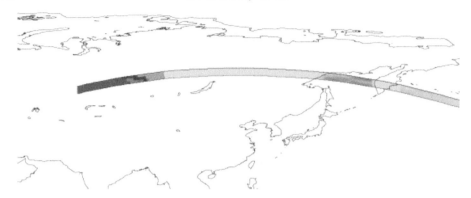

Figure 4. This figure presents the western limb results of the Apophis risk path. The path stretches from Kazakhstan, through northern Russia and into the Pacific Ocean. Cells shaded dark represent higher casualty figures generated from the impact into the centre of the cell.

Figure 5. This figure presents the eastern limb of the Apophis path of risk stretching from the Pacific Ocean across Central America to the Cape Verde Islands in the Atlantic Ocean. Dark shaded cells are those which generated most casualties from the impact into the centre of that cell.

Figure 5. These casualties are generated along the populated coastal regions of Central America.

Land impacts in this region across Central America show some interesting features. In general these impacts generate fewer casualties than the surrounding ocean impacts. However, particular impacts in Venezuela produce casualty estimates comparable to these oceanic events. This variability is due to the location dependence of land impacts. For example a direct impact into Caracas will produce many more casualties than an impact only a few kilometres outside the city. Overall the land impacts in Columbia and Venezuela generate more casualties than those in the Central American countries of Nicaragua, Costa Rica and Panama.

The most easterly potential impact sites approach the Cape Verde Islands off the western coast of Africa. Impacts at this end of the path generate relatively high casualty figures. The generated tsunami affects the islands themselves as well as coastal communities of Mauritania, Senegal and Gambia.

4.3. Conclusions

NEOSim demonstrates that, at both ends of the risk path, there exists a region. Deflection of the asteroid impact site in either direction will increase the threat to either one of these two regions. Thus an assessment would need to be made to determine which of these populations would most feasibly accept the increasing risk. In either case the mitigation mission launched needs to be fully capable of deflecting the asteroid's orbit to prevent the object 'just hitting' and thus catastrophically affecting either of these two populations.

Of interest are the relatively low casualty estimates for Pacific and Atlantic Ocean impacts far out to sea. These impacts, while appearing to have little consequence, still generate many casualties and are a significant hazard. However, relative to impacts around Central America, these ocean events represent a significantly lower threat. Impacts close to Hawaii are highlighted as generating more casualties. Warning time is a factor that will dramatically reduce the number of casualties resulting from an ocean impact. Accurate prior knowledge of the impact site combined with the long travel time of the tsunami wave (of the order of hours) will aid in evacuation and reduce the total casualty figure. This will help prevent scenes such as those witnessed during the Sumatran earthquake induced tsunami of 2004.

5. The Application of the NEOSim tool to aid the Decision Making Process

5.1. *The UK Risk Case Study*

Studying a specific region on the Earth allows for an assessment of what form the risk to the region or a particular country in that region takes. The NEOSim output indicates whether the country's population is at greater risk from ocean or land impacts and which impacts will generate most casualties. In the case of the UK, the threat is greatest from ocean impacts and the generated tsunami waves. Many factors are involved in determining this threat, particularly the amount of coastline exposed to deep ocean, the number of coastal populations and the local topography and bathymetry.

The UK is protected from many possible Atlantic Ocean impacts by the presence of Ireland. However, it is impossible to segregate the study to one particular country as, while the majority of Wales is protected by the presence of Ireland, the western communities of Ireland will be severely affected. Therefore it is impossible to study the consequences for a whole region as many countries will be affected by any impact.

The threat mitigation decision making process for the UK concerns populations are protected from the NEO impact hazard. NEOSim demonstrates that the greatest threat is faced by tsunami inundation from Atlantic Ocean impacts. Thus, to mitigate this threat, research is required into methods for protecting the major cities (indicated previously) from this inundation. One feasible method would be through a tsunami warning system with evacuation strategy put in place for each major city. Thus, in the event of an impact, the most densely populated regions at risk places could be evacuated to reduce the overall loss of life.

5.2. *The Apophis Risk Case Study*

Despite the recent orbit refinement of the asteroid Apophis essentially removing its impact threat for 2036, the experience highlights the potential for Earth to be hit unexpectedly by a small NEO. Thus there is a real need for rapid assessment of the impact consequences and determination of the populations at greatest risk. Examining the Apophis path of risk demonstrates the dramatic consequences for an impact. The study highlights regions around Central America as being at greatest risk from the asteroid. Therefore any mitigation attempt would most likely focus at moving the asteroid's orbit away from this region. However, any mitigation manoeuvre will always have the effect of increasing the risk to one region over another. In this example, large populations at risk are highlighted at each end of the path. Thus any mitigation mission will move the asteroid towards one of these regions.

Decisions regarding the relative importance of one community over another are difficult to make, but need to be made before any mitigation scenario is attempted in terms of modifications of a NEO's orbit, with resultant change in impact site. Such discussion would be required at an international level as the consequences are truly multi-national. Application of the NEOSim tool enables easy assessment of various scenarios quickly and cheaply to inform this discussion as well as highlighting which countries will be affected.

5.3. *Future Work*

Work has been ongoing to develop an advanced impact simulator called NEOimpactor which improves on NEOSim in three major areas:

- The inclusion of an infrastructure damage model to determine the economic cost of an impact event,
- The advancement of the tsunami model from a ray tracing method to a neural network approach to cater for diffraction around coastlines, and
- The development of a database software architecture to enable manipulation of data layers to power new and novel investigations. One application of this database model is to provide feedback about the effects suffered by only one pre-selected country.

Early results have been compared to the outputs of NEOSim for equivalent impact events and also compared to real world scenarios including the Sumatra Indian Ocean tsunami.

Acknowledgements

We would like to thank Professor R. Crowther, the principal sponsor of the project, for his useful input and advice through out.

References

United Nations Office of the Special Envoy for Tsunami Recovery, [Online] available from: http://www.tsunamispecialenvoy.org/country/humantoll.asp [19 July 2006]

Chyba, C.F., Thomas, P.J. & Zahnle, K.J. 1993, *Nature* 361, 40

Lyne, J.E., Tauber, M. & Fought, R. 1996, *J. Geophys. Res.* 101, 23207

Borovička, J., Popova, O.P., Nemtchinov, I.V., Spurný, P. & Ceplecha, Z. 1998, *Astron. Astrophys.* 334, 713

Baldwin, B. & Sheaffer, Y. 1971, *J. Geophys. Res.* 76, 4653

Foschini, L. 1998, *Astron. Astrophys.* 337, 5

Klinkrad, H., Fritsche, B. & Lips, T. 2004, *International Astronautical Congress 2004*, vol. IAC-04-IAA.5.12.2.07

Collins, G.S., Melosh, H.J. & Marcus, R.A. 2004, *Lunar and Planetary Laboratory*, [Online] www.lpl.arizona.edu/ImpactEffects [28 January 2005]

Glasstone S. & Dolan P.J. 1977, The Effects of Nuclear Weapons, 3rd Edition, *United States Department of Defence*

Collins, G.S., Melosh, H.J. & Marcus, R.A. 2005, Meteoritics & Planetary Science 40, 817

Chesley, S.R. & Ward, S.N. 2003, *AAS/Division for Planetary Sciences Meeting Abstracts*, vol. 35

Ward, S.N. & Asphaug, E. 2000, *Icarus* 145, 64

National Aeronautics and Space Administration 2003, [Online] http://visibleearth.nasa.gov/view_rec.php?id=116 [15 March 2005]

Epilogue

Near Earth Objects, Our Celestial Neighbors: Opportunity And Risk
Proceedings IAU Symposium No. 236, 2006
A. Milani, G.B. Valsecchi & D. Vokrouhlický, eds.
© 2007 International Astronomical Union
doi:10.1017/S1743921307003638

Concluding remarks:
15 years ago, now and in the near future

Andrea Milani

Department of Mathematics, University of Pisa, Piazza Pontecorvo 5, 56127 Pisa, Italy
email: milani@dm.unipi.it

1. Highlights and conclusions

For conclusions, the traditional approach in IAU meetings is to have a highlights speech, in which the most important points of the meeting are stressed. After consulting with the co-chairs of the Scientific Organizing Committee, I decided I would rather discuss the spirit of this meeting from a more conceptual point of view.

We think this meeting has been good, in that it has well outlined the state of the art on the subject of Near Earth Objects (NEOs), and this state of the art is in turn very good. However, to appreciate how good it is, we need to consider the evolution of the state of the art over some significant time span. In the opening remarks by Valsecchi we have been informed on the state of the art 236 years ago, when for the first time an object, comet D/1770 L1 Lexell, was discovered during a close approach to the Earth. We do not need to go that far. I would like to ask you to make the effort to go back, with your memory if you are as old as me, by reading the references if you are young, to 15 years ago, at the time of the Spaceguard Survey Report.

2. Looking back

At that time we had a lot of apparently good ideas, and most of them were wrong. Just a couple of examples: one is transport mechanisms. We have remembered in this meeting George Wetherill: he was the one who stated the problem of trasport of meteorites and Near Earth Asteroids in the modern, scientific way, and in all his otherwise very successful scientific career he was never able to solve it, because he did not yet have the tools.

At that time, and I am speaking of 1991, not of the Middle Ages, we did not have any idea that a main belt asteroid, billions of years old, could nevertheless be on a chaotic orbit. We did not have any idea that the non gravitational perturbations, well known for artificial satellites, could play a significant role in the dynamical evolution of an asteroid. As for comets, we did not have the faintest idea that there was a source population, the Trans Neptunian Objects, of which we knew only one example, Pluto, which of course by itself is not a very good source of comets. We did not have an idea that asteroids could end up in the Sun, which we now know to be the main "sink" of NEOs.

Thus all our models on the origin of meteorites and NEOs were wrong. I do not want to offend anybody, of course many of you may claim to have written a precursor paper presenting some of the ideas later found to be right, but if we have to be honest, we have to count the times we have been right and also the ones in which we have been wrong, and the fact is, at that time we were more often wrong than right.

Other examples of things we did not know, at about that time: we did not have any example of binary asteroid, apart from some who turned out to be false (let us politely say "later not confirmed"). We did have radar ranging to asteroids, not radar images.

We did not have very much in the way of spacecraft images either (Gaspra images were obtained in that same year 1991), thus we had to use Phobos images or, more often, Bill Hartmann's paintings to figure out what an asteroid would look like. Or maybe use, as Paolo Farinella *et al.* did, a Walt Disney comic strip to illustrate the concept of rubble pile.

On the subject of impact risk, we were aware of its existence, and we had already an idea of the order of magnitude of the probability of a catastrophic impact. Our population models, even for the largest NEA (> 1 km), were only accurate within a factor 2, which was not bad for the tools we had at the time. The completeness levels were of the order of 10%, and the surveys for discovering NEOs and other solar system objects were covering the sky, in a very incomplete way, up to a limiting magnitude 17.

Moreover, even after NEOs had been discovered, we did not have the mathematical tools to decide if they could impact the Earth in the medium terms, say 50 years. Still halfway through this period, in 1998, we could have a discussion among the best specialists of orbit determination which would not end with a consensus view on the possibility of an impact of a given object†.

3. Progress achieved

This was the level we were at the time: it was a young and immature science, which made rapid progress, as it is normal when you start from a low level of knowledge. If we look at the list of problems which have been solved in these 15 years we find it is very impressive. I have already mentioned the transport of NEAs and meteorites, the binaries and more in general the possible spin states of small bodies, the spacecraft and radar images and the consequent possibility to describe asteroids also in geomorphologic terms.

There has been an enormous increase of available data, especially after the beginning of operations of the automated CCD survey starting in 1998, but also because of spacecraft visits and of the availability of more powerful telescopes and instruments for physical observations. We can describe the increase by saying that we have about two orders of magnitude more objects known, and the number of objects for which we have other information besides the discovery has grown roughly in proportion.

We have also introduced the conceptual tools which have allowed to solve the problems. Our science has become mature: we can make predictions testing the theories and they turn out to be right, in most of the cases. There are a number of connections which were not even suspected and are now well understood. As an example, the connection between asteroid families, zodiacal light, meteorites and even traces of cosmic dust in geologic layers, was not even discussed much in this meeting because is now taken for granted. Another example is the transneptunians as source region for comets, including the Near Earth ones like Lexell: even after the discoveries of TNO started flowing in, at the beginning they were not considered a possible source.

An even more extreme example is meteor astronomy. Not so long ago it was regarded by most of our colleague astronomers as more witchcraft than science. The very idea that you could wait for many nights for something which may not even happen, and anyway could involve just grams of material, would be considered a waste of astronomical resources. Now even this is a mature science, which can successfully predict meteor showers and provide information on some comets and asteroids which is not available otherwise, short of a dedicated space mission; this is acheieved by using advanced technology, of which we

† 1997 XF$_{11}$.

even had a glimpse in one exibition in the hall of this General Assembly‡. Thus we have had plenty of new data because we worked hard, sometimes against a lot of difficulties, to obtain them, and we have used them overall very well.

There is also the issue of recognition within the astronomical community. We are having a significant recognition by the IAU. To have a NEO symposium at the IAU General assembly is not a small thing: there has never been an IAU meeting on such topic, and there has never been a Symposium on an even vaguely related topic at the previous General Assemblies.

4. Complacency warning

So, all is well in this brave new world for NEO science: no, this is not what I would like to say to conclude this Symposium. As people of my age know, being mature is a mixed blessing. By being mature, you tend to be more competent, but also to become complacent. To use a common, even abused, terminology, from the scientific revolution our science has undergone in the last 15 years we are passing to a stage of normal science. I have to tell you, even taking the risk of being unpleasant, that I have perceived, also in this meeting, symptoms of complacency, like a feeling of normal science. That is, very good science, using sound concepts, improving continously on what had been done before, but I do have the impression that the rate of discovery is slackening. Thus please look into yourself and ask yourself whether you are being complacent.

There are some symptoms of denial of annoying problems and discrepancies, generally small things which may or may not point to some weakness in our theoretical understanding and our models. There is the frequency of Tunguska-class events, with the additional outlier of the statistically too close approach of (99942) Apophis, which keeps coming back after we think we had solved the problem already. There are the size distribution of long period comets, and the physical aging of comets, still not properly constrained, with implications on the impact risk estimation which are small in relative sense, but may become large if the asteroid risk is taken care of. There is the issue of a suspected distribution of impacts on the Earth which may be statistically inconsistent with uncorrelated impacts. There are still ideas on the internal structure of asteroids which are at best to be considered theories, certainly not established facts. All minor disturbances with respect to the general success of the theories we have developed and carefully tested in recent years, still indications that we may have new and possibly important things to be discovered.

The reason for this is that we are in a transition phase. We have had a large increase in the availability of data, and we have had new conceptual tools, and this has fueled very rapid progress. However, unless our science receives a large inflow of fresh data, allowing us to find that our theories may be false, as well as true, we are at risk of decline.

5. The foreseable future

For our good luck, among the things which happened recently, there has been an important change of context for our specific discipline, and this is why we wanted to have the IAU Secretary General at the last session of this meeting. There has been widespread recognition, among the astronomical community but also among the political authorities in charge of funding science, of two important facts. One, NEOs are a subject

‡ The exhibition showing the automated meteor observing station by the Ondřejov Observatory, Czech Republic.

of science, and more specifically astronomical science. Two, the problem of NEO risk is responsability of the astronomers. Which means that the astronomers are entitled to ask for supplementary resources, on top of their normal reserach grants, to complete an activity which is useful to mankind, namely protecting our planetary environment from the imapct risk, but it is they who have to do it, to ask for the money and to use it to build the necessary astronomical resources. And if the astronomers are clever enough, they can build them in such a way that they can be efficiently used for multiple purposes, thus getting also a lot of science out of it.

This leads us to what is going to happen next, that is the next generation surveys, of which the first one, Pan-STARRS is expected to be operational in 2007. Thus the current transition period is going to be very short, it will be over soon: this is why this is the right moment for soul searching, to make sure we understand what are the things which are not clear, and which will be cast into doubt by the next very large inflow of data.

Three of these next generation surveys have been presented in this meeting† and apparently they have all acted very well to exploit this opportunity, being able to apply for sponsorship justified also by the mitigation of the NEO risk, but designing to perform multiple science with the same instrument and on the same images. In this way, NEOs are indeed a risk and an opportunity at the same time, as in the title of this meeting.

Thus I think there will soon be the opportunity to have a big boost in our science, and I would like to conclude by expressing my hope: that this period of normal science will just be a pause between two revolutions, and we will have the chance, together with younger people, to be involved a new phase of rapid progress as we have been actors in the previous one.

† Pan-STARRS, LSST and Discovery Channel Telescope.

Index

494

Object Index

Subject Index

Printed in the United States
by Baker & Taylor Publisher Services